Strategies for Tree Improvement under Stress Conditions

Strategies for Tree Improvement under Stress Conditions

Editors

Jie Luo
Wentao Hu

MDPI

Basel • Beijing • Wuhan • Barcelona • Belgrade • Novi Sad • Cluj • Manchester

Editors

Jie Luo
College of Horticulture and Forestry Sciences
Huazhong Agricultural University
Wuhan
China

Wentao Hu
College of Forestry and Landscape Architecture
South China Agricultural University
Guangzhou
China

Editorial Office
MDPI
St. Alban-Anlage 66
4052 Basel, Switzerland

This is a reprint of articles from the Special Issue published online in the open access journal *Forests* (ISSN 1999-4907) (available at: www.mdpi.com/journal/forests/special_issues/tree_stress_mycorrhizal).

For citation purposes, cite each article independently as indicated on the article page online and as indicated below:

Lastname, A.A.; Lastname, B.B. Article Title. *Journal Name* **Year**, *Volume Number*, Page Range.

ISBN 978-3-0365-8495-9 (Hbk)
ISBN 978-3-0365-8494-2 (PDF)
doi.org/10.3390/books978-3-0365-8494-2

Contents

forests

MDPI

Editorial

Strategies for Tree Improvement under Stress Conditions

Jie Luo [1] and Wentao Hu [2,*]

1 College of Horticulture and Forestry Sciences, Hubei Engineering Technology Research Center for Forestry Information, Huazhong Agricultural University, Wuhan 430070, China; luojie@mail.hzau.edu.cn or jie.luo@vip.163.com
2 State Key Laboratory of Conservation and Utilization of Subtropical Agro-Bioresources, Guangdong Laboratory for Lingnan Modern Agriculture, Guangdong Key Laboratory for Innovative Development and Utilization of Forest Plant Germplasm, College of Forestry and Landscape Architecture, South China Agricultural University, Guangzhou 510642, China
* Correspondence: hwt@scau.edu.cn

1. Introduction

Forests are vital ecosystems, covering a significant portion of the Earth's land area and providing essential ecological services and valuable products for human society [1]. However, forests face numerous challenges, including climate change, water deficits, nutrient limitations, and the emergence of pests and diseases [2]. To ensure the resilience and sustainability of forests, it is crucial to understand the genetic and molecular mechanisms underlying tree responses to these stressors. This Special Issue on "Strategies for tree improvement under stress conditions" presents a collection of 19 research papers that contribute to the recent advances of physiological and molecular mechanisms in woody plants in adapting to stress conditions.

2. Drought

Research papers in this collection explore various aspects of tree genetics and molecular responses to drought stresses. In one study, the effects of water deficit on artificially bred poplar hybrids were investigated, revealing that male siblings exhibit better protection than female siblings under water-deficient conditions [3]. Li et al. investigated the impact of nitrogen (N) supply on water uptake, drought resistance, and hormone regulation in *Populus simonii* seedlings under PEG-induced drought stress and reveal that increasing N supply may enhance drought tolerance by reducing transpiration rate and oxidative stress while improving water uptake and antioxidant activity [4]. The overexpression of the *SpsNAC005* gene promotes growth, development, and stress tolerance in transgenic plants under drought conditions [5]. The expression levels of *MaTCP2*, *MaTCP4-1*, *MaTCP8*, *MaTCP9-1*, and *MaTCP20-2* exhibited a significant correlation with the process of root development, suggesting their involvement in regulating root growth under drought conditions [6]. Furthermore, these identified MaTCP transcription factors hold potential implications for enhancing the drought tolerance of mulberry plants [6]. Liu et al. identified and analyzed 18 auxin response factors (ARFs) in *Santalum album* and tissue-specific expression and drought-induced expression patterns were observed, with six genes overexpressed in haustorium and three genes overexpressed under drought stress [7]. These findings provide insights into the functions of *S. album* ARF genes, particularly in haustorium formation and response to drought stress [7]. *Robinia pseudoacacia* can access shallow soil water in wet years and utilize deeper soil water in dry years to maintain growth and resistance to drought stress, which provide the solutions for rainwater resource planning and management in forest plantations [8].

Citation: Luo, J.; Hu, W. Strategies for Tree Improvement under Stress Conditions. *Forests* **2023**, *14*, 1320. https://doi.org/10.3390/f14071320

Received: 16 June 2023
Accepted: 26 June 2023
Published: 27 June 2023

3. Heavy Mental

Heavy mental accumulation in soils affects the normal growth of plants; however, the exogenous application of nutrients may mitigate the toxic effects of heavy metals. Here, Li et al. suggest that high-dose Fe mitigates Cd-induced growth suppression, promotes Cd transport to aboveground tissues, and enhances antioxidant capacity in poplar, which provides insights for Cd-contaminated soil remediation using poplar [9]. Wang et al. investigated the effects of selenium (Se) on cadmium (Cd) accumulation and the physiological mechanisms underlying Se-mediated regulation of Cd-induced oxidative stress in *Juglans regia*, suggesting that the exogenous application of Se, especially at 200 μM, reduces Cd accumulation, enhances antioxidant enzyme activities, and alleviates Cd-induced stress in walnut roots [10]. Another study focused on the transcriptomic response to zinc stress in mulberry, uncovering organ-specific differences in gene expression [11].

4. Salinity

Saline is one of the most serious abiotic stresses that affecting plant growth and development worldwide [12]. Zhang et al. found that grafting can ameliorate the inhibition of salinity on the photosynthetic capacity of *Hibiscus syriacus*, mainly resulting from alleviated limitations on photosynthetic pigments, photochemical efficiency, and the Calvin-Benson–Bassham cycle [13]. Additionally, the overexpression of the poplar *WRKY51* transcription factor was found to enhance salt tolerance in *Arabidopsis thaliana*, demonstrating the potential of genetic engineering for improving tree resilience [14]. Pang et al. showed that highly expressed transcription factor genes were correlated with key salt tolerance indices, suggesting their potential as genetic resources for salt tolerance breeding in *Salix matsudana* [15]. Moreover, the overexpression of *SpsNAC005* from *Salix psammophila* in poplar significantly improved its tolerance to salt stress [5].

5. Nutrient

N is one of the most important macronutrients for growth and development in woody plants, and applications of N can significantly increase productivity [16,17]. Responses of fine root traits and soil nitrogen to fertilization methods and N application amounts in a poplar plantation were investigated, shedding light on the interactions between tree roots and soil nutrient availability [18]. NH_4^+-N and NO_3^--N distributions have different impacts on the root morphology and growth of *Cunninghamia lanceolata* and *Schima superba* seedlings, in which tailoring N application based on N form and plant species is recommended for seedling cultivation [19]. The influence of trace elements on the traits and active compounds of *Camellia oleifera* in nutrient-poor forests was also examined, revealing that exogenous applications of zinc and Se could significantly improve the qualities of its fruits [20]. The trace element boron has been shown to be essential for woody plants overcoming stress conditions [21]. Liu et al. uncovered the effects of *Funneliformis mosseae* inoculation on *C. oleifera* seedlings under normal and boron deficient conditions and found that AMF inoculation improves boron deficiency resistance and that AMF colonization is influenced by boron availability [22].

6. Other Strategies for Tree Improvement

The genome-wide identification of the *PP2C* gene family and expression-level analyses of the PP2Cs in *Paulownia fortunei* in response to rifampicin and methyl methanesulfonate treatments were studied, providing insights into their potential roles in stress responses [23]. The chemical composition of walnut oil, including fatty acids, micronutrients, and secondary metabolites, was analyzed in different walnut species and hybrids cultivated at various sites [24]. Significant variations in composition and content were observed between species and sites, which could be valuable for site selection and improving the nutritional quality of walnut oil [24]. Differentially methylated regions (DMRs) and associated genes (DMGs) after gtafting in pecan were analyzed, which identified the key genes involved in hormone response, suggesting their crucial roles in graft growth regulation [25]. This study provides valuable insights into the epigenetic mechanisms underlying rootstock-induced growth changes in pecan, paving the way for tree improvement using grafting in this plant species [25].

7. Conclusions and Prospects

These research papers collectively highlight the significance of molecular genetics and genomics in addressing the challenges faced by forest ecosystems. By unraveling the genetic basis of tree responses to various stressors, we can develop targeted strategies for tree improvement, conservation, and sustainable forest management. Furthermore, these studies encompass a wide range of tree species, including economically important species, rare and endangered species, and ecologically significant trees, broadening our understanding of forest species beyond model species.

As guest editors, we believe that this Special Issue presents a valuable compilation of research findings that contribute to tree improvements under stress conditions. The diversity of species, conditions and genetic traits investigated underscores the potential of molecular genetics to enhance forest health, ecosystem services, and sustainable production. However, more research is needed to further expand our knowledge and address the complex challenges faced by forests in the face of climate change and other stressors. By continuing to explore the intricacies of tree physiology and genetics, we can pave the way for a more resilient and sustainable future for our forests and the multitude of benefits they provide.

Funding: The study was supported by the Fundamental Research Funds for the Central Universities (Grant no. 2262022YLYJ007).

Conflicts of Interest: The authors declare no conflict of interest.

References

1. Gamfeldt, L.; Snäll, T.; Bagchi, R.; Jonsson, M.; Gustafsson, L.; Kjellander, P.; Ruiz-Jaen, M.C.; Fröberg, M.; Stendahl, J.; Philipson, C.D.; et al. Higher levels of multiple ecosystem services are found in forests with more tree species. *Nat. Commun.* **2013**, *4*, 1340. [CrossRef] [PubMed]
2. Arora, N.K.; Fatima, T.; Mishra, I.; Verma, M.; Mishra, J.; Mishra, V. Environmental sustainability: Challenges and viable solutions. *Environ. Sustain.* **2018**, *1*, 309–340. [CrossRef]
3. Chen, F.; Gong, Y.; Liu, S.; Wang, Y.; Luo, L.; Zhu, G.; Zhao, H. More effective protection supports male better than female siblings over water deficit in artificially bred poplar hybrids. *Forests* **2023**, *14*, 995. [CrossRef]
4. Li, Z.; Wang, X.; Liu, Y.; Zhou, Y.; Qian, Z.; Yu, Z.; Wu, N.; Bian, Z. Water uptake and hormone modulation responses to nitrogen supply in *Populus simonii* under PEG-induced drought stress. *Forests* **2022**, *13*, 907. [CrossRef]
5. Yang, H.; Fan, L.; Yu, X.; Zhang, X.; Hao, P.; Wei, D.; Zhang, G. Analysis of the NAC gene family in *Salix* and the identification of *SpsNAC005* gene contributing to salt and drought tolerance. *Forests* **2022**, *13*, 971. [CrossRef]
6. Wei, W.; He, J.; Luo, Y.; Yang, Z.; Xia, X.; Han, Y.; He, N. Root-growth-related matcp transcription factors responsive to drought stress in mulberry. *Forests* **2023**, *14*, 143. [CrossRef]
7. Liu, X.; Liu, Y.; Wang, S.; Qin, F.; Wang, D.; Chen, Y.; Hu, L.; Meng, S.; Lu, J. Complete genome expression analysis of the Auxin Response Factor gene family in sandalwood and their potential roles in drought stress. *Forests* **2022**, *13*, 1934. [CrossRef]
8. Ma, C.; Meng, H.; Xie, B.; Li, Q.; Li, X.; Zhou, B.; Wang, Q.; Luo, Y. In situ rainwater collection and infiltration system alleviates the negative effects of drought on plant-available water, fine root distribution and plant hydraulic conductivity. *Forests* **2022**, *13*, 2082. [CrossRef]
9. Li, M.; Liu, C.; Zhang, D.; Wang, B.; Ding, S. The influence of iron application on the growth and cadmium stress tolerance of poplar. *Forests* **2022**, *13*, 2023. [CrossRef]
10. Wang, B.; Zhang, D.; Wang, W.; Song, Y.; Lu, M.; Ding, S. Foliar application of selenium reduces cadmium accumulation in walnut seedlings. *Forests* **2022**, *13*, 1493. [CrossRef]
11. Huang, S.; Kang, X.; Yu, T.; Yidilisi, K.; Zhang, L.; Cao, X.; Chao, N.; Liu, L. Comparative RNA-Seq analysis reveals the organ-specific transcriptomic response to zinc stress in mulberry. *Forests* **2023**, *14*, 842. [CrossRef]
12. Luo, J.; Shi, W.; Li, H.; Janz, D.; Luo, Z.-B. The conserved salt-responsive genes in the roots of *Populus* × *canescens* and *Arabidopsis thaliana*. *Environ. Exp. Bot.* **2016**, *129*, 48–56. [CrossRef]
13. Zhang, S.; Yu, W.; Lu, Z.; Xiang, P.; Wang, Z.; Hua, J.; Gu, C.; Cai, J.; Lu, Y. *Hibiscus hamabo* rootstock-grafting improves photosynthetic capacity of *Hibiscus syriacus* under salt stress. *Forests* **2023**, *14*, 1226. [CrossRef]
14. Zhou, Y.; Li, Q.; Zhang, Y. Overexpression of the poplar *WRKY51* transcription factor enhances salt tolerance in *Arabidopsis thaliana*. *Forests* **2023**, *14*, 191. [CrossRef]

15. Pang, Y.; Guo, L.; Wang, T.; Liu, W.; Mao, P.; Cao, X.; Geng, Y.; Cao, B. Screening of key indices and the gene transcriptional regulation analysis related to salt tolerance in *Salix matsudana* seedlings. *Forests* **2022**, *13*, 754. [CrossRef]
16. Li, Z.; Deng, S.; Zhu, D.; Wu, J.; Zhou, J.; Shi, W.; Fayyaz, P.; Luo, Z.-B.; Luo, J. Proteomic reconfigurations underlying physiological alterations in poplar roots in acclimation to changing nitrogen availability. *Environ. Exp. Bot.* **2023**, *211*, 105367. [CrossRef]
17. Luo, J.; Zhou, J.-J. Growth performance, photosynthesis, and root characteristics are associated with nitrogen use efficiency in six poplar species. *Environ. Exp. Bot.* **2019**, *164*, 40–51. [CrossRef]
18. Yan, X.; Dai, T.; Gao, Y.; Di, N.; Jia, L. Responses of fine root traits and soil nitrogen to fertilization methods and nitrogen application amounts in a poplar plantation. *Forests* **2023**, *14*, 282. [CrossRef]
19. Liang, H.; Wang, L.; Wang, Y.; Quan, X.; Li, X.; Xiao, Y.; Yan, X. Root development in *Cunninghamia lanceolata* and *Schima superba* seedlings expresses contrasting preferences to nitrogen forms. *Forests* **2022**, *13*, 2085. [CrossRef]
20. Dai, Q.; Deng, Z.; Pan, L.; Nie, L.; Yang, Y.; Huang, Y.; Huang, J. Effects of trace elements on traits and functional active compounds of *Camellia oleifera* in nutrient-poor forests. *Forests* **2023**, *14*, 830. [CrossRef]
21. Luo, J.; Liang, Z.; Wu, M.; Mei, L. Genome-wide identification of *BOR* genes in poplar and their roles in response to various environmental stimuli. *Environ. Exp. Bot.* **2019**, *164*, 101–113.
22. Liu, J.; Zhang, M.; Fan, J.; Ding, W.; Chen, L.; Luo, J.; Liu, Y.; Mei, L. The synergistic effects of AMF inoculation and boron deficiency on the growth and physiology of *Camellia oleifera* seedlings. *Forests* **2023**, *14*, 1126. [CrossRef]
23. Zhao, Z.; Zhang, P.; Deng, M.; Cao, Y.; Fan, G. Genome-wide identification and expression analyses of the *PP2C* gene family in *Paulownia fortunei*. *Forests* **2023**, *14*, 207. [CrossRef]
24. Yang, H.; Xiao, X.; Li, J.; Wang, F.; Mi, J.; Shi, Y.; He, F.; Chen, L.; Zhang, F.; Wan, X. Chemical compositions of walnut (*Juglans* Spp.) oil: Combined effects of genetic and climatic factors. *Forests* **2022**, *13*, 962. [CrossRef]
25. Liu, Z.; Tan, P.; Liang, Y.; Shang, Y.; Zhu, K.; Peng, F.; Li, Y. Grafting with different rootstocks induced DNA methylation alterations in pecan [*Carya illinoinensis* (Wangenh.) K. Koch]. *Forests* **2023**, *14*, 4.

forests

Article

Hibiscus hamabo Rootstock-Grafting Improves Photosynthetic Capacity of *Hibiscus syriacus* under Salt Stress

Shuqing Zhang [1,2,†], Wanwen Yu [1,†], Zhiguo Lu [2,†], Peng Xiang [1,2], Zhiquan Wang [2], Jianfeng Hua [2], Chunsun Gu [2], Jinfeng Cai [1] and Yan Lu [2,*]

1 Co-Innovation Center for the Sustainable Forestry in Southern China, Nanjing Forestry University, Nanjing 210037, China; shuqing20210226@163.com (S.Z.); youeryuww@163.com (W.Y.); xpeng0501@163.com (P.X.); caijinfeng1984@njfu.edu.cn (J.C.)

2 Jiangsu Key Laboratory for the Research and Utilization of Plant Resources, Institute of Botany, Jiangsu Province and Chinese Academy of Sciences (Nanjing Botanical Garden Mem. Sun Yat-Sen), Nanjing 210014, China; lzgjs@cnbg.net (Z.L.); wangzhiquan@cnbg.net (Z.W.); jfhua@cnbg.net (J.H.); chunsungu@cnbg.net (C.G.)

* Correspondence: luy666888@cnbg.net

† These authors contributed equally to this work.

Abstract: *Hibiscus syriacus*, a woody ornamental plant with great economic value, is vulnerable to salinity. Hence, its cultivation in saline areas is severely restricted. Although grafting *H. syriacus* onto *H. hamabo* rootstock can greatly improve *H. syriacus*'s salt resistance, the photosynthetic response of *H. syriacus* to grafting and salt stress remains largely unknown. To address this question, self-rooted (Hs), self-grafted (Hs/Hs), and *H. hamabo*-grafted (Hs/Hh) *H. syriacus* were exposed to 0 or 300 mM NaCl. Salt significantly reduced the net and maximum photosynthetic rates, chlorophyll content, and maximum (Fv/Fm) and actual (ΦPSII) photochemical quantum yield of photosystem II (PSII), as well as the apparent electron transport rate, in Hs and Hs/Hs. However, these reductions were largely alleviated when *H. syriacus* was grafted onto *H. hamabo*. In line with the changes in the chlorophyll fluorescence parameters, the expression of genes encoding subunits of PSII and PSI in Hs/Hh was higher than that in Hs and Hs/Hs under saline conditions. Moreover, *H. hamabo* rootstock grafting upregulated the genes involved in the Calvin–Benson–Bassham cycle in *H. syriacus* under salt conditions. These results indicate that grafting can ameliorate the inhibition of salinity on the photosynthetic capacity of *H. syriacus*, mainly resulting from alleviated limitations on photosynthetic pigments, photochemical efficiency, and the Calvin–Benson–Bassham cycle.

Keywords: grafting; salt stress; gas exchange; chlorophyll fluorescence; *Hibiscus syriacus* Linn.; *Hibiscus hamabo* Sieb. et Zucc

Citation: Zhang, S.; Yu, W.; Lu, Z.; Xiang, P.; Wang, Z.; Hua, J.; Gu, C.; Cai, J.; Lu, Y. *Hibiscus hamabo* Rootstock-Grafting Improves Photosynthetic Capacity of *Hibiscus syriacus* under Salt Stress. *Forests* **2023**, *14*, 1226. https://doi.org/10.3390/f14061226

Academic Editor: Cate Macinnis-Ng

Received: 16 May 2023
Revised: 4 June 2023
Accepted: 11 June 2023
Published: 13 June 2023

1. Introduction

Salinity is one of the major critical environmental constraints that limits the growth, yield, and quality of trees, crops, and horticultural plants [1–3]. Nowadays, over 20% of the arable land throughout the world is salt influenced [4], and improper fertilization and irrigation further exacerbate salinization [5]. It is estimated that approximately half of the cultivated land will be affected by salinity in 2050 [6]. The declining productivity of plants grown in saline environments is usually correlated with a reduction in their photosynthesis. The reduction in gas exchange is mainly due to the following factors: (i) stomatal closure, which restricts the intercellular CO_2 concentration (Ci) for carboxylation; (ii) impairment of the photosynthetic apparatus, resulting in the limitation of photosystem II (PSII) efficiency; and (iii) reduction in the activities of ribulose-1,5-bisphosphate carboxylase/oxygenase (Rubisco) for the Calvin–Benson–Bassham (CBB) cycle [7]. Photosynthesis is recognized as a good indicator of a plant's response to salt stress [8]. Thus, exploring the response of photosynthesis to NaCl stress could lay a physiological foundation for promoting plants' salt tolerance.

Grafting is a well-established agronomic practice to promote the salt tolerance of plants by transferring its branch or bud to the stem or root of a salt-tolerant species [2,9]. Owing to its low cost and ease of operation, grafting is widely applied in vegetable, fruit, and woody crop production to improve their salt resistance [5,10,11]. For instance, grafting the scions of Thompson Seedless, a salt-sensitive variety of grapevine (*Vitis vinifera* L.), onto the rootstocks of the 110R variety, which has a higher salt-tolerant capacity, can greatly enhance the growth and salt resistance of the scions [10]. Grafting can also alleviate the damage of salt stress to mulberry (*Morus alba* L.) and avocado (*Persea americana* Mill.) [12,13]. Grafting the scions of citrus (*Citrus aurantium* L.) onto salt-tolerant rootstocks reveals higher salinity tolerance than grating onto salt-sensitive rootstocks [14]. Moreover, previous studies have revealed that salt stress significantly decreases the net photosynthetic rate in self-rooted and/or self-grafted salt-sensitive watermelon (*Citrullus lanatus* Matsum. Et Nakai) and cucumber (*Cucumis sativus* L.), whereas salt-tolerant rootstock-grafted plants exhibit a much higher net photosynthetic rate under NaCl stress [7]. Currently, little information is available on the mechanisms underlying grafting that improve the photosynthetic capacity of plants under salt stress.

Hibiscus syriacus L., which belongs to the genus Hibiscus, is an important woody ornamental plant. There are more than 350 varieties with distinct flower colors and forms distributed around the world [15]. The flowers of this species have a long blooming period. Aside from its ornamental value, *H. syriacus* also has certain medicinal properties, including antifungal, antihypertensive, and anti-inflammatory [16]. However, *H. syriacus* plants grown in saline areas exhibit leaf shrinkage and plant withering, indicating their susceptibility to salt stress. This seriously restricts the cultivation and application of *H. syriacus* in saline areas. *H. hamabo* Sieb. et Zucc., also a member of the genus Hibiscus, is a significant semi-mangrove shrub with golden yellow flowers [17]. However, its flowers are sparse and short-lived, which severely limits its ornamental value [18]. This plant is halophytic and can survive in habitats where the NaCl concentration varies from 1.1 to 1.5% [18]. Owing to its superior salt tolerance, *H. hamabo* has been used as a rootstock to graft *H. syriacus*. The grafted *H. syriacus* plants can grow well in coastal and inland saline lands, indicating that grafting has greatly enhanced the salt tolerance of *H. syriacus*. However, there is currently little information available regarding the photosynthetic response of grafted *H. syriacus* to soil salinity.

In this study, self-rooted (Hs), self-grafted (Hs/Hs), and *H. hamabo*-grafted (Hs/Hh) *H. syriacus* are exposed to either 0 (−Na, serving as the control) or 300 (+Na) mM NaCl, respectively. We hypothesize that grafting *H. syriacus* scions onto *H. hamabo* rootstocks would markedly enhance the photosynthetic capacity and induce the transcriptional regulation of key genes that participate in photosynthesis under salt stress. To test this hypothesis, we characterize the changes in the photosynthetic rates, photosynthetic pigment contents, chlorophyll fluorescence, and expression of key genes (genes encoding subunits of PSII and PSI and involved in the CBB cycle) involved in photosynthesis in the leaves of Hs, Hs/Hs, and Hs/Hh in response to salt stress.

2. Materials and Methods

2.1. Plant Cultivation and Salt Treatment

Cuttings (length, 15 cm; diameter, 0.5 cm) of *Hibiscus syriacus* Linn. and *Hibiscus hamabo* Sieb.et zucc. were planted in plastic pots (2.6 L) containing soil (categorized as Alfisol based on USDA soil taxonomy; total nitrogen, 842 mg kg^{-1}; available phosphorus, 42 mg kg^{-1}; available potassium, 152 mg kg^{-1}; 1.30% organic matter; pH, 6.9). When the stems of the one-year-old cuttings were 6 mm thick, 48 *H. syriacus* and 24 *H. hamabo* plants were selected for grafting. Among them, 24 *H. syriacus* plants were selected as self-rooted plants (Hs). The remaining *H. syriacus* and *H. hamabo* plants were used for grafting using the splice-grafting technique to obtain self-grafted *H. syriacus* (Hs/Hs) and heterografted *H. syriacus* (*H. syriacus* grafted on *H. hamabo* rootstock, Hs/Hh) plants, respectively. All plants were grown in a greenhouse (natural light and temperature; relative humidity, 75%). Seventy-five days after grafting, the growth of Hs/Hs and Hs/Hh became vigorous, and each of the three types of plants (Hs, Hs/Hs, and Hs/Hh) was equally divided into two groups. The plants in each group were irrigated with either 0 (−Na, serving as control) or 300 (+Na) mM NaCl. The irrigation was performed every other day. To avoid salt shock, NaCl was added to obtain the desired concentration (300 mM, 29.2 dS m^{-1}) in increments of 50 mM at a time.

2.2. Harvesting

After 23 days of salt treatment, distinct morphological differences appeared among the groups. The leaf samples (leaf plastochron index = 5 to 10) were collected and frozen immediately in liquid nitrogen. Frozen samples were ground into fine powder in liquid nitrogen using a ball mill (GT300, Beijing Grinder Instrument Co. Ltd., Beijing, China) and stored at −80 °C for further analysis. For further biochemical analysis, equal amounts of frozen samples acquired from two plants that received the same treatment in each plant type were pooled and thoroughly mixed. As a result, six mixed samples were obtained for each treatment per plant type.

2.3. Determination of Chlorophyll and Carotenoids

The concentrations of photosynthetic pigments were determined using the ethanol extraction procedure, as described elsewhere [19]. Briefly, about 50 mg of the sample was extracted using 10 mL of 95% ethanol. The mixture was left to soak in the dark overnight until the samples turned white. The concentrations of chlorophyll a, chlorophyll b, and carotenoid in the supernatant were determined spectrophotometrically at 665, 649, and 470 nm, respectively.

2.4. Determination of Gas Exchange

Prior to harvesting, two mature leaves (LPI = 7–8) from six plants per group were selected for photosynthetic measurements. The net photosynthetic rate (Pn), intercellular CO_2 concentration (Ci), transpiration rate (E), stomatal conductance (Gs), and photosynthetic water-use efficiency (WUE, calculated as Pn/E) were measured using a portable photosynthesizer (CIRAS-3, PP-Systems, Amesbury, MA, USA) according to the manufacturer's instructions. The photosynthetic active radiation gradient (PAR) was set at 1000 μmol m^{-2} s^{-1}, and the external CO_2 concentration was maintained at 400 ± 10 μmol mol^{-1}. The environmental temperature was maintained at 32 ± 5 °C.

For the diurnal variation of photosynthesis, the gas exchange was measured every 2 h from 8:00 to 18:00. The measured leaves were marked to ensure that the same leaf was measured at different time points.

For the determination of the light-response curves of Pn, the following PAR levels were set: 2000, 1800, 1600, 1400, 1200, 1000, 800, 600, 400, 300, 200, 100, 50, 0 μmol m^{-2} s^{-1}. The Pn value under each PAR was recorded every 2 min.

2.5. Determination of Chlorophyll Fluorescence Parameters

The chlorophyll fluorescence parameters of the leaves were determined using a portable pulse-modulated fluorometer (FMS-2, Hansatech, King's Lynn, Norfolk, UK) between 9:00 and 11:00 am according to the manufacturer's instructions. Briefly, endogenous actinic light was activated to measure the minimum fluorescence (Fs), maximum fluorescence (Fm′), and actual photochemical quantum yield of PSII (ΦPSII) after light adaptation, as suggested by Lu et al. [20]. After dark adaptation for 0.5 h, the initial fluorescence (Fo), maximum fluorescence (Fm), steady-state fluorescence (Fs), and maximum photochemical quantum yield of PSII (Fv/Fm) were determined, as described previously [20]. The non-photochemical quenching coefficient (NPQ), photochemical quenching coefficient (qP), and apparent electron transport rate (ETR) were calculated according to the following formula: $NPQ = Fm/Fm' - 1$, $qP = \Phi PSII/(Fv/Fm)$, $ETR = \Phi PSII \times PPFD \times f \times \alpha$ [21,22].

2.6. Analysis of Transcript Levels

Total RNA was isolated from the fine, powdered fresh samples (ca. 50 mg) using an OminiPlant RNA Kit (CW2598S, CVBIO, Taizhou, China). RNA concentration was determined using a Colibri ultramicro spectrophotometer (LB915, Corrine Technology, Beijing, China). RNA integrity was checked using 1% agarose gel electrophoresis. The RNA (ca. 1 μg) was used to synthesize the first strand of cDNA using a PrimeScript RT reagent kit (RR047A, TaKaRa, Dalian, China). The synthesized cDNA was used for quantitative real-time PCR (qRT-PCR), as described elsewhere [23]. qRT-PCR was conducted using a TB Green Premix Ex Taq II (RR820, TaKaRa) in a StepOnePlus Real-Time PCR System (Thermo Fisher Scientific, Waltham, MA, USA). Primers were designed specifically for each gene, and 18S rRNA was selected as the reference gene (Table S1).

2.7. Statistical Analysis

Statistical tests were performed using Statgraphics (STN, St Louis, MO, USA), as described elsewhere [24], with minor modifications. Briefly, prior to statistical analysis, the data were tested for normality. Two-way ANOVAs were used with grafting and NaCl as two factors. Differences in means were considered significant if the *p*-value was less than 0.05. The light-response curve was fitted using the rectangular hyperbola model, as described elsewhere [25]. Principal component analysis (PCA) was performed using the command prcomp in R (http://www.r-project.org/, accessed on 9 January 2023).

3. Results

3.1. Morphology and Photosynthetic Pigments

After 23 days of 300 mM NaCl treatment, the leaves of Hs and Hs/Hs turned yellow but the leaves of Hs/Hh remained green (Figure 1a). Consistently, the chlorophyll a concentration was decreased by 22% and 31%, and the chlorophyll (a + b) concentration was reduced by 19% and 23% in the leaves of Hs and Hs/Hs, whereas no marked differences were found for these two parameters in Hs/Hh leaves grown under NaCl treatment compared to the control (Figure 1b,d). The carotenoid and chlorophyll a/b concentrations were reduced by 29% and 38% in the leaves of Hs/Hs, but no marked differences were found in Hs/Hh leaves grown under NaCl conditions compared to those grown under control conditions (Figure 1e,f). Except for the carotenoid content in Hs/Hs under control conditions, the concentrations of chlorophyll a, chlorophyll b, chlorophyll (a + b), and carotenoid were all significantly higher in Hs/Hh than in Hs and Hs/Hs, regardless of the NaCl treatment (Figure 1b–e).

3.2. Gas-Exchange Parameters

The Pn, E, and Gs were significantly decreased in the leaves of all the plants exposed to NaCl compared to those grown under control conditions. In addition, the reduced levels of these parameters observed in Hs/Hh were significantly lower than those in Hs and Hs/Hs (Figure 2a–c). The Ci was not markedly altered by salt in the leaves of any of the plants (Figure 2d). The E/Gs was increased by 177%, 147%, and 95%, respectively, in the leaves of Hs, Hs/Hs, and Hs/Hh treated with NaCl compared to those under control conditions (Figure 2e). The WUE was reduced by 39% in the leaves of Hs, but no significant differences were observed in the leaves of Hs/Hs and Hs/Hh treated with NaCl compared to those grown under control conditions (Figure 2f). Except for the Pn in Hs/Hs, no significant differences were found in the Pn, E, Gs, Ci, WUE, and E/Gs in all plants under control conditions (Figure 2a–f). However, the Pn, E, Gs, and WUE in Hs/Hh leaves were all significantly higher, and the E/Gs was lower compared to the levels observed in Hs and Hs/Hs under salt conditions (Figure 2a–f). For instance, the Pn in Hs/Hh (11.90 μmol CO_2 m^{-2} s^{-1}) leaves was, respectively, 243% and 225% higher than that in Hs (3.47 μmol CO_2 m^{-2} s^{-1}) and Hs/Hs (3.67 μmol CO_2 m^{-2} s^{-1}) leaves under salt treatment (Figure 2a).

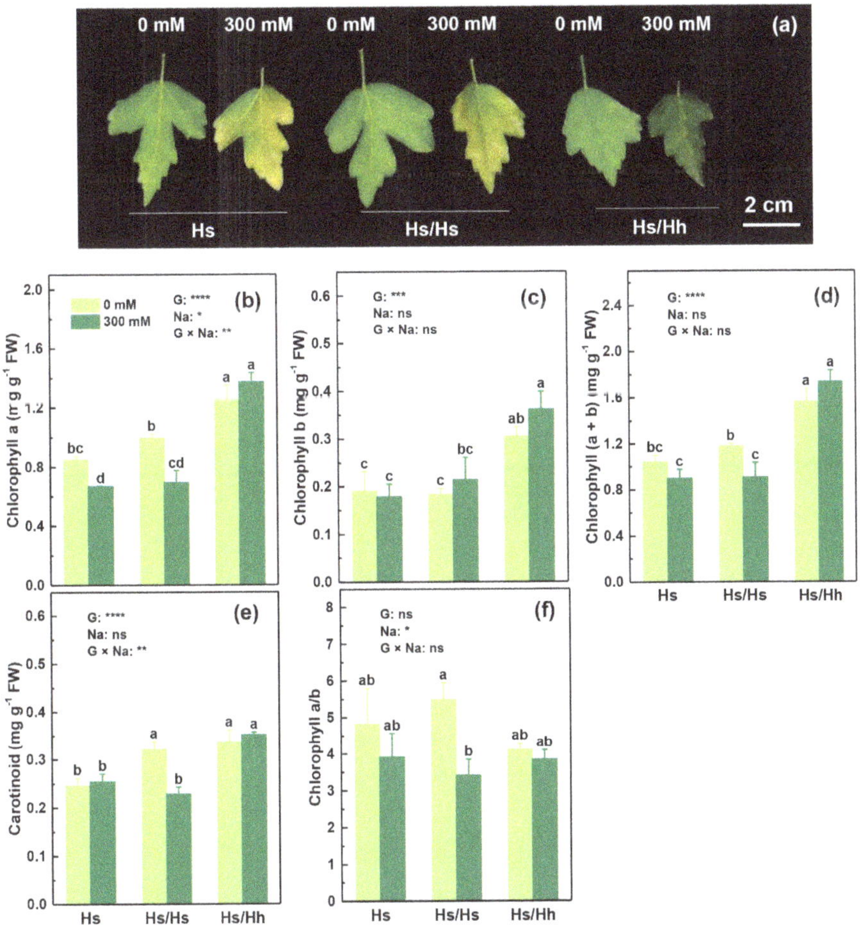

Figure 1. Morphology (**a**), and concentrations of chlorophyll a (**b**), chlorophyll b (**c**), chlorophyll (a + b) (**d**), carotenoid (**e**), and chlorophyll a/b (**f**) of grafted *H. syriacus* leaves grown under either 0 (control) or 300 mM NaCl conditions. Hs: self-rooted *H. syriacus*; Hs/Hs: self-grafted *H. syriacus*; Hs/Hh: heterografted *H. syriacus* with *H. hamabo* as the rootstock. Data indicate means ± SE (n = 6). Different letters on the bars indicate significant differences between the treatments. *p*-values based on the ANOVA tests for grafting (G), NaCl (Na), and their interaction (G × Na) are indicated: * $p < 0.05$; ** $p < 0.01$; *** $p < 0.001$; **** $p < 0.0001$; ns: not significant.

The diurnal changes in the Pn in all the plants showed single-peak curve change patterns, and no mid-day depression was evident (Figure 3a). The highest daily Pn observed in Hs/Hh (10:00) grown under control and salt conditions occurred later than that in Hs (14:00) grown under control and salt conditions, as well as Hs/Hs grown under salt conditions (Figure 3a). The Pn was decreased in all the plants grown under NaCl conditions compared to those grown under control conditions at each time point (Figure 3a). The Pn in the leaves of Hs/Hh was consistently higher than that of Hs and Hs/Hs grown under salt treatment between 8:00 and 16:00 (Figure 3a).

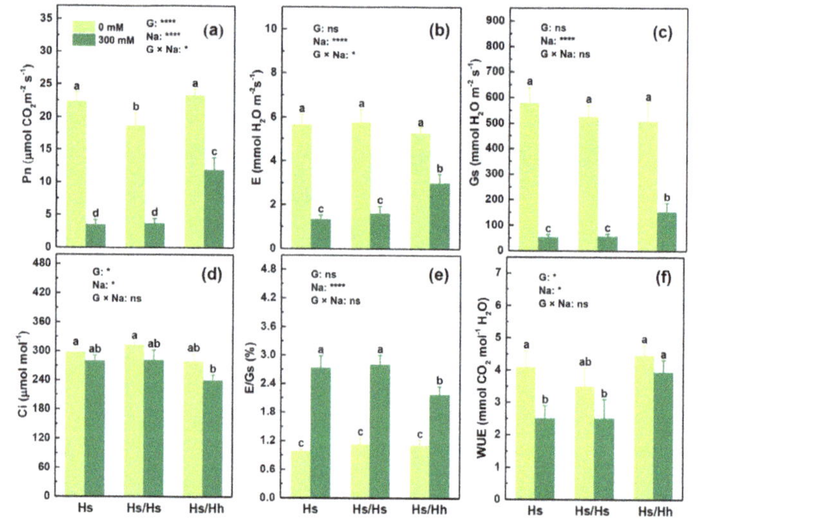

Figure 2. (**a**) Net photosynthetic rate (Pn), (**b**) transpiration rate (E), (**c**) stomatal conductance (Gs), (**d**) intercellular CO_2 concentration (Ci), (**e**) E/Gs, and (**f**) photosynthetic water-use efficiency (WUE) of grafted *H. syriacus* leaves grown under either 0 (control) or 300 mM NaCl conditions. Hs: self-rooted *H. syriacus*; Hs/Hs: self-grafted *H. syriacus*; Hs/Hh: heterografted *H. syriacus* with *H. hamabo* as the rootstock. Data indicate means ± SE (n = 6). The information about the treatments and statistical analysis is the same as that mentioned in Figure 1.

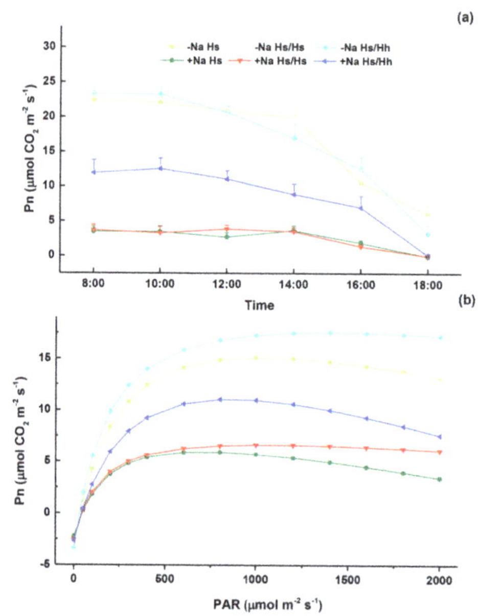

Figure 3. Diurnal variation of photosynthesis (**a**) and light-response curves (**b**) of grafted *H. syriacus* leaves grown under either 0 (−Na, serving as the control) or 300 mM NaCl (+Na) conditions. Hs: self-rooted *H. syriacus*; Hs/Hs: self-grafted *H. syriacus*; Hs/Hh: heterografted *H. syriacus* with *H. hamabo* as the rootstock. In (**a**), bars indicate means ± SE (n = 6); in (**b**), numbers represent the averages of six biological replicates.

Figure 3b shows the fitted light-response curves of the Pn in the leaves of the three plants. The Pn–PAR curves of all the plants are parabolic. The patterns of variation in the curves are quite similar for Hs grown under both control and salt conditions and Hs/Hh grown under salt conditions. In other words, as the PAR increased, the Pn exhibited a trend of first increasing and then slowly decreasing. The light-response curves of the Pn in Hs/Hs grown under control and salt conditions and Hs/Hh grown under control conditions exhibited a trend of first increasing and subsequently plateauing. Specifically, when the PAR was less than 200 μmol m^{-2} s^{-1}, the Pn increased rapidly as the PAR increased. When the PAR ranged from 200 to 800 (for plants under NaCl treatment) or 1000 (for plants under control conditions) μmol m^{-2} s^{-1}, the rate of increase of the Pn gradually slowed down. When the PAR further increased, the Pn in the Hs grown under control and salt conditions and the Hs/Hh grown under salt conditions gradually decreased, and the curves in the Hs/Hs grown under control and salt conditions and the Hs/Hh grown under control conditions became flat. Except for 0 μmol m^{-2} s^{-1}, the Pn in the three plants grown under NaCl conditions decreased compared to those grown under control conditions, and the Pn in Hs/Hh was higher than in Hs and Hs/Hs grown under salt treatment.

The maximum net photosynthetic rate (Pnmax) and apparent quantum efficiency (AQY) were significantly reduced in the leaves of all the plants grown under NaCl conditions compared to those grown under control conditions (Table 1). The Pnmax value was markedly higher in Hs/Hh (11.30 μmol CO_2 m^{-2} s^{-1}) than in Hs (5.97 μmol CO_2 m^{-2} s^{-1}) under salt treatment (Table 1). The AQY value in Hs/Hh (0.14 mol mol^{-1}) was 27% and 21% higher than in Hs (0.11 mol mol^{-1}) and Hs/Hs (0.12 mol mol^{-1}) under control conditions (Table 1). The dark respiration rate (Rd) was decreased by 34% in the leaves of Hs grown under NaCl conditions compared to those grown under control conditions (Table 1).

Table 1. Simulated parameters in light-response curves of grafted *H. syriacus* leaves grown under either 0 (control) or 300 mM NaCl conditions. Hs: self-rooted *H. syriacus*; Hs/Hs: self-grafted *H. syriacus*; Hs/Hh: heterografted *H. syriacus* with *H. hamabo* as the rootstock. Data indicate means ± SE (n = 6). The different letters beside the values indicate significant differences. The *p*-values based on the ANOVA tests for grafting (G), NaCl (Na), and their interaction (G × Na) are indicated: * $p < 0.05$; *** $p < 0.001$; **** $p < 0.0001$; ns: not significant. Pnmax: the maximum net photosynthetic rate; LSP: the light saturation point; LCP: the light compensation point; AQY: the apparent quantum efficiency; Rd: the dark respiration rate.

Grafting	NaCl	Pnmax (μmol CO_2 m^{-2} s^{-1})	LSP (μmol m^{-2} s^{-1})	LCP (μmol m^{-2} s^{-1})	AQY (mol mol^{-1})	Rd (mmol CO_2 m^{-2} s^{-1})
Hs	0 mM	15.32 ± 1.42 a	1057.32 ± 186.86 ab	36.09 ± 4.26 a	0.11 ± 0.01 b	3.39 ± 0.15 a
	300 mM	5.97 ± 0.99 d	713.27 ± 63.24 b	45.17 ± 9.88 a	0.06 ± 0.01 c	2.25 ± 0.34 b
Hs/Hs	0 mM	12.99 ± 1.30 ab	1274.99 ± 366.86 a	35.11 ± 2.46 a	0.12 ± 0.00 b	3.36 ± 0.09 a
	300 mM	7.47 ± 1.74 cd	818.93 ± 29.76 ab	47.97 ± 0.20 a	0.08 ± 0.00 c	2.98 ± 0.05 ab
Hs/Hh	0 mM	15.28 ± 0.44 a	1051.27 ± 6.70 ab	44.86 ± 4.28 a	0.14 ± 0.00 a	3.53 ± 0.52 a
	300 mM	11.30 ± 1.26 bc	899.88 ± 149.53 ab	43.00 ± 6.96 a	0.08 ± 0.01 c	2.79 ± 0.12 ab
p-values	G	ns	ns	ns	*	ns
	Na	****	*	ns	****	***
	G × Na	ns	ns	ns	ns	ns

3.3. Chlorophyll Fluorescence Parameters

Although the Fm, Fm′, ΦPSII, qP, and ETR were significantly decreased in the leaves of Hs, Hs/Hs, and Hs/Hh exposed to NaCl compared to those grown under control conditions, the reduced levels of these indexes in Hs/Hh were significantly lower than those in Hs and Hs/Hs (Figure 4a,b,d,f,g). The Fv/Fm was decreased by 26% and 33% in the leaves of Hs and Hs/Hs, respectively, whereas it was not significantly altered in Hs/Hh leaves grown under NaCl conditions compared to those grown under control conditions (Figure 4c). The NPQ was increased by 55% and 276% in the leaves of Hs/Hs and Hs/Hh, and the ETR/Pn was increased by 292%, 185%, and 69%, respectively, in the leaves of Hs, Hs/Hs, and Hs/Hh treated with NaCl compared to those grown under control conditions (Figure 4e,g). The Fm, Fm′, Fv/Fm, ΦPSII, qP, and ETR in Hs/Hh were markedly higher than those in Hs and Hs/Hs under NaCl treatment, but no such differences were found in the leaves grown under control conditions (Figure 4a–d,f,g). The ETR/Pn in Hs/Hh (8.95) leaves was lower than that in Hs (21.92) and Hs/Hh (20.01) leaves under salt treatment, whereas no such difference was observed in the leaves grown under control conditions (Figure 4h). Grafting and salt stress had no effect on the Fo and Fs (Figure S1).

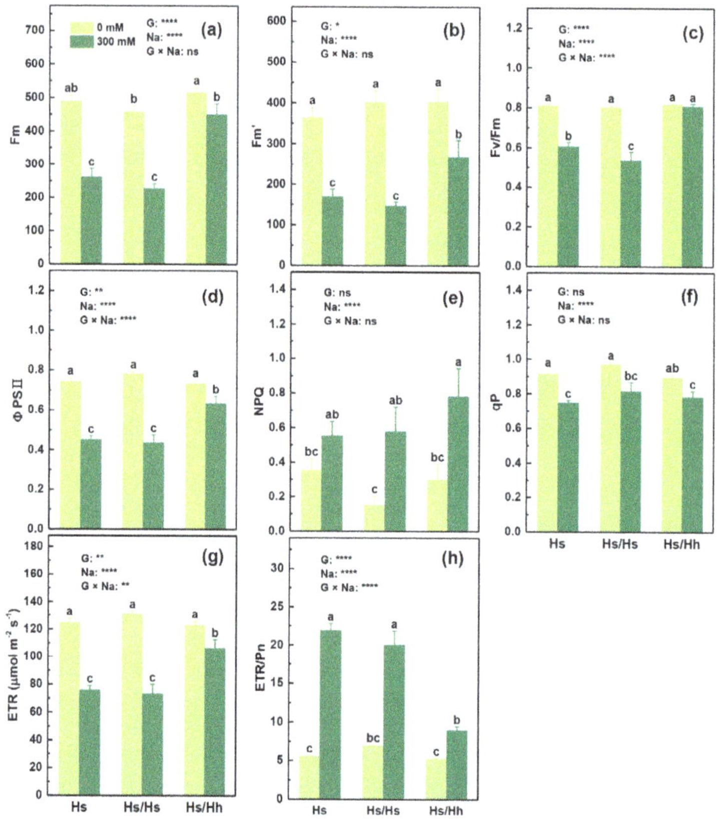

Figure 4. Fm (**a**), Fm′ (**b**), Fv/Fm (**c**), ΦPSII (**d**), NPQ (**e**), qP (**f**), ETR (**g**), and ETR/Pn (**h**) of grafted *H. syriacus* leaves grown under either 0 (control) or 300 mM NaCl conditions. Hs: self-rooted *H. syriacus*; Hs/Hs: self-grafted *H. syriacus*; Hs/Hh: heterografted *H. syriacus* with *H. hamabo* as the rootstock. Data indicate means ± SE (n = 6). The information about the treatments and statistical analysis is the same as that mentioned in Figure 1.

3.4. PCA Analysis of Photosynthetic Parameters

In order to investigate the relationship between varying the Pn and grafting and saline treatments, PCA was performed (Figure 5a, Table S2). PC1 and PC2 accounted for 48% and 22% of the total variation, respectively. PC1 clearly distinguished the effects of salt, and PC2 revealed the effects of grafting when the plants were exposed to NaCl treatments. Moreover, the differences between the control and salt-stress treatments in Hs/Hh were smaller than those in Hs and Hs/Hs, which indicated that Hs/Hh was more tolerant to saline stress than Hs and Hs/Hs. The variables Fo, Fs, qP, Gs, and E/Gs were key contributors to PC1, and the variables Fm, ΦPSII, ETR, chlorophyll a, chlorophyll (a + b), and Fm′ were essential factors for PC2. These results demonstrate that grafting induced a higher Pn capacity under salt conditions, which can be attributed mainly to the chlorophyll concentration, Fm, ΦPSII, and ETR.

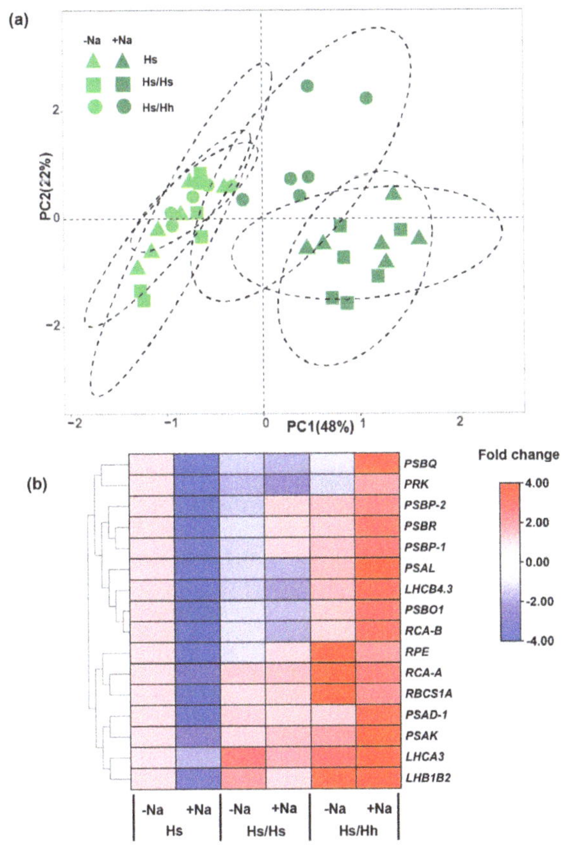

Figure 5. Principal component analysis (PCA) plots of photosynthetic parameters (**a**), cluster analysis of transcriptional fold changes of photosynthetic genes (**b**) of grafted *H. syriacus* leaves grown under either 0 (−Na, serving as control) or 300 mM NaCl (+Na) conditions. Hs: self-rooted *H. syriacus*; Hs/Hs: self-grafted *H. syriacus*; Hs/Hh: heterografted *H. syriacus* with *H. hamabo* as the rootstock.

3.5. Changes in mRNA Levels of Key Genes Participating in Photosynthesis

The transcript levels of representative genes involved in photosynthesis were assessed in the three plants (Figure 5b). These genes included *Photosystem I subunit D-1* (*PSAD-1*), *PSAK*, and *PSAL*, which encode complexes of PSI, and *Photosystem II oxygen-evolving complex 1* (*PSBO1*), *Photosystem II subunit P-1* (*PSBP-1*), *PSBP-2*, *PSBQ*, and *PSBR*, which encode the oxygen-evolving complexes of eukaryotic PSII. The expression of these genes was markedly lower in Hs leaves but was upregulated in Hs/Hh leaves treated with salt compared to those grown under control conditions. The expression of these genes was significantly higher in the leaves of Hs/Hh compared to Hs and Hs/Hh compared to Hs/Hs grown under salt treatment.

Photosystem I light-harvesting complex gene 3 (LHCA3), Photosystem II light-harvesting complex gene B1B2 (LHB1B2), and *light-harvesting complex Photosystem II (LHCB4.3)* are involved in capturing light energy in PSI and PSII. Except for *LHCB4.3* in Hs/Hs, the expression of these three genes was significantly lower in the leaves of Hs and Hs/Hs but was markedly higher in Hs/Hh leaves grown under salt conditions compared to those grown under control conditions. The mRNA levels of *LHCA3* and *LHB1B2* were upregulated in the leaves of Hs/Hh compared to Hs grown under control conditions. The transcript levels of *LHCA3*, *LHB1B2*, and *LHCB4.3* were markedly higher in the leaves of Hs/Hh compared to Hs and Hs/Hh compared to Hs/Hs grown under salt conditions.

Ribulose bisphosphate carboxylase small subunit 1A (RBCS1A), *rubisco activase-A (RCA-A)*, *RCA-B*, *ribulose-phosphate 3-epimerase (RPE)*, and *phosphoribulokinase (PRK)* participate in the Calvin–Benson–Bassham (CBB) cycle, which is a vital pathway of photosynthesis in plants. The expression of these five genes was markedly lower in Hs leaves grown under salt conditions compared to those grown under control conditions. The mRNA levels of *RPE*, *RCA-A*, and *RBCS1A* were downregulated, but the transcript levels of *RCA-B* and *PRK* were significantly higher in Hs/Hh leaves grown under salt conditions compared to those grown under control conditions. The expression of *RPE*, *RCA-A*, and *RBCS1A* was markedly higher in the leaves of Hs/Hh compared to Hs and Hs/Hh compared to Hs/Hs, regardless of the NaCl treatment. The mRNA levels of RCA-B and PRK were higher in the leaves of Hs/Hh compared to Hs and Hs/Hh compared to Hs/Hs grown under salt conditions.

4. Discussion

4.1. Grafting onto H. hamabo Rootstock Ameliorates the Inhibition of Saline on the Photosynthetic Capacity of H. syriacus Leaves

Photosynthesis provides the primary material and energy for plant growth. The photosynthetic capacity of most plants has been reported to be reduced by salt stress [5,20,26], and saline-tolerant plants can better acclimate to NaCl stress with a lesser decrease in the Pn [27]. Consistently, the Pn and Pnmax were markedly decreased in the leaves of Hs, Hs/Hs, and Hs/Hh exposed to salt compared to those grown under control conditions. More importantly, Hs/Hh exhibited a higher Pn and Pnmax than Hs and Hs/Hs under saline conditions, demonstrating that grafting onto *H. hamabo* rootstock could largely enhance the salt-tolerance of *H. syriacus* by alleviating the limitations of salt stress on its photosynthetic capacity. This result is consistent with previous studies on grafting in salt-treated watermelon, cucumber, and tomato (*Lycopersicon esculentum* Mill.) [5,28,29].

In general, salt-induced restrictions on the Pn can result from either stomatal closure or non-stomatal-related limitations [30]. For instance, a marked decrease in the Pn, Gs, and Ci has been observed in NaCl-exposed non-grafted and grafted apple (*Malus pumila* Mill.), tomato, and cucumber, implying that stomatal limitation is mainly responsible for the salt-induced reduction in the Pn [28,30,31]. In contrast, we found that the Pn and Gs were markedly decreased, whereas the Ci was not altered by salt in Hs, Hs/Hs, and Hs/Hh, indicating that non-stomatal limitation was likely the dominant factor for the reduced Pn under salt stress. Similar results were observed in 200 mM NaCl-treated *H. syriacus* [20]. Moreover, Hs/Hh exhibited a higher WUE than Hs and Hs/Hs when grown under saline treatment. Since the WUE is calculated as the ratio of Pn to E, the higher WUE in the salt-exposed Hs/Hh may have been caused by the grafting-induced improvement in photosynthetic performance. Having a higher WUE is significant for a plant's salt tolerance because it helps reduce the absorption of salt and alleviates water deficiency under salt stress [32].

4.2. Grafting Alleviates the Inhibition of Salt Stress on Photosynthetic Capacity through Amelioration of Photosynthetic Pigment Reduction

Chlorophyll is the main photosynthetic pigment in higher plants. The light energy harvested by chlorophyll is converted to stable chemical energy under a series of complex reactions in the chloroplast, and plant photosynthesis will be limited due to the damage and degradation of chlorophyll [20]. A reduced chlorophyll content has been observed in many plants suffering from salt stress [20,33,34]. In this study, the concentrations of chlorophyll a and chlorophyll (a + b) were significantly decreased by salt stress in Hs and Hs/Hs. Since chlorophyll b was not markedly altered by salt stress, the decreased chlorophyll (a + b) content can mainly be attributed to the reduced concentration of chlorophyll a in salt-exposed Hs and Hs/Hs. Notably, chlorophyll a is the primary photosynthetic pigment involved in determining the photosynthetic rate [8]. Thus, the reduction in the Pn was likely mainly due to the decreased concentration of chlorophyll a in salt-treated Hs and Hs/Hs. Moreover, the chlorophyll a and chlorophyll (a + b) concentrations were not altered in Hs/Hh leaves exposed to salt compared to those grown under control conditions. This indicates that grafting effectively ameliorated the salinity-induced reduction in chlorophyll concentrations. Similar findings of grafting alleviating the salt-induced decline in the chlorophyll content have also been reported in watermelon, tomato, and apple [5,7,9,30]. These results suggest that the inhibition of salt stress on photosynthetic capacity through grafting is caused by the alleviation of reductions in the concentrations of chlorophyll a and chlorophyll (a + b).

4.3. Grafting Alleviates the Inhibition of Salt Stress on Photosynthetic Capacity through Amelioration of Limitations on Photochemical Efficiency

The decline in photochemical activity is known to be one of the non-stomatal factors that limits photosynthesis [35]. Therefore, the photochemical efficiency was determined in this study. The Fv/Fm, the maximum quantum use efficiency of PSII, is a classic parameter related to PSII functioning [36]. In the current study, the Fv/Fm was decreased by 26% and 33% in Hs and Hs/Hs under 300 mM NaCl treatment, indicating that photoinhibition occurred, which could be the result of damage to PSII [37]. Nonetheless, the constant Fv/Fm in Hs/Hh demonstrates that the salt-induced photoinhibition of PSII was repaired by the *H. hamabo* rootstock. Similar results were observed in salt-exposed grafted watermelon, cucumber, and tomato [7,28,38]. In this study, salt stress led to significant reductions in the ΦPSII and ETR in Hs, Hs/Hs, and Hs/Hh, likely owing to the decrease in the open PSII reaction centers (qP). Since the ΦPSII and ETR represent the photochemical conversion efficiency of PSII [38], their reductions indicate that the electron transporting for carbon fixation was limited by salt in the leaves of Hs, Hs/Hs, and Hs/Hh. This would harm the PSII complexes and cause the production of excess reactive oxygen species [7]. However, Hs/Hh exhibited a higher ΦPSII and ETR than Hs and Hs/Hs under saline treatment, suggesting that the *H. hamabo* rootstock can alleviate the salt-induced inhibition of the photochemical conversion efficiency of PSII in H. syriacus. Moreover, the ΦPSII and ETR were reduced, whereas the Fv/Fm was not altered in Hs/Hh, which was likely a result of the improvement in the NPQ under saline conditions (Figure 4e), implying that more energy was safely dissipated as heat in PSII [28].

In fact, photosynthesis is a very complex process that depends on the synergistic functioning of large multi-subunit pigment–protein complexes of PSII and PSI, which are embedded in specific regions of the thylakoid membrane [39]. The genes responsible for encoding the subunits of PSII and PSI are closely associated with the photochemical efficiency in plants [40,41]. For instance, the suppression of *PSBO1* and *PSBP2* through RNA interference (RNAi) led to a marked reduction in the Fv/Fm and ΦPSII and an accumulation of PSII core subunits [42–44]. Arabidopsis *psad1-1* mutant plants exhibited retarded growth and lower PSI activity, Fv/Fm, and ΦPSII, as well as downregulation of most genes participating in the light phase of photosynthesis compared to wild-type (WT) plants [41]. The transcript levels of *PSBO1*, *PSBP-1*, *PSBP-2*, *PSBQ*, *PSBR*, *PSAD-1*, *PSAK*, and *PSAL* were significantly reduced by salt in Hs, and the transcript levels of these genes in Hs/Hh were higher than in Hs and Hs/Hs under saline conditions, suggesting that *H. hamabo* rootstock can alleviate the salt-induced inhibition of mRNA levels of genes encoding the subunits of PSII and PSI in *H. syriacus*. Thus, these results indicate that the amelioration of the inhibition of salt on photosynthetic capacity through grafting can be attributed to the *H. hamabo* rootstock-enhanced expression of genes encoding the subunits of PSII and PSI, as well as the stimulation of the photochemical efficiency in *H. syriacus* under saline conditions.

4.4. Grafting Alleviates the Limitation of Salt Stress on Photosynthetic Capacity through Amelioration of Inhibition on mRNA Levels of Genes Involved in the CBB Cycle

The CBB cycle is responsible for CO_2 fixation in photosynthesis [45]. In this cycle, ribulose 1,5-bisphosphate carboxylase/oxygenase (Rubisco) is a primary rate-limiting enzyme that catalyzes the binding of CO_2 to ribulose-1,5-bisphosphate (RuBP) to form 3-phosphoglycerate [46,47]. Ribulose-phosphate 3-epimerase (RPE) catalyzes the conversion of xylulose-5-P to form ribulose-5-phosphate (Ru5P) [48], and phosphoribulokinase (PRK) catalyzes the ATP-dependent phosphorylation of Ru5P to form RuBP [45]. The de-activation of the Rubisco and a reduced Pn have often been observed in long-term NaCl stress [29]. The mRNA levels of *RBCS1A*, which encodes a Rubisco small subunit that constitutes Rubisco [49], as well as *RCA-A* and *RCA-B*, which code for Rubisco activase responsible for activating Rubisco [50], and *RPE* and *PRK* were downregulated by saline in Hs. The transcript levels of these genes in Hs/Hh were significantly higher than those in Hs and Hs/Hh under salt conditions. Similar results have been found in salt-exposed grafted watermelon [7]. T-DNA insertion mutants of *rbcs1a* revealed significant reductions in the Rubisco activity and net photosynthetic rates compared to wild-type (WT) *Arabidopsis* [49]. RCA uses the energy from ATP hydrolysis to restore catalytic competence to Rubisco [47]. A previous study revealed that the reduction in the net CO_2 assimilation was related to Rubisco deactivation owing to the inhibition of RCA under moderate heat stress in *A. thaliana* [51]. Therefore, these results suggest that the amelioration of the inhibition of salt stress on photosynthetic capacity through grafting can be attributed to the higher expression of *RBCS1A*, *RCA*, *RPE*, and *PRK* under saline conditions.

5. Conclusions

Taken together, our results show that after 23 days of NaCl treatment, the leaves of Hs and Hs/Hs exhibited chlorosis, whereas the leaves Hs/Hh remained green. Saline stress markedly decreased the Pn, Pnmax, E, Gs, WUE, chlorophyll a, chlorophyll (a + b), Fm, Fm′, Fv/Fm, ΦPSII, and ETR in the leaves of Hs and Hs/Hs. In contrast, the reductions in these parameters were greatly alleviated when *H. syriacus* was grafted onto *H. hamabo* rootstock. Since the Ci was not markedly altered by 300 mM NaCl in Hs, Hs/Hs, and Hs/Hh, it indicates that these three plants experienced significant non-stomatal limitations to photosynthesis under salt conditions. Consistent with the changes in the Fv/Fm, ΦPSII, and ETR, genes encoding subunits of PSII and PSI were downregulated by salt in Hs, and the transcript levels of these genes were higher in Hs/Hh than in Hs and Hs/Hs under salt conditions. Moreover, *H. hamabo* rootstock grafting upregulated *RBCS1A*, *RCA*, *RPE*, and *PRK*, which were involved in the CBB cycle in *H. syriacus* scions under saline treatment. These results demonstrate that grafting can alleviate the inhibition of salt on the photosynthetic capacity of *H. syriacus*, which is mainly attributed to ameliorated limitations on photosynthetic pigments, photochemical efficiency, and the CBB cycle.

Supplementary Materials: The following supporting information can be downloaded at https://www.mdpi.com/article/10.3390/f14061226/s1, Figure S1. Fs and Fo of grafted *H. syriacus* leaves; Table S1. Primers used for RT-qPCR; Table S2. PCA loadings of physiological parameters.

Author Contributions: S.Z., W.Y. and Z.L.: visualization, investigation, data curation, methodology, software, and writing—original draft preparation. P.X., Z.W., J.H., C.G. and J.C.: methodology, visualization, investigation, and data curation. Y.L.: conceptualization, supervision, and writing—reviewing and editing. All authors have read and agreed to the published version of the manuscript.

Funding: This study was jointly supported by the Jiangsu Special Fund on Technology Innovation of Carbon Dioxide Peaking and Carbon Neutrality (BE2022420), the Jiangsu Institute of Botany Talent Fund (JIBTF202208), and the Jiangsu Science and Technology Plan Project (BE2021367).

Institutional Review Board Statement: Not applicable.

Informed Consent Statement: Not applicable.

Data Availability Statement: All data supporting the findings of this study are available within this paper and its Supplementary Materials published online.

Conflicts of Interest: The authors declare no conflict of interest.

References

1. Behera, T.K.; Krishna, R.; Ansari, W.A.; Aamir, M.; Kumar, P.; Kashyap, S.P.; Pandey, S.; Kole, C. Approaches involved in the vegetable crops salt stress tolerance improvement: Present status and way ahead. *Front. Plant Sci.* **2022**, *12*, 3104. [CrossRef] [PubMed]
2. Li, W.; Meng, R.; Liu, Y.; Chen, S.; Jiang, J.; Wang, L.; Zhao, S.; Wang, Z.; Fang, W.; Chen, F. Heterografted chrysanthemums enhance salt stress tolerance by integrating reactive oxygen species, soluble sugar, and proline. *Hortic. Res.* **2022**, *9*, uhac073. [CrossRef] [PubMed]
3. Luo, J.; Shi, W.; Li, H.; Janz, D.; Luo, Z.-B. The conserved salt-responsive genes in the roots of *Populus* × *canescens* and *Arabidopsis thaliana*. *Environ. Exp. Bot.* **2016**, *129*, 48–56. [CrossRef]
4. Li, H.; Lin, J.; Yang, Q.-S.; Li, X.-G.; Chang, Y.-H. Comprehensive analysis of differentially expressed genes under salt stress in pear (*Pyrus betulaefolia*) using RNA-Seq. *Plant Growth Regul.* **2017**, *82*, 409–420. [CrossRef]
5. Yan, Y.; Wang, S.; Wei, M.; Gong, B.; Shi, Q. Effect of Different Rootstocks on the Salt Stress Tolerance in Watermelon Seedlings. *Hortic. Plant J.* **2018**, *4*, 239–249. [CrossRef]
6. Rahman, A.; Hossain, M.S.; Mahmud, J.-A.; Nahar, K.; Hasanuzzaman, M.; Fujita, M. Manganese-induced salt stress tolerance in rice seedlings: Regulation of ion homeostasis, antioxidant defense and glyoxalase systems. *Physiol. Mol. Biol. Plants* **2016**, *22*, 291–306. [CrossRef]
7. Yang, Y.; Yu, L.; Wang, L.; Guo, S. Bottle gourd rootstock-grafting promotes photosynthesis by regulating the stomata and non-stomata performances in leaves of watermelon seedlings under NaCl stress. *J. Plant Physiol.* **2015**, *186–187*, 50–58. [CrossRef]
8. Barhoumi, Z.; Atia, A.; Hussain, A.A.; Albinhassan, T.H.; Saleh, K.A. Effects of high salinity on photosynthesis characteristics, leaf histological components and chloroplasts ultrastructure of *Avicennia marina* seedlings. *Acta Physiol. Plant.* **2022**, *44*, 85. [CrossRef]
9. Feng, X.; Guo, K.; Yang, C.; Li, J.; Chen, H.; Liu, X. Growth and fruit production of tomato grafted onto wolfberry (*Lycium chinense*) rootstock in saline soil. *Sci. Hortic.* **2019**, *255*, 298–305. [CrossRef]
10. Patil, S.; Shinde, M.; Prashant, R.; Kadoo, N.; Upadhyay, A.; Gupta, V. Comparative Proteomics Unravels the Differences in Salt Stress Response of Own-Rooted and 110R-Grafted Thompson Seedless Grapevines. *J. Proteome Res.* **2020**, *19*, 583–599. [CrossRef]
11. Niu, M.; Huang, Y.; Sun, S.; Sun, J.; Cao, H.; Shabala, S.; Bie, Z. Root respiratory burst oxidase homologue-dependent H_2O_2 production confers salt tolerance on a grafted cucumber by controlling Na^+ exclusion and stomatal closure. *J. Exp. Bot.* **2018**, *69*, 3465–3476. [CrossRef] [PubMed]
12. Zhang, H.H.; Li, X.; Zhang, S.B.; Yin, Z.P.; Zhu, W.X.; Li, J.B.; Meng, L.; Zhong, H.X.; Xu, N.; Wu, Y.N. Rootstock Alleviates Salt Stress in Grafted Mulberry Seedlings: Physiological and PSII Function Responses. *Front. Plant Sci.* **2018**, *9*, 1806. [CrossRef]
13. Lazare, S.; Yasuor, H.; Yermiyahu, U.; Kuhalskaya, A.; Brotman, Y.; Ben-Gal, A.; Dag, A. It takes two: Reciprocal scion-rootstock relationships enable salt tolerance in 'Hass' avocado. *Plant Sci.* **2021**, *312*, 111048. [CrossRef] [PubMed]
14. Simpson, C.R.; Nelson, S.D.; Melgar, J.C.; Jifon, J.; Schuster, G.; Volder, A. Effects of salinity on physiological parameters of grafted and ungrafted citrus trees. *Sci. Hortic.* **2015**, *197*, 483–489. [CrossRef]
15. Magdalita, P.M.; San Pascual, A.O. Hibiscus (*Hibiscus rosa-sinensis*): Importance and Classification. In *Floriculture and Ornamental Plants*; Datta, S.K., Gupta, Y.C., Eds.; Springer: Singapore, 2020; pp. 1–44. [CrossRef]
16. Punasiya, R.; Devre, K.; Pillai, S. Pharmacognostic and Pharmacological overview on *Hibiscus syriacus* L. *Int. J. Pharm. Life Sci.* **2014**, *5*, 3617–3621.
17. Wang, Z.; Xue, J.-Y.; Hu, S.-Y.; Zhang, F.; Yu, R.; Chen, D.; Van de Peer, Y.; Jiang, J.; Song, A.; Ni, L.; et al. The genome of *Hibiscus hamabo* reveals its adaptation to saline and waterlogged habitat. *Hortic. Res.* **2022**, *9*, uhac067. [CrossRef] [PubMed]
18. Sakhanokho, H.F.; Islam-Faridi, N.; Babiker, E.M.; Nelson, C.D.; Stringer, S.J.; Adamczyk Jr, J.J. Determination of nuclear DNA content, ploidy, and FISH location of ribosomal DNA in *Hibiscus hamabo*. *Sci. Hortic.* **2020**, *264*, 109167. [CrossRef]
19. He, J.; Ma, C.; Ma, Y.; Li, H.; Kang, J.; Liu, T.; Polle, A.; Peng, C.; Luo, Z.-B. Cadmium tolerance in six poplar species. *Environ. Sci. Pollut. Res.* **2013**, *20*, 163–174. [CrossRef] [PubMed]
20. Lu, W.; Wei, G.; Zhou, B.; Liu, J.; Zhang, S.; Guo, J. A comparative analysis of photosynthetic function and reactive oxygen species metabolism responses in two hibiscus cultivars under saline conditions. *Plant Physiol. Biochem. PPB* **2022**, *184*, 87–97. [CrossRef]
21. Maxwell, K.; Johnson, G.N. Chlorophyll fluorescence—A practical guide. *J. Exp. Bot.* **2000**, *51*, 659–668. [CrossRef]
22. Toral-Juárez, M.A.; Avila, R.T.; Cardoso, A.A.; Brito, F.A.L.; Machado, K.L.G.; Almeida, W.L.; Souza, R.P.B.; Martins, S.C.V.; DaMatta, F.M. Drought-tolerant coffee plants display increased tolerance to waterlogging and post-waterlogging reoxygenation. *Environ. Exp. Bot.* **2021**, *182*, 104311. [CrossRef]
23. Cao, X.; Jia, J.; Zhang, C.; Li, H.; Liu, T.; Jiang, X.; Polle, A.; Peng, C.; Luo, Z.B. Anatomical, physiological and transcriptional responses of two contrasting poplar genotypes to drought and re-watering. *Physiol. Plant.* **2014**, *151*, 480. [CrossRef]
24. Lu, Y.; Deng, S.; Li, Z.; Wu, J.; Zhu, D.; Shi, W.; Zhou, J.; Fayyaz, P.; Luo, Z.B. Physiological Characteristics and Transcriptomic Dissection in Two Root Segments with Contrasting Net Fluxes of Ammonium and Nitrate of Poplar Under Low Nitrogen Availability. *Plant Cell Physiol.* **2022**, *63*, 30–44. [CrossRef] [PubMed]
25. Ye, Z.P.; Yu, Q. A coupled model of stomatal conductance and photosynthesis for winter wheat. *Photosynthetica* **2008**, *46*, 637–640. [CrossRef]

26. Penella, C.; Landi, M.; Guidi, L.; Nebauer, S.G.; Pellegrini, E.; San Bautista, A.; Remorini, D.; Nali, C.; Lopez-Galarza, S.; Calatayud, A. Salt-tolerant rootstock increases yield of pepper under salinity through maintenance of photosynthetic performance and sinks strength. *J. Plant Physiol.* **2016**, *193*, 1–11. [CrossRef] [PubMed]
27. Yan, K.; Wu, C.W.; Zhang, L.H.; Chen, X.B. Contrasting photosynthesis and photoinhibition in tetraploid and its autodiploid honeysuckle (*Lonicera japonica* Thunb.) under salt stress. *Front. Plant Sci.* **2015**, *6*, 227. [CrossRef]
28. He, Y.; Zhu, Z.; Yang, J.; Ni, X.; Zhu, B. Grafting increases the salt tolerance of tomato by improvement of photosynthesis and enhancement of antioxidant enzymes activity. *Environ. Exp. Bot.* **2009**, *66*, 270–278. [CrossRef]
29. Liu, Z.; Bie, Z.; Huang, Y.; Zhen, A.; Niu, M.; Lei, B. Rootstocks improve cucumber photosynthesis through nitrogen metabolism regulation under salt stress. *Acta Physiol. Plant.* **2013**, *35*, 2259–2267. [CrossRef]
30. Li, C.; Wei, Z.; Liang, D.; Zhou, S.; Li, Y.; Liu, C.; Ma, F. Enhanced salt resistance in apple plants overexpressing a *Malus vacuolar* Na^+/H^+ antiporter gene is associated with differences in stomatal behavior and photosynthesis. *Plant Physiol. Biochem. PPB* **2013**, *70*, 164–173. [CrossRef]
31. Zhen, A.; Bie, Z.; Huang, Y.; Liu, Z.; Lei, B. Effects of salt-tolerant rootstock grafting on ultrastructure, photosynthetic capacity, and H_2O_2-scavenging system in chloroplasts of cucumber seedlings under NaCl stress. *Acta Physiol. Plant.* **2011**, *33*, 2311–2319. [CrossRef]
32. Karaba, A.; Dixit, S.; Greco, R.; Aharoni, A.; Trijatmiko, K.R.; Marsch-Martinez, N.; Krishnan, A.; Nataraja, K.N.; Udayakumar, M.; Pereira, A. Improvement of water use efficiency in rice by expression of *HARDY*, an Arabidopsis drought and salt tolerance gene. *Proc. Natl. Acad. Sci. USA* **2007**, *104*, 15270–15275. [CrossRef] [PubMed]
33. Yang, Y.; Tang, R.J.; Jiang, C.M.; Li, B.; Kang, T.; Liu, H.; Zhao, N.; Ma, X.J.; Yang, L.; Chen, S.L.; et al. Overexpression of the *PtSOS2* gene improves tolerance to salt stress in transgenic poplar plants. *Plant Biotechnol. J.* **2015**, *13*, 962–973. [CrossRef] [PubMed]
34. Dos Santos Araujo, G.; de Oliveira Paula-Marinho, S.; de Paiva Pinheiro, S.K.; de Castro Miguel, E.; de Sousa Lopes, L.; Camelo Marques, E.; de Carvalho, H.H.; Gomes-Filho, E. H_2O_2 priming promotes salt tolerance in maize by protecting chloroplasts ultrastructure and primary metabolites modulation. *Plant Sci. Int. J. Exp. Plant Biol.* **2021**, *303*, 110774. [CrossRef]
35. Souza, R.; Machado, E.; Silva, J.; Lagôa, A.; Silveira, J. Photosynthetic gas exchange, chlorophyll fluorescence and some associated metabolic changes in cowpea (*Vigna unguiculata*) during water stress and recovery. *Environ. Exp. Bot.* **2004**, *51*, 45–56. [CrossRef]
36. Baker, N.R. Chlorophyll fluorescence: A probe of photosynthesis in vivo. Annu. Rev. *Plant Biol.* **2008**, *59*, 89–113. [CrossRef]
37. Lucini, L.; Rouphael, Y.; Cardarelli, M.; Canaguier, R.; Kumar, P.; Colla, G. The effect of a plant-derived biostimulant on metabolic profiling and crop performance of lettuce grown under saline conditions. *Sci. Hortic.* **2015**, *182*, 124–133. [CrossRef]
38. Liu, Z.X.; Bie, Z.L.; Huang, Y.; Zhen, A.; Lei, B.; Zhang, H.Y. Grafting onto *Cucurbita moschata* rootstock alleviates salt stress in cucumber plants by delaying photoinhibition. *Photosynthetica* **2012**, *50*, 152–160. [CrossRef]
39. Ilikova, I.; Ilik, P.; Opatikova, M.; Arshad, R.; Nosek, L.; Karlicky, V.; Kucerova, Z.; Roudnicky, P.; Pospisil, P.; Lazar, D.; et al. Towards spruce-type photosystem II: Consequences of the loss of light-harvesting proteins LHCB3 and LHCB6 in Arabidopsis. *Plant Physiol.* **2021**, *187*, 2691–2715. [CrossRef] [PubMed]
40. Che, Y.; Kusama, S.; Matsui, S.; Suorsa, M.; Nakano, T.; Aro, E.M.; Ifuku, K. Arabidopsis PsbP-Like Protein 1 Facilitates the Assembly of the Photosystem II Supercomplexes and Optimizes Plant Fitness under Fluctuating Light. *Plant Cell Physiol.* **2020**, *61*, 1168–1180. [CrossRef]
41. Ihnatowicz, A.; Pesaresi, P.; Varotto, C.; Richly, E.; Schneider, A.; Jahns, P.; Salamini, F.; Leister, D. Mutants for *photosystem I subunit D* of *Arabidopsis thaliana*: Effects on photosynthesis, photosystem I stability and expression of nuclear genes for chloroplast functions. *Plant J. Cell Mol. Biol.* **2004**, *37*, 839–852. [CrossRef]
42. Yi, X.; Hargett, S.R.; Frankel, L.K.; Bricker, T.M. The effects of simultaneous RNAi suppression of PsbO and PsbP protein expression in photosystem II of Arabidopsis. *Photosynth. Res.* **2008**, *98*, 439–448. [CrossRef] [PubMed]
43. Yi, X.; McChargue, M.; Laborde, S.; Frankel, L.K.; Bricker, T.M. The manganese-stabilizing protein is required for photosystem II assembly/stability and photoautotrophy in higher plants. *J. Biol. Chem.* **2005**, *280*, 16170–16174. [CrossRef] [PubMed]
44. Ifuku, K.; Yamamoto, Y.; Ono, T.-a.; Ishihara, S.; Sato, F. PsbP Protein, But Not PsbQ Protein, Is Essential for the Regulation and Stabilization of Photosystem II in Higher Plants. *Plant Physiol.* **2005**, *139*, 1175–1184. [CrossRef]
45. Yu, A.; Xie, Y.; Pan, X.; Zhang, H.; Cao, P.; Su, X.; Chang, W.; Li, M. Photosynthetic Phosphoribulokinase Structures: Enzymatic Mechanisms and the Redox Regulation of the Calvin-Benson-Bassham Cycle. *Plant Cell* **2020**, *32*, 1556–1573. [CrossRef] [PubMed]
46. Andersson, I.; Backlund, A. Structure and function of Rubisco. *Plant Physiol. Biochem.* **2008**, *46*, 275–291. [CrossRef] [PubMed]
47. Carmo-Silva, A.E.; Salvucci, M.E. The regulatory properties of Rubisco activase differ among species and affect photosynthetic induction during light transitions. *Plant Physiol.* **2013**, *161*, 1645–1655. [CrossRef]
48. Favery, B.; Lecomte, P.; Gil, N.; Bechtold, N.; Bouchez, D.; Dalmasso, A.; Abad, P. *RPE*, a plant gene involved in early developmental steps of nematode feeding cells. *EMBO J.* **1998**, *17*, 6799–6811. [CrossRef]
49. Izumi, M.; Tsunoda, H.; Suzuki, Y.; Makino, A.; Ishida, H. *RBCS1A* and *RBCS3B*, two major members within the Arabidopsis *RBCS* multigene family, function to yield sufficient Rubisco content for leaf photosynthetic capacity. *J. Exp. Bot.* **2012**, *63*, 2159–2170. [CrossRef]

50. Kim, S.Y.; Stessman, D.J.; Wright, D.A.; Spalding, M.H.; Huber, S.C.; Ort, D.R. Arabidopsis plants expressing only the redox-regulated Rca-α isoform have constrained photosynthesis and plant growth. *Plant J.* **2020**, *103*, 2250–2262. [CrossRef]
51. Kurek, I.; Chang, T.K.; Bertain, S.M.; Madrigal, A.; Liu, L.; Lassner, M.W.; Zhu, G. Enhanced thermostability of Arabidopsis Rubisco activase improves photosynthesis and growth rates under moderate heat stress. *Plant Cell* **2007**, *19*, 3230–3241. [CrossRef]

forests

MDPI

Article

The Synergistic Effects of AMF Inoculation and Boron Deficiency on the Growth and Physiology of *Camellia oleifera* Seedlings

Junying Liu [1], Mengxue Zhang [1], Jie Fan [1], Wenna Ding [1], Longsheng Chen [2], Jie Luo [1], Yongzhong Liu [1] and Li Mei [1,*]

[1] College of Horticulture and Forestry Sciences, Hubei Engineering Technology Research Center for Forestry Information, Huazhong Agricultural University, Wuhan 430070, China; liujunying@webmail.hzau.edu.cn (J.L.)

[2] Research Institute of Oil Tea Camellia, Hunan Academy of Forestry, Changsha 410004, China

* Correspondence: meili@mail.hzau.edu.cn; Tel.: +86-13871496862

Abstract: Arbuscular mycorrhizal fungi (AMF) symbiosis has been shown to improve the ability to obtain nutrients and resist adverse environmental conditions. However, there are few studies on the functions of AMF in the absorption and accumulation of boron (B). Moreover, it is still unclear whether the root colonization rates of AMF are limited by B deficiency. In this study, *Camellia oleifera* seedlings were planted in normal and boron-deficient substrates, and the seedlings were inoculated with *Funneliformis mosseae* or left uninoculated. The growth and physiological indices of *C. oleifera* seedlings were determined. The results of this experiment indicate that AMF inoculation increased the plant biomass, B content, B accumulation, and antioxidant enzyme activity in both normal and boron-deficient *C. oleifera* seedlings. Furthermore, boron deficiency resulted in a decrease in the AMF root colonization efficiency and the inhibition of *C. oleifera* seedlings' growth and physiological activity. These findings suggest that AMF inoculation could improve the resistance to B-deficiency stress. Additionally, the colonization efficiency of AMF was adversely affected by B deficiency; thus, AMF play a cooperative role with B in the growth and physiological functions of plants. The results provide a theoretical basis for taking measures to solve B-deficiency stress in *C. oleifera* and other plants' cultivation.

Keywords: *Camellia oleifera*; boron deficiency; arbuscular mycorrhizal fungi (AMF); root morphology; physiological parameters

Citation: Liu, J.; Zhang, M.; Fan, J.; Ding, W.; Chen, L.; Luo, J.; Liu, Y.; Mei, L. The Synergistic Effects of AMF Inoculation and Boron Deficiency on the Growth and Physiology of *Camellia oleifera* Seedlings. *Forests* **2023**, *14*, 1126. https://doi.org/10.3390/f14061126

Academic Editor: Stefan Arndt

Received: 27 March 2023
Revised: 23 May 2023
Accepted: 23 May 2023
Published: 30 May 2023

1. Introduction

Boron (B), as one of the essential trace elements, plays an important role in sugar transportation and hormone and phenolic metabolism for the growth and development of higher plants [1,2]. B deficiency induces the accumulation of reactive oxygen species and accelerates peroxidation in the root membrane system. B deficiency also affects the levels of phenylalanine and tyrosine in the roots, promotes the biosynthesis of salicylic acid, caffeic acid, and ferulic acid, and increases lignin contents in plant roots, which can lead to structural and morphological changes [3]. The typical symptoms of B deficiency in crops include bud development without flowering in *Gossypium hirsutum* and flowering without seed production in *Brassica napus* [4,5]. Boron application, on the other hand, has been shown to increase the grain size, reduce ear sterility, and increase the yield of *Oryza sativa* [6], as well as reduce crop sensitivity to abiotic stresses, such as drought, salinity, or heavy-metal toxicity [7]. However, the excessive application of B can lead to B toxicity due to the narrow suitable range of B demand in plants. Therefore, alleviating the adverse effects of B deficiency by improving the root absorption and utilization efficiency of B in plants is very crucial.

Arbuscular mycorrhizal fungi (AMF) are symbiotic with approximately 80% of terrestrial plants, increasing the efficiency of soil water and nutrient absorption while mitigating the negative effects of stress on plants [8,9]. AMF colonization can improve the plant's uptake and effective utilization of mineral nutrients, such as nitrogen and phosphorus. A few studies have demonstrated that AMF can alleviate excess B toxicity symptoms in plants by increasing biomass and decreasing the tissue B concentration [10,11]. AMF inoculation could also be an effective method to promote the phytoremediation of B-contaminated soils under salt and drought stresses [10]. However, the roles of AMF in B uptake and utilization under B-deficient conditions are poorly understood [12].

Previous research has confirmed that AMF can obtain carbohydrates from host plants [8,13], but it is still unclear whether the symbiosis between AMF and host plants is affected by the levels of mineral nutrients such as B. Recent studies have shown that B deficiency can negatively affect plant growth and the AMF colonization rate. For example, B deficiency reduced root growth and the AMF colonization rate of Norway spruce (*Picea abies*) [4,14,15]. These findings suggest that B may be necessary for effective symbiosis between AMF and host plants [16]. Moreover, some fungi can transport B in their hyphae, indicating that B may play an important role in the relationship between fungi and plants [11]. However, the specific effects of B-deficiency stress on plant AMF colonization in B-deficient environments and the physiological mechanism of the interaction between B deficiency and AMF remain unclear.

The plantations of *Camellia oleifera* cover an expansive area of over 5 million hectares in southern China. These plantations serve crucial functions in resolving the conflict between the demand for and supply of edible oil, boosting the income of farmers, and preserving the ecological balance of the region. The *C. oleifera* planting area is located in the area with red and yellow soil in southern China. The soil's B content in this region, affected by heavy rainfall and leaching, was generally below the normal range of 0.5–1 mg kg^{-1} because of the heavy rainfall leaching influence [17]. B deficiency results in *C. oleifera* plantations exhibiting symptoms such as flowering without bearing fruit, buds without blossoming, and cracks in fruit shells. These symptoms significantly impact the quality and productivity of plantations and serve as the primary inhibitors of the industry's growth.

It has been proven that *C. oleifera* roots can live in symbiosis with AMF [18]. To what extent AMF infection alleviates the symptoms of boron deficiency in *C. oleifera* and whether the AMF infection efficiency is limited by boron deficiency, however, are not clear yet. Clarifying the effects of AMF on B absorption and accumulation, as well as the effects of B deficiency on the AMF colonization efficiency, will provide a basis for understanding the interaction between B and AMF and its physiological mechanism. Additionally, the insights from this study will underpin the development of effective cultivation strategies and address the issue of B-deficiency stress in the *C. oleifera* industry.

2. Materials and Methods

2.1. Experimental Materials and Treatments

C. oleifera 'Chang Lin 40′ (CL40) seedlings were planted in the greenhouse with a plastic board roof, with no additional temperature and humidity control measures other than natural ventilation with open windows. The roots of 2-year-old seedlings were sterilized with 10% hydrogen peroxide (H_2O_2) for 15 s and then repeatedly washed with distilled water before transplanting them to bigger pots (5 L). The culture substrate was a mixture of pearlite and sand (pearlite–sand = 3:1, v/v), sterilized by high-pressure steam (121 °C, 0.11 Mpa, 2 h). *C. oleifera* seedlings were inoculated with or without *Funneliformis mosseae* on 1 July 2021. The inoculant was composed of fungal spores, mycelia, an inoculated root segment and other propagules, and a mixed matrix, which was purchased from the Institute of Root Research, Yangtze University. The active spore count in the inoculant was 70–80 per gram. To ensure full contact between the AMF agent and the root system, 25 g of the dried fungal agent was spread evenly onto the rhizosphere per pot of inoculant-treated *C. oleifera* seedlings. The same amount of sterilized inactive AMF agent was added in the same way in the non-AMF-inoculated treatments.

Inoculated and uninoculated *C. oleifera* seedlings were irrigated daily with 40 mL of Hoagland nutrient solution with normal B (2.86 mg L^{-1}) or deficient B (0 mg L^{-1}) per plant once a day. Four treatments were applied: AMF inoculation and normal B (+AM+B), AMF inoculation and B deficiency (+AM−B), no AMF inoculation and normal B (−AM+B), and no AMF inoculation and B deficiency (−AM−B). There were 30 replicate seedlings for each treatment.

2.2. Sampling and Plant Growth Parameters

Three replicate seedlings of *C. oleifera* were randomly selected for the measurement of plant height, ground diameter, and other growth indices 60 days after treatment. All plant samples were separated into roots, stems, and leaves and then killed at 105 °C for 30 min and dried at 75 °C until reaching constant weight. Subsequently, the biomass and root-to-shoot (R/S) ratio were calculated. The dried samples were ground into powder for the determination of B content. Three more replicate plants were chosen at random and separated into roots, stems, and leaves. The fresh roots were scanned using an Epson scanner (12000 XL), and the root length, surface area, and volume were examined using the WinRHIZO analysis software (Pro 32-bit 2019a). Some fresh roots were isolated from three plants to determine root vitality and the mycorrhizal infection rate, while root activity was evaluated using 2,3,5-triphenyl-tetrazolium chloride staining [19]. The fresh samples were put in buffer solution and then stored at −80°C in the refrigerator for the determination of physiological indicators.

The AMF root colonization rate was determined using the method outlined by Vierheilig et al. [20]. Specifically, 0.2 g of a fresh, fine root was cut into 1 cm sections, fixed in FAA until transparent, and then stained. Fifteen stained root segments were arranged neatly on glass slides, covered with lactic acid and a slide, and gently squeezed. Pictures were taken using a light microscope to visualize invasion points, mycelium, arbuscules, vesicles, and spores. The colonization rate of each root was determined using a grading score of 1, 10, 20, 30, 40, 50, 60, 70, 80, 90, or 100, and the number of root segments at each level was recorded to calculate the AMF root colonization rate for each treatment.

$$AMF\ colonization\ rate\ (\%) = \sum(0 \times N1 + 10 \times N2 + \ldots\ldots + 100 \times N11)/N \times 100\% \quad (1)$$

2.3. Boron Analysis and Physiological Parameter Evaluation

Three plants were mixed into one sample, and three duplicate samples were used for boron and physiological parameter determination. The boron contents were determined using the curcumin colorimetric method as described by Bingham [21]. Each dry sample weighed 0.533 g and was dry-ashed in a muffle furnace at 550 °C for 5 h. It was then dissolved in a 0.1 mol L^{-1} HCL solution. The supernatant was absorbed with a curcumin-oxalic acid solution, dried using a water bath, dissolved with alcohol, and filtered with dry filter paper into a light contrast groove. Finally, the boron content and accumulation were calculated.

The malondialdehyde (MDA) content was determined with the thiobarbituric acid reaction [22]. Into a graduated test tube, 2 mL of crude enzyme solution and 3 mL of a 5% trichloroacetic acid solution of 0.5% sulfobarbituric acid were added, and the mixture was heated in a boiling water bath for 10 min. After quickly cooling, it was centrifuged at 4500 r min^{-1} for 10 min. The supernatant was then measured for absorbance using a spectrophotometer at 600 nm and 532 nm (UV-4802), with distilled water as a 100% blank transmittance.

The soluble sugar content was determined using the method outlined by Zhang et al. [23]. Fresh and clean plant samples weighing 0.1 g were extracted two times in boiling water for 30 min each time with 5 mL of distilled water. After filtration into a 25 mL volumetric flask, 0.5 mL of the sample extract was combined with 1.5 mL of distilled water in a 10 mL graduated tube, and the absorbance was measured with a spectrophotometer (UV-4802) at a 630 nm wavelength to determine the soluble sugar content.

The activities of superoxide dismutase (SOD), peroxidase (POD), and catalase (CAT) were determined using the method described by Zhou and Leul [24]. Specifically, 0.5 g of the plant sample was placed in a pre-cooled mortar for enzyme solution extraction. Then, 1 mL of phosphate buffer was added to an ice bath, and the volume was increased to 5 mL, after which it was centrifuged at 4000 r.p.m. for 10 min.

SOD activity was determined by adding 0.05 mol L^{-1} phosphate buffer, 130 mmol L^{-1} methionine solution, 750 μmol L^{-1} tetrazole solution, 100 μmol L^{-1} EDTA-Na_2 solution, 20 μmol L^{-1} riboflavin solution, and crude enzyme solution to a test tube containing phosphate buffer. The reaction mixture was then exposed to 4000 lux daylight for 20 min. The absorbance of the reaction mixture was measured at 560 nm using a spectrophotometer (UV-4802).

POD activity was determined by adding 3 mL of 25 mmol L^{-1} guaiacol solution ($C_6H_4(OH)(OCH_3)$) to a centrifuge tube, followed by the addition of 0.2 mL of 250 mmol L^{-1} hydrogen peroxide solution. The reaction mixture was mixed thoroughly to make a homogeneous solution. Then, 0.1 mL of crude enzyme solution was added to the reaction mixture as a substrate and introduced evenly into the spectrophotometer (UV-4802). The absorbance of the reaction mixture was measured at 470 nm every 1 min for 10 min.

CAT activity was determined by adding 3 mL of 20 mmol L^{-1} hydrogen peroxide solution to a 10 mL centrifuge tube. Then, 50 μL of crude enzyme solution was quickly mixed with the reaction mixture and transferred to a quartz cuvette. The absorbance of the reaction mixture was continuously measured at 240 nm using a spectrophotometer (UV-4802) for 3 min, and the initial and final values were recorded.

2.4. Statistical Analysis

All data were analyzed using IBM SPSS Statistics 26 software. Before statistical analysis, data were tested for normality using appropriate tests. If normality assumption was not met, data were transformed before being subjected to significance analysis. Significance was analyzed using the Duncan method with a significance level of $p < 0.05$. The effect of B-deficiency stress treatment on morphological and physiological parameters was determined using two-way ANOVA.

3. Results

3.1. The Effect of Boron Deficiency and AMF Inoculation on the Growth Parameters

Among the four treatments, +AM+B-treated seedlings showed the highest height elongation, ground diameter thickening, shoot and root biomass, total biomass, and R/S ratio. They were followed by those obtained with +AM−B and −AM+B treatments, while the lowest values were observed in −AM−B-treated seedlings. The biomass and R/S ratio showed significant differences ($p < 0.05$) between the +AM+B and −AM−B treatments. Statistical analysis of the data presented in Table 1 indicated that AMF colonization significantly benefits the growth parameters under B-deficiency stress treatment.

Table 1. Effects of boron deficiency and AMF inoculation on growth parameters (n = 3).

Treatments	+AM+B	+AM−B	−AM+B	−AM−B
Height elongation (cm)	3.45 ± 0.21^{a}	2.81 ± 0.29^{a}	1.35 ± 0.08^{b}	1.26 ± 0.18^{b}
Ground diameter thickening (mm)	1.28 ± 0.14^{a}	1.04 ± 0.18^{a}	1.05 ± 0.11^{a}	0.42 ± 0.04^{b}
Root biomass (g)	5.15 ± 0.51^{a}	4.85 ± 0.40^{a}	4.57 ± 1.00^{a}	3.40 ± 0.50^{b}
Shoot biomass (g)	10.03 ± 0.36^{a}	9.47 ± 0.16^{a}	9.64 ± 0.16^{a}	7.29 ± 0.46^{b}
Total biomass (g)	16.02 ± 0.81^{a}	13.47 ± 0.74^{ab}	15.56 ± 0.82^{a}	11.29 ± 0.97^{b}
R/S ratio	0.69 ± 0.01^{a}	0.69 ± 0.01^{a}	0.67 ± 0.03^{a}	0.43 ± 0.05^{b}

Note: +AM+B: AMF inoculation and normal B; +AM−B: AMF inoculation and B deficiency; −AM+B: no AMF inoculation and normal B; −AM−B: no AMF inoculation and B deficiency. The same applies to tables below. Values are expressed as means ± SE. Significant differences between treatments were determined using Tukey's honestly significant difference test. Different lowercase letters indicate significant differences between treatments, $p < 0.05$.

3.2. Boron-Deficiency Effect on AMF Root Colonization Rate, Activity, and Morphology

The AMF colonization rate of +AM+B-treated seedlings was 46.1%, significantly higher than that of seedlings in the +AM−B treatment (35.6%). In contrast, no AMF colonization was observed in the roots of AMF non-inoculated *C. oleifera* seedlings. This result indicates that B deficiency could inhibit the development of the AMF symbionts (Figure 1). The root sections of AMF-inoculated plants showed the presence of mycelia, follicles, arbuscular, and spores of AMF. The inoculated roots were closely attached to the surface by hyphae and kept growing along the root surface and branches, forming multiple invasion points. The hyphae grew in the form of a binary fork, forming a dendritic structure. Root cortical cells produced multiple round or oval vesicles filled with cellular chambers (Figure 1B–D).

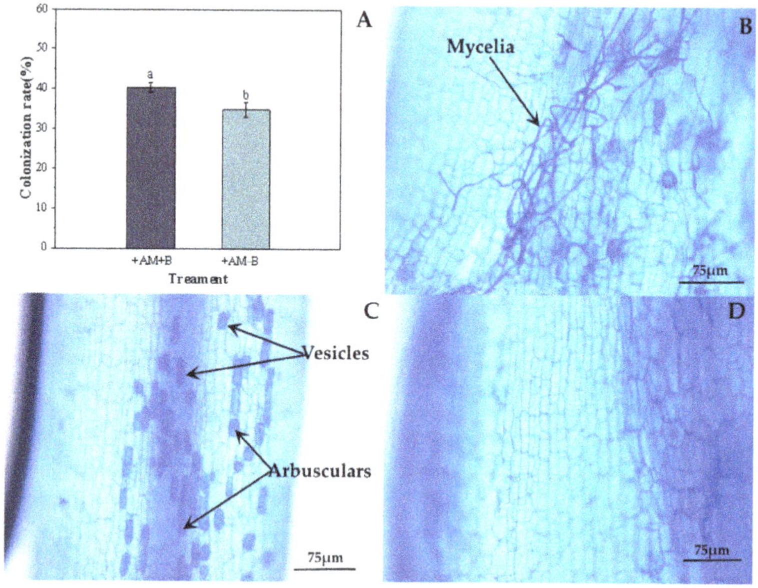

Figure 1. AMF root colonization rate (**A**), light micrographs of AMF-inoculated root (**B**,**C**), and AMF non-inoculated root (**D**) of *C. oleifera* seedlings. +AM+B: AMF inoculation and normal B; +AM−B: AMF inoculation and B deficiency. Values (**A**) are expressed as means ± SE. Significant differences between means were determined using significant difference T-test. Different lowercase letters indicate significant differences between treatments, $p < 0.05$.

The +AM+B-treated seedlings had the maximum root activity after treatment (Figure 2). AMF inoculation improved the root activity in both B-deficient and normal-B-supplied *C. oleifera* seedlings ($p < 0.05$).

The +AM+B-treated *C. oleifera* seedlings had the maximum total root length, root surface area, and root volume (Figure 2). The total root length, root surface area, and root volume of −AM+B-treated seedlings decreased by 5.9%, 33.6%, and 23%, respectively, and those in the +AM−B treatment decreased by 7.2%, 34.9%, and 36.2%, respectively, compared with +AM+B. Under B-deficient conditions, the total root length, root surface area, and root volume of AMF-inoculated *C. oleifera* seedlings were significantly higher than those of uninoculated seedlings 60 days after treatment ($p < 0.05$).

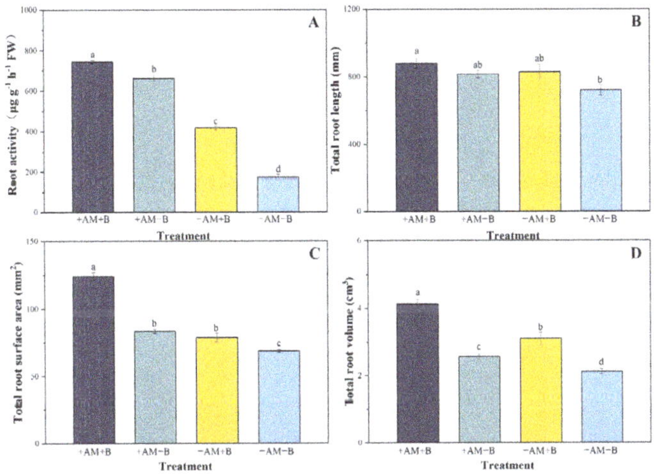

Figure 2. Effects of boron deficiency and AMF inoculation on root activity and root morphology of *C. oleifera* seedlings (n = 3). Root activity (**A**), total root length (**B**), total root surface area (**C**) and total root volume (**D**) of *C. oleifera* seedlings. +AM+B: AMF inoculation and normal B; +AM−B: AMF inoculation and B deficiency; −AM+B: no AMF inoculation and normal B; −AM−B: no AMF inoculation and B deficiency. Values are expressed as means ± SE. Significant differences between means were determined using Tukey's honestly significant difference test. Different lowercase letters indicate significant differences between treatments ($p < 0.05$).

3.3. Effect of Boron Deficiency and AMF Inoculation on Boron Content and Accumulation

AMF inoculation improved the B content in stems and leaves in both B-deficiency- and normal-B-supply-treated *C. oleifera* seedlings ($p < 0.05$) (Figure 3B,C). Under normal B treatment conditions, the B content in the roots, stems, and leaves of uninoculated seedlings decreased by 18.7%, 12.5%, and 12.1%, respectively, compared with AMF-inoculated treatments (Figure 3A–C). B accumulation in normal-B-treated plants was higher than that in B-deficiency-treated *C. oleifera* seedlings ($p < 0.05$). AMF inoculation significantly increased the plant's B accumulation in normal-B treatments (Figure 3D).

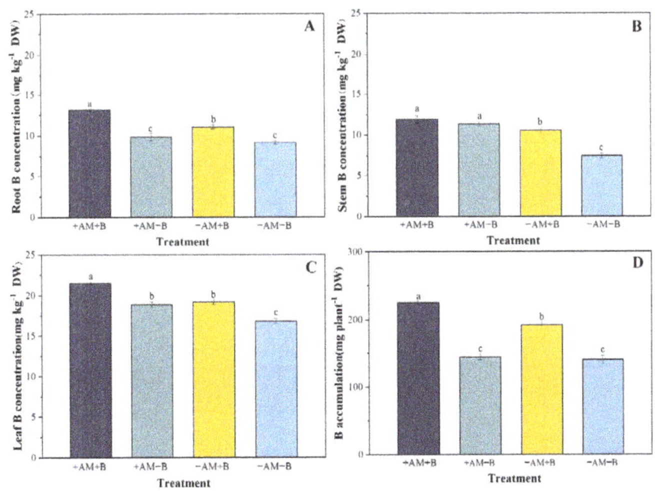

Figure 3. Effects of boron deficiency and AMF inoculation on boron concentration and accumulation in *C. oleifera* seedlings (n = 3). Root B concentration (**A**), stem B concentration (**B**), leaf B concentration

(**C**) and B accumulation (**D**) of *C. oleifera* seedlings. +AM+B: AMF inoculation and normal B; +AM−B: AMF inoculation and B deficiency; −AM+B: no AMF inoculation and normal B; −AM−B: no AMF inoculation and B deficiency. Values are expressed as means ± SE. Significant differences between means were determined using Tukey's honestly significant difference test. Different lowercase letters indicate significant differences between treatments ($p < 0.05$).

3.4. Effect of Boron Deficiency and AMF Inoculation on Antioxidant Enzyme Activity

AMF inoculation significantly increased soluble sugar contents in roots and leaves under different boron treatments ($p < 0.05$) of *C. oleifera* seedlings (Figure 4A,B). The MDA contents in roots and leaves significantly differed among the four treatments, and the MDA contents in leaves were significantly higher than that in the roots of *C. oleifera* seedlings ($p < 0.05$). B deficiency increased the MDA content, while AMF inoculation significantly decreased MDA contents in both the leaves and roots of *C. oleifera* seedlings (Figure 4C,D).

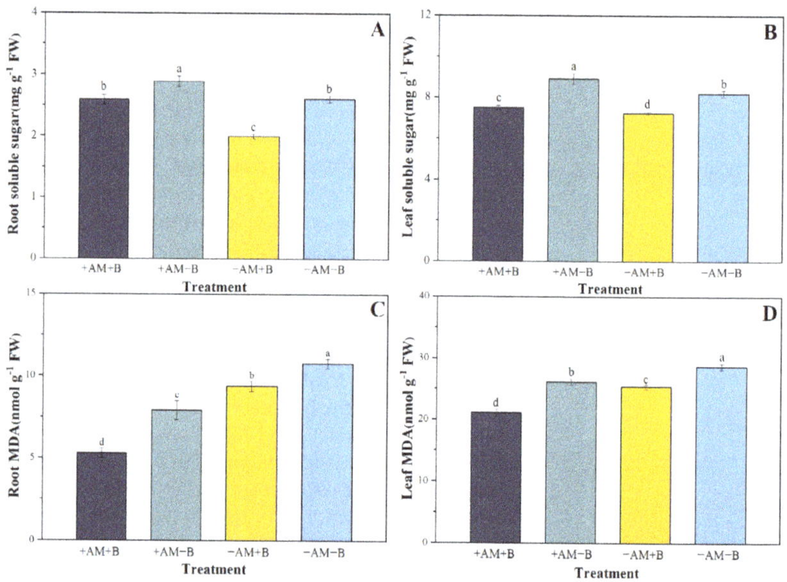

Figure 4. Effects of boron deficiency and AMF inoculation on soluble sugar and MDA contents of *C. oleifera* seedlings (n = 3). Root soluble sugar (**A**), leaf soluble sugar (**B**), Root MDA (**C**) and leaf MDA (**D**) of *C. oleifera* seedlings. +AM+B: AMF inoculation and normal B; +AM−B: AMF inoculation and B deficiency; −AM+B: no AMF inoculation and normal B; −AM−B: no AMF inoculation and B deficiency. Values are expressed as means ± SE. Significant differences between means were determined using Tukey's honestly significant difference test. Different lowercase letters indicate significant differences between treatments, $p < 0.05$.

The SOD activities in roots and leaves significantly differed among the four treatments of *C. oleifera* seedlings ($p < 0.05$). The +AM−B-treated plants had the highest SOD activity in roots and leaves (Figure 5A). The B-deficiency treatment significantly improved the SOD, POD, and CAT activities in the roots and leaves in both AMF-inoculated and non-inoculated *C. oleifera* seedlings. AMF inoculation enhanced SOD, POD, and CAT activities in roots and leaves in B-deficiency-treated *C. oleifera* seedlings ($p < 0.05$).

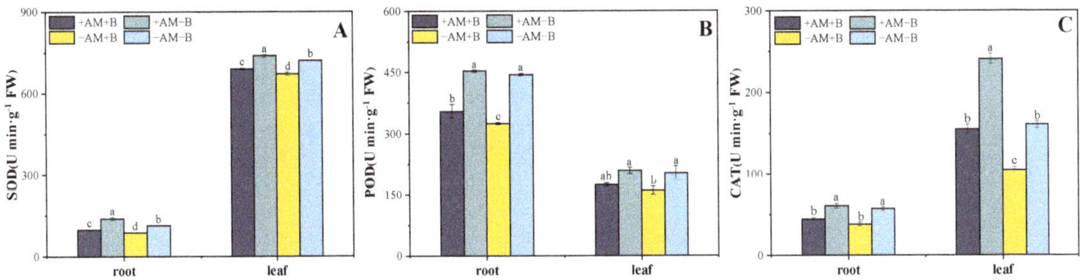

Figure 5. Effects of boron deficiency and AMF inoculation on the SOD, POD, and CAT activities of *C. oleifera* seedlings (n = 3). Root and leaf SOD (**A**), root and leaf POD (**B**), root and leaf CAT (**C**) of *C. oleifera* seedlings. +AM+B: AMF inoculation and normal B; +AM−B: AMF inoculation and B deficiency; −AM+B: no AMF inoculation and normal B; −AM−B: no AMF inoculation and B deficiency. Values are expressed as means ± SE. Significant differences between means were determined using Tukey's honestly significant difference test. Different lowercase letters indicate significant differences between treatments of roots or leaves ($p < 0.05$).

3.5. Principal Component Analysis

A principal component analysis of 14 indicators was performed for dimension reduction. There are two components with eigenvalues greater than 1, and PCA scores for the four treatments are shown in Figure 6. The first principal component eigenvalue is 8.610, and the larger contributing indexes are total root length and soluble sugar; the second principal component eigenvalue is 5.614, and the larger contributors are SOD and CAT. The main variables with high loads on axis 1 are the R/S ratio, root length, root surface area, root volume, root vitality, and malondialdehyde. The main variables with high loads on axis 2 are mainly soluble sugar, soluble protein, SOD, POD, CAT, and boron accumulation. There is a positive correlation between root activity and the mycorrhizal infection rate and R/S ratio. The four treatments are distributed in different quadrants, indicating significant differences between treatments.

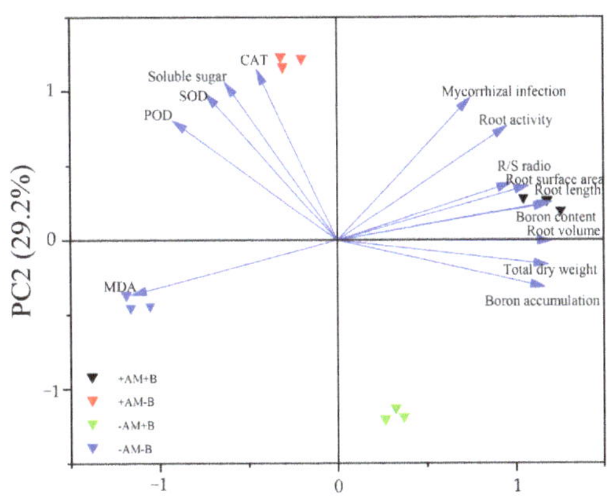

Figure 6. Principal component analysis of variables. The arrow lines of the variables are plotted as the correlation coefficients between them and the first two principal components. CAT, SOD, POD, and MDA are catalase, superoxide dismutase, peroxidase, and malondialdehyde, respectively, in *C. oleifera* plants.

According to the principal component analysis, a comprehensive evaluation model of *C. oleifera* quality under different treatments was established (F = 0.66F1 + 0.29F2), from which the comprehensive scores of four treatments (Table 2) were calculated, and the scores were ranked from high to low: +AM+B, +AM−B, −AM+B, and −AM−B.

Table 2. Scores and sorting of the principal components of all detected indexes in 4 treatments.

Treatment	Principal Component Score		Scores	Sort
	F1	F2		
+AM+B	3.58	0.54	2.52	1
+AM−B	−0.96	2.50	0.09	2
−AM+B	0.99	−2.30	−0.01	3
−AM−B	−3.61	−0.74	−2.59	4

Note: +AM+B: AMF inoculation and normal B; +AM−B: AMF inoculation and B deficiency; −AM+B: no AMF inoculation and normal B; −AM−B: no AMF inoculation and B deficiency.

The relationships between 14 indexes and 4 treatments were analyzed by principal component analysis. The AMF colonization rate, total biomass, root activity, soluble sugar, root length, root surface area, and root volume made significant contributions to PC1, while there was a large negative correlation between POD and PC1 (Figure 6).

4. Discussion

4.1. Effect of Boron Deficiency on AMF Colonization

The colonization of AMF depends on the host plant species, the classification of AMF, and the soil environment. Several studies have reported that soil environmental factors affect the growth of the AMF symbiont, such as low phosphorus levels, which inhibit plant AMF root colonization [25–27]. This result showed that B deficiency significantly reduced the AMF colonization rate, which expanded our knowledge by showing that the soil B level is also an important factor affecting AMF colonization.

Boron deficiency significantly decreases the plant's total root length, which will limit the contact chances between the root system and AMF in the soil to a certain extent. Many researchers have reported that B deficiency damages the root cell structure [28], which may hinder the invasion and decrease the colonization rate of AMF. In addition, B has been proven to be transported in the hyphae of most fungi, and B also plays a role in maintaining the cell wall structure of hyphae [11]. Under the condition of B deficiency, the stability of the AMF fungal vesicle envelope was impaired, resulting in the production of nitrogenase, and nitrogen fixation was almost non-existent [29]. Therefore, B deficiency inhibits hyphal growth, which is also the main reason that AMF reduce their ability to infect plants [16]. These results further confirm that B is necessary for the establishment of effective symbiosis between AMF and plants.

4.2. Effect of AMF on Plant Growth, Boron Absorption, and Utilization

B deficiency has been proven to inhibit the growth and elongation of plant roots [30]. These research results also showed that the total root length, total root surface area, total root volume, and root biomass of AMF-inoculated *C. oleifera* seedlings were significantly higher than those of non-inoculated seedlings, which is consistent with other research results showing that B deficiency limited the growth and development of plant roots [28,30]. AMF inoculation significantly increased the total root length, surface area, volume, and biomass of *C. oleifera* seedlings under B-deficient conditions, which is consistent with published studies [31,32]. The symbiotic relationship between AMF and plant roots effectively increased the root length, root surface area, root volume, and biomass, which facilitates plant adaptation to B deficiency and other stresses. AMF inoculation increased the B accumulation in both normal-B-supply and B-deficient conditions. The colonization of AMF promotes B absorption and accumulation and effectively alleviates growth and other physiological symptoms in B-deficient environments.

4.3. Effect of AMF and Boron Deficiency on Plant Physiological Characteristics

Plants under stress conditions will produce a series of physiological responses to reduce the damage to themselves. Soluble sugars, as important carbohydrates in plants, increase plant resistance to environmental stress through their osmoregulatory function [33]. Our results showed that AMF inoculation significantly increased the contents of soluble sugars and enhanced the cell stability of *C. oleifera* seedlings under different boron-supply treatments, the same result as in *C. sinensis* [34]. Moreover, the soluble sugar content was significantly higher in boron-deficient than normal-boron-treated *C. oleifera* seedlings. It has been speculated that the accumulation of soluble sugars may result in an increase in osmotic regulators and carbohydrate transportation to improve the adaptation to B deficiency and other stresses [35].

MDA is the final decomposition product of membrane lipid peroxidation, which suppresses the activity of cell-protective enzymes and antioxidant contents, resulting in membrane lipid peroxidation and severe plant cell death [36]. The root and leaf MDA contents of *C. oleifera* seedlings inoculated with AMF were significantly lower than those of non-AMF-inoculated seedlings, indicating that AMF could protect the plant membrane system and thus relieve the stress symptoms of B deficiency in *C. oleifera* seedlings.

The plant exoenzyme protection system consists of SOD, POD, and CAT [22]. CAT is a redox enzyme that converts H_2O_2 into H_2O and O_2, thus preventing H_2O_2 accumulation in the cell, and SOD acts as a key defensive enzyme against free radicals in catalyzing superoxide dismutation to H_2O_2 and O_2 [37], thus functioning to eliminate excess superoxide anions and sustain normal physiological metabolism under stress [38]. POD has a wide variety of physiological roles, including lignification, the crosslinking of polymers in cell walls, the formation of suberin, and resistance to stress [39]. The SOD, POD, and CAT activities were higher in the roots and leaves of AMF-inoculated seedlings than in those without AMF, indicating that AMF could improve antioxidant enzyme activity and protect plants from adversity [40,41]. The SOD, POD, and CAT activities in roots and leaves were significantly higher in B-deficiency-treated seedlings than in normal-B-treated *Brassica* seedlings, which is consistent with the findings of Pandey and Archana [42]. Another study also confirmed that AMF can effectively improve the activities of SOD, POD, and CAT in the leaves of tobacco (*Nicotiana tabacum* L.) and further improve the activities of plant antioxidant enzymes to alleviate toxic effects and enhance plant stress resistance [43].

5. Conclusions

The results demonstrated that AMF inoculation improved both seedling growth and B absorption and utilization efficiency, in addition to enhancing the physiological functions of *C. oleifera* seedlings under both normal and B-deficiency stress conditions. B deficiency had a direct and negative impact on growth and physiological parameters and also reduced the AMF root colonization rate of *C. oleifera* seedlings. These conclusions provide an important theoretical basis for understanding the effect of the interaction between B and AMF on plant growth, providing ideas for solving the problem of B deficiency in *C. oleifera* by AMF inoculation.

Author Contributions: Conceptualization, L.M. and J.L. (Jie Luo); methodology, J.L. (Junying Liu); software, J.F. and Y.L.; validation, J.L. (Junying Liu), M.Z. and W.D.; formal analysis, J.F.; investigation, J.L. (Junying Liu); resources, L.C.; data curation, J.L. (Junying Liu); writing—original draft preparation, J.L. (Junying Liu).; writing—review and editing, L.M.; visualization, J.L. (Junying Liu); supervision, L.M.; project administration, L.M.; funding acquisition, L.M. All authors have read and agreed to the published version of the manuscript.

Funding: This research was supported by grants from the National Key Research Program (2017YFC0505503) and the earmarked fund for CARS26.

Data Availability Statement: Not applicable.

Conflicts of Interest: The authors declare no conflict of interest.

References

1. Yan, L.; Li, S.; Muhammad, R.; Jiang, C.C. Proline metabolism and biosynthesis behave differently in response to boron-deficiency and toxicity in *Brassica napus*. *Plant Physiol. Biochem.* **2021**, *167*, 529–540. [CrossRef] [PubMed]
2. Mehboob, N.; Yasirasir, T.A.; Ul-Allah, S.; Nawaz, A.; Ahmad, N.; Hussain, M. Interactive effect of boron application methods and boron-tolerant bacteria (*Bacillus* sp. MN54) improves nodulation, grain yield, profitability and biofortification of kabuli chickpea grown under irrigated and rainfed conditions. *J. Soil Sci. Plant Nutr.* **2022**, *22*, 5068–5085. [CrossRef]
3. Wu, X.W.; Riaz, M.; Yan, L.; Jiang, C.C. Distribution and mobility of foliar-applied boron (^{10}B) in citrange rootstock under different boron conditions. *Plant Growth Regul.* **2020**, *39*, 575–582. [CrossRef]
4. Dell, B.; Huang, L.B. Physiological response of plants to low boron. *Plant Soil* **1997**, *193*, 103–120. [CrossRef]
5. Subedi, K.D.; Gregory, P.J.; Gooding, M.J. Boron accumulation and partitioning in wheat cultivars with contrasting tolerance to boron deficiency. *Plant Soil* **1999**, *214*, 141–152. [CrossRef]
6. Atique-ur-Rehman; Farooq, M.; Nawaz, A.; Rehman, A.; Iqbal, S. Soil application of boron improves the tillering, leaf elongation, panicle fertility, yield and its grain enrichment in fine-grain aromatic rice. *J. Plant Nutr.* **2015**, *38*, 338–354. [CrossRef]
7. García-Sánchez, F.; Simón-Grao, S.; Martínez-Nicolás, J.J.; Alfosea-Simón, M.; Liu, C.G.; Chatzissavvidis, C.; Pérez-Pérez, J.G.; Cámara-Zapata, J.M. Multiple stresses occurring with boron toxicity and deficiency in plants. *J. Hazard. Mater.* **2020**, *397*, 122713. [CrossRef]
8. Smith, S.E.; Read, D.J. *Mycorrhizal Symbiosis*, 3rd ed.; Academic Press: London, UK, 2008.
9. Lehto, T.; Zwiazek, J.J. Ectomycorrhizas and water relations of trees, a review. *Mycorrhiza* **2011**, *21*, 71–90. [CrossRef]
10. Liu, C.G.; Dai, Z.; Cui, M.Y.; Lu, W.K.; Sun, H.W. Arbuscular mycorrhizal fungi alleviate boron toxicity in *Puccinellia tenuiflora* under the combined stresses of salt and drought. *Environ. Pollut.* **2018**, *240*, 557–565. [CrossRef]
11. Simón-Grao, S.; Nieves, M.; Martínez-Nicolás, J.J.; Alfosea-Simón, M.; Cámara-Zapata, J.M.; Fernández-Zapata, J.C.; García-Sánchez, F. Arbuscular mycorrhizal symbiosis improves tolerance of Carrizo citrange to excess boron supply by reducing leaf B concentration and toxicity in the leaves and roots. *Ecotox. Environ. Saf.* **2019**, *173*, 322–330. [CrossRef]
12. Lavola, A.; Aphalo, P.J.; Lehto, T. Boron and other elements in sporophores of ectomycorrhizal and saprotrophic fungi. *Mycorrhiza* **2001**, *21*, 155–165. [CrossRef]
13. Nehls, U. Mastering ectomycorrhizal symbiosis: The impact of carbohydrates. *J. Exp. Bot.* **2008**, *59*, 1097–1108. [CrossRef]
14. Möttönen, M.; Lehto, T.; Aphalo, P.J. Growth dynamics and mycorrhizas of Norway spruce (*Picea abies*) seedings in relation to boron supply. *Trees* **2001**, *15*, 319–326. [CrossRef]
15. Räisänen, M.; Repo, T.; Lehto, T. Cold acclimation was partially impaired in boron deficient Norway spruce seedings. *Plant Soil* **2007**, *292*, 271–282. [CrossRef]
16. Bolaños, L.; Redondo-Nieto, M.; Bonilla, I.; Wall, L.G. Boron requirement in the *Discaria trinervis* (Rhamnaceae) and Frankia symbiotic relationship. Its essentiality for Frankia BCU110501 growth and nitrogen fixation. *Physiol. Plant.* **2002**, *115*, 563–570. [CrossRef]
17. Zhang, Z.Q.; Mao, P.; Yu, D.S.; Xu, L. Typical soil pH change characteristics in red soil area in the past 25 years—Taking Yujiang County of Jiangxi Province as an example. *Act. Pedol. Sin.* **2018**, *55*, 1545–1553. (In Chinese)
18. Lin, Y.L.; Li, Z.Y.; Zhang, L.P.; Wu, F.; Yang, Y.; Tan, M.X.; Hu, D.N. Effects of organic phosphorus and AM fungi on growth, root morphology and photosynthetic characteristics of *Camellia oleifera*. *Non-Wood For. Res.* **2021**, *39*, 121–128+210. (In Chinese)
19. Wang, J.E.; Li, D.W.; Zhang, Y.L.; Zhao, Q.; He, Y.M.; Gong, Z.H. Defence responses of pepper (*Capsicum annuum* L.) infected with incompatible and compatible strains of *Phytophthora capsici*. *Eur. J. Plant Pathol.* **2013**, *136*, 625–638. [CrossRef]
20. Vierheilig, H.; Coughlan, A.P.; Wyss, U.; Piche, Y. Ink and vinegar, a simple staining technique for arbuscular mycorrhizal fungi. *Appl. Environ. Microbiol.* **1998**, *64*, 5004–5007. [CrossRef] [PubMed]
21. Bingham, F.T. Boron. In *Methods of Soil Analysis Part 2: Chemical and Microbiological Properties*; American Society of Agronomy: Madison, WI, USA, 1982; pp. 431–446.
22. Carmak, I.; Horst, J.H. Effects of aluminum on lipid peroxidation, superoxide dismutase, catalase, and peroxidase activities in root tips of soybean (*Glycine max*). *Physiol. Plant.* **1991**, *83*, 463–468. [CrossRef]
23. Zhang, P.; Zhang, H.; Liu, J.N.; Huan, X.J.; Wang, Q.C.; Li, L.; Liu, Y.J.; Qin, P. Effects of drought and re-watering treatment on physiological and biochemical indexes of different drought-resistant wheat varieties/lines at seedling stage. *Acta Agric. Boreali-Occident. Sin.* **2020**, *29*, 1795–1802. (In Chinese)
24. Zhou, W.; Leul, M. Uniconazole-induced tolerance of rape plants to heat stress in relation to changes in hormonal levels, enzyme activities and lipid peroxidation. *Plant Growth Regul.* **1999**, *27*, 99–104. [CrossRef]
25. Tarafdar, J.C.; Marschner, H. Phosphatase activity in the rhizosphere and hyphosphere of VA mycorrhizal wheat supplied with inorganic and organic phosphorus. *Soil Biol. Biochem.* **1994**, *26*, 387–395. [CrossRef]
26. Tanaka, Y.; Yano, K. Nitrogen delivery to maize via mycorrhizal hyphae depends on the form of N supplied. *Plant Cell Environ.* **2005**, *28*, 1247–1254. [CrossRef]
27. Kazadi, T.A.; Lwalaba, J.L.W.; Ansey, B.K.; Muzulukwau, J.M.; Katabe, G.M.; Karul, M.I.; Baert, G.; Haesaert, G.; Mundende, R.P.M. Effect of phosphorus and arbuscular mycorrhizal fungi (AMF) inoculation on growth and productivity of maize (*Zea mays* L.) in a tropical ferralsol. *Gesunde Pflanz.* **2022**, *74*, 159–165. [CrossRef]

28. Camacho-Cristóbal, J.J.; Martín-Rejano, E.M.; Herrera-Rodríguez, M.B.; Navarro-Gochicoa, M.T.; Rexach, J.; González-Fontes, A. Boron deficiency inhibits root cell elongation via an ethylene/auxin/ros-dependent pathway in Arabidopsis seedlings. *J. Exp. Bot.* **2015**, *66*, 3831–3840. [CrossRef] [PubMed]
29. Zarea, M.J.; Karimi, N.; Goltapeh, E.M.; Ghalavand, A. Effect of cropping systems and arbuscular mycorrhizal fungi on soil microbial activity and root nodule nitrogenase. *J. Saudi Soc. Agric. Sci.* **2011**, *10*, 109–120. [CrossRef]
30. Chen, L.S.; Han, S.; Qi, Y.P.; Yang, L.T. Boron stresses and tolerance in citrus. *Afr. J. Biotechnol.* **2012**, *11*, 5961–5969.
31. Schellenbaum, L.; Berta, G.; Ravolanirina, F.; Tisserant, B.; Gianinazzi, S.; Fitter, A.H. Influence of endomycorrhizal infection on root morphology in a micropropagated woody plant species (*Vitis vinifera* L.). *Ann. Bot.* **1991**, *68*, 135–141. [CrossRef]
32. Borkowska, B. Growth and photosynthetic activity of micropropagated strawberry plants inoculated with endomycorrhizal fungi (AMF) and growing under drought stress. *Acta Physiol. Plant.* **2002**, *24*, 365–370. [CrossRef]
33. Tisarum, R.; Theerawitaya, C.; Samphumphuang, T.; Singh, H.P.; Chaum, S. Foliar application of glycinebetaine regulates soluble sugars and modulates physiological adaptations in sweet potato (*Ipomoea batatas*) under water deficit. *Protoplasma* **2020**, *257*, 197–211. [CrossRef] [PubMed]
34. Singh, S.; Pandey, A.; Kumar, B.; Palni, L.M.S. Enhancement in growth and quality parameters of tea [*Camellia sinensis* (L.) O. Kuntze] through inoculation with arbuscular mycorrhizal fungi in an acid soil. *Biol. Fertil. Soils* **2010**, *46*, 427–433. [CrossRef]
35. Huang, Z.; Zou, Z.; He, C.; He, Z.; Zhang, Z.; Li, J. Physiological and photosynthetic responses of melon (*Cucumis melo* L.) seedlings to three *Glomus* species under water deficit. *Plant Soil* **2011**, *339*, 391–399. [CrossRef]
36. Prasad, T.K. Role of catalase in inducing chilling tolerance in pre-emergent maize seedlings. *Plant Physiol.* **1997**, *114*, 1369–1376. [CrossRef] [PubMed]
37. Sreenivasulu, N.; Grimm, B.; Wobus, U.; Weschke, W. Differential response of antioxidant compounds to salinity stress in salt tolerant and salt-sensitive seedlings of foxtail millet (*Setaria italica*). *Physiol. Plant.* **2000**, *109*, 435–442. [CrossRef]
38. Singh, K.; Yadav, S.K.; Mishra, R.K. Differential effect of salinity on thermotolerance of SOD isoforms in seven varieties of *Vigna mungo* (L.) Hepper. *Plant Physiol. Rep.* **2019**, *24*, 279–288. [CrossRef]
39. Mazza, G.; Charles, C.; Ouchet, M.; Ricard, J.; Raynaud, J. Isolation, purification and physico-chemical properties of turnip peroxidases. *Biochim. Biophys. Acta* **1968**, *167*, 89–98. [CrossRef]
40. Akensous, F.Z.; Anli, M.; Boutasknit, A.; Ben-Laouane, R.; Ait-Rahou, Y.; Ahmed, H.B.; Nasri, N.; Hafidi, M.; Meddich, A. Boosting date palm (*Phoenix dactylifera* L.) growth under drought stress: Effects of innovative biostimulants. *Gesunde Pflanz.* **2022**, *74*, 961–982. [CrossRef]
41. Wang, J.P.; Yuan, J.H.; Ren, Q.; Zhang, B.; Zhang, J.C.; Huang, R.Z.; Wang, G.G. Arbuscular mycorrhizal fungi enhanced salt tolerance of *Gleditsia sinensis* by modulating antioxidant activity, ion balance and P/N ratio. *Plant Growth Regul.* **2022**, *97*, 33–49. [CrossRef]
42. Pandey, N.; Archana. Antioxidant responses and water status in *Brassica* seedlings subjected to boron stress. *Acta Physiol. Plant.* **2013**, *35*, 697–706. [CrossRef]
43. Begum, N.; Akhtar, K.; Ahanger, M.A.; Iqbal, M.; Wang, P.P.; Mustafa, N.S.; Zhang, L.X. Arbuscular mycorrhizal fungi improve growth, essential oil, secondary metabolism, and yield of tobacco (*Nicotiana tabacum* L.) under drought stress conditions. *Environ. Sci. Pollut. Res.* **2021**, *28*, 45276–45295. [CrossRef] [PubMed]

forests

MDPI

Article

More Effective Protection Supports Male Better Than Female Siblings over Water Deficit in Artificially Bred Poplar Hybrids

Fugui Chen [1], Yu Gong [1], Shuangyan Liu [1], Yiyun Wang [1], Linjie Luo [1], Guoping Zhu [1] and Han Zhao [2,*]

1 Anhui Provincial Key Laboratory of Molecular Enzymology and Mechanism of Major Diseases, College of Life Sciences, Anhui Normal University, Wuhu 241000, China; chenfg@ahnu.edu.cn (F.C.)
2 Research Institute of Non-Timber Forestry, Chinese Academy of Forestry, Zhengzhou 450003, China
* Correspondence: zhaohan@caf.ac.cn

Abstract: Sexually dimorphic response to stress has been observed in assorted natural dioecious plants. Up to now, few studies have focused on the difference of stress responses between artificially bred siblings. To determine the sexual dimorphism between artificially bred sibling poplar trees, we conducted a study comparing the response to water deficit between male and female *Populus* × *euramericana* siblings. This pair of hybrids was analyzed in terms of growth, photosynthesis, membrane injury and repair systems, as well as gene regulation patterns. The female and male siblings presented distinct responses to water deficit, with greater inhibition in females' growth and photosynthesis. The results also displayed that in females, relative electrolyte leakage and malonaldehyde content were higher than those in males under water deficit conditions. On the other hand, water deficit caused a greater increase in both SOD activity and POD activity in males than those in females. Consistent with these physiological differences, the expression of several stress-related genes, including *SOD*, *GST*, *bHLH35*, and *PsbX1*, was regulated differently between female and male hybrids by water deficit stress. Higher expression of *SOD* in moderate-water-deficit-treated females and higher *GST*, *bHLH35* expression in both moderate- and severe-water-deficit-treated females suggest that the female sib is more sensitive, whilst higher expression of *SOD* in severe-water-deficit-treated males and higher *PsbX1* expression in water-deficit-treated males testify that males protect cells better. To achieve an integrated view, all these variables were analyzed through the use of a principal component analysis and a total discrepancy between the sexes in their response to water deficit was demonstrated. The results indicate that, compared with male poplar sibs, females are more sensitive, but deploy a weaker protective apparatus to deal with water deficit.

Keywords: sexual dimorphism; siblings; water deficit; gene regulation; ROS scavenging; *Populus* × *euramericana*

Citation: Chen, F.; Gong, Y.; Liu, S.; Wang, Y.; Luo, L.; Zhu, G.; Zhao, H. More Effective Protection Supports Male Better Than Female Siblings over Water Deficit in Artificially Bred Poplar Hybrids. *Forests* **2023**, *14*, 995. https://doi.org/10.3390/f14050995

Academic Editor: Stefan Arndt

Received: 11 April 2023
Revised: 6 May 2023
Accepted: 10 May 2023
Published: 11 May 2023

1. Introduction

There are 15,600 dioecious plant species which have been identified to date, accounting for 5%–6% of the angiosperm plants on Earth [1]. Under environmental selection, dioecious plants separate females from male individuals, which has been hypothesized to increase outcrossing for facilitating species evolution and resolving intralocus sexual conflicts over the allocation of resources [2–5]. Despite their autosomal genetic similarity, females and males of various dioecious plants are different in morphological, physiological, and ecological features, including possessing different responses to environmental stresses [6,7]. In dioecious plants, sexual-related stress resistance bias is selected and evolves naturally along with sexual selection on both autosomes and sexual chromosomes over several decades to million years [2,8]. Artificially bred female and male siblings are generated as hybrids from a specific pair of parents, and are produced through carefully controlled fertilization, precluding paternal input except from the selected male parent. These hybrids are quite similar in autosomes, and this is reflected in similarities in morphological, physiological, and ecological features. However, these similarities do not include the similarity in responses to environmental stresses [9,10]. The distinct physiological responses to stress between artificially bred female and male siblings may be linked to gene expression or sex determination biases [3].

Water is indispensable to plants in retaining the balance of cell turgor, osmotic potential, photosynthesis, respiration, etc. [11–13]. Drought has induced huge pecuniary losses in global crop production in the last decade and will increasingly be a misfortune for agriculture, humanity, and livestock alike [14]. Water deficit, another meaning for drought for plants, greatly restricts plant distribution, growth, development, and productivity [15]. To perceive, avoid, and compensate for drought-induced harm, plants have evolved various defense strategies. The perception of stress by plants is through initiating several complex signaling networks, such as phytohormone level change, kinase/phosphatase signaling cascade regulation, stress-related gene expression, and reactive oxygen species (ROS) production [15,16]. However, the excess ROS induced by water deficit is harmful to cellular integrity and biomacromolecules. To survive, plants have to detoxify excess ROS toxicity through enzymatic and/or nonenzymatic mechanisms [17–19]. It can be considered that the capability of detoxification of ROS might be correlated with the ability of plants to resist stress. Under water deficit stress, resistant and sensitive plants respond differently, spanning from morphology and physiology to biochemistry and gene regulation. Typically, the sensitive plant responds quickly to water deficit but is exposed to more harm due to a weaker protective strategy for water deficiency than a tolerant plant at either cell, tissue, or overall plant level [14,20,21]. This theory is supported in many native dioecious species, whereby it appears that the female is the sensitive sex and is harmed more severely by stress compared to male plants [22–28]. Few research studies have focused on whether and how the sexual-specific response to water deficit between sexes is manifest between cultured female and male sibs [9,10].

Several gene expression mechanisms, especially stress response and resistance-related-gene regulation, endue plants with different responses and resistance to stress [14,15,29,30]. The response- and resistance-related genes' products, for sensing and resisting drought, are classified into functional and regulatory groups [30]. In the first group, there are water channels and transporters, detoxification enzymes, protection factors, and osmolyte biosynthesis enzymes and proteases, and in the regulatory groups, there are transcription factors, such as DREBs, AREB, MYC, MYB, bZIP, bHLH, and NAC, and protein kinases and phosphatases, phospholipid metabolism and ABA biosynthetic pathway components [30]. Superoxide dismutase (SOD) is a crucial antioxidant enzyme responsible for ROS scavenging and its expression was higher in tolerant plants compared with sensitive ones [31–33]. Glutathione S-transferase (GST) reduces hydroperoxides produced during oxidative stress and was upregulated in drought-tolerance-enhanced transgenetic poplar [34,35]. Basic helix–loop–helix (bHLH) transcription factors are involved in plant growth and development, secondary metabolite biosynthesis, photomorphogenesis, signal transduction, and stress response. *bHLH*-gene-overexpressed poplar show higher resistance to drought [36–38]. Photosystem II subunit X (PsbX) protein maintains efficient electron transport in PS II and safeguards PSII integrity for photosynthesis [39–41]. It is sensitive to stresses and greater PsbX expression, the greater integrity of PSII, and higher photosynthesis capacity, even facing stress [42].

Populus, a dioecious tree with a fully sequenced genome, is a typical model for surveying both physiological and genetic sexual-specific response to stress in woody plants [43–45]. Previous studies have revealed different responses between female and male plants in many native poplars to various stresses [28,46–51]. Nevertheless, these poplars investigated in these studies were natural populations or cuttings (clones) from either male or female plants. The different responses between these native sexes developed under sex-specific evolutionary selection from several decades to million years, which means that the genomes between native female and male poplars are greatly different in both sex chromosomes and autosomes. However, the female and male poplar sibs, which are hybrids from the same pair of parents, have been selected in only several years and are much more similar between females and males in morphological, physiological, and biochemical traits and gene expression, except for the sexes [9,10]. Here, to determine whether and what sexual dimorphism is present in artificial hybrids, the sibs of *P.* × *euramericana*, the female line 'Nanlin-895', and the male line 'Nanlin-1388', were chosen for a direct or integrated study of morphological variability, alteration of physiological and biochemical parameters, gene expression analysis, etc., under water deficit conditions.

2. Materials and Methods

2.1. Plant Materials and Water Deficit Treatment

The stalks of brother and sister sibling *P.* × *euramericana* were collected from the clones 'Nanlin-895' (*P.* × *euramericana* cv. 'Nanlin-895') and 'Nanlin-1388' (*P.* × *euramericana* cv. 'Nanlin-1388'), respectively, which are brother–sister sibs of the maternal clone 'I-69' (*P. deltoides* Bartr. cv. 'Lux') and the paternal clone 'I-45' (*P.* × *euramericana* (Dode) Guineir cv. 'I-45/51'). Stalks were planted in 5 L plastic pots filled with 3 kg of homogenized soil and 4 g of slow-release fertilizer (N:P:K = 13:10:14). Following two months of growth, eighteen male and eighteen female seedlings, similar in both stature and height, were chosen for the water deficit study. The seedlings were grown in a greenhouse with a glass shelter at Anhui Normal University in Wuhu, China. The experimental design was completely randomized with two factors: sex and water deficit stress. Water deficit treatment was set at three levels: control (70%–80% of soil water holding capacity (SWHC)), moderate water deficit stress (50%–60% of SWHC), and severe water deficit stress (30%–40% of SWHC). Each treatment involved six male and six female seedlings (three biological replicates per sex with two cuttings per biological replicate). To maintain the soil water content, each of the seedlings was watered with an adjustable volume water into the pots according to Li et al. [52]. The treatment was ongoing for three months from the 1 June to the 1 September 2019.

2.2. Morphology and Photosynthesis Assay

Along with the treatments, the seedlings' heights were measured every two days, and the shoot basal diameters (diameters of the boles at the soil surface) were measured every ten days. The growth curves for both height and basal diameter were generated using the binomial regression method in the SPSS 23.0 package (Chicago, IL, USA). At the end of the treatments, photographs of the plants were taken. Four cuttings from each treatment and sex were randomly selected for gas exchange rate assays. Light response curves were generated on the third or fourth fully expanded leaf. The parameters measured included the net photosynthesis rate (A), stomatal conductance (g_s), intercellular CO_2 concentration (C_i), transpiration rate (E), primary photochemical efficiency of PSII (Fv/Fm), sum of the quantum yields of PSII photochemistry (Φ_{PSII}), photochemical quenching (qP), and non-photochemical quenching (NPQ). These measurements were obtained using the Li-COR 6400 system (LI-COR, Inc., Lincoln, NE, USA) under the following conditions: leaf temperature of 25 ± 2 °C, photosynthetic photon flux (PPF) of 1400 μmol m^{-2} s^{-1}, relative air humidity of 70%, and ambient CO_2 concentration of 400 μmol mol^{-1}. Light response curves were generated using PPF values of 0, 100, 250, 500, 750, 1000, 1200, 1300, 1400, 1500, 1600, and 1800 μmol m^{-2} s^{-1}, at an ambient CO_2 concentration of 400 μmol mol^{-1}, leaf temperature of 25 ± 2 °C, and relative air humidity 70%. Gas exchange parameters and light response were measured between 10:00 a.m. and 1:00 p.m. The modified rectangular hyperbola model was used to generate a regression analysis for light response data, and the corresponding formula is as follows [53,54]:

$$A = \phi PAR \frac{1 - \beta\phi}{1 + \gamma\phi} - R_d$$

where φ represents the initial apparent quantum efficiency, PAR refers to photosynthetic photon flux density, R_d stands for the rate of dark respiration, and β and γ represent the corresponding coefficients. The maximum net photosynthetic rate (A_{max}) can be calculated using the following formula:

$$A_{max} = \phi\left(\frac{\sqrt{\beta+\gamma}-\sqrt{\beta}}{\gamma}\right) - R_d$$

Nonlinear regression analysis was performed using the SPSS version 23.0 (Chicago, IL, USA) software to generate the light response curve regression.

2.3. Chlorophyll Pigment Content Assay

To determine the chlorophyll pigment content, leaves from cuttings of each gender and treatment were cut into approximately 0.2 cm strips. About 0.1 g of each sample was then ground using a mortar and pestle with 5 mL of 80% acetone, and the mixture was transferred to a 50 mL Falcon tube, with the mortar and pestle being washed several times using 80% acetone. A total of 25 mL of mixture was obtained by adding 80% acetone. The mixture was incubated in darkness at room temperature for 12 h, with gentle stirring every 4–5 h. After filtration with cheese cloth, the absorbance of the extractive was measured at 645 and 663 nm with a spectrophotometer (SMA5000, Merinton, StellarNet, Inc., Tampa, FL, USA). The pigment contents were calculated using the follows [55]:

$$\begin{aligned} Chla &= 12.25 * A_{663} - 2.79 * A_{645} \\ Chlb &= 21.50 * A_{645} - 5.10 * A_{663} \\ TC &= 7.15 * A_{663} + 18.71 * A_{645} \end{aligned}$$

where *Chl a*, *Chl b*, and *TC* represent chlorophyll a, chlorophyll b, and total chlorophyll contents, respectively, and A_{645}, A_{663} are the absorbance values of the extracted solution at 645 and 663 nm, respectively. The unit of chlorophyll is expressed in mg mL^{-1}.

2.4. Relative Electrolyte Leakage Assay

The relative electrolyte leakage (REL) was measured according to the method described by Liao et al. [56]. Leaves from four cuttings of each sex and treatment were selected and washed with deionized water several times. Leaf discs (0.3 g) were prepared by avoiding the main veins and then incubated at room temperature for 30 min with gentle shaking every several minutes. The electrical conductivity (C1) of the bathing solution was measured using a conductivity detector (DDB-303A, INESA Analytical Instrument Co., Ltd., Shanghai, China). The glass tubes containing the bathing solution and leaf discs were then boiled for 10 min and allowed to cool to room temperature. The electrical conductivity (C2) of the boiled solution was measured, and the relative electrolyte leakage was calculated as follows:

$$\mathrm{REL}(\%) = (\mathrm{C1}/\mathrm{C2}) * 100$$

where C1 is the electrical conductivity of the bathing solution before boiling, and C2 is the electrical conductivity of the boiled solution. The REL is expressed as a percentage.

2.5. ROS Scavenging Enzyme Activity Assay

Four fully expanded leaves were randomly chosen from each replicate to conduct the ROS scavenging enzyme activity assay. The extraction of superoxide dismutase (SOD: EC 1.15.1.1) and peroxidase (POD: EC 1.11.1.7) was carried out according to a previously published method, and the activity was measured [57]. To extract the enzymes, 0.5 g of fresh leaves was ground in 5 mL of iced 50 μM phosphate buffer (pH 7.8) containing 1% *w/v* polyvinyl pyrrolidone (PVP) and then centrifuged at 12,000× *g*, 4 °C for 15 min. After that, 10 μL of the supernatant was mixed with 4 mL of reaction system (50 μM phosphate buffer (pH 7.8), 77.12 μM nitroblue tetrazolium chloride (NBT), 13.37 mM methionine, 0.1 mM ethylene diamine tetraacetic acid (EDTA), and 80.2 μM riboflavin). The reaction was initiated by illuminating the mixture with a white fluorescent lamp

(4000 Lux). After 20 min of illumination, the absorbance at 560 nm was measured using a UV spectrophotometer (SMA5000, Merinton). A negative control system without enzymes was used for baseline measurements. One unit activity of SOD (U) was defined as the amount of enzyme necessary to inhibit 50% of NBT reduction [58]. The POD activity was initiated by adding 20 μL of the supernatant to the POD reaction system which contained 50 μM of phosphate buffer (pH 6.0), 50 μM of guaiacol, and 50 μM of H_2O_2, and mixing by inverting and righting the tube three times. The rate of absorbance change at 470 nm was monitored immediately with a UV spectrophotometer (SMA5000, Merinton) for three minutes. POD activity was defined as the ability to convert guaiacol to tetraguaiacol and was evaluated from the change in absorbance value per minute [59].

2.6. Malondialdehyde Content Assay

To assess the level of membrane lipid peroxidation, the malondialdehyde (MDA) content was determined following a previously established method [23]. For each treatment, four fully expanded leaves were randomly selected, and 0.5 g of fresh leaves was homogenized in an ice bath using 5 mL of phosphate buffer (pH 7.8). The homogenate was then centrifuged at 12,000× *g* for 20 min at 4 °C. Next, 1 mL of the supernatant was mixed with 2 mL of the reaction mixture containing 0.6% (*w*/*v*) TBA and 10% (*v*/*v*) TCA; then, the mixture was incubated in boiling water for 15 min and quickly cooled in an ice bath. Subsequently, the mixture was centrifuged at 12,000× *g* for 10 min and the absorbance of the supernatant was measured at 450, 532, and 600 nm using a UV spectrophotometer (SMA5000, Merinton). Finally, the MDA content was calculated:

$$C(\mu M) = 6.45 * (A_{532} - A_{600}) - 0.56 * A_{450}$$

where A_{450}, A_{532}, and A_{600} denote the absorption of the supernatant at the wavelengths of 450 nm, 532 nm, and 600 nm respectively.

2.7. RNA Extraction and qRT-PCR Assay

Poplar leaves were subjected to total RNA extraction using the Trizol Total RNA Extractor Kit (Sangon Biotech, Shanghai, China) following the manufacturer's instructions. Subsequently, complementary DNA (cDNA) was synthesized using the Monscript™ RTIII Super Mix Kit (Monad Biotech Co., Ltd., WuHan, China) according to the manufacturer's guidelines. The primers for *SOD*, *GST*, *bHLH35*, and *PsbX1* transcript were developed using SnapGene software 2.3.2 and are listed in Supplementary Table S4. PCR was performed with the MonAmp™ ChemoHS qPCR Mix Kit (Monad Biotech Co., Ltd. WuHan, China) on Roche LightCycler™ 96, using SYBRGreen as the fluorescent detection dye. The internal control used was UBQ, and the $2^{-\Delta\Delta CT}$ method was employed for determining the relative expression of each gene [60]. The entire experiment was conducted in triplicate to ensure the accuracy of the results.

2.8. Statistical Analysis

To test the different responses, data of growth, gas exchange, chlorophyll pigment content, chlorophyll fluorescence, REL, MDA content, ROS enzyme activities, and gene expression were compared between the sexes and among the treatments using generalized linear models in IBM SPSS 23.0 package (Chicago, IL, USA). Two-way analysis of variance (ANOVA) with a post hoc Duncan multiple comparison was used for statistically significantly differing means at a $p < 0.05$ level. All data were tested for, and validated to be, both normally distributed and with a homogeneity of variance before comparisons were performed.

To assess and compare the composite different response between female and male siblings, the principal component analysis (PCA) was carried out based on the growth, photosynthetic and biochemical parameters, and gene expression of *P.* × *euramericana* siblings. Prior to PCA, the correlation between the traits was examined and is documented in Supplemental Table S5. The PCA analysis was conducted using SIMCA 13.0 software (Umetrics AB, Umea, Sweden).

3. Results

3.1. Water Deficit Had Different Effects on Growth, Net Photosynthesis, and Chlorophyll Fluorescence in Female and Male Poplar Sibs

Water deficit induced more serious leaf curling and chlorosis in females than in males (Figure 1A). The height and basal diameter increments were significantly reduced in both sexes by water deficit, and the reduction was more severe in females' basal diameter than that in males under severe water deficit (Figure 1B,D, Table S1). Furthermore, water deficit resulted in a greater depression in female height and basal diameter growth than in males (Figure 1C,E, Table S1). Water deficit decreased *A* in both female and male sibs and this decrease was statistically significantly greater in females (Table 1). However, other gas exchange parameters such as *gs*, *Ci*, and *E* were not greatly affected by water deficit and showed no significant difference between sexes of poplar cuttings (Table S2). Moreover, *Fv/Fm*, Φ_{PSII}, and *qP* were all statistically significantly lower in female sibs than in male sibs under water deficit conditions (Table 1).

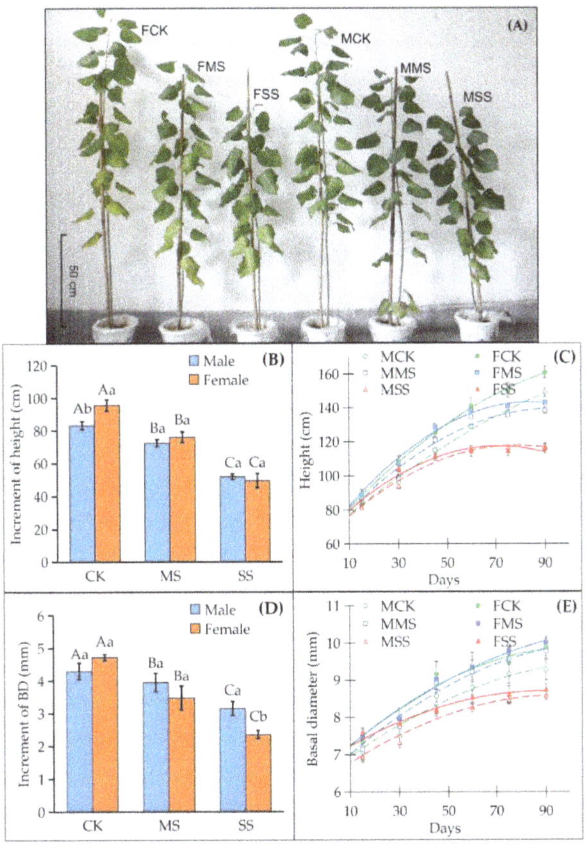

Figure 1. Phenotypic symptom and growth of female and male sibs of *P.* × *euramericana* under different irrigation conditions. (**A**) Phenotypic differences between sib lines; (**B**) increment of height growth; (**C**) height growth curves; (**D**) increment of basal diameter; (**E**) basal diameter growth curves.

MCK and FCK, male and female lines under control treatment (70%–80% of SWHC); MMS and FMS, male and female lines under moderate water deficit stress (50%–60% of SWHC); MSS and FSS, male and female lines under severe water deficit stress (30%–40% of SWHC). Different uppercase letters above the bars denote significant differences among the control and water-deficit-treated female and male lines separately, and different lowercase letters denote significant differences between the sexes of each treatment at the level of $p \leq 0.05$ according to Duncan post hoc tests. BD, basal diameter. Values are means ± SE ($n \geq 3$).

Table 1. Photosynthesis and chlorophyll fluorescence parameters of female and male *P.* × *euramericana* sibs under different irrigation conditions.

Parameters		A (μmol m^{-2} s^{-1})	Fv/Fm	Φ_{PSII}	qP
Male	CK	9.976 ± 0.208 Aa	0.824 ± 0.002 Aa	0.057 ± 0.003 Aa	0.116 ± 0.005 Aa
	MS	9.135 ± 0.391 ABa	0.823 ± 0.003 Aa	0.047 ± 0.001 Ba	0.096 ± 0.001 Ba
	SS	8.008 ± 0.752 Ba	0.808 ± 0.003 Ba	0.036 ± 0.001 Ca	0.075 ± 0.002 Ca
Female	CK	8.367 ± 0.562 Ab	0.825 ± 0.002 Aa	0.053 ± 0.001 Aa	0.116 ± 0.004 Aa
	MS	6.945 ± 0.494 ABb	0.817 ± 0.002 Ba	0.039 ± 0.001 Bb	0.083 ± 0.003 Bb
	SS	5.897 ± 0.676 Bb	0.799 ± 0.002 Cb	0.026 ± 0.001 Cb	0.050 ± 0.004 Cb
P-level	*P*stress	0.001	0.000	0.000	0.000
	*P*sex	0.000	0.018	0.000	0.000

Note: *A*, net photosynthesis rate; *Fv*/*Fm*, maximal PSII quantum yield; Φ_{PSII}, quantum yield in PSII; *qP*, photochemical quenching parameter. CK, control treatment (70%–80% of SWHC); MS, moderate water deficit stress (50%–60% of SWHC); SS, severe water deficit stress (30%–40% of SWHC). Within a column, different uppercase letters following values denote significant differences among the control and water-deficit-treated female and male lines separately, and different lowercase letters denote significant differences between the sexes of each treatment according to Duncan posthoc tests. P_{stress}, water deficit treatment effect; P_{sex}, sex effect. Values are means ± SE ($n \geq 3$).

3.2. *Female and Male Poplar Sibs Responded Differently to Light and the Chlorophyll Content Was Reduced Differently by Water Deficit*

Female and male hybrids showed different light response curve, especially under water deficit conditions (Figure 2). Regression analysis using the modified rectangular hyperbola model showed that the regressed *A* in males was higher than in females when the photosynthetic photon flux increased to above 500 μmol m^{-2} s^{-1}, and this difference was more pronounced under water deficit conditions. The coefficients of the fitted curves are presented in Table S3. Furthermore, *Amax*, regressed by the modified rectangular hyperbola model, was reduced in both males and females under water deficit stress, with a more noticeable reduction in females than that in males under severe water deficit (Table 2). Additionally, the chlorophyll a, b, and total chlorophyll content decreased in both female and male sibs under water deficit. The chlorophyll content was lower in females than males under both control and water deficit conditions (Table 2).

Table 2. The maximum net photosynthetic rate and chlorophyll pigments content of female and male *P.* × *euramericana* under different irrigation conditions.

Parameters		*Amax* (μmol m^{-2} s^{-1})	*Chl a* (mg mL^{-1})	*Chl b* (mg mL^{-1})	*TC* (mg mL^{-1})
Male	CK	11.045 ± 0.501 Aa	26.086 ± 0.194 Aa	47.599 ± 0.358 Aa	73.686 ± 0.552 Aa
	MS	9.473 ± 0.195 Ba	24.721 ± 0.176 Ba	45.086 ± 0.328 Ba	69.806 ± 0.504 Ba
	SS	7.444 ± 0.635 Ca	21.290 ± 0.139 Ca	38.787 ± 0.255 Ca	60.077 ± 0.394 Ca
Female	CK	10.684 ± 0.400 Aa	23.134 ± 0.261 Ab	42.155 ± 0.478 Ab	65.289 ± 0.738 Ab
	MS	8.522 ± 0.371 Ba	18.008 ± 0.601 Bb	32.802 ± 1.096 Bb	50.810 ± 1.697 Bb
	SS	6.059 ± 0.084 Cb	12.382 ± 0.399 Cb	22.569 ± 0.706 Cb	34.951 ± 1.105 Cb
P-level	*P*stress	0.019	0.000	0.000	0.000
	*P*sex	0.000	0.000	0.000	0.000

Note: A_{max}, maximum net photosynthetic rate; *Chl a*, chlorophyll a content; *Chl b*, chlorophyll b content; and *TC*, total chlorophyll content. CK, control treatment (70%–80% of SWHC); MS, moderate water deficit stress (50%–60% of SWHC); SS, severe water deficit stress (30%–40% of SWHC). Within a column, different uppercase letters following values denote significant differences among the control and water-deficit-treated female and male lines separately, and different lowercase letters denote significant differences between the sexes of each treatment according to Duncan post hoc tests. The *p* values for water deficit, sex, and their combined effects are denoted. P_{stress}, water deficit treatment effect; P_{sex}, sex effect. Values are means ± SE ($n \geq 3$).

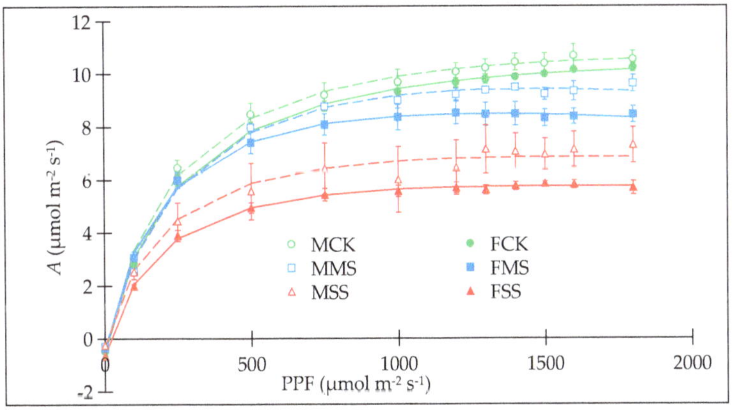

Figure 2. Light response curves of female and male *P.* × *euramericana* hybrids under different irrigation conditions. MCK and FCK, male and female lines under control treatment (70%–80% of SWHC); MMS and FMS, male and female lines under moderate water deficit stress (50%–60% of SWHC); MSS and FSS, male and female lines under severe water deficit stress (30%–40% of SWHC). Values are means ± SE (n = 3).

3.3. Water Deficit Damaged Cell Membrane Differently and Induced MDA Contents Differently between Female and Male Sibs

The parameter REL is used to measure plant membrane damage, and its analysis was conducted on both female and male *P.* × *euramericana* under different irrigation conditions. It was observed that the water deficit resulted in an increase in REL in both males and females. Additionally, when subjected to severe water deficit conditions, the REL levels were significantly higher in females than in males (Figure 3A). Another key observation was related to MDA, which is the final product of lipid oxidation caused by ROS and is known to be harmful to other macromolecules within the cell. The content of MDA was examined in female and male *P.* × *euramericana* under different irrigation conditions. It was found that after water deficit treatments, MDA contents increased in both sexes (Figure 3B). Notably, under severe water deficit conditions, the MDA content in females was significantly higher than that in males (as shown in Figure 3B).

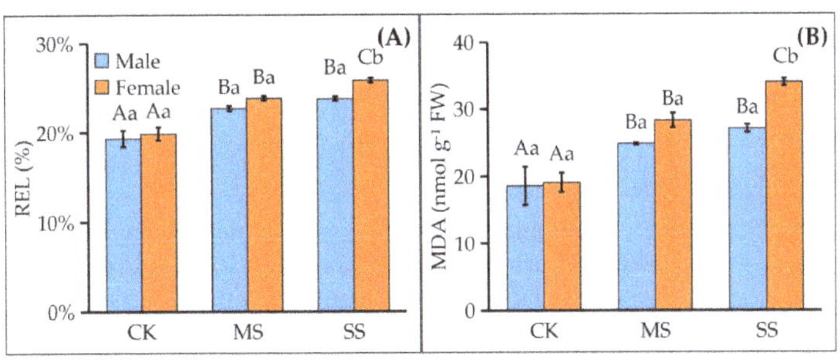

Figure 3. Relative electrolyte leakage (**A**) and MDA content (**B**) of female and male *P.* × *euramericana* lines under different irrigation conditions. CK, control treatment (70%–80% of SWHC); MS, moderate water deficit stress (50%–60% of SWHC); SS, severe water deficit stress (30%–40% of SWHC). Different uppercase letters above the bars denote significant differences among the control and water-deficit-treated female and male lines separately, and different lowercase letters denote significant differences between the sexes of each treatments according to at the level of $p \leq 0.05$ according to Duncan post hoc tests. Values are means ± SE ($n \geq 3$).

3.4. Water Deficit Induced Different Activities of ROS Scavenging Enzymes

It is widely recognized that a plant's ability to scavenge ROS is positively correlated with its resistance to stress. Therefore, in this study, the activities of ROS scavenging enzymes, SOD and POD, were measured in female and male cuttings subjected to different irrigation treatments. As anticipated, the enzyme activities in control females and males were quite similar. However, the response to water deficit treatment was distinct between the two genders. Specifically, the activities of SOD (Figure 4A) and POD (Figure 4B) were significantly higher in males following exposure to water deficit.

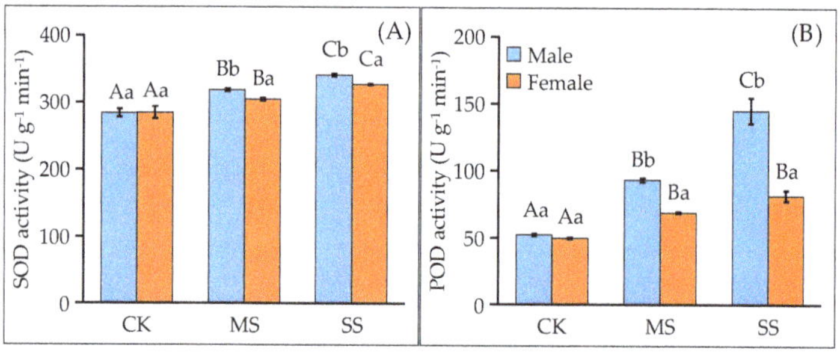

Figure 4. Activities of SOD (**A**) and POD (**B**) of female and male *P.* × *euramericana* lines under different irrigation conditions. CK, control treatment (70%–80% of SWHC); MS, moderate water deficit stress (50%–60% of SWHC); SS, severe water deficit stress (30%–40% of SWHC). Different uppercase letters above the bars denote significant differences among the control and water-deficit-treated female and male lines separately, and different lowercase letters denote significant differences between the sexes of each treatments according to at the level of $p \leq 0.05$ according to Duncan post hoc tests. Values are means ± SE (n = 3).

3.5. Water Deficit Regulated Stress-Related Gene Expression Differently in Male and Female Poplar Hybrids

In order to clarify the different regulation patterns in response to water deficit, we investigated the expression of four stress-related genes in female and male *P.* × *euramericana* sibs. It was found that under moderate water deficit conditions, the expression of *SOD* and *GST* was significantly higher in females than in males (Figure 5A,B). However, under severe water deficit, *SOD* expression was higher in male sibs (Figure 5A). The gene *bHLH35* was upregulated in females under water deficit stress but downregulated in males, and the expression level in females was higher than that in males under both moderate and severe water deficit (Figure 5C). In contrast, *PsbX1* gene was downregulated in females but upregulated in males in response to water deficit (Figure 5D).

3.6. PCA Displayed Comprehensively Different Responses between Male and Female Sibs to Water Deficit

To gain a comprehensive understanding of the differential responses to water deficit between female and male sibs of *P.* × *euramericana*, we performed a principal component analysis (PCA) based on growth, physiological and biochemical parameters, and gene expression data of stress-related genes after different irrigation treatments (Figure 6, Table S6). The first principal component (PC1), which accounted for 58.6% of the total variation, was identified as the stress axis, effectively separating the control group from the water deficit treatment groups, except for the moderate-water-deficit-treated males in the control group. The second principal component (PC2), accounting for 19.4% of the total variation, was identified as the sex axis. PC2 separated the male and female sibs into upper and lower regions, except for MCK3, which was located in the lower region (Figure 6A). The weights of the variables' contribution to PCA revealed that the distribution of REL, MDA content, and NPQ aligned with the direction of severe-water-deficit-treated females, whereas the distribution of SOD activity and gene expression of *GST* and *bHLH35* (in Figure 6B) aligned with the direction of severe-water-deficit-treated males (Figure 6). Other variables, such as gas exchange, chlorophyll content, and growth, contributed to PCA in distinguishing control from water-deficit-treated samples (Figure 6B).

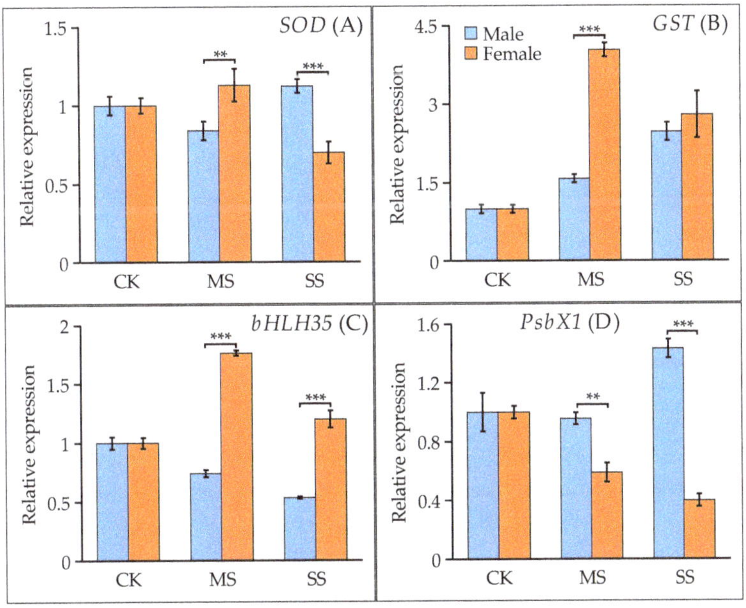

Figure 5. Expression of stress-related genes in female and male *P.* × *euramericana* sibs under different irrigation conditions. (**A**) *SOD*, (**B**) *GST*, (**C**) *bHLH35*, (**D**) *PsbX1*. CK, control treatment (70%–80% of SWHC); MS, moderate water deficit stress (50%–60% of SWHC); SS, severe water deficit stress (30%–40% of SWHC). Duncan post hoc tests at the following levels, **: $0.001 < p < 0.01$; ***: $p < 0.001$. Values are means ± SE ($n \geq 3$).

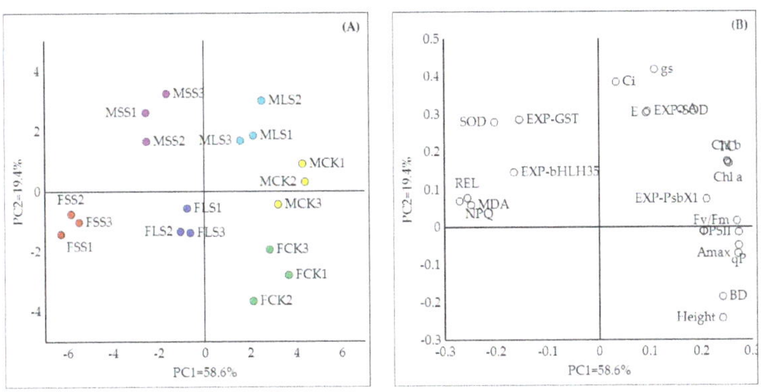

Figure 6. PCA of physiological, biochemical, and the gene expression parameters response to water deficit of female and male *P.* × *euramericana* lines under different irrigation conditions. (**A**) Score scatter plot and (**B**) loading scatter plot of all parameters. MCK and FCK, male and female lines under control treatment (70%–80% of SWHC); MMS and FMS, male and female lines under moderate water deficit stress (50%–60% of SWHC); MSS and FSS, male and female lines under severe water deficit stress (30%–40% of SWHC); EXP with hyphen in front of each gene in (**B**) indicates gene expression value.

4. Discussion

We present evidence of sexual dimorphism in the response to water deficit in artificially bred female and male poplar siblings. Our study investigated the physiological, biochemical, and gene regulation responses of female and male poplar lines to water deficit, and the results indicated that artificially bred male poplar siblings were stronger in resisting water deficit due to more effective ROS scavenging over female siblings.

4.1. Females Were More Vulnerable to Water Deficit in Growth and Photosynthesis Than Males

Water deficit causes plant yield loss through stunted plant growth or even mortality [14,61,62]. Plants that are more resistant to drought are typically able to maintain better growth under stress conditions [11]. Studies have shown that female plants tend to suffer more in terms of growth than males under suboptimal environmental conditions [10,56,63,64]. This sexual bias in growth response may be the result of millions of years of natural and sexual selection [65]. However, dioecious populations that have been artificially bred do not have the benefit of millions of years of selection and are, therefore, more similar in their secondary sexual traits [9,10]. Despite this, we observed sexually dimorphic responses to water deficit in the growth of female and male hybrid poplars in our study.

One mechanism that may contribute to growth depression under water deficit is the reduction in photosynthesis. Stress conditions are known to inhibit photosynthesis by decreasing the net photosynthesis rate and the capacity for photosynthesis, as well as reducing light use efficiency [12,66]. Our results show that water deficit significantly reduced the photosynthetic capacity of *P.* × *euramericana* sibs, and this inhibition was more pronounced in females under severe water deficit stress. This result is consistent with previous findings in natural dioecious plants but has never been found before in artificially bred sibling plants [27,49]. Chlorophyll fluorescence parameters have been used to examine photosynthetic performance under stress conditions [67–70]. A previous study found that freezing reduced Fv/Fm, Φ_{PSII}, and qP in *P. euphratica*, but these parameters were higher in transgenic lines with higher tolerance compared to wild lines under freezing conditions [71]. Our results showed lower Fv/Fm, Φ_{PSII}, and qP in female sib *P.* × *euramericana* compared to males, suggesting that the maximal and actual quantum yield of PSII in females was lower than in males during water deficit. This result is consistent with the transgenic study and natural population observations, indicating that males can acquire efficient photosynthesis ability even under water stress conditions [71,72].

4.2. Female Was Damaged More Severely by Water Deficit Due to Its Weaker ROS Scavenging Enzyme Activities

The cell membrane is highly susceptible to stress, and its integrity and stability are key indicators of a plant's ability to tolerate stress [73,74]. Two important markers of damage caused by stress in plants are REL and MDA content, which are products of membrane lipid peroxidation [74,75]. Higher membrane permeability, as measured by REL, is positively correlated with cell membrane injury induced by drought [74,76]. In the present study, REL was found to be elevated in both sexes of *P.* × *euramericana* under water deficit, with higher levels observed in females. Similarly, higher levels of MDA were observed in female poplar lines than in males under severe water deficit. This suggests that the cell membrane is more intact in artificially bred male poplar lines during water deficit, in line with previous studies of natural dioecious plants [48,56,77]. Excessive production of ROS under stress can harm cellular macromolecules, raise REL, and lead to excess MDA accumulation, which ultimately result in plant death [13,17,19]. To combat this, plants possess enzymatic systems, including superoxide dismutase, peroxidase, catalase, ascorbate peroxidase, and glutathione reductase, that scavenge ROS and provide protection against oxidative stress [19,78]. SODs, the first defensive line against ROS, convert superoxide into stable H_2O_2, whereas PODs detoxify H_2O_2 to H_2O [13,17,78]. Previous studies have found that drought led to a rapid increase in ROS in plant cells and a subsequent rise in SOD and

POD activity [19,72,79,80]. In the present study, SOD and POD activity increased under water deficit conditions in both sexes of poplar, indicating activation of the enzymatic system to avert ROS damage. Notably, the activities of SOD and POD were significantly higher in males than in females under severe water deficit, suggesting that males are more efficient at converting superoxide into H_2O_2 and then detoxifying them through POD.

4.3. Water Deficit Regulated Genes Encoding Stress-Related Proteins Differently in Female and Male Lines of *P.* × *euramericana*

Plants regulate gene expression to respond to and resist stresses. Different regulatory patterns of genes which encode stress-related proteins endow plants with different levels of stress resistance. For example, in poplar, the *SOD* gene, which affects the balance of ROS, is upregulated by salt and drought stress [34]. In tetraploid *Poncirus trifoliata*, the tolerant genotype has higher levels of *SOD* gene expression than its diploid progenitors under drought stress [32]. In our research, the *SOD* gene was expressed higher in male poplar than that in females under water deficit conditions, and its transcript levels correlated with higher SOD activity in males. These results suggest that male poplar may possess more effective ROS scavenging abilities than their female siblings. *GST* genes confer plant tolerance by increasing the activities of enzymes that scavenge ROS to maintain ROS homeostasis and decrease cellular damage [81–83]. Previous research has demonstrated that improved GST activity enhances drought tolerance in *P.* × *euramericana* by eliminating excess ROS [34]. Similarly, in female *P.* × *euramericana* under moderate water deficit conditions, *GST* expression was significantly higher than in male siblings, indicating females have higher sensitivity to stress. Despite the sensitive response to water deficit, the final SOD and POD activity levels were lower in females, indicating weaker ROS scavenging capacities. bHLH genes participate in various biological processes, including plant stress responses [36,84]. *bHLH* genes promote plant tolerance to drought by regulating photosynthesis, ROS scavenging, and growth [37,85]. Similar to the *GST* expression pattern, in male *P.* × *euramericana*, the expression of *bHLH35* was lower than in females under water deficit conditions. We attribute these results to the fact that females showed a more sensitive response to water deficit, whereas males exhibited higher levels of photosynthesis, growth, and ROS scavenging, thus indicating that males are more tolerant to water deficit than females. PsbX, a low-molecular-weight protein in PSII, is a key component affecting the integrity of PSII and subsequently regulating photosynthesis [39–41]. In both female and male cuttings of poplar siblings, the expression of *PsbX1* increased significantly under water deficit conditions. Importantly, the degree of upregulation in males was significantly higher than females. These results are consistent with the physiological results of higher A and Fv/Fm ratio in water-deficit-treated males. Physiological and photosynthesis gene expression results indicate that male poplar siblings possess a higher photosynthesis capacity than their female siblings under water deficit conditions, likely due to higher PSII integrity and greater protection in males under water deficit conditions.

4.4. PCA Showed an Overall Difference in Water Deficit Response between Female and Male Poplar Sibs

PCA is a powerful tool for analyzing large datasets as it depicts the relationships among variables and observations. It enables us to identify which variables contribute unique or similar information to the model [86]. In our study, PCA revealed that males and females responded differently to water deficit. Principal component 1 (PC1) clearly separated control and water-deficit-treated lines, except for the moderate-water-deficit-treated males in the control samples, suggesting that females were more sensitive to water deficit than males. The second component, PC2 separated males from females under both control and water deficit conditions. However, the separation between water-deficit-treated males and females was more pronounced than that of the control-treated lines. These results confirmed that female and male poplar siblings showed striking differences in physiological processes and gene regulation under stress conditions. The loading plot demonstrated the variables' contribution to the PCA of observations. The distribution of REL, MDA content, and *NPQ* was in line with the distribution of severe water deficit females in the score plot. In contrast, gas exchange parameters grouped moderate-water-deficit-treated males into control-treated poplar. These results align with previous studies showing native females' higher sensitivity to stress than males [49,57,87].

5. Conclusions

Our findings indicate that male and female *P.* × *euramericana* siblings show different responses to water deficit, which is especially evident in artificially cultivated female and male pairs. Females exhibit greater inhibition of growth and photosynthesis, with higher REL and MDA levels indicating more severe damage caused by water deficit. In contrast, males show higher levels of SOD and POD activity, as well as greater expression of the *SOD* and *PsbX1* genes, suggesting that they are better equipped to handle water deficit by more effectively scavenging ROS. These results confirm that sexually dimorphic response to stress is present in both natural and artificially cultivated dioecious plants. Based on our findings and the work of others, we suggest examining the correlation between reduced growth and survival of stress in sibling plants. While our research shows that males may maintain better growth during drought stress for various reasons, it is possible that female plants may be less susceptible to death in the long run due to their ability to reduce their growth and conserve energy in times of stress. It is important to reconsider the consequences of response to different degrees of stresses and the short-term/long-term consequences of stress resistance in plants. Furthermore, more research should be conducted to investigate whether female plants are hedging their long-term survival by responding poorly to drought stress and limiting their growth. This may lead to a paradigm shift in scientists' understanding of plant adaptation to stress and open the door for further investigation into the survival strategies of dioecious plants under stress. Moreover, identifying and characterizing genes that influence this sex-specific response to water deficit stress would be an invaluable pursuit for future studies.

Supplementary Materials: The following supporting information can be downloaded at: https://www.mdpi.com/article/10.3390/f14050995/s1, Table S1: Formulas of height and basal diameter growth fitting curves; Table S2: Gas exchange parameters; Table S3: Coefficients of light response curves; Table S4: Primers used in fluorescence quantitative PCR; Table S5: Correlation matrix of parameters in all samples for PCA. Table S6: Variables used in PCA.

Author Contributions: Conceptualization, F.C.; Methodology, Y.G., Y.W. and F.C.; Formal analysis, Y.G. and F.C.; Writing—Original Draft Preparation, Y.G.; Writing—Review and Editing, F.C., Y.G., L.L. and S.L.; Visualization, F.C. and Y.G.; Supervision, F.C.; Project Administration, F.C.; Funding acquisition, F.C., G.Z. and H.Z. All authors have read and agreed to the published version of the manuscript.

Funding: This research was funded by National Natural Science Foundation of China, 31200469 and Natural Science Foundation of Anhui Province, 1308085QC62.

Data Availability Statement: The data are available on request from the corresponding author and supporting data can be found in the Supplementary Materials.

Acknowledgments: We express our gratitude to the National Natural Science Foundation of China (grant 31200469) and the Natural Science Foundation of Anhui Province (grant 1308085QC62) for providing the funding for this research. We extend our appreciation to the Anhui Provincial Key Laboratory of the Conservation and Exploitation of Biological Resources for their support in facilitating this study. Special thanks go to Professor Luozhong Tang from the Nanjing Forestry University for providing us with the female and male *Populus* × *euramericana* stalks. Finally, we acknowledge ChatGPT, Academic Phrasebank, and Grammarly for refining the language and polishing the manuscript.

Conflicts of Interest: The authors declare no conflict of interest.

References

1. Renner, S.S. The relative and absolute frequencies of angiosperm sexual systems: Dioecy, monoecy, gynodioecy, and an updated online database. *Am. J. Bot.* **2014**, *101*, 1588–1596. [CrossRef] [PubMed]
2. Bachtrog, D.; Mank, J.E.; Peichel, C.L.; Kirkpatrick, M.; Otto, S.P.; Ashman, T.-L.; Hahn, M.W.; Kitano, J.; Mayrose, I.; Ming, R.; et al. Sex Determination: Why So Many Ways of Doing It? *PLoS Biol.* **2014**, *12*, e1001899. [CrossRef] [PubMed]
3. Zemp, N.; Tavares, R.; Muyle, A.; Charlesworth, D.; Marais, G.A.B.; Widmer, A. Evolution of sex-biased gene expression in a dioecious plant. *Nat. Plants* **2016**, *2*, 16168. [CrossRef] [PubMed]
4. Henry, I.M.; Akagi, T.; Tao, R.; Comai, L. One hundred ways to invent the sexes: Theoretical and observed paths to dioecy in plants. *Annu. Rev. Plant Biol.* **2018**, *69*, 553–575. [CrossRef]
5. Darwin, C. *The Different Forms of Flowers on Plants of the Same Species*; Murray: London, UK, 1877. [CrossRef]
6. Álvarez-Cansino, L.; Diaz Barradas, M.C.; Zunzunegui, M.; Paz Esquivias, M.; Dawson, T.E. Gender-specific variation in physiology in the dioecious shrub *Corema album* throughout its distributional range. *Funct. Plant Biol.* **2012**, *39*, 968–978. [CrossRef]
7. Liu, X.; Dong, T.; Zhang, S. Adaptation Responses of Dioecious and Hermaphroditic Tree Species to Abiotic Stress. *Forests* **2023**, *14*, 383. [CrossRef]
8. Barrett, S.C.; Hough, J. Sexual dimorphism in flowering plants. *J. Exp. Bot.* **2013**, *64*, 67–82. [CrossRef]
9. Sun, Z.; Shen, Y.; Niinemets, Ü. Responses of isoprene emission and photochemical efficiency to severe drought combined with prolonged hot weather in hybrid *Populus*. *J. Exp. Bot.* **2020**, *71*, 7364–7381. [CrossRef]
10. Lin, T.; Tang, J.; He, F.; Chen, G.; Shi, Y.; Wang, X.; Han, S.; Li, S.; Zhu, T.; Chen, L. Sexual differences in above- and belowground herbivore resistance between male and female poplars as affected by soil cadmium stress. *Sci. Total Environ.* **2022**, *803*, 150081. [CrossRef] [PubMed]
11. Choat, B.; Brodribb, T.J.; Brodersen, C.R.; Duursma, R.A.; López, R.; Medlyn, B.E. Triggers of tree mortality under drought. *Nature* **2018**, *558*, 531–539. [CrossRef]
12. Basu, S.; Ramegowda, V.; Kumar, A.; Pereira, A. Plant adaptation to drought stress. *F1000Research* **2016**, *5*, 1554. [CrossRef] [PubMed]
13. Miller, G.; Suzuki, N.; Ciftci-Yilmaz, S.; Mittler, R. Reactive oxygen species homeostasis and signalling during drought and salinity stresses. *Plant Cell Environ.* **2010**, *33*, 453–467. [CrossRef] [PubMed]
14. Gupta, A.; Rico-Medina, A.; Caño-Delgado, A.I. The physiology of plant responses to drought. *Science* **2020**, *368*, 266–269. [CrossRef] [PubMed]
15. Zhang, H.; Zhu, J.; Gong, Z.; Zhu, J. Abiotic stress responses in plants. *Nat. Rev. Genet.* **2022**, *23*, 104–119. [CrossRef] [PubMed]
16. Devireddy, A.R.; Zandalinas, S.I.; Fichman, Y.; Mittler, R. Integration of reactive oxygen species and hormone signaling during abiotic stress. *Plant J.* **2021**, *105*, 459–476. [CrossRef]
17. Apel, K.; Hirt, H. Reactive oxygen species: Metabolism, oxidative stress, and signal transduction. *Annu. Rev. Plant Biol.* **2004**, *55*, 373–399. [CrossRef]
18. Mittler, R. Oxidative stress, antioxidants and stress tolerance. *Trends Plant Sci.* **2002**, *7*, 405–410. [CrossRef]
19. Sachdev, S.; Ansari, S.A.; Ansari, M.I.; Fujita, M.; Hasanuzzaman, M. Abiotic Stress and Reactive Oxygen Species: Generation, Signaling, and Defense Mechanisms. *Antioxidants* **2021**, *10*, 277. [CrossRef]
20. Mukarram, M.; Choudhary, S.; Kurjak, D.; Petek, A.; Khan, M.M.A. Drought: Sensing, signalling, effects and tolerance in higher plants. *Physiol. Plant.* **2021**, *172*, 1291–1300. [CrossRef]
21. Wahab, A.; Abdi, G.; Saleem, M.H.; Ali, B.; Ullah, S.; Shah, W.; Mumtaz, S.; Yasin, G.; Muresan, C.C.; Marc, R.A. Plants' Physio-Biochemical and Phyto-Hormonal Responses to Alleviate the Adverse Effects of Drought Stress: A Comprehensive Review. *Plants* **2022**, *11*, 1620. [CrossRef]
22. Wang, X.; Curtis, P.S. Gender-specific responses of *Populus tremuloides* to atmospheric CO_2 enrichment. *New Phytol.* **2001**, *150*, 675–684. [CrossRef]
23. Xu, X.; Yang, F.; Xiao, X.; Zhang, S.; Korpelainen, H.; Li, C. Sex-specific responses of *Populus cathayana* to drought and elevated temperatures. *Plant Cell Environ.* **2008**, *31*, 850–860. [CrossRef] [PubMed]
24. Chen, L.; Zhang, S.; Zhao, H.; Korpelainen, H.; Li, C. Sex-related adaptive responses to interaction of drought and salinity in *Populus yunnanensis*. *Plant Cell Environ.* **2010**, *33*, 1767–1778. [CrossRef] [PubMed]
25. Lei, Y.B.; Jiang, Y.L.; Chen, K.; Duan, B.; Zhang, S.; Korpelainen, H.; Niinemets, Ü.; Li, C. Reproductive investments driven by sex and altitude in sympatric *Populus* and *Salix* trees. *Tree Physiol.* **2017**, *37*, 1503–1514. [CrossRef]
26. Melnikova, N.V.; Borkhert, E.V.; Snezhkina, A.V.; Kudryavtseva, A.V.; Dmitriev, A.A. Sex-Specific Response to Stress in *Populus*. *Front. Plant Sci.* **2017**, *8*, 1827. [CrossRef]
27. Xia, Z.; He, Y.; Yu, L.; Lv, R.; Korpelainen, H.; Li, C. Sex-specific strategies of phosphorus (P) acquisition in *Populus cathayana* as affected by soil P availability and distribution. *New Phytol.* **2020**, *225*, 782–792. [CrossRef]
28. Yu, L.; Huang, Z.; Tang, S.; Korpelainen, H.; Li, C.J.E.; Botany, E. *Populus euphratica* males exhibit stronger drought and salt stress resistance than females. *Environ. Exp. Bot.* **2023**, *205*, 105114. [CrossRef]
29. Yamaguchi-Shinozaki, K.; Shinozaki, K. Transcriptional regulatory networks in cellular responses and tolerance to dehydration and cold stresses. *Annu. Rev. Plant Biol.* **2006**, *57*, 781–803. [CrossRef]

30. Shinozaki, K.; Yamaguchi-Shinozaki, K. Gene networks involved in drought stress response and tolerance. *J. Exp. Bot.* **2006**, *58*, 221–227. [CrossRef]
31. Choudhury, F.K.; Rivero, R.M.; Blumwald, E.; Mittler, R. Reactive oxygen species, abiotic stress and stress combination. *Plant J.* **2017**, *90*, 856–867. [CrossRef]
32. Wei, T.; Wang, Y.; Xie, Z.; Guo, D.; Chen, C.; Fan, Q.; Deng, X.; Liu, J. Enhanced ROS scavenging and sugar accumulation contribute to drought tolerance of naturally occurring autotetraploids in *Poncirus trifoliata*. *Plant Biotechnol. J.* **2019**, *17*, 1394–1407. [CrossRef] [PubMed]
33. Zhang, H.; Wang, Z.; Li, X.; Gao, X.; Dai, Z.; Cui, Y.; Zhi, Y.; Liu, Q.; Zhai, H.; Gao, S.; et al. The IbBBX24–IbTOE3–IbPRX17 module enhances abiotic stress tolerance by scavenging reactive oxygen species in sweet potato. *New Phytol.* **2022**, *233*, 1133–1152. [CrossRef] [PubMed]
34. Wei, H.; Movahedi, A.; Xu, C.; Sun, W.; Li, L.; Wang, P.; Li, D.; Zhuge, Q. Overexpression of *PtHMGR* enhances drought and salt tolerance of poplar. *Ann. Bot.* **2019**, *125*, 785–803. [CrossRef]
35. Edwards, R.; Dixon, D.P. Plant Glutathione Transferases. In *Methods Enzymology*; Sies, H., Packer, L., Eds.; Academic Press: Cambridge, MA, USA, 2005; Volume 401, pp. 169–186.
36. Gao, S.; Li, C.; Chen, X.; Li, S.; Liang, N.; Wang, H.; Zhan, Y.; Zeng, F. Basic helix-loop-helix transcription factor *PxbHLH02* enhances drought tolerance in *Populus* (*Populus simonii* × *P. nigra*). *Tree Physiol.* **2023**, *43*, 185–202. [CrossRef]
37. Liang, B.; Wan, S.; Ma, Q.; Yang, L.; Hu, W.; Kuang, L.; Xie, J.; Huang, Y.; Liu, D.; Liu, Y. A Novel bHLH Transcription Factor PtrbHLH66 from Trifoliate Orange Positively Regulates Plant Drought Tolerance by Mediating Root Growth and ROS Scavenging. *Int. J. Mol. Sci.* **2022**, *23*, 15053. [CrossRef] [PubMed]
38. Abe, H.; Urao, T.; Ito, T.; Seki, M.; Shinozaki, K.; Yamaguchi-Shinozaki, K. Arabidopsis AtMYC2 (bHLH) and AtMYB2 (MYB) function as transcriptional activators in abscisic acid signaling. *Plant Cell* **2003**, *15*, 63–78. [CrossRef] [PubMed]
39. García-Cerdán, J.G.; Sveshnikov, D.; Dewez, D.; Jansson, S.; Funk, C.; Schröder, W.P. Antisense Inhibition of the PsbX Protein Affects PSII Integrity in the Higher Plant *Arabidopsis thaliana*. *Plant Cell Physiol.* **2008**, *50*, 191–202. [CrossRef]
40. Xiao, L.; Liu, X.; Lu, W.; Chen, P.; Quan, M.; Si, J.; Du, Q.; Zhang, D. Genetic dissection of the gene coexpression network underlying photosynthesis in *Populus*. *Plant Biotechnol. J.* **2020**, *18*, 1015–1026. [CrossRef]
41. Biswas, S.; Eaton-Rye, J.J. PsbX maintains efficient electron transport in Photosystem II and reduces susceptibility to high light in Synechocystis sp. PCC 6803. *Biochim. Biophys. Acta (BBA)—Bioenerg.* **2022**, *1863*, 148519. [CrossRef]
42. Xiao, Y.; Huang, G.; You, X.; Zhu, Q.; Wang, W.; Kuang, T.; Han, G.; Sui, S.-F.; Shen, J.-R. Structural insights into cyanobacterial photosystem II intermediates associated with Psb28 and Tsl0063. *Nat. Plants* **2021**, *7*, 1132–1142. [CrossRef]
43. Jansson, S.; Douglas, C.J. *Populus*: A model system for plant biology. *Annu. Rev. Plant Biol.* **2007**, *58*, 435–458. [CrossRef]
44. Wullschleger, S.; Weston, D.; Davis, J. *Populus* responses to edaphic and climatic cues: Emerging evidence from systems biology research. *Crit. Rev. Plant Sci.* **2009**, *28*, 368–374. [CrossRef]
45. Wullschleger, S.D.; Jansson, S.; Taylor, G. Genomics and Forest Biology: *Populus* Emerges as the Perennial Favorite. *Plant Cell* **2002**, *14*, 2651–2655. [CrossRef] [PubMed]
46. Chen, F.; Shen, J.; Min, D.; Ke, L.; Tian, X.; Korpelainen, H.; Li, C. Male *Populus cathayana* than female shows higher photosynthesis and less cellular injury through ABA-induced manganese transporting inhibition under high manganese condition. *Trees* **2018**, *32*, 255–263. [CrossRef]
47. Song, H.; Cai, Z.; Liao, J.; Zhang, S. Phosphoproteomic and Metabolomic Analyses Reveal Sexually Differential Regulatory Mechanisms in Poplar to Nitrogen Deficiency. *J. Proteome Res.* **2020**, *19*, 1073–1084. [CrossRef] [PubMed]
48. Miao, L.F.; Yang, F.; Han, C.Y.; Pu, Y.J.; Ding, Y.; Zhang, J. Sex-specific responses to winter flooding, spring waterlogging and post-flooding recovery in *Populus deltoides*. *Sci. Rep.* **2017**, *7*, 2534. [CrossRef]
49. Zhang, R.; Liu, J.Y.; Liu, Q.S.; He, H.G.; Xu, X.; Dong, T.F. Sexual differences in growth and defence of *Populus yunnanensis* under drought stress. *Can. J. For. Res.* **2019**, *49*, 491–499. [CrossRef]
50. Zhang, Y.D.; Virjamo, V.; Sobuj, N.; Du, W.; Yin, Y.; Nybakken, L.; Guo, H.; Julkunen-Tiitto, R. Sex-related responses of European aspen (*Populus tremula* L.) to combined stress: TiO2 nanoparticles, elevated temperature and CO_2 concentration. *J. Hazard. Mater.* **2018**, *352*, 130–138. [CrossRef]
51. Stromme, C.B.; Julkunen-Tiitto, R.; Olsen, J.E.; Nybakken, L. The dioecious *Populus tremula* displays interactive effects of temperature and ultraviolet-B along a natural gradient. *Environ. Exp. Bot.* **2018**, *146*, 13–26. [CrossRef]
52. Li, C.; Ren, J.; Luo, J.; Lu, R. Sex-specific physiological and growth responses to water stress in *Hippophae rhamnoides* L. populations. *Acta Physiol. Plant.* **2004**, *26*, 123. [CrossRef]
53. Yang, Z.; Ma, L.; Jia, Z.; Sun, Y. Application on Spss Software in Photosynthetic Light-Response Curve of *Quercus Variabilis* Leaf. In *Recent Advances in Computer Science and Information Engineering*; Qian, Z., Cao, L., Su, W., Wang, T., Yang, H., Eds.; Springer: Berlin/Heidelberg, Germany, 2012; Volume 3, pp. 567–573.
54. Ye, Z.; Suggett, D.J.; Robakowski, P.; Kang, H. A mechanistic model for the photosynthesis–light response based on the photosynthetic electron transport of photosystem II in C3 and C4 species. *New Phytol.* **2013**, *199*, 110–120. [CrossRef]
55. Lichtenthaler, H.K. Chlorophylls and carotenoids: Pigments of photosynthetic biomembranes. In *Methods Enzymology*; Academic Press: Cambridge, MA, USA, 1987; Volume 148, pp. 350–382.
56. Liao, J.; Song, H.F.; Tang, D.T.; Zhang, S. Sexually differential tolerance to water deficiency of *Salix paraplesia*—A female-biased alpine willow. *Ecol. Evol.* **2019**, *9*, 8450–8464. [CrossRef]

57. Chen, F.; Zhang, S.; Zhu, G.; Korpelainen, H.; Li, C. *Populus cathayana* males are less affected than females by excess manganese: Comparative proteomic and physiological analyses. *Proteomics* **2013**, *13*, 2424–2437. [CrossRef]
58. Maehly, A.C. The Assay of Catalases and Peroxidases. *Methods Biochem. Anal.* **1954**, *1*, 357–424.
59. Giannopolitis, C.N.; Ries, S.K. Superoxide Dismutases: I. Occurrence in Higher Plants. *Plant Physiol.* **1977**, *59*, 309–314. [CrossRef]
60. Livak, K.J.; Schmittgen, T.D. Analysis of Relative Gene Expression Data Using Real-Time Quantitative PCR and the $2^{-\Delta\Delta CT}$ Method. *Methods* **2001**, *25*, 402–408. [CrossRef]
61. Canning, C.M.; Mood, B.J.; Bonsal, B.; Howat, B.; Laroque, C.P. Comparison of tree-growth drought legacies of three shelterbelt species in the Canadian prairies. *Agric. For. Meteorol.* **2023**, *330*, 109317. [CrossRef]
62. Brodribb, T.J.; Powers, J.; Cochard, H.; Choat, B. Hanging by a thread? Forests and drought. *Science* **2020**, *368*, 261–266. [CrossRef]
63. Hultine, K.R.; Bush, S.E.; Ward, J.K.; Dawson, T.E. Does sexual dimorphism predispose dioecious riparian trees to sex ratio imbalances under climate change? *Oecologia* **2018**, *187*, 921–931. [CrossRef]
64. Morales, M.; Pinto-Marijuan, M.; Munne-Bosch, S. Seasonal, sex- and plant size-related effects on photoinhibition and photoprotection in the dioecious Mediterranean dwarf palm, *Chamaerops humilis*. *Front. Plant Sci.* **2016**, *7*, 1116. [CrossRef]
65. Korgiopoulou, C.; Bresta, P.; Nikolopoulos, D.; Karabourniotis, G. Sex-specific structural and functional leaf traits and sun-shade acclimation in the dioecious tree *Pistacia vera* (Anacardiaceae). *Funct. Plant Biol.* **2019**, *46*, 649–659. [CrossRef]
66. Hu, H.; He, B.; Ma, L.; Chen, X.; Han, P.; Luo, Y.; Liu, Y.; Fei, X.; Wei, A. Physiological and transcriptome analyses reveal the photosynthetic response to drought stress in drought-sensitive (Fengjiao) and drought-tolerant (Hanjiao) *Zanthoxylum bungeanum* cultivars. *Front. Plant Sci.* **2022**, *13*, 968714. [CrossRef] [PubMed]
67. Baker, N.R. Chlorophyll fluorescence: A probe of photosynthesis in vivo. *Annu. Rev. Plant Biol.* **2008**, *59*, 89–113. [CrossRef] [PubMed]
68. Helm, L.T.; Shi, H.; Lerdau, M.T.; Yang, X. Solar-induced chlorophyll fluorescence and short-term photosynthetic response to drought. *Ecol. Appl.* **2020**, *30*, e02101. [CrossRef] [PubMed]
69. Donaldson, L. Autofluorescence in Plants. *Molecules.* **2020**, *25*, 2393. [CrossRef]
70. Sherstneva, O.; Khlopkov, A.; Gromova, E.; Yudina, L.; Vetrova, Y.; Pecherina, A.; Kuznetsova, D.; Krutova, E.; Sukhov, V.; Vodeneev, V. Analysis of chlorophyll fluorescence parameters as predictors of biomass accumulation and tolerance to heat and drought stress of wheat *Triticum aestivum*. *Funct. Plant Biol.* **2022**, *49*, 155–169. [CrossRef]
71. He, F.; Li, H.; Wang, J.; Su, Y.; Wang, H.; Feng, C.; Yang, Y.; Niu, M.; Liu, C.; Yin, W.; et al. PeSTZ1, a C2H2-type zinc finger transcription factor from *Populus euphratica*, enhances freezing tolerance through modulation of ROS scavenging by directly regulating *PeAPX2*. *Plant Biotechnol. J.* **2019**, *17*, 2169–2183. [CrossRef]
72. Zhang, S.; Chen, L.; Duan, B.; Korpelainen, H.; Li, C. *Populus cathayana* males exhibit more efficient protective mechanisms than females under drought stress. *For. Ecol. Manag.* **2012**, *275*, 68–78. [CrossRef]
73. Bowler, C.; Van Montagu, M.; Inzé, D. Superoxide dismutase and stress tolerance. *Annu. Rev. Plant Biol.* **1992**, *43*, 83–116. [CrossRef]
74. Bajji, M.; Kinet, J.-M.; Lutts, S. The use of the electrolyte leakage method for assessing cell membrane stability as a water stress tolerance test in durum wheat. *Plant Growth Regul.* **2002**, *36*, 61–70. [CrossRef]
75. Hodges, D.M.; DeLong, J.M.; Forney, C.F.; Prange, R.K. Improving the thiobarbituric acid-reactive-substances assay for estimating lipid peroxidation in plant tissues containing anthocyanin and other interfering compounds. *Planta* **1999**, *207*, 604–611. [CrossRef]
76. Whitlow, T.H.; Bassuk, N.L.; Ranney, T.G.; Reichert, D.L. An improved method for using electrolyte leakage to assess membrane competence in plant tissues. *Plant Physiol.* **1992**, *98*, 198–205. [CrossRef]
77. Qin, F.; Liu, G.; Huang, G.Q.; Dong, T.F.; Liao, Y.M.; Xu, X. Zinc application alleviates the adverse effects of lead stress more in female *Morus alba* than in males. *Environ. Exp. Bot.* **2018**, *146*, 68–76. [CrossRef]
78. Kumar, V.; Khare, T.; Sharma, M.; Wani, S.H. ROS-Induced Signaling and Gene Expression in Crops Under Salinity Stress. In *Reactive Oxygen Species and Antioxidant Systems in Plants: Role and Regulation under Abiotic Stress*; Khan, M.I.R., Khan, N.A., Eds.; Springer: Singapore, 2017; pp. 159–184.
79. Melandri, G.; Abd Elgawad, H.; Floková, K.; Jamar, D.C.; Asard, H.; Beemster, G.T.S.; Ruyter-Spira, C.; Bouwmeester, H.J. Drought tolerance in selected aerobic and upland rice varieties is driven by different metabolic and antioxidative responses. *Planta* **2021**, *254*, 13. [CrossRef]
80. Li, C.; Wan, Y.; Shang, X.; Fang, S. Responses of Microstructure, Ultrastructure and Antioxidant Enzyme Activity to PEG-Induced Drought Stress in *Cyclocarya paliurus* Seedlings. *Forests* **2022**, *13*, 836. [CrossRef]
81. Hasanuzzaman, M.; Nahar, K.; Anee, T.I.; Fujita, M. Glutathione in plants: Biosynthesis and physiological role in environmental stress tolerance. *Physiol. Mol. Biol. Plants* **2017**, *23*, 249–268. [CrossRef]
82. Nianiou-Obeidat, I.; Madesis, P.; Kissoudis, C.; Voulgari, G.; Chronopoulou, E.; Tsaftaris, A.; Labrou, N.E. Plant glutathione transferase-mediated stress tolerance: Functions and biotechnological applications. *Plant Cell Rep.* **2017**, *36*, 791–805. [CrossRef]
83. Yang, G.; Wang, Y.; Xia, D.; Gao, C.; Wang, C.; Yang, C. Overexpression of a GST gene (*ThGSTZ1*) from *Tamarix hispida* improves drought and salinity tolerance by enhancing the ability to scavenge reactive oxygen species. *Plant Cell Tissue Org.* **2014**, *117*, 99–112. [CrossRef]
84. Toledo-Ortiz, G.; Huq, E.; Quail, P.H. The Arabidopsis basic/helix-loop-helix transcription factor family. *Plant Cell* **2003**, *15*, 1749–1770. [CrossRef]

85. Dong, Y.; Wang, C.; Han, X.; Tang, S.; Liu, S.; Xia, X.; Yin, W. A novel bHLH transcription factor PebHLH35 from *Populus euphratica* confers drought tolerance through regulating stomatal development, photosynthesis and growth in *Arabidopsis*. *Biochem. Biophys. Res. Commun.* **2014**, *450*, 453–458. [CrossRef]
86. Ringnér, M. What is principal component analysis? *Nat. Biotechnol.* **2008**, *26*, 303–304. [CrossRef] [PubMed]
87. Rozas, V.; Le Quesne, C.; Rojas-Badilla, M.; Gonzalez-Reyes, A.; Donoso, S.; Olano, J.M. Climatic cues for secondary growth and cone production are sex-dependent in the long-lived dioecious conifer Araucaria araucana. *Agric. For. Meteorol.* **2019**, *274*, 132–143. [CrossRef]

forests

MDPI

Article

Comparative RNA-Seq Analysis Reveals the Organ-Specific Transcriptomic Response to Zinc Stress in Mulberry

Shuai Huang [1], Xiaoru Kang [1], Ting Yu [1], Keermula Yidilisi [1], Lin Zhang [1,2], Xu Cao [1,2], Nan Chao [1,2,*] and Li Liu [1,2,*]

[1] Jiangsu Key Laboratory of Sericultural Biology and Biotechnology, School of Biotechnology, Jiangsu University of Science and Technology, Zhenjiang 212018, China

[2] Key Laboratory of Silkworm and Mulberry Genetic Improvement, Ministry of Agriculture and Rural Affairs, Sericultural Research Institute, Chinese Academy of Agricultural Sciences, Zhenjiang 212018, China

* Correspondence: chaonan1989@126.com (N.C.); morusliu@126.com (L.L.)

Abstract: Mulberry (*Morus, Moraceae*) is an important economic plant that is considered zinc-rich. Zinc (Zn) is a micronutrient that plays vital roles in various bio-processes in plants and animals. In the present study, a comparative transcriptome analysis associated with physiological indicators was performed to reveal the potential mechanism in different organs in response to zinc toxicity in mulberry. Physiological indicators in mulberry plants treated with increasing concentrations of zinc were monitored to reveal the tolerance limits to zinc concentration. Transcriptome analysis of different organs in mulberry under excess zinc stress was performed to reveal the spatial response to zinc stress. The results show that the hormone signaling pathway and secondary metabolism including lignin biosynthesis, flavonoid biosynthesis and sugar metabolism are important for excess zinc treatment responses. In addition, the organ-based spatial response of these pathways is indicated. Lignin biosynthesis mainly responds to zinc stress in lignified tissues or organs such as stems, flavonoid biosynthesis is the main response to zinc stress in leaves, and sugar metabolism is predominant in roots. Further co-expression network analysis indicated candidate genes involved in the organ-based spatial response. Several transcription factors and genes involved in phenylpropanoid biosynthesis, cell wall biogenesis and sugar metabolism were further validated and designed as organ-based response genes for zinc stress.

Keywords: mulberry; physiology; organ-specific; secondary metabolism; transcriptome; zinc stress

Citation: Huang, S.; Kang, X.; Yu, T.; Yidilisi, K.; Zhang, L.; Cao, X.; Chao, N.; Liu, L. Comparative RNA-Seq Analysis Reveals the Organ-Specific Transcriptomic Response to Zinc Stress in Mulberry. *Forests* **2023**, *14*, 842. https://doi.org/10.3390/f14040842

Academic Editor: Chikako Honda

Received: 28 February 2023
Revised: 14 April 2023
Accepted: 17 April 2023
Published: 20 April 2023

1. Introduction

Zinc (Zn) is a micronutrient that is necessary for higher plants, animals and humans. Zinc is known to bind to various proteins and works as a cofactor involved in metabolic processes in the plant and animal kingdoms. Zinc is important for people to maintain their fitness level by affecting physical growth, the immune system, reproductive health and brain development [1,2]. Zinc deficiency in humans is a critical nutritional and health problem in the world. It affects, on average, one-third of the world's population in different countries [3]. Dietary modification with zinc-rich foods is a recommended intervention strategy to improve zinc intake for humans to keep normal healthy growth and reproduction [4].

Plants are the main sources of many food products including staples, such as rice, wheat and maize, and non-staple food, such as vegetables and fruits [5]. In addition, some plants provide specific nutrients to benefit our health. Zinc-deficient plants generally have low tissue zinc concentrations and therefore, in addition to reduced crop yields, the crop products from these plants make a lower contribution to the zinc content in the human diet [5]. Soil-plant relationships also affected the zinc content in animal food products, another human dietary source [5]. Therefore, plant zinc content is one of the main sources of human zinc intake. Increasing plant zinc content can alleviate zinc deficiency in the human body. Zinc deficiency in crops is found in many countries and regions around the world [3] The average total zinc concentration in soils was reported to be around 55 mg/kg [5]. On the other hand, zinc concentration in soils has gradually increased in the last decades as a consequence of human activities [6–8] including industrial processes, agriculture, increasing use of biosolids and metal mining [9,10]. The maximum zinc concentration permitted in sewage sludge-amended soils (pH 6–7) among different European countries is within the range of 100 mg/kg to 300 mg/kg zinc [11]. Zinc-polluted soils also lead to zinc toxicity in plants, which affects growth and yield [6–8].

It is possible to clean up zinc-polluted soils with zinc hyperaccumulator plants along with the production of zinc-rich foods. In plants, zinc modulates the activity of a large number of enzymes involved in the maintenance of biomembrane integrity, participates in carbohydrate metabolism and protein synthesis and plays an important role in indoleacetic acid metabolism [12,13]. Further, zinc protects cells from the damage caused by reactive oxygen species [14]. Some studies also indicated that a relatively high concentration of zinc in soil could help to reduce the cadmium content in crops [15,16]. It has been proved that the application of zinc can effectively relieve lead toxicity in *Lactuca sativa* and *Houttuynia cordata* [7,17,18]. However, beyond certain concentrations (100 mg/kg~500 mg/kg), zinc is toxic to vascular plants [19]. Mulberry is a woody plant with resistance to heavy metals, such as iron and cadmium, and is also used in medicine and food. Mulberry is capable of taking up small amounts of heavy metals and was reported to have the ability to clean up zinc-polluted soils. Mulberry planted in mines was measured to migrate 254,532.8 mg zinc every square meter of plough layer soil [20]. The contents of zinc in mulberry showed spatial differences with quite different zinc concentrations in different organs (leaf, root, bark and stem) [20]. In mulberry, the leaves and fruits are known as sites rich in zinc and are used to produce zinc-rich food [21].

The availability of the *Morus notabilis* genome and chromosome-level genome of *Morus alba* promote the transcriptome analysis of mulberry in response to various stresses [22,23]. However, little knowledge is available in terms of the response to zinc toxicity in mulberry. *M. alba* variety *Fengchi* is a new variety created by the Sericultural Research Institute, Chinese Academy of Agricultural Sciences, that is expected to spread and grow in extreme environment conditions as forage. Given the great potential of *M. alba* variety *Fengchi* to improve the ecological environment, it is also expected to be used as a heavy metal hyperaccumulator in mines. Our previous studies assessed the potential roles of lignin biosynthetic genes in response to zinc stress in *Fengchi* [24]. In the present study, we performed physiological analysis of mulberry plants under zinc stress and validated the limit of zinc concentration for mulberry tolerance. Transcriptome analysis of different organs in mulberry under excess zinc stress was performed to reveal the spatial response to zinc stress. Specifically, we address the following questions: (i) what is the concentration of zinc in soil that causes zinc stress in mulberry, (ii) how does mulberry respond to zinc stress at the transcription level and (iii) what are the organ-specific responses to zinc stress in different mulberry organs?

2. Materials and Methods

2.1. Plant Materials

The materials used in this study were obtained from the National Germplasm Resource Nursery of the Institute of Sericulture, Chinese Academy of Agricultural Sciences. Zinc treatments were reported in our previous studies [24,25]. In brief, one-year-old seedlings of *M. alba* variety *Fengchi* were transplanted into plastic pots with soil, and the potted plants were irrigated with 400 mL/kg of Murashige and Skoog (MS) medium to provide nutrients [26]. Zinc sulfate powder was applied near the roots of the mulberry trees as the excess zinc stress treatment, in which the zinc ion concentration ranges from 0 to 450 mg/kg with a gradient set every 50 mg/kg. The root, stem and leaf tissues were quickly frozen in liquid nitrogen and stored at −80 °C. This experiment was performed using three mulberry seedlings with a similar growth status as biological replicates. These collected samples were used for both RNA-Seq and RT-qPCR (quantitative real-time PCR) analysis.

2.2. Determining the Contents of Physiological Indicators Related to Zinc Toxicity

The samples collected on day 15 were used for the determination of physiological indexes. Chlorophyll contents including Chlorophyll a (Chl a), Chlorophyll b (Chl b) and total Chlorophyll (Chl a + b) were determined with ethanol extraction followed by spectrophotometry [27]. The MDA contents in roots and stems were measured using the thiobarbituric acid method as described by Sairam and Srivastava (2001) [28]. Then, 0.5 g of the plant samples were extracted in 4.0 mL of 10% trichloroacetic acid (TCA) and centrifuged at 10,000× *g* for 10 min at 4 °C. The absorbance of the supernatant was determined at 532 and 600 nm with a spectrometer (BioTek ®Epoch 2, BioTek Instruments, Inc., Winooski, VT, USA) [28]. Proline (Pro) content was measured according to Silva et al. (2016) [29]. SOD activity was determined using a SOD measurement kit (Suzhou Keming Technology, Suzhou, China) according to the manufacturer's instructions. The MDA, proline and SOD contents in the leaves, roots and stems were measured, respectively. The zinc supply concentration resulting in zinc toxicity was determined based on both physiological indicators and plant phenotypes. Graphpad Prism8.0 was used to perform ANOVA and visualize the results. $p < 0.05$ was considered significant. All the above measurements were carried out with three biological replications.

2.3. RNA-Seq and Data Processing

The RNA-Seq dataset of mulberry containing different organs under the excess zinc treatment (450 mg/kg) was obtained using the Illumina sequencing system, and the trim galore (version-0.6.4) was used to remove the adapters and perform a quality control of the reads. The trimmed reads were further aligned to the Morus alba genome released by Jiao et al. (2020) using bowtie2 (version-2.3.2) [30]. Samtools was used to operate the bam files. The genome annotation file (.gff3) was used to calculate the expression matrix using StringTie v2.15 [31]. Differentially expressed genes (DEGs) were obtained using DEseq2 by comparing the expression levels of sample pairs [32]. A weighted correlation network analysis (WGCNA) was performed to screen the co-expressed DEGs [33]. R version 4.1.2 was used for R-package-based analyses.

2.4. Workflow for Comprehensive Transcriptome Analysis

The sample correlation was assessed using DEGs with both Principal Component Analysis (PCA) and Pearson correlation analysis. All DEGs were annotated to their orthologs in *Arabidopsis thialiana* using Blast. The DEGs involved in the response to zinc stress were further compared between different organs to reveal the possible organ-specific responses. Classification of DEGs was performed based on Venn analysis, and KEGG and GO analysis of the specific class of DEGs was performed using DAVID online tools [34]. WGCNA was performed using the DEGs as input and the treatments as conditions, and cytoscape 3.01 was used to visualize the co-expression network. TBtools v1.09876 and R version 4.1.2 were used to perform the above analysis and visualize the results [35].

2.5. RT-qPCR Analysis of Key Genes Involved in the Response to Zinc Toxicity

Total RNA extraction and cDNA synthesis were performed as in our previous report [35]. Quantitative real-time PCR (RT-qPCR) was performed using an ABI StepOnePlus™ Real-Time PCR System (USA) with actin as the reference gene [36]. Reactions were prepared in a total volume of 10 μL containing 5 μL 2× ChamQ SYBR Color qPCR Master Mix (High ROX Premixed) (Vazyme, Nanjing, China), 1 μL cDNA template and 0.3 μM of each primer. The program was set at 95 °C for 15 min, 45 cycles of 20 s at 95 °C and 60 s at 60 °C. The relative expression level was calculated using $2^{-\delta Ct}$, and the fold change (treatment/control) was calculated using $2^{-\delta\delta Ct}$. The primers are listed in Table S1. Graphpad Prism8.0 was used to perform an ANOVA and visualize the RT-qPCR results. $p < 0.05$

was considered significant. Three biological replicates, each with three technical replicates, were used for RT-qPCR.

3. Results

3.1. Mulberry Organ-Specific Physiological Responses to Zinc Treatment

As both zinc deficiency and zinc excess can lead to plant abnormal growth, several physiological indicators were selected to assess their effect on physiological processes. Malondialdehyde (MDA) contents changed slowly along with increasing zinc concentrations and showed a sharp increase at 350 mg/kg followed by a sharp decrease in MDA contents (Figure 1A). Although MDA contents showed a similar trend in the stems and roots with minor differences, the stems showed a significantly higher level than in roots. A similar bell-pattern trend in proline content was also observed in different organs, with the summit of proline content at the zinc concentration of 200 mg/kg. The proline content in leaves showed a significantly higher level than in the stems or roots (Figure 1B). SOD activity in different organs showed quite different trends with increasing zinc concentrations (Figure 1C). SOD activity increased sharply at 50 mg/kg zinc with a significant difference compared with the control, which thereafter kept stable in stems. The highest SOD activity was found in the leaves and roots at 350 and 250 mg/kg zinc, respectively, which thereafter decreased with an increase in zinc concentration. A significant difference can be observed at the summits compared with the controls. The chlorophyll content including chlorophyll A and B in the leaves showed an increase–decrease-increase fluctuation with a summit at 150 mg/kg zinc (Figure 1D). The above bell-pattern change in physiological indicators along with increasing zinc concentration was also reported in a previous study [37]. The bell-pattern change might indicate that plants experience a process including benefiting from suitable supply of zinc, suffering excess zinc toxicity and damage in the physiological response mechanism. It is obvious that a 350 mg/kg zinc supply completely induced physiologically adverse effects in all organs of mulberry. In addition, the mulberry seedlings under the >350 mg/kg zinc treatment showed growth retardation (Figure S1).

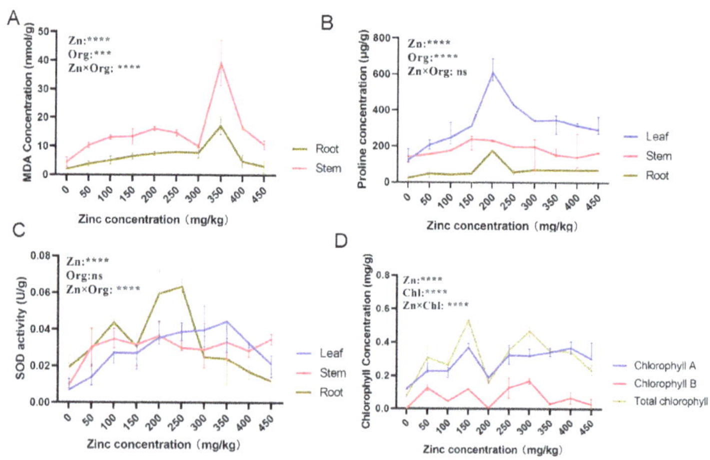

Figure 1. Change in physiological indicators under the gradient zinc concentration treatment on different organs in mulberry. (**A**). MDA contents in different organs under different zinc concentration treatments; (**B**). SOD activity in different organs under different zinc concentration treatments; (**C**). proline contents in different organs under different zinc concentration treatments; and (**D**). chlorophyll contents in different organs under different zinc concentration treatments. A two-way ANOVA was performed to analyze the difference resulting from zinc treatments and organs or types of chlorophyll. ***, $p < 0.001$; ****, $p < 0.0001$. In addition, significance analysis was also performed for every two points using ANOVA, and the results are available in Table S2.

3.2. Transcriptomic Analysis Showed Organ-Specific Differences in Response to Excess Zinc Treatments

Given the results of spatial physiological responses to zinc treatments at different concentrations, a zinc concentration >350 mg/kg would be toxic to mulberry plants and inhibit growth, as we observed (Figure S1). Therefore, the leaves, stems and roots from mulberry plants exposed to 450 mg/kg zinc were collected for a further study to explore mulberry spatial responses at transcription levels under excess zinc. A summary of the RNA-Seq datasets is available in Table S3. The Q30 values of the RNA-Seq clean data from all samples were >91%. Both the PCA and Pearson correlation analysis-based heatmaps using Fragments Per Kilobase Million values (FPKM) of DEGs showed that the replications of each sample set are clustered together (Figure 2A,B). In addition, the organ-specific difference was indicated by both the PCA and Pearson correlation analysis-based heatmaps. The difference resulting from different organs seems to be a dominant factor for distinguishing samples. Transcription-level disturbances due to excess zinc treatments in different organs are also quite different (Figure 2A,C,D). Maximum DEGs were identified in the stems (572), which also included the most number of down-regulated genes (349), while DEGs in the leaves (total DEGs: 319) and roots (total DEGs: 215) are mainly up-regulated genes (Figure 2C,D). Therefore, organ-specific differences should be considered when further analyses of DEGs involved in zinc stress responses are performed.

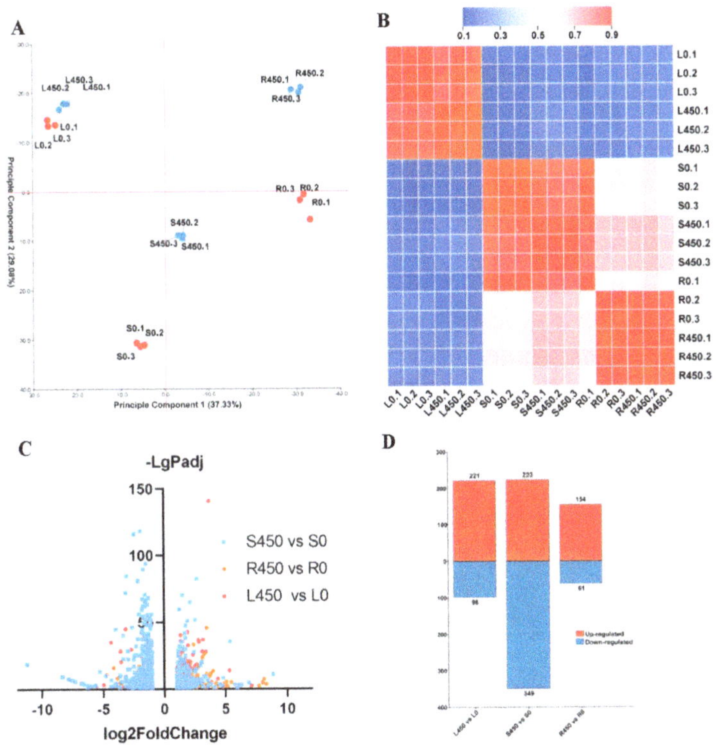

Figure 2. Comparative transcriptome analysis in different organs under excess zinc treatment. (**A**). PCA showing different organ samples under zinc treatments and control; (**B**). heatmap showing different organ samples under zinc treatments and the control based on Pearson correlation coefficients; (**C**). volcano diagram showing DEGs in different organs under zinc treatments and the control; and (**D**). up-regulated and down-regulated genes in different organs under zinc treatments and the control. L0, S0 and R0 indicate the leaves, stems and roots of mulberry without zinc treatment (controls), respectively. L450, R450 and S450 indicate the leaves, stems and roots of mulberry under the 450 mg/kg zinc treatment, respectively.

3.3. Organ-Specific DEGs in Mulberry under Zinc Stress

The Venn diagram showed that quite a large number of DEGs are organ-specific under excess zinc treatment, indicative of the organ-specific responses to zinc toxicity in mulberry (Figure 3A). The GO and KEGG pathway enrichment analysis using organ-specific DEGs showed that different biological processes were involved in the organ-specific response to zinc toxicity in mulberry. In the leaves, the hormone response and signal transduction pathways are well enriched (Figure 3B). In addition, the KEGG pathway analysis also showed that the secondary metabolic biosynthesis pathways including flavonoid biosynthesis and the indole alkaloid biosynthesis pathway are also significantly enriched (Figure 3B). Different pathways are enriched for root-specific DEGs including the jasmonic acid biosynthesis pathway and processes involved in sugar metabolites. Quite a lot of sugar metabolite-related processes including the starch and sucrose biosynthesis pathway, the carbohydrate metabolic process and the glucan catabolic process are significantly enriched (Figure 3C). These results indicate that JA and sugar metabolism participate in the response to zinc toxicity in roots. In the stems, hormone-related pathways are also significantly enriched. Phenylpropanoid biosynthesis and the following lignin biosynthesis and flavonoid biosynthesis are enriched in the stems (Figure 3D). Obviously, these results imply that there is quite a different disturbance in different organs in response to zinc toxicity in mulberry. It is possible that a more complex regulation network of genes involved in zinc stress exists in mulberry.

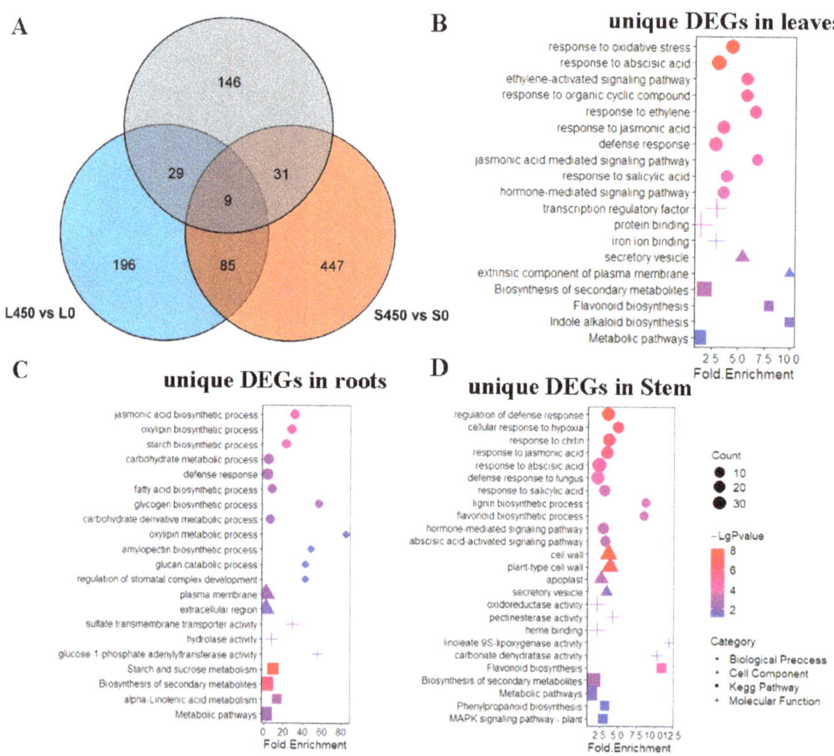

Figure 3. Organ-specific DEGs in mulberry under zinc stress. (**A**). Venn diagram showing DEGs in different organs in mulberry under zinc stress; (**B**). enrichment analysis showing unique DEGs in leaves in mulberry under zinc stress; (**C**). enrichment analysis showing unique DEGs in roots in mulberry under zinc stress; and (**D**). enrichment analysis showing unique DEGs in stems in mulberry under zinc stress.

A total of 81 transcription factors were annotated as the DEGs involved in the response to zinc toxicity including ERF (17), MYB (11), NAC (9) and WRKY (7). It is interesting that these TFs also showed organ-specific preferences. These 17 zinc toxicity-responsive ERFs mainly showed significant differences in the leaves and stems (16/17) than in the roots (1/17) (Table 1, Figure S2). The fold-change in these genes is also available in Table S5. Since most ERFs are annotated as factors involved in hormone responses and stress responses, the results match our GO and KEGG pathway enrichment analysis. In addition, zinc toxicity-responsive MYBs also showed organ-specific preferences. Among the eleven differently expressed MYBs, seven MYBs were identified as DEGs unique to stems, and two MYBs were identified as DEGs in stems and other organs (leaves or roots). Stem-specific, differentially expressed MYBs are mainly annotated as factors involved in phenylpropanoid biosynthesis or cell wall biogenesis (Table 1 and Figure S2). For example, stem-specific DEGs *MYB7* and *MYB52* were proposed to be involved in lignin biosynthesis, and *MYB66* and *MYB123* were proposed to be involved in flavonoid biosynthesis in Arabidopsis [38]. Several WRKYs, including *WRKY40*, *WRKY50* and *WRKY51* and bHLH *MYC4*, that were annotated as JA-related genes showed significantly different expressions in the roots under excess zinc treatment (Table 1). These results further suggested that organ-specific transcriptional regulation networks might be important in the response to zinc toxicity in mulberry.

Table 1. Zinc toxicity-responsive transcription factors in mulberry.

Gene ID	Organ	Ortholog	Gene Name	TF Type	Annotation
M.alba_G0008931	L	AT3G07340	*CIB3*	bHLH	Photoperiodic flowering
M.alba_G0005811	L	AT2G22850	*bZIP6*	bZIP	Vascular development
M.alba_G0006171	L	AT1G75390	*bZIP44*	bZIP	Stress response and development
M.alba_G0017047	L	AT1G27730	*STZ*	C2H2	Stress response
M.alba_G0013540	L	AT4G29190	*AtC3H49*	C3H	Cold response
M.alba_G0018475	L	AT3G47500	*CDF3*	Dof	Nitrogen responses
M.alba_G0006380	L	AT4G39780	*ERF60*	ERF	Defense response and light stimulus
M.alba_G0020260	L	AT2G44840	*ERF13*	ERF	Ethylene-activated signaling pathway
M.alba_G0020348	L	AT4G17500	*ERF-1*	ERF	Ethylene-activated signaling pathway
M.alba_G0003406	L	AT5G51190	*ERF105*	ERF	Response to cold stress
M.alba_G0016087	L	AT2G47520	*ERF71*	ERF	Response to hypoxia stress
M.alba_G0016369	L	AT1G50640	*ERF3*	ERF	Ethylene-activated signaling pathway
M.alba_G0001958	L	AT3G02550	*LBD41*	LBD	Stress response
M.alba_G0017051	L	AT3G49940	*LBD38*	LBD	Cellular metal ion homeostasis
M.alba_G0012646	L	AT4G37260	*MYB73*	MYB	Stress responses and leaf senescence
M.alba_G0006600	L	AT1G01720	*ANAC002*	NAC	Response to wounding and abscisic acid
M.alba_G0007251	L	AT1G69490	*ANAC029*	NAC	Leaf senescence
M.alba_G0013712	L	AT1G01720	*ANAC002*	NAC	Response to wounding and abscisic acid
M.alba_G0019459	L	AT4G27410	*ANAC072*	NAC	ABA-mediated dehydration response
M.alba_G0007498	L	AT1G13260	*EDF4*	RAV	Response to low temperature
M.alba_G0011100	L	AT1G80840	*WRKY40*	WRKY	Fungus defense
M.alba_G0009278	R	AT4G01500	*NGA4*	B3	Leaf, stigma development
M.alba_G0005063	R	AT4G17880	*MYC4*	bHLH	Activate JA-responses
M.alba_G0010656	R	AT4G20970	*NA*	bHLH	Dehydration stress memory
M.alba_G0018432	R	AT3G47640	*PYE*	bHLH	Regulating response to iron deficiency

Table 1. *Cont.*

Gene ID	Organ	Ortholog	Gene Name	TF Type	Annotation
M.alba_G0015523	R	AT5G28770	*bZIP63*	Bzip	Circadian phase in response to sugars
M.alba_G0001737	R	AT5G39660	*CDF2*	Dof	Photoperiodic flowering response
M.alba_G0003407	R	AT4G17500	*ERF-1*	ERF	Ethylene-activated signaling pathway
M.alba_G0004215	R	AT3G13040	*γ-YB2*	G2-like	Phosphate starvation
M.alba_G0016414	R	AT2G31180	*MYB14*	MYB	Cold or wound stress
M.alba_G0006667	R	AT5G64530	*ANAC104*	NAC	Xylem development
M.alba_G0013320	R	AT1G80840	*WRKY40*	WRKY	Photosynthesis and Iron Homeostasis
M.alba_G0013615	R	AT5G26170	*WRKY50*	WRKY	Defense response to fungus, JA response
M.alba_G0018966	RLS	AT3G18960	*REM7*	B3	Tissue development
M.alba_G0018963	RLS	AT5G23090	*NF-YB13*	NF-YB	NA
M.alba_G0002019	RS	AT5G16770	*MYB9*	MYB	Suberin biosynthesis and transport
M.alba_G0006489	RS	AT5G64810	*WRKY51*	WRKY	Jasmonic acid-inducible defense responses
M.alba_G0013321	RS	AT1G80840	*WRKY40*	WRKY	Jasmonic acid-inducible defense responses
M.alba_G0001053	S	AT4G29930	*NA*	bHLH	NA
M.alba_G0004848	S	AT1G32640	*MYC2*	bHLH	Activate JA-responses
M.alba_G0012659	S	AT1G72210	*BHLH96*	bHLH	Regulation of RNA polymerase II
M.alba_G0018293	S	AT4G20970	*NA*	bHLH	Defense response to fungus
M.alba_G0019112	S	AT4G25440	*ZFWD1*	C3H	Development
M.alba_G0003582	S	AT2G40140	*SZF2*	C3H	Response to biotic and abiotic stresses
M.alba_G0005725	S	AT4G38960	*BBX19*	DBB	Photomorphogenesis and flowering
M.alba_G0005394	S	AT5G51990	*CBF4*	ERF	Drought stress and abscisic acid treatment
M.alba_G0012399	S	AT1G21910	*DREB26*	ERF	Response to JA and SA, abiotic stress
M.alba_G0016407	S	AT2G31230	*ERF15*	ERF	Stress response
M.alba_G0005395	S	AT5G51990	*CBF4*	ERF	Drought stress and abscisic acid treatment
M.alba_G0014394	S	AT1G50420	*SCL-3*	GRAS	Response to gibberellin
M.alba_G0013518	S	AT2G22840	*GRF1*	GRF	Leaf development.
M.alba_G0019709	S	AT3G61890	*HB-12*	HD-ZIP	Leaf and stem development
M.alba_G0013465	S	AT4G37540	*LBD39*	LBD	Cell wall biogenesis
M.alba_G0011537	S	AT5G35550	*MYB123*	MYB	Anthocyanin biosynthesis
M.alba_G0012042	S	AT1G17950	*MYB52*	MYB	Lignin, xylan and cellulose biosynthesis
M.alba_G0018463	S	AT5G61420	*MYB28*	MYB	Seed development and aliphatic glucosinolate biosynthesis
M.alba_G0018447	S	AT3G47600	*ATMYB94*	MYB	Cuticular wax biosynthesis
M.alba_G0011536	S	AT2G16720	*MYB7*	MYB	General phenylpropanoid and lignin R2R3-MYB repressors
M.alba_G0018280	S	AT5G14750	*MYB66*	MYB	Anthocyanin production and differentiation of trichome cells
M.alba_G0013188	S	AT1G75250	*ATRL6*	MYB	Signal transduction
M.alba_G0019458	S	AT3G15510	*ANAC056*	NAC	System development
M.alba_G0009218	S	AT3G04070	*ANAC047*	NAC	Response to flooding
M.alba_G0009713	S	AT5G63790	*ANAC102*	NAC	Mediating response to low oxygen stress

Table 1. *Cont.*

Gene ID	Organ	Ortholog	Gene Name	TF Type	Annotation
M.alba_G0011705	S	AT4G14540	*NF-YB3*	NF-YB	Response to heat, response to water deprivation
M.alba_G0006285	S	AT4G24660	*ATHB22*	ZF-HD	Embryo development ending in seed dormancy
M.alba_G0007224	S	AT1G69600	*ATHB29*	ZF-HD	Early responsive to dehydration stress.
M.alba_G0013466	SL	AT1G27730	*STZ*	C2H2	Stress response
M.alba_G0015192	SL	AT3G46080	*NA*	C2H2	Transient stress
M.alba_G0003253	SL	AT5G52020	*DREB*	ERF	Glucosinolate metabolic process
M.alba_G0003254	SL	AT5G51990	*CBF4*	ERF	Drought stress and abscisic acid treatment
M.alba_G0003536	SL	AT2G40340	*AtERF48*	ERF	Response to abscisic and acid stress
M.alba_G0005396	SL	AT5G51990	*CBF4*	ERF	Response to drought stress and abscisic acid treatment
M.alba_G0017242	SL	AT4G34410	*ERF109*	ERF	Retarding programmed cell death under salt stress
M.alba_G0019814	SL	AT1G19210	*ERF17*	ERF	JA, defense to biotic stresses
M.alba_G0000389	SL	AT5G48150	*PAT1*	GRAS	Callus formation, photomorphogenesis, red, far-red light phototransduction
M.alba_G0011734	SL	AT4G17230	*SCL13*	GRAS	Cellular response to hypoxia, heat
M.alba_G0004071	SL	AT5G04760	*DIV2*	MYB	Negative roles in salt stress and is required for ABA signaling in Arabidopsis
M.alba_G0014170	SL	AT3G44350	*ANAC061*	NAC	Response to salt stress
M.alba_G0005182	SL	AT4G11070	*WRKY41*	WRKY	ABA defense response
M.alba_G0014899	SL	AT2G38470	*WRKY33*	WRKY	Stress response
M.alba_G0019631	SL	AT2G46400	*WRKY46*	WRKY	ABA signaling and auxin homeostasis in response to abiotic stress

Note: The symbols were the gene names given to the orthologs in *Arabidopsis thaliana* and the function annotation refers to the information in TAIR.

3.4. Network of Genes Involved in the Response to Zinc Toxicity in Different Organs

WGCNA showed that modules MEbrown, MEgreen and MEturquoise had significant correlations with different organs, which indicated that DEGs in these modules are organ-specific responses under zinc stress (Figure 4A). Further analysis showed that the top co-expressed genes (nodes of the top 500 connections with correlation coefficients > 0.6) in the leaves are involved in hormone signaling. Several transcription factors including *EFR-1*, *EFR13*, *EFR71*, *ANAC002* and *WRKY40* and jasmonate signaling-related genes *JAZ1* and *JAZ8* were co-expressed (Figure 4C). Co-expressed genes in the stems are also hormone signaling-related genes (Figure 4B). The co-expression network of zinc-response genes in the roots was quite different from the networks in the leaves and stems. Several sugar metabolism-related genes such as *CWINV1*, *BE3*, *GBSS1* and *ADG1* showed a significant correlation and interacted with other stress response genes including *PYE*, *JMT* and *AVP* (Figure 4D). In addition, another module, MEblues, showed a significant correlation with the treatments, which indicated these DEGs might be generally important genes involved in the response to zinc toxicity. Repressor *BAN* and activator *MYB52* together with a series of genes involved in lignin biosynthesis and flavonoid biosynthesis comprised a regulation network of phenylpropanoid biosynthesis and cell wall biogenesis involved in the response to excess zinc. In addition, *ANAC104* involved in suberin biosynthesis and *MYB9* involved in xylem development were also proposed to be involved in cell wall biogenesis. The connections between phenylpropanoid biosynthesis and other stress response genes were also shown in the network. *ANAC072* and *PIP2;1* were reported to be involved in the ABA-mediated dehydration response.

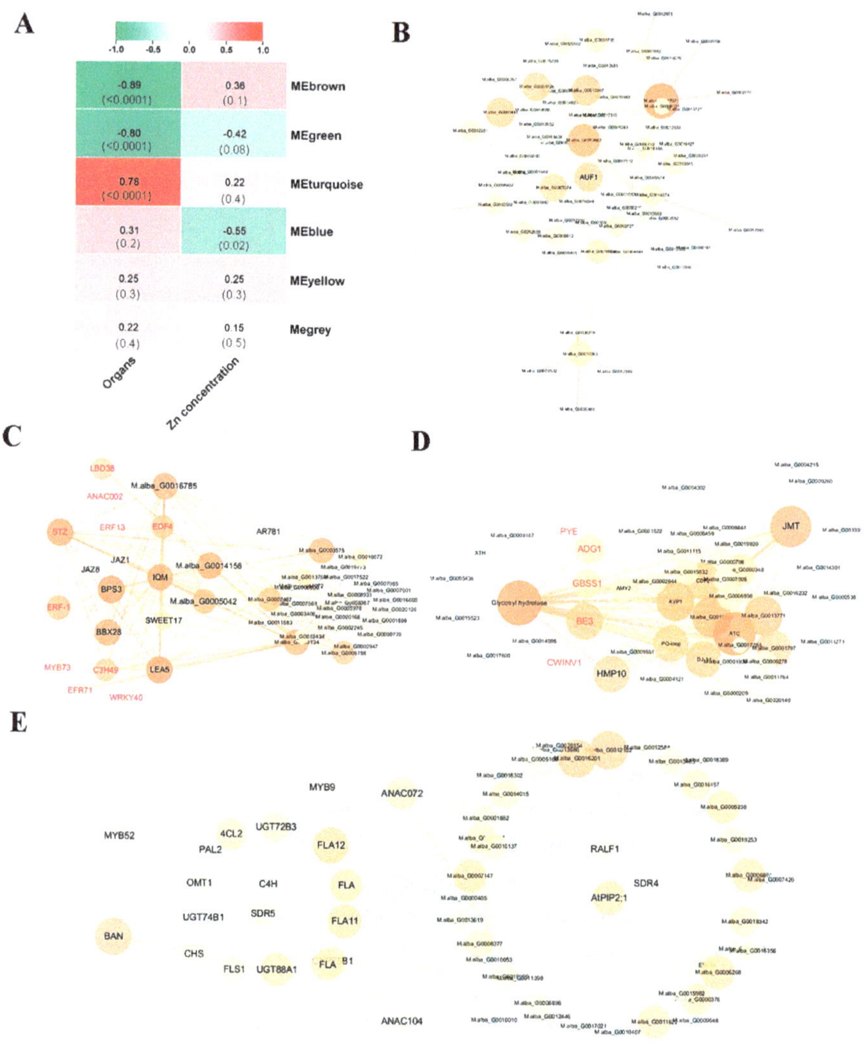

Figure 4. Co-expression of DEGs in different organs in mulberry under zinc stress. (**A**). Heatmap of module-trait associations based on WGCNA; (**B**). co-expression network showing DEGs in the stems of mulberry under zinc stress; (**C**). co-expression network showing DEGs in the leaves of mulberry under zinc stress; (**D**). co-expression network showing DEGs in the roots of mulberry under zinc stress; and (**E**). co-expression network showing DEGs in both the stems and roots of mulberry under zinc stresses. Co-expressed genes were clustered in the same-colored modules such as MEbrown, MEblue, etc.

3.5. Validation of Transcription Levels of Key Genes Responsive to Zinc Toxicity

Several key genes responsive to zinc toxicity based on our WGCNA and co-expression networks were further validated using RT-qPCR. The detailed information for these selected genes is provided in Table S4. The RT-qPCR results correspond well with the RNA-Seq results except for *BAN*, which showed no significantly different expression in all detected organs of mulberry exposed to zinc stress (Figure 5 and Table S4). Genes involved in phenylpropanoid biosynthesis, such as *PAL2*, *CHS* and *4CL2* (Figure 5A–C), genes involved in sugar metabolism, such as *CWINV1*, *GBSS1* and *ADG1* (Figure 5D–F), and transcription factors involved in cell wall biogenesis, such as *MYB9* and *ANAC 104* (Figure 5H,I), show quite similar organ-specific expression change comparing with the results of the transcriptome analysis. These genes are considered organ-specific zinc stress response genes.

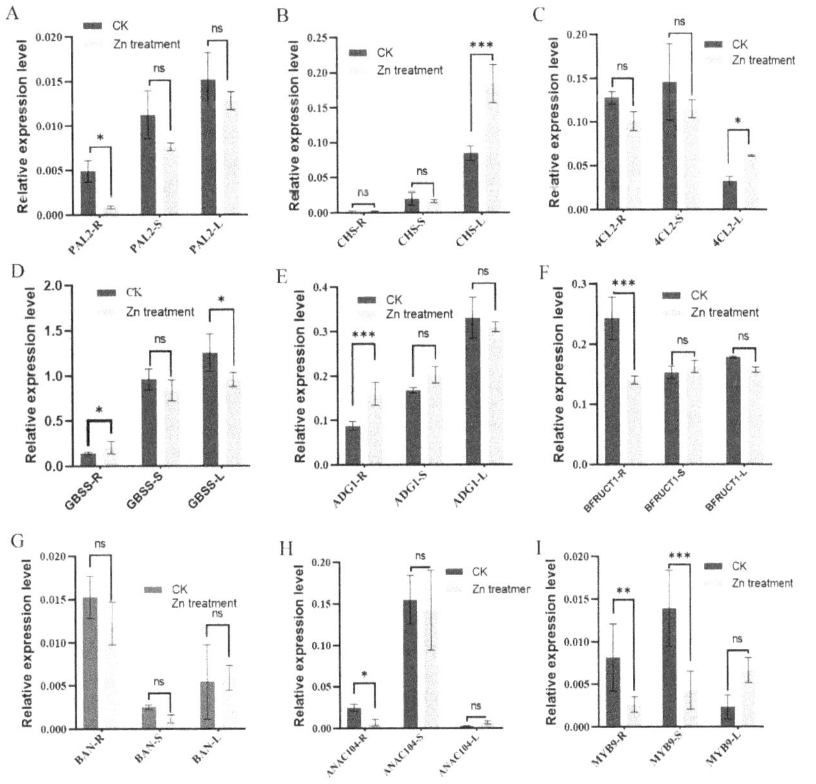

Figure 5. RT-qPCR showing the selected DEGs in response to zinc stress in mulberry. (**A–I**), relative expression levels of selected DEGs in different organs under zinc stresses. the gene names were marked in each subfigure and R, S and L indicated roots, stems and leaves respectively. Error bars represent SE. Asterisks indicate significant differences as determined with Student's *t*-test (* $p < 0.05$; ** $p < 0.01$; *** $p < 0.001$).

4. Discussion

4.1. Limitation of Excess Zinc Supply in Soil for Mulberry

Some heavy metals such as zinc are known as micronutrients that are essential to plant growth and survival [1]. However, plants would suffer zinc toxicity if the zinc concentration in the soil is beyond the limitation of zinc concentration in the soil. The threshold total zinc values from the literature for zinc in sensitive plant species is 150~200 mg/kg zinc, and 100–500 mg/kg zinc is regarded as the range of zinc contents at which the yield of many crops might be reduced by 25% due to toxicity [39]. Zinc toxicity function disrupts key enzymatic reactions in many cellular processes including carbon fixation and metabolism [40–42]. Different plants had quite different capacities to deal with zinc toxicity, and the limitation in zinc supply that resulted in zinc toxicity varied for different plant species [39]. Toxicity symptoms usually become visible at >300 mg/kg zinc in leaf DW, although some crops show toxicity symptoms at <100 mg/kg zinc in leaf DW [5,43,44]. Mandarin orange (*Citrus reticulata* Blanco) seedlings supplied with 5 mM (~325 mg/kg) zinc are considered zinc sufficient and induced prolific growth and sprouted abundantly, while plants that received more than 10 mM (~650 mg/kg) of zinc suffered zinc toxicity with prime features of growth retardation, defoliation and sluggish root growth [27]. Wheat under 14 mg/kg zinc treatment in the soil can help to deal with drought stress and result in zinc-mediated alleviation of drought stress [45]. Some plants such as *Thlaspi caerulescens* are zinc hyperaccumulators that show zinc-hypertolerant capacity. Mulberry is also reported to have an outstanding ability to uptake zinc and survive in zinc-polluted mines [20]. A previous study revealed that a 50 mg/kg zinc treatment can not only promote plant growth but also alleviates the adverse effects of lead stress in *Morus alba* [46]. Heavy metal toxicity also can be reflected by the change in physiological indicators levels such as MDA, SOD, proline and chlorophyll [7,27,37,46]. In the present study, the limitation of excess zinc supply in the soil for mulberry was identified by referring to physiological indicators. A 350 mg/kg zinc supply completely induced physiological adverse effects in all organs of mulberry, and beyond this limitation, possible damages that cannot be alleviated by physiological responses occurred when supplying excess zinc (Figure 1). The bell-pattern change might indicate that plants experience a process including benefiting from a suitable supply of zinc, suffering excess zinc toxicity and damage in the physiological response mechanism. Similar physiological indicator change patterns were also reported in the previous study [37]. In fact, mulberry plants with 400 mg/kg or 450 mg/kg zinc supply can still survive but only with slow growth (Figure S1). The maximum zinc concentration permitted in sewage sludge-amended soils (pH 6–7) is in the range of 100 mg/kg to 300 mg/kg zinc in some countries [11]. Therefore, mulberry is a potential plant that can be used as a bio-cleaner in zinc-polluted areas.

4.2. Organ-Specific Responses to Zinc Toxicity

Zinc homeostasis showed a spatial-temporal distribution, and a spatial response to acute zinc deficiency in Sorghum has been reported [47]. The distribution of zinc in mulberry also showed spatial differences with quite different zinc concentrations in different organs (leaf, root, bark and stem) [20]. Therefore, organ-specific responses to zinc toxicity should be evaluated. Our results and previous reports have indicated that hormone signaling pathways and several secondary metabolite-related pathways, such as lignin and flavonoids, are important for zinc stress responses [8,19,24,47–49]. These pathways were also identified in the present study, and organ-specific responses were also revealed. Hormone signal pathways participate in the response to zinc toxicity in all organs, but some hormones may play dominant roles in specific organs of mulberry. For example, in the roots, only JA-related pathways were enriched, including its biosynthesis pathway and alpha-linolenic acid metabolism. Secondary metabolism is another important biological process involved in the response to zinc toxicity in all organs, but genes involved in specific secondary metabolites were enriched in specific organs such as the flavonoid biosynthesis pathway in leaves, sugar metabolism in roots and lignin biosynthesis in stems. The causes of organ-specific responses to zinc toxicity in mulberry are various. The distribution of zinc in organs, organ preference expression of key genes involved in pathways and roles of different organs in response to stresses can together result in organ-specific responses. For example, genes involved in the lignin biosynthesis pathway prefer expression in lignified tissues, which may lead to this pathway mainly responding to zinc toxicity in stems or lignified roots [24]. Sugar homeostasis in vascular tissue is important for the response to various stresses including drought, oxidative stress and stresses resulting from heavy metals [50,51]. Our results also indicated that the sugar metabolic pathway participates in the response to zinc toxicity in the roots of mulberry.

4.3. Molecular Regulation Network of Genes in Response to Zinc Exposure

The current understanding of plant zinc homeostasis regulation mechanisms is mainly based on studies on *Arabidopsis*. The transcriptional level regulation of zinc homeostasis based on *bZIP19* and *bZIP23* in response to changes in cellular zinc status was reported in *Arabidopsis* [8]. However, the molecular mechanism of the transcriptional regulation network in response to zinc toxicity is little reported for woody plants. In the present study, a total of 88 transcription factors including ERFs, MYBs bHLHs and WRKYs were screened as possible regulators in response to zinc toxicity using transcriptome analysis. AP2/ERF proteins have important functions in the transcriptional regulation of a variety of biological processes and are known as regulators involved in hormone responses such as the ABA response, JA response in chlorophyll degradation and both the biotic and abiotic stress responses [52–57]. ERFs were the most predominant TFs identified in this study. These ERFs mainly show different expression levels in the leaves and participate in regulating a network involved in hormone-signaling pathway-mediated stress responses. EFR-1 and EFR13 were reported to be involved in the ethylene-activated signaling pathway, and ERF71 was identified as a regulator involved in the response to hypoxia stress [55]. The present study indicated that these EFRs were possibly important regulators involved in the response to zinc toxicity through the hormone signaling pathway in mulberry leaves. Other hormone response genes such as *JAZ1* and *JAZ8* were also co expressed with these *ERF*s to comprise the regulation network in mulberry leaves (Figure 4B) In addition to the hormone signaling pathway, the secondary metabolic pathway is also important for the response to zinc toxicity. In mulberry roots, a possible regulation network including sugar metabolism-related genes was also built, indicating their important roles in response to zinc toxicity (Figure 4D). Our previous study reported the genes involved in the lignin biosynthesis pathway positively respond to zinc toxicity in lignifying tissues [24]. In the present study, a possible regulation network of phenylpropanoid biosynthesis and cell wall biogenesis was identified, and their roles in the response to zinc toxicity were primarily revealed. Repressor BAN, activator MYB52 as well as ANAC104 and MYB9 were identified as regulators in this network.

5. Conclusions

Finally, a schematic diagram was summarized (Figure 6). Mulberry seedlings show different growth statuses under different concentrations of zinc treatments. Mulberry seedlings show growth retardation when suffering > 350 mg/kg zinc supply in the soil, and the physiological indicators show a bell-pattern change with a summit at high zinc concentrations. Different pathways are enriched in different organs in response to excess zinc treatment indicating an organ-specific response in mulberry. Flavonoids in leaves, lignin and cell wall in stems and sugar metabolism in roots are important for response to zinc toxicity in specific organs.

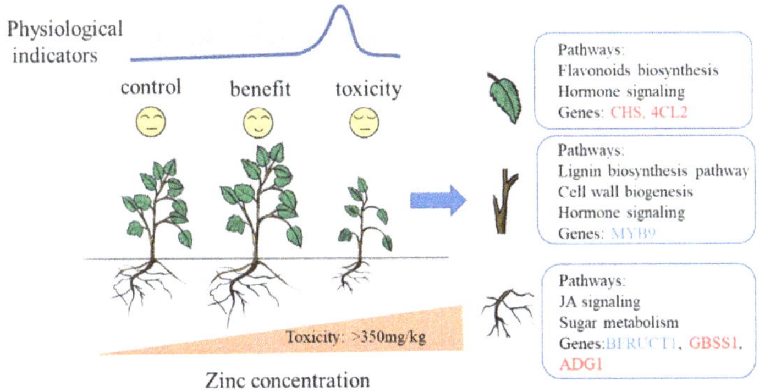

Figure 6. Schematic diagram showing the organ-based response to zinc toxicity in mulberry.

Mulberry seedlings show different growth statuses under different concentrations of zinc treatments. Plants show growth retardation when suffering > 350 mg/kg zinc supply in the soil, and the physiological indicators show a bell-pattern change with a summit at high zinc concentrations. Different pathways are enriched in different organs in response to excess zinc treatment, indicating an organ-specific response in mulberry. The up-regulated genes are colored red, and the down-regulated genes are colored blue.

Supplementary Materials: The following supporting information can be downloaded at: https://www.mdpi.com/article/10.3390/f14040842/s1, Table S1: Primers used in this study; Table S2: Detailed statistics significance analysis results for the change in physiological indicators; Table S3: Summary of the RNA-Seq dataset; Table S4: Detailed information and fold change in selected genes for RT-qPCR; Table S5: DEGs in different organs; Figure S1: Growth status of mulberry seedlings after 15-day zinc treatments; Figure S2: Distribution of different expressed transcription factors (DETFs) in different organs. R, unique DETFs in roots; S, unique DETFs in stems; L; unique DETFs in leaves; RS, DETFs in both roots and stems; SL, DETFs in both stems and leaves.

Author Contributions: L.L. and N.C. guided this work and provided advice; L.Z. design the experiments; S.H., X.K., T.Y. and K.Y. performed the experiments and analyzed the data; N.C. analyzed the data, organized the figures and wrote the manuscript. X.C. revised the manuscript. All authors have read and agreed to the published version of the manuscript.

Funding: This work was jointly supported by the Natural Science Foundation of Jiangsu Province (BK20210879 to Nan Chao), the National Natural Science Foundation (NSF 32201526 to Nan Chao), the Crop Germplasm Resources Protection Project of the Ministry of Agriculture and Rural Affairs of the People's Republic of China (19190172), the National Infrastructure for Crop Germplasm Resources (NICGR-2019-43), and the China Agriculture Research System of MOF and MARA.

Acknowledgments: We thank Feng Jiao at Northwest University of Agriculture and Forestry who provided us with the genome annotation file for *M. alba*.

Conflicts of Interest: The authors declare no conflict of interest.

References

1. Hambidge, M. Human Zinc Deficiency. *J. Nutr.* **2000**, *130* (Suppl. S5), 1344S. [CrossRef]
2. Cherasse, Y.; Urade, Y. Dietary Zinc Acts as a Sleep Modulator. *Int. J. Mol. Sci.* **2017**, *18*, 2334. [CrossRef]
3. Hotz, C.; Brown, K.H. Assessment of the risk of zinc deficiency in populations and options for its control. *Food Nutr. Bull.* **2004**, *25*, S130–S162.
4. Gibson, R.S.; King, J.C.; Lowe, N. A Review of Dietary Zinc Recommendations. *Food Nutr. Bull.* **2016**, *37*, 443–460. [CrossRef]
5. Alloway, B.J. *Zinc in Soils and Crop Nutrition*; International Zinc International Association: Brussels, Belgium; Fertilizer Industry Association: Paris, France, 2008.
6. Tsonev, T.; Cebola Lidon, F.J. Zinc in plants-an overview. *Emir. J. Food Agric.* **2012**, *24*, 322–333.
7. Qin, F.; Liu, G.; Huang, G.; Dong, T.; Liao, Y.; Xu, X. Zinc application alleviates the adverse effects of lead stress more in female Morus alba than in males. *Environ. Exp. Bot.* **2018**, *146*, 68–76. [CrossRef]
8. Zlobin, I.E. Current understanding of plant zinc homeostasis regulation mechanisms. *Plant Physiol. Biochem.* **2021**, *162*, 327–335. [CrossRef]
9. Fuentes, D.; Disante, K.B.; Valdecantos, A.; Cortina, J.; Vallejo, V.R. Sensitivity of Mediterranean woody seedlings to copper, nickel and zinc. *Chemosphere* **2007**, *66*, 412–420. [CrossRef]
10. Bose, S.; Bhattacharyya, A.K. Heavy metal accumulation in wheat plant grown in soil amended with industrial sludge. *Chemosphere* **2008**, *70*, 1264–1272. [CrossRef]
11. McGrath, S.; Chang, A.; Page, A.; Witter, E. Land application of sewage sludge: Scientific perspectives of heavy metal loading limits in Europe and the United States. *Environ. Rev.* **1994**, *2*, 108–118. [CrossRef]
12. Disante, K.B.; Fuentes, D.; Cortina, J. Response to drought of Zinc-stressed *Quercus suber* L. seedlings. Environ. *Exp. Bot.* **2011**, *70*, 96–103. [CrossRef]
13. Pandey, N.; Pathak, G.C.; Singh, A.K.; Sharma, C.P. Enzymic changes in response to zinc nutrition. *J. Plant Physiol.* **2002**, *159*, 1151–1153. [CrossRef]
14. Chakmak, I. Possible roles of zinc in protecting plant cells from damage by reactive oxygen. *New Phytol.* **2000**, *146*, 185–205. [CrossRef]
15. Choudhary, M.; Bailey, L.; Grant, C.; Leisle, D. Effect of Zinc on the concentration of Cd and Zinc in plant tissue of two durum wheat lines. *Can. J. Plant Sci.* **1995**, *75*, 445–448. [CrossRef]
16. Lin, Y.F.; Aarts, M.G. The molecular mechanism of zinc and cadmium stress response in plants. *Cell. Mol. Life Sci.* **2012**, *69*, 3187–3206. [CrossRef] [PubMed]
17. He, P.P.; Lv, X.Z.; Wang, G.Y. Effects of Se and Zinc supplementation on the antagonism against Pb and Cd in vegetables. *Environ. Int.* **2004**, *30*, 167–172. [CrossRef]

18. Li, C.; Xu, G.; Zang, R.; Korpelainen, H.; Berninger, F. Sex-related differences in leaf morphological and physiological responses in Hippophae rhamnoides along an altitudinal gradient. *Tree Physiol.* **2007**, *27*, 399–406. [CrossRef]
19. Zhang, Y.; Wang, Y.; Ding, Z.; Wang, H.; Song, L.; Jia, S.; Ma, D. Zinc stress affects ionome and metabolome in tea plants. *Plant Physiol. Biochem.* **2017**, *111*, 318–328. [CrossRef]
20. Jiang, Y.; Huang, R.; Yan, X.; Jia, C.; Long, T. Mulberry for environmental protection. *Pak. J. Bot.* **2017**, *49*, 781–788.
21. Srivastava, S.; Kapoor, R.; Thathola, A.; Srivastava, R.P. Nutritional quality of leaves of some genotypes of mulberry (*Morus alba*). *Int. J. Food Sci. Nutr.* **2006**, *57*, 305–313. [CrossRef]
22. Cao, X.; Du, W.; Shang, C.; Shen, Q.; Liu, L.; Cheng, J. Comparative transcriptome reveals circadian and hormonal control of adventitious rooting in mulberry hardwood cuttings. *Acta Physiol. Plant.* **2018**, *40*, 197. [CrossRef]
23. Shang, C.; Yang, H.; Ma, S.; Shen, Q.; Liu, L.; Hou, C.; Cao, X.; Cheng, J. Physiological and Transcriptomic Changes during the Early Phases of Adventitious Root Formation in Mulberry Stem Hardwood Cuttings. *Int. J. Mol. Sci.* **2019**, *20*, 3707. [CrossRef] [PubMed]
24. Chao, N.; Yu, T.; Hou, C.; Liu, L.; Zhang, L. Genome-wide analysis of the lignin toolbox for morus and the roles of lignin related genes in response to zinc stress. *PeerJ* **2021**, *9*, e11964. [CrossRef] [PubMed]
25. Chao, N.; Wang, R.F.; Hou, C.; Yu, T.; Miao, K.; Cao, F.Y.; Fang, R.J.; Liu, L. Functional characterization of two chalcone isomerase (CHI) revealing their responsibility for anthocyanins accumulation in mulberry. *Plant Physiol. Biochem.* **2021**, *161*, 65–73. [CrossRef] [PubMed]
26. Susheelamma, B.N.; Shekar, K.R.; Sarkar, A.; Rao, M.R.; Datta, R.K. Genotype and hormonal effects on callus formation and regeneration in mulberry. *Euphytica* **1996**, *90*, 25–29. [CrossRef]
27. Subba, P.; Mukhopadhyay, M.; Mahato, S.K.; Bhutia, K.D.; Mondal, T.K.; Ghosh, S.K. Zinc stress induces physiological, ultra-structural and biochemical changes in mandarin orange (*Citrus reticulata Blanco*) seedlings. *Physiol. Mol. Biol. Plants* **2014**, *20*, 461–473. [CrossRef]
28. Arora, A.; Sairam, R.; Srivastava, G. Oxidative stress and antioxidative system in plants. *Curr. Sci.* **2002**, *82*, 1227–1238.
29. Silva, P.; Matos, M. Assessment of the impact of Aluminum on germination, early growth and free proline content in *Lactuca sativa* L. Ecotoxicol. *Environ. Saf.* **2016**, *131*, 151–156. [CrossRef]
30. Langdon, W.B. Performance of genetic programming optimised Bowtie2 on genome comparison and analytic testing (GCAT) benchmarks. *BioData Min.* **2015**, *8*, 1. [CrossRef]
31. Pertea, M.; Pertea, G.M.; Antonescu, C.M.; Chang, T.C.; Mendell, J.T.; Salzberg, S.L. StringTie enables improved reconstruction of a transcriptome from RNA-seq reads. *Nat. Biotechnol.* **2015**, *33*, 290–295. [CrossRef]
32. Anders, S.; Huber, W. Differential expression analysis for sequence count data. *Nat. Preced.* **2010**, *2010*, 5. [CrossRef]
33. Langfelder, P.; Horvath, S. WGCNA: An R package for weighted correlation network analysis. *BMC Bioinform.* **2008**, *9*, 559. [CrossRef] [PubMed]
34. Huang, D.W.; Sherman, B.T.; Tan, Q.; Kir, J.; Liu, D.; Bryant, D.; Guo, Y.; Stephens, R.; Baseler, M.W.; Lane, H.C. DAVID Bioinformatics Resources: Expanded annotation database and novel algorithms to better extract biology from large gene lists. *Nucleic Acids Res.* **2007**, *35* (Suppl. S2), W169–W175. [CrossRef] [PubMed]
35. Chen, C.; Chen, H.; Zhang, Y.; Thomas, H.R.; Frank, M.H.; He, Y.; Xia, R. TBtools: An Integrative Toolkit Developed for Interactive Analyses of Big Biological Data. *Mol. Plant* **2020**, *13*, 1194–1202. [CrossRef] [PubMed]
36. Shukla, P.; Reddy, R.A.; Ponnuvel, K.M.; Rohela, G.K.; Shabnam, A.A.; Ghosh, M.K.; Mishra, R.K. Selection of suitable reference genes for quantitative real-time PCR gene expression analysis in Mulberry (*Morus alba* L.) under different abiotic stresses. *Mol. Biol. Rep.* **2019**, *46*, 1809–1817. [CrossRef] [PubMed]
37. Kumar Tewari, R.; Kumar, P.; Nand Sharma, P. Morphology and physiology of zinc-stressed mulberry plants. *J. Plant Nutr. Soil Sci.* **2008**, *171*, 286–294. [CrossRef]
38. Liu, L.; Chao, N.; Yidilisi, K.; Kang, X.; Cao, X. Comprehensive analysis of the MYB transcription factor gene family in Morus alba. *BMC Plant Biol.* **2022**, *22*, 281. [CrossRef]
39. Kabata-Pendias, A. *Trace Elements in Soils and Plants*; CRC Press: Boca Raton, FL, USA, 2000.
40. Monnet, F.; Vaillant, N.; Vernay, P.; Coudret, A.; Sallanon, H.; Hitmi, A. Relationship between PSII activity, CO_2 fixation, and Zinc, Mn and Mg contents of Lolium perenne under zinc stress. *J. Plant Physiol.* **2001**, *158*, 1137–1144. [CrossRef]
41. Roitto, M.; Rautio, P.; Julkunen-Tiitto, R.; Kukkola, E.; Huttunen, S. Changes in the concentrations of phenolics and photosynthates in Scots pine (*Pinus sylvestris* L.) seedlings exposed to nickel and copper. *Environ. Pollut.* **2005**, *137*, 603–609. [CrossRef]
42. Burzyński, M.; Żurek, A. Effects of copper and cadmium on photosynthesis in cucumber cotyledons. *Photosynthetica* **2007**, *45*, 239–244. [CrossRef]
43. Broadley, M.R.; White, P.J.; Hammond, J.P.; Zelko, I.; Lux, A. Zinc in plants. *New Phytol.* **2007**, *173*, 677–702. [CrossRef] [PubMed]
44. Marschner, H. *Marschner's Mineral Nutrition of Higher Plants*; Academic Press: Cambridge, MA, USA, 2011.
45. Ma, D.; Sun, D.; Wang, C.; Ding, H.; Qin, H.; Hou, J.; Xin, H.; Xie, Y.; Guo, T. Physiological Responses and Yield of Wheat Plants in Zinc-Mediated Alleviation of Drought Stress. *Front. Plant Sci.* **2017**, *8*, 860. [CrossRef] [PubMed]
46. Fan, W.; Guo, Q.; Liu, C.; Liu, X.; Zhang, M.; Long, D.; Xiang, Z.; Zhao, A. Two mulberry phytochelatin synthase genes confer zinc/cadmium tolerance and accumulation in transgenic Arabidopsis and tobacco. *Gene* **2018**, *645*, 95–104. [CrossRef] [PubMed]

47. Li, Y.; Zhang, Y.; Shi, D.; Liu, X.; Qin, J.; Ge, Q.; Xu, L.; Pan, X.; Li, W.; Zhu, Y.; et al. Spatial-temporal analysis of zinc homeostasis reveals the response mechanisms to acute zinc deficiency in Sorghum bicolor. *New Phytol.* **2013**, *200*, 1102–1115. [CrossRef] [PubMed]
48. Colebrook, E.H.; Thomas, S.G.; Phillips, A.L.; Hedden, P. The role of gibberellin signalling in plant responses to abiotic stress. *J. Exp. Biol.* **2014**, *217 Pt 1*, 67–75. [CrossRef]
49. Azevedo, H.; Azinheiro, S.G.; Munoz-Merida, A.; Castro, P.H.; Huettel, B.; Aarts, M.G.; Assuncao, A.G. Transcriptomic profiling of Arabidopsis gene expression in response to varying micronutrient zinc supply. *Genome Data* **2016**, *7*, 256–258. [CrossRef]
50. Sami, F.; Yusuf, M.; Faizan, M.; Faraz, A.; Hayat, S. Role of sugars under abiotic stress. *Plant Physiol. Biochem.* **2016**, *109*, 54–61. [CrossRef]
51. Rosa, M.; Prado, C.; Podazza, G.; Interdonato, R.; Gonzalez, J.A.; Hilal, M.; Prado, F.E. Soluble sugars–metabolism, sensing and abiotic stress: A complex network in the life of plants. *Plant Signal. Behav.* **2009**, *4*, 388–393. [CrossRef]
52. Illgen, S.; Zintl, S.; Zuther, E.; Hincha, D.K.; Schmulling, T. Characterisation of the ERF102 to ERF105 genes of Arabidopsis thaliana and their role in the response to cold stress. *Plant Mol. Biol.* **2020**, *103*, 303–320. [CrossRef]
53. Je, J.; Chen, H.; Song, C.; Lim, C.O. Arabidopsis DREB2C modulates ABA biosynthesis during germination. *Biochem. Biophys. Res. Commun.* **2014**, *452*, 91–98. [CrossRef]
54. Lee, S.B.; Lee, S.J.; Kim, S.Y. AtERF15 is a positive regulator of ABA response. *Plant Cell Rep.* **2015**, *34*, 71–81. [CrossRef] [PubMed]
55. Nakano, T.; Suzuki, K.; Fujimura, T.; Shinshi, H. Genome-wide analysis of the ERF gene family in Arabidopsis and rice. *Plant Physiol.* **2006**, *140*, 411–432. [CrossRef] [PubMed]
56. Yin, X.R.; Xie, X.L.; Xia, X.J.; Yu, J.Q.; Ferguson, I.B.; Giovannoni, J.J.; Chen, K.S. Involvement of an ethylene response factor in chlorophyll degradation during citrus fruit degreening. *Plant J.* **2016**, *86*, 403–412. [CrossRef] [PubMed]
57. Zhang, H.; Huang, L.; Dai, Y.; Liu, S.; Hong, Y.; Tian, L.; Huang, L.; Cao, Z.; Li, D.; Song, F. Arabidopsis AtERF15 positively regulates immunity against Pseudomonas syringae pv. tomato DC3000 and Botrytis cinerea. *Front. Plant Sci.* **2015**, *6*, 686. [CrossRef] [PubMed]

forests

MDPI

Article

Effects of Trace Elements on Traits and Functional Active Compounds of *Camellia oleifera* in Nutrient-Poor Forests

Qiuyue Dai [1,2,†], Zheng Deng [1,†], Lan Pan [1], Lang Nie [1], Yunyuan Yang [1], Yongfang Huang [1] and Jiuxiang Huang [1,*]

1 College of Forests and Landscape Architecture, South China Agricultural University, Guangzhou 510507, China; 15968826856@163.com (Q.D.); zheng_1816@stu.scau.edu.cn (Z.D.); panlan@scau.edu.cn (L.P.); 13928749780@163.com (L.N.); yunyuan@stu.scau.edu.cn (Y.Y.); hyfang@scau.edu.cn (Y.H.)

2 Beilingshan State-Owned Forest Farm, Zhaoqing 526000, China

* Correspondence: jxhuang@scau.edu.cn

† These authors contributed equally to this work.

Abstract: *Camellia oleifera* is a major woody oilseed species in China, but it is typically cultivated in nutrient-poor soils and may be affected by various trace elements. This study examined how spraying selenium, boron, and zinc trace elements affected the traits and functional active compounds of *C. oleifera* under nutrient deficiency. The results revealed significant variations in the effects of different trace element combinations on *C. oleifera*. Optimal concentrations of zinc and selenium are critical for promoting the growth and development of *C. oleifera* fruit. The transverse diameter of the fruit, the single fruit weight, the number of seeds per fruit, the single fresh seed weight, the oil content in the fruit, and the oil yield per plant of other treatments can be increased by up to 3.07%, 10.57%, 23.66%, 30.23%, 7.94%, and 21.95%, respectively, at most, compared to the control group. Diluting zinc from 1000 to 1500 times and maintaining a selenium concentration from 100 to 200 mg/L has been found to be beneficial for fruit growth. While low concentrations of selenium may promote an increase in fruit transverse diameter, high concentrations of selenium, along with high dilutions of zinc, can have the opposite effect, leading to a reduction in fruit diameter. However, a high concentration of selenium can positively impact the number of seeds per fruit. The most effective combination was found to be a selenium concentration of 0 mg/L, a boron concentration of 4 mg/L, and a zinc dilution of 1500. Interestingly, lower concentrations of selenium and boron, as well as lower dilutions of zinc, were found to increase the oil yield per plant. This suggests that a careful balance of trace elements is required to promote both fruit growth and oil content. The total sterol, squalene, total flavonoid, and polyphenol content of other treatments can be increased by up to 28.81%, 32.07%, 188.04%, and 92.61%, respectively, at most, compared to the control group. Selenium fertilizer and boron fertilizer increased the total sterol content in *Camellia* oil and had a significant positive correlation at the 0.01 level, but zinc fertilizer had little influence on it. High concentration selenium fertilizer generally increased the squalene, total flavonoid, and polyphenol content in *Camellia* oil, but boron and zinc fertilizers had little effect on these components. The results suggested that choosing appropriate fertilizer combinations could improve nutrient deficiency in *C. oleifera* and enhance the functional active compounds of its oil, thereby enhancing its value.

Keywords: *Camellia* oil; foliar fertilizer; nutrient-poor

Citation: Dai, Q.; Deng, Z.; Pan, L.; Nie, L.; Yang, Y.; Huang, Y.; Huang, J. Effects of Trace Elements on Traits and Functional Active Compounds of *Camellia oleifera* in Nutrient-Poor Forests. *Forests* **2023**, *14*, 830. https://doi.org/10.3390/f14040830

Academic Editor: Cate Macinnis-Ng

Received: 21 March 2023
Revised: 9 April 2023
Accepted: 13 April 2023
Published: 18 April 2023

1. Introduction

Camellia oleifera, a crucial oil crop in southern China, is often cultivated in arid and infertile soil [1]. The primary value of *C. oleifera* lies in its oil production. However, this production is often limited by soil nutrient deficiencies. *Camellia* oil contains antioxidant and functional active components that make it a valuable resource for the medical and cosmetic industries. These components are known to have balanced and healthy effects in reducing the risk of obesity, cancer, and heart disease [2,3]. *Camellia* oil is known for its nutritional value and health benefits, which are attributed to its high content of unsaturated fatty acids, antioxidants, and other bioactive compounds [4]. For example, the unsaturated fatty acids found in *Camellia* oil, such as oleic acid, have been shown to have cardioprotective effects and improve blood lipid profiles [5]. Tea polyphenols, and flavonoids, have potent antioxidant and anti-inflammatory properties, which may help protect against chronic diseases such as cancer and cardiovascular disease [6]. Squalene, a natural antioxidant found in *Camellia* oil, and saponins have been found to have antimicrobial and anti-inflammatory properties [7]. Overall, the combination of these beneficial compounds in *Camellia* oil makes it a valuable resource for the food, cosmetic, and pharmaceutical industries. *Camellia* flowers are abundant in phenolic compounds and serve as important natural sources of active ingredients for the food industry [8]. Previous studies have indicated that the levels of stigmasterol, tocopherol, beta-carotene, and lutein in *Camellia* oil are significantly higher than those found in olive oil [9]. The functional and active compounds found in *Camellia* oil play a crucial role in enhancing its quality. However, due to its tolerance in barren soil, *C. oleifera* is often planted in barren areas, which results in a reduction in the active compounds of *Camellia* oil and a decrease in nutritional value. Therefore, fertilization is needed to supplement nutrients.

The soil nutrient levels in *C. oleifera* forests in China exhibit substantial variability, and the growth of *C. oleifera* is restricted by different elements [10]. Stress induced by zinc and boron results in a significant decrease in the growth and relative water content of *C. oleifera* [11]. Selenium, on the other hand, can enhance the nutritional value of *Camellia* oil and reduce the potential for heavy metal contamination [12]. Trace elements, including zinc, boron, and selenium, play a crucial role in the growth and nutrient content of *C. oleifera*. Zinc is an essential micronutrient for plants, but an excess or deficiency of it can cause oxidative stress, which can damage plant cells. When *Camellia* plants are exposed to zinc stress, they experience an increase in reactive oxygen species, which can cause oxidative damage to cells. In response to this stress, *Camellia* plants increase the activity of antioxidant enzymes to prevent oxidative damage [13]. However, zinc can improve the water use efficiency of tea plants under drought stress and also reduce oxidative stress caused by drought by increasing the activity of antioxidant enzymes [14]. Studies have shown that the application of boron fertilizer can improve the yield and quality of *C. oleifera* by enhancing soil fertility, balancing nutrient availability, and preventing nutrient deficiencies [15]. The use of selenium fertilizer can significantly increase the selenium content in *Camellia* oil, which in turn can improve its antioxidant and other beneficial functional components [16]. Previous studies have demonstrated that both soil and foliar fertilizer applications can significantly increase the selenium content in *C. oleifera* oil. However, the enrichment effect of foliar applications is superior to that of root applications [17]. Additionally, applying an active selenium compound fertilizer to soil can increase the nutrient content in mature *C. oleifera* forests, and the foliar spraying of selenium compound fertilizer can improve both the quality and yield of *C. oleifera* [18]. Therefore, by understanding the role of trace elements and using fertilizer effectively, it may be possible to increase both the production and quality of *C. oleifera* in barren conditions. However, there are limited studies on the effects of trace elements on the functional active compounds of *Camellia* oil, such as total sterols, squalene, total flavonoids, and polyphenols.

The objective of this study was to investigate the effects of trace elements such as zinc, boron, and selenium on the indicators of *C. oleifera* forests grown in nutrient-poor conditions and on the functional and active components of *Camellia* oil. The results of this research will provide a theoretical basis for improving the level and quality of *Camellia* oil and its functional active compounds, which would improve economic benefits.

2. Materials and Methods

2.1. Plant Material and Study Area

Camellia oleifera 'Cenruan 3', 8 years old, was used as the experimental species. The study area was located in Chini Town, Guangzhou City, China, and had a subtropical monsoon climate, warm and rainy, with long summers and short winters. The average monthly temperature was about 23.4 °C in October but, in exceptional months, was 13.1 °C. The average monthly precipitation was 98.5 mm in October [19]. The pH value of the soil was 5.6, the organic matter content was 21.26 mg/kg, the basic nitrogen content was 46.82 mg/kg, the available phosphorus content was 6.08 mg/kg, the available potassium content was 56.09 mg/kg, the available boron content was 0.22 mg/kg, the available zinc content was 1.76 mg/kg, and the selenium content was 0.159 mg/kg.

2.2. Experimental Design

An orthogonal experimental design was used to apply various levels of selenium (0, 100, 150, and 200 mg/L), boron (0, 2, 4, and 6 mg/L), and zinc (at dilutions of 0, 1000, 1500, and 2000) as a foliar fertilizer spray. The raw material for selenium is $K_3SeP_3O_{10}$. The raw material for boron is $Na_2B_4O_7 \cdot 10H_2O$. The raw material for zinc is $C_{10}H_{12}N_2O_8ZnNa_2$, and the concentration of the original zinc solution before dilution was 145 mg/L. Mix the above raw materials with water and stir evenly. Each treatment (16 in total) equaled 30 L of fertilizer. The fertilizer was applied every two months from November 2020. Thirty trees were used for each fertilizer combination, representing a total of 480 trees. The fertilization plan is shown in Table 1.

Table 1. Orthogonal design implementation.

Treatment Number	Selenium (mg/L)	Boron (mg/L)	Zinc (Dilution Factor)
CK	0	0	-
Y1	0	2	1000
Y2	0	4	1500
Y3	0	6	2000
Y4	100	0	1000
Y5	100	2	0
Y6	100	4	2000
Y7	100	6	1500
Y8	150	0	1500
Y9	150	2	2000
Y10	150	4	0
Y11	150	6	1000
Y12	200	0	2000
Y13	200	2	1500
Y14	200	4	1000
Y15	200	6	0

2.3. Measurement of Camellia Fruit Characteristics and Economic Indicators

In October 2021, 15 trees were randomly selected for treatment and their yields were measured. Thirty fruits were randomly selected from thirty trees for each treatment, and the characteristics of the fruits were quantified. The longitudinal and transverse diameters of each fruit were measured with vernier calipers, and each fruit was weighed. Each fruit was then crushed with a hammer to release the peel and seeds. The peel thickness was measured with vernier calipers, and the number of seeds was counted. The fresh seeds from each fruit were also weighed. Seeds were dried to constant weight at 60 °C and stored in a refrigerator at 4 °C to measure the oil content.

Grinding the seeds into a powder. Two grams of dried seed powder per sample were wrapped in filter paper, baked to a constant weight at 105 °C, and then weighed on the filter paper and soaked overnight in ether. After 3 h at 65 °C in a crude fatty acid analyzer, the sample was dried again at 105 °C and weighed. The other fruit quality parameters were calculated using the following equations [20]: ratio of fresh seed per fruit (RFS) (%)= total mass of fresh seeds (g)/total mass of fresh fruit weight (g) × 100%; ratio of dry seed to fresh seed (RDS) (%) = total mass of dry seeds (g)/total mass of fresh seeds (g) × 100%; ratio of kernels in dry seed (RKS) (%) = total mass of fresh kernels (g)/total mass of dry seeds (g) × 100%; ratio of dry to fresh kernels (RDK) (%) = total mass of dry kernels (g)/total mass of fresh kernels (g) × 100%; oil content in dry kernels (OCK) (%) = total mass of oil (g)/total mass of dry kernels (g) × 100%; oil content in fruit (OCF) (%) = RFS × RDS × RKS × RDK × OCK.

2.4. Measurement of Functional Active Compounds of Camellia Oil

For each treatment, 5 kg of fruit was picked from 30 trees for oil extraction, and red or yellow peels were randomly selected. The fruit was dried indoors in the shade for a week and then left in the sun until it cracked, and then the peeled seeds baked at 105 °C to a constant weight. We then ground the seeds into powder and press oil and stored in a refrigerator at 4 °C.

Measurement of the total sterol content in *Camellia* oil according to standard GB/T 25223-2010 [21]. An amount of 250 mg of oil was placed in a 25 mL flask, and the unsaponifiable substances were extracted via an alumina column. Then, the unsaponifiable substances was separated on the thin layer chromatographic plate, and the total sterol content was obtained by gas chromatographic analysis.

Measurement of the squalene content in *Camellia* oil according to standard LS/T 6120-2017 [22]. An amount of 300 μL of internal standard squalene solution was accurately absorbed in a 250 mL round-bottomed flask and then dried with a nitrogen blower, and 0.2–2 g of the oil was added to the flask. An amount of 50mL of potassium hydroxide–ethanol solution was added for saponification and extraction, and the resulting concentrate was injected into a gas chromatograph for analysis of the squalene content.

The total flavonoid content of *Camellia* oil was determined using the following method [23]. An amount of 1 mL of oil from the solution to be tested was placed in a 10 mL centrifuge tube, and 0.4 mL 5% sodium nitrite was added. The solution was shaken well and left to stand for 6 min. Then, 0.4 mL 10% aluminum nitrate was added, shaken well, and left to stand for 6 min, followed by 4 mL 4% sodium hydroxide, which was again shaken well, and 4 mL 60% anhydrous ethanol. After 15 min, the absorbance was measured at 510 nm with a spectrophotometer after centrifugation at 3500 r/min for 10 min. The standard rutin, 10 mg, was added to a 10 mL volumetric bottle with 60% anhydrous ethanol to obtain a 1 mg/mL standard solution, and an absorbance of 0.025–0.4 mg/mL standard solution was determined. According to standard curve of rutin, the total flavonoid content was calculated.

The polyphenols content of *Camellia* oil was determined using the following method, according to standard LS/T 6119-2017 [24]. An amount of 1 mL of oil from the solution to be tested was put into a 10 mL colorimetric tube, and 5 mL 10% foline-phenol was added. It was then left to stand for 3–5 min, 4 mL 7.5% Na_2CO_3 was added, and it was again left to stand for 1 h. Absorbance was measured at 625 nm with a spectrophotometer, alongside a blank control. A standard curve was created using 30 mg gallic acid as a standard in a 100 mL volumetric bottle with distilled water added to obtain a 300 μg/mL solution. The absorbance value of 10–50 μg/mL of the standard solution was determined. According to standard curve of gallic acid, the polyphenols content was calculated.

2.5. Statistical Analysis

Microsoft Excel 2019 (Microsoft Corp, Redmond, Washington, DC, USA) was used for data processing and mapping. SPSS 24.0 (SPSS Inc., Chicago, IL, USA) was used for multiple comparisons, variance analyses, and correlation analyses.

3. Results

3.1. The Effect of Trace Elements on Fruit Characteristics

As shown in Table 2, there were slight differences between treatments in peel thickness, fruit transverse diameter, fruit longitudinal diameter, single fruit weight, the number of seeds per fruit, and single fresh seed weight. The transverse diameter of the fruit in treatment Y3 (0 mg/L selenium, 6 mg/L boron, and 2000 dilution of zinc) was significantly lower than that in the control group ($p < 0.05$), suggesting that high concentrations of selenium and high dilutions of zinc have the effect of reducing fruit transverse diameter, while the other treatments showed no significant difference ($p > 0.05$). The transverse diameter of fruit in treatment Y11 was the highest, and 9.1% higher than in treatment Y3. Both had received 6 mg/L boron, suggesting that boron had no direct effect on fruit transverse diameter. Treatment Y11 included 150 mg/L selenium, while treatment Y3 included 0 mg/L, and treatments Y4, Y5, Y7, and Y8 included 100, 100, 100, and 150 mg/L, respectively, indicating that low concentrations of selenium could perhaps promote an increase in fruit transverse diameter. However, there were no significant differences between the 15 treatments and the control group ($p > 0.05$). Treatment Y3 had the lowest fruit longitudinal diameter, while treatment Y2 had the highest fruit longitudinal diameter, which was 6.86% higher than that of treatment Y3. The transverse diameter of fruit of treatments Y2 can be increased by up to 3.07% compared to the control group. With a 1500 dilution of zinc, the fruit longitudinal diameter was higher, suggesting that this dilution had a positive effect on longitudinal diameter growth.

Table 2. Multiple comparison of the effects of trace elements on *Camellia oleifera* fruit characteristics.

Treatment Number	Longitudinal Diameter of Single Fruit (mm)	Transverse Diameter of Single Fruit (mm)	Peel Thickness of Single Fruit (mm)	Single Fruit Weight (g)	Number of Seeds Per Fruit	Single Fresh Seed Weight (g)
	Mean ± Standard Error	Mean ± Standard Error	Mean ± Standard Error	Mean	Mean	Mean
CK	33.84 ± 2.91 ab	35.29 ± 3.65 ab	3.80 ± 0.55 bcd	22.81	6.17	1.29
1	33.27 ± 3.28 ab	35.00 ± 3.80 ab	3.63 ± 0.83 cd	22.16	4.93	1.68
2	34.88 ± 2.80 a	35.22 ± 3.22 ab	3.77 ± 0.65 bcd	22.66	5.57	1.27
3	32.64 ± 2.91 b	32.97 ± 3.11 c	3.67 ± 0.68 cd	18.88	5.60	1.15
4	34.19 ± 2.68 ab	35.58 ± 3.46 ab	3.77 ± 0.54 bcd	23.29	5.30	1.47
5	34.27 ± 3.18 ab	35.49 ± 3.50 ab	4.21 ± 0.77 a	22.4	6.03	1.27
6	34.11 ± 2.79 ab	35.28 ± 3.01 ab	4.00 ± 0.61 abc	22.35	5.90	1.26
7	34.51 ± 3.48 a	35.37 ± 3.38 ab	3.93 ± 0.63 abc	23.37	5.37	1.45
8	33.92 ± 3.56 ab	35.40 ± 4.12 ab	4.02 ± 0.63 abc	23.06	6.13	1.22
9	33.34 ± 2.96 ab	33.65 ± 3.87 bc	3.97 ± 0.64 abc	20.39	5.50	1.27
10	34.20 ± 2.97 ab	34.8 ± 2.79 abc	3.52 ± 0.63 d	21.79	5.90	1.24
11	34.14 ± 3.60 ab	35.97 ± 3.93 a	4.07 ± 0.69 ab	25.22	7.63	1.21
12	33.41 ± 1.85 ab	34.0 ± 2.14 abc	3.65 ± 0.50 cd	20.17	6.17	1.11
13	34.40 ± 2.94 ab	35.24 ± 2.95 ab	3.8 ± 0.56 abcd	22.85	7.30	1.22
14	34.18 ± 1.96 ab	34.6 ± 2.36 abc	3.8 ± 0.69 abcd	21.95	7.03	1.02
15	33.12 ± 2.98 ab	34.0 ± 3.41 abc	3.97 ± 0.52 abc	20.99	6.20	1.14

Data indicate mean ± SE ($n = 30$). Different letters in the same column indicate significant differences ($p < 0.05$).

The peel thickness in treatment Y5 was significantly higher than that of the control group ($p < 0.05$), while the thickness in treatment Y10 was significantly lower than that of the control ($p < 0.05$) and 16.39% lower than that of treatment Y5. However, there were no significant differences in peel thickness between the other treatments and the control ($p > 0.05$). Both treatments Y5 and Y10 contained a 0 dilution of zinc, suggesting that zinc had no direct effect on peel thickness. Treatment Y10 (150 mg/L selenium, 4 mg/L boron, and 0 dilution of zinc) had the best effect on reducing peel thickness.

The single fruit weight of treatment 11 was the highest and 10.57% higher than in the control group, and that of treatment 3 was the lowest and 17.23% lower than in the control group. The single fruit weights of treatments 11, 7, 4, 8, and 13 were higher than in the control group. When the dilution of zinc is 1000–1500 times and the concentration of selenium is 100–200 mg/L, the growth of *C. oleifera* fruit is promoted.

Treatment Y11 yielded the most seeds per fruit, while treatment Y1 yielded the least grains per fruit. The single seed weight for treatment Y1 was the heaviest, while that of treatment Y14 was the lightest. Treatment Y11 had the highest fruit weight, 10.57% higher than the control group, while treatment Y3 had the lowest fruit weight, at 17.23% lower than the control group. The number of seeds per fruit in treatment 11, 14, 13, and 15 was higher than in control group. The amount of selenium applied was between 150–200 mg/L, suggesting that a high concentration of selenium can promote an increase in the number of seeds per fruit.

3.2. The Effect of Trace Elements on Economic Indicators

As shown in Table 3, different micronutrient treatments had variable effects on the oil yield of the fresh seeds. Treatment Y1 had the highest fresh seed ratio per fruit, 2.74% higher than the control, while treatment Y8 had the lowest fresh seed ratio. Treatment Y3 had the highest ratio of dry seed to fresh seed, while treatment Y4 had the lowest ratio of dry seed to fresh seed. Treatment Y2 had the highest ratio of kernels in the dried seeds, while treatment Y3 had the lowest ratio of kernels. Treatment Y6 had the most ratio of dried kernels to fresh kernels, while treatment Y9 had the least. Treatment Y2 had the highest yield of oil from both kernel and fruit, while treatment Y6 had the lowest yield. The oil content in fruit of treatments Y2 can be increased by up to 7.94% compared to the control group. Overall, the best effect was achieved with a selenium concentration of 0 mg/L, a boron concentration of 4 mg/L, and a zinc dilution of 1500, as demonstrated by the oil content of the fruits.

Table 3. Effects of trace elements on economic indicators for *Camellia oleifera*.

Treatment Number	Ratio of Fresh Seed Per Fruit (%)	Ratio of Dry Seed to Fresh Seed (%)	Ratio of Kernels in Dry Seed (%)	Ratio of Dry to Fresh Kernels (%)	Oil Content in Dry Kernels (%)	Oil Content in Fruit (%)	Fruit Yield Per Plant (kg)	Oil Yield Per Plant (kg)
CK	34.76	52.86	69.61	98.79	54.83	6.93	11.81	0.82
1	37.50	56.44	67.09	98.03	53.41	7.44	13.46	1.00
2	33.54	57.94	70.07	98.97	55.47	7.48	13.22	0.99
3	34.17	61.46	58.16	98.54	51.39	6.19	10.00	0.62
4	33.38	49.19	65.12	98.52	51.41	5.42	14.27	0.77
5	34.23	49.79	64.46	98.47	51.93	5.62	17.05	0.96
6	33.17	49.99	58.50	99.10	48.64	4.68	13.23	0.62
7	33.37	54.93	66.63	98.66	51.48	6.20	9.93	0.62
8	32.44	52.58	67.54	98.24	52.58	5.95	10.75	0.64
9	34.15	54.76	67.66	97.51	51.72	6.38	12.12	0.77
10	33.53	58.45	62.85	99.77	49.50	6.08	11.92	0.72
11	35.53	49.86	62.45	99.36	48.94	5.38	13.03	0.70
12	33.91	51.14	67.03	98.26	51.43	5.87	10.99	0.65
13	34.27	49.74	64.80	98.75	50.15	5.47	15.65	0.86
14	32.82	52.15	65.93	98.63	51.24	5.70	15.90	0.91
15	33.55	55.02	67.81	98.95	51.86	6.42	12.29	0.79

Treatment Y5 had the highest fruit yield per plant at 17.05 kg, which was 44.41% higher than the control, followed by treatment Y13, while treatment Y7 had the lowest average yield. In combination with the oil content in fruit, the oil yield per plant for each treatment was ranked as Y1 > Y2 > Y5 > Y14 > Y13 > CK > Y15 > Y4 = Y9 > Y10 > Y11 > Y12 > Y8 > Y3 = Y6 = Y7. The oil yield per plant decreases to the minimum when the treatment is Y3, Y6, and Y7. The highest oil yield per plant for treatments Y1 (0 mg/L selenium, 2 mg/L boron, and 1000 dilution of zinc) and Y2 (0 mg/L selenium, 4 mg/L boron, and 1500 dilution of zinc) were 1.00 and 0.99 kg, respectively. The oil yield per plant of treatments Y1 can be increased by up to 21.95% compared to the control group. The results showed that a low concentration of selenium, a low concentration of boron, and a low dilution of zinc could increase the oil yield per plant.

3.3. *The Effects of Trace Elements on Functional and Active Components of Camellia Oil*

3.3.1. Effects of Trace Elements on Total Sterol Content

As shown in Figure 1, applying foliar fertilizer could both significantly increase and significantly decrease the total sterol content of *Camellia* oil. The total sterol content under treatments Y11, Y6, Y15, Y9, and Y14 was significantly higher than that of the control group ($p < 0.05$), by 28.81%, 12.96%, 10.52%, and 7.62%, respectively. However, the total sterol content under treatment Y1 was significantly lower than that of the control group ($p < 0.05$). The total sterol content under the other treatments showed no significant difference with the control group ($p > 0.05$). The application rates of selenium were 150 mg/L, 100 mg/L, and 0 mg/L in treatments Y11, Y6, and Y1, respectively, the application rates of boron were 6 mg/L, 4 mg/L, and 2 mg/L, respectively, and the dilutions of zinc were 1000, 2000, and 1000, respectively, indicating that the high concentrations of selenium and boron could promote an increase in total sterol content, while zinc had no direct effect.

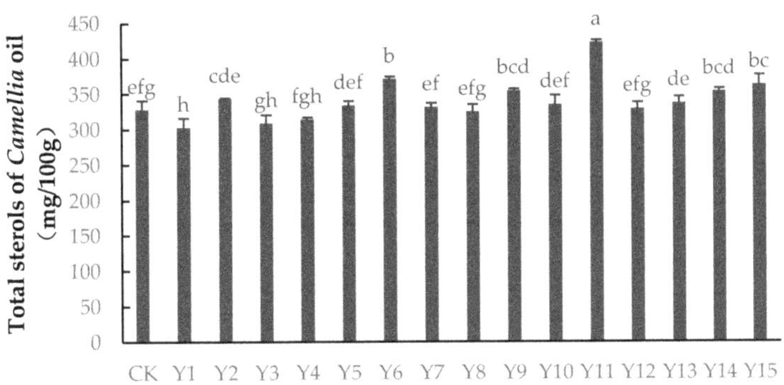

Figure 1. The content of total sterols in *Camellia* oil under different treatments. Note: Data indicate mean ± SE ($n = 3$). Different letters indicate significant differences ($p < 0.05$) between different treatments.

3.3.2. Effects of Trace Elements on Squalene Content

As shown in Figure 2, applying foliar fertilizer could both significantly increase and significantly decrease the squalene content in *Camellia* oil. The squalene content of treatments Y12, Y13, Y14, Y11, Y15, Y2, Y6, Y8, Y9, and Y10 was significantly higher than that of the control group ($p < 0.05$). Of these, the squalene content was highest under Y12 and Y13, being 32.07% and 31.36% more, respectively, than in the control group. The squalene content under Y7 and Y4 was lower than the control group, but not significantly ($p > 0.05$), while the squalene content under Y1 and Y3 was significantly lower than in the control group ($p < 0.05$), by 4.66% and 5.68%, respectively. For treatments Y12, Y13, Y1, and Y3, the selenium dosage was 200 mg/L, 200 mg/L, 0 mg/L, and 0 mg/L, respectively, the boron dosage was 0 mg/L, 2 mg/L, 2 mg/L, and 6 mg/L, respectively, and the dilution of zinc was 2000, 1500, 1000, and 2000, respectively. This indicated that a high concentration of selenium could promote the squalene content of *Camellia* oil, while boron and zinc had no direct impact.

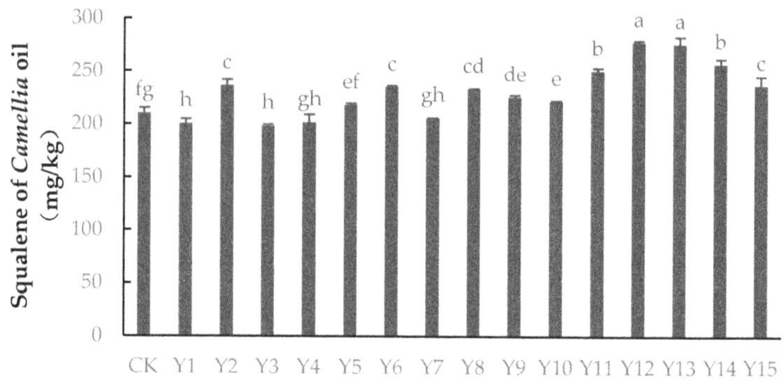

Figure 2. The content of squalene in *Camellia* oil under different treatments. Note: Data indicate mean ± SE (n = 3). Different letters indicate significant differences ($p < 0.05$) between different treatments.

3.3.3. Effects of Trace Elements on Total Flavonoid Content

As shown in Figure 3, applying foliar fertilizer could both significantly increase and significantly decrease the total flavonoid content of *Camellia* oil. The total flavonoid content under treatments Y2, Y4, Y7, Y12, Y13, Y14, and Y15 was significantly higher than that of the control group ($p < 0.05$). Of these, Y4 had the highest total flavonoid content, being 188.04% higher than in the control group, followed by Y14 and Y13, which were 104.23% and 110.2% higher, respectively. The total flavonoid content under treatments Y1, Y3, Y5, Y6, Y8, Y9, and Y11 was significantly lower than in the control group ($p < 0.05$), with Y3 being 44.59% lower. The total flavonoid content under treatment Y10 was not significantly different from that of the control group ($p > 0.05$). Treatment Y4 received 100 mg/L of selenium, 0 mg/L of boron, and a 1000 dilution of zinc, indicating that this combination of elements maximized the total flavonoid content of *Camellia* oil. For treatments Y12, Y13, Y14, and Y15, the selenium application was 200 mg/L, indicating that 200 mg/L selenium could significantly increase the total flavonoid content of *Camellia* oil. The boron application was 0, 2, 4, and 6 mg/L, respectively, and the dilution of zinc 2000, 1500, 1000, and 0, respectively, indicating that boron and zinc had no direct effect on the total flavonoid content of *Camellia* oil.

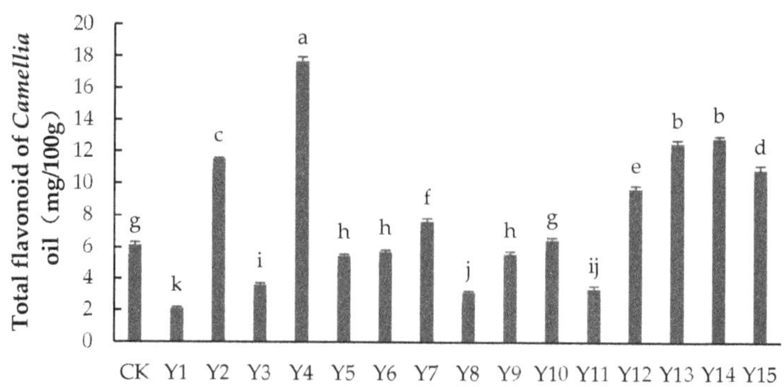

Figure 3. The content of total flavonoid in *Camellia* oil under different treatments. Note: Data indicate mean ± SE (n = 3). Different letters indicate significant differences ($p < 0.05$) between different treatments.

3.3.4. Effects of Trace Elements on Polyphenol Content

As shown in Figure 4, spraying foliar fertilizer could both significantly increase and significantly decrease the polyphenol content of *Camellia* oil. The polyphenol content under treatments Y2, Y13, Y14, Y12, Y15, and Y4 was significantly higher than that of the control group ($p < 0.05$), while the polyphenol content under treatments Y1, Y5, Y6, Y7, Y8, and Y11 was significantly lower ($p < 0.05$). There was no significant difference between treatments Y3, Y9, and Y10 and the control group ($p > 0.05$). Treatment Y2 had the highest polyphenol content, being 92.61% higher than in the control group, followed by treatments Y13 and Y14, which were 66.5% and 51.07% higher. The polyphenol content of treatment Y1 was the lowest, being 60.92% lower than in the control group, followed by Y6, which was 30.71% lower. Under treatment Y2, the selenium application was 0 mg/L, boron 4 mg/L, and the dilution of zinc 1500, indicating that this combination could maximize the polyphenol content of *Camellia* oil. For treatments Y12, Y13, Y14, and Y15, the selenium application was 200 mg/L, indicating that 200 mg/L selenium could significantly increase the polyphenol content of *Camellia* oil.

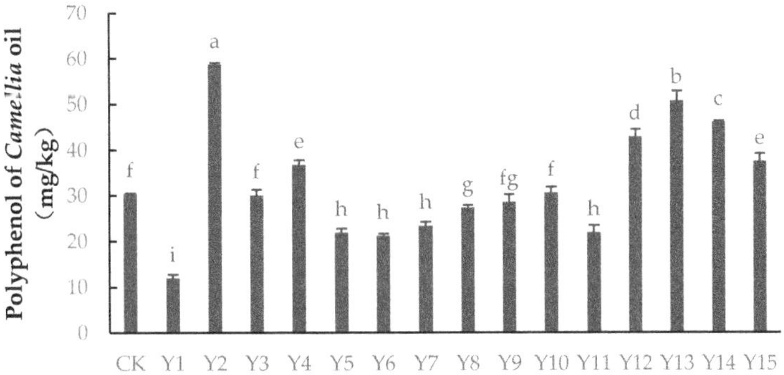

Figure 4. The content of polyphenol in *Camellia* oil under different treatments. Note: Data indicate mean ± SE (n = 3). Different letters indicate significant differences ($p < 0.05$) between different treatments.

3.4. Correlation Analysis between Trace Elements and Components in Camellia Oil

As shown in Table 4, the total sterol content of *Camellia* oil had a very significant positive correlation with selenium and boron at the level of 0.01. The squalene content of *Camellia* oil had a very significant positive correlation with selenium at the level of 0.01, and there was a significant positive correlation between total flavonoids and selenium at the level of 0.05. There was no significant correlation between the polyphenol content of *Camellia* oil and the trace elements.

Table 4. Correlation analysis between trace elements and components in *Camellia oleifera*.

Trace Element	Total Sterol Content of *Camellia* Oil	Squalene Content of *Camellia* Oil	Total Flavonoid Content of *Camellia* Oil	Polyphenol Content of *Camellia* Oil
Selenium	0.403 **	0.713 **	0.328 *	0.249
Boron	0.458 **	−0.072	−0.149	−0.072
Zinc	−0.03	0.223	−0.05	0.122

* Significant at the 0.05 level. ** Significant at the 0.01 level.

As shown in Table 5, the variance analysis showed that selenium had a very significant effect on the squalene content of *Camellia* oil ($p < 0.01$) but no significant effect on the total sterol content. Boron and zinc had no significant effect on the squalene and total sterol content of *Camellia* oil. The range (R value) analysis showed that the relative weighting for the effect of the trace elements on the squalene content of *Camellia* oil was selenium > zinc > boron, and, for the total sterol content, selenium > boron > zinc. Selenium, boron, and zinc had no significant effect on the total flavonoid and polyphenol content of *Camellia* oil ($p > 0.05$), although selenium had a greater effect than zinc. The relative weighting for the effect of the trace elements on the polyphenol content of *Camellia* oil was selenium > boron > zinc.

Table 5. Analysis of variance and extreme value analysis of trace elements and components in *Camellia* oil.

Source of Variation	Degrees of Freedom	Total Sterol Content		Squalene Content		Total Flavonoid Content		Polyphenol Content	
		p Value	*R* Value	*p* Value	*R* Value	*p* Value	*R* Value	*p* Value	*R* Value
Selenium	3	0.315	38.250	0.003	50.688	0.085	6.850	0.117	18.500
Boron	3	0.361	35.500	0.903	14.650	0.702	2.775	0.577	10.890
Zinc	3	0.937	27.375	0.842	16.188	0.806	2.868	0.612	10.860

$p < 0.05$ indicates a significant difference, <0.01 indicates an extremely significant difference; *R* represents extreme differences, indicating the weight of influence of different factors.

4. Discussion

Camellia plants that are in a nutrient-poor environment for an extended period of time will experience a significant reduction in yield, and the content of quality-related metabolites will be affected [25]. Previous research has shown that there are complex interactions between various elements in soil and *C. oleifera*, so the interaction between them should be fully considered to achieve precise fertilization and further improve fertilizer utilization [26]. In this study, 16 kinds of microelement combinations were designed to fertilize the *C. oleifera* forest under the environment of nutrient deficiency. The results showed that Y10 (150 mg/L selenium, 4 mg/L boron, and 0 diluted zinc) had the best effect on reducing the thickness of fruit peels. A low concentration of selenium might promote the increase in fruit transverse diameter. When the dilution of zinc was 1500, it had a positive effect on the growth of fruit longitudinal diameter, while boron concentration had little effect on fruit size. Xiang et al.'s research on tea plants aligns with the findings of this study, indicating that an appropriate concentration of selenium can result in a substantial increase in tea plant biomass and yield [27]. According to the research conducted by Hajiboland et al., the growth of tea plants was not found to be inhibited by a lack of boron, a critical micronutrient for many plants [28]. Recent studies, including one by Lu et al., have suggested that the application of a 0.05% zinc fertilizer concentration can significantly enhance the growth of *C. oleifera* [29]. Jiang et al. have shown that different fertilization methods, and the interaction of phosphorus and potassium fertilizers, have a significant impact on peel thickness. Nitrogen fertilizer applied at 50 g/plant, phosphorus fertilizer at 50 g/plant, and potassium fertilizer at 50 g/plant can all significantly reduce the peel thickness of *C. oleifera* [30]. The increase in pericarp thickness ends with the completion of lignin accumulation in the pericarp [31]. Cch NST1 and Cch BLH6 are lignin-regulated nuclear genes in *C. oleifera* fruit [32]. The change in the size of *C. oleifera* fruit after foliar spraying with selenium, boron and zinc may be due to the promotion or inhibition of *CchNST1* and *CchBLH6* gene expression.

The application of selenium, boron, and zinc had no significant effect on the fresh seed ratio or the number of seeds in *C. oleifera* fruit. The rate of seed inclusion during the accumulation of seed lignin and the number of ovules remaining in the fruit at harvest determine the fresh seed ratio and the number of seeds in *C. oleifera* fruit. *C. oleifera* seed inclusions accumulate from July to September, and fertilization during this period could increase the number of inclusions and improve the fresh seed ratio [33,34]. Therefore, applying appropriate fertilizer during this period could be beneficial to the number of seeds in *C. oleifera* fruit, but the physiological and molecular mechanisms that regulate ovule development have not yet been studied. In light of the selenium-poor soil conditions observed in the study area, the present research sheds new light on the crucial role that this trace element plays in promoting plant reproduction. Notably, the study's results demonstrate that a concentration of selenium above a certain threshold can be a potent factor in boosting the number of seeds produced per fruit.

Spraying with selenium, boron, and zinc can also increase the oil yield of *C. oleifera*. The results showed that Y1 and Y2 treatments could increase the oil content of fresh fruits and the oil yield per plant. Genetic quality, seed maturity, fruiting quantity, and the soil quality of *C. oleifera* forest land all affect the oil content of *C. oleifera*. For example, lime soil and calcium-rich purple sand soil are beneficial to the seed oil content of *C. oleifera* [33]. Fertilizer spraying and the application of amino acids can increase the oil content of *C. oleifera* [35]. Similarly, when selenium fertilizer is applied at 200–300 g/hm^2, the oil content of sunflower kernels can be significantly improved [36]. In another study the application of 0.1% boron increased the oil content of *C. oleifera* to 8.25% more than that of the control [37]. The beneficial impact of selenium, boron, and zinc on the quality of *Camellia* oil could potentially be explained by their role in promoting the uptake and assimilation of other essential mineral elements by the plant. By facilitating nutrient absorption, these trace elements may improve the taste, aroma, and nutritional properties of the oil [8,12]. Lin et al. have discovered alleles of three enzymes that play important roles in enhancing the yield and quality of seed oil during *C. oleifera* domestication [38].

The trace elements can also significantly increase the content of functionally active components in *Camellia* oil. The study's findings suggest that selenium and boron may have a synergistic effect on the total sterol content of *Camellia* oil, while zinc has no direct impact. Specifically, the application of selenium at a concentration of 150 mg/L, in conjunction with 6 mg/L of boron, can significantly enhance the total sterol content of *Camellia* oil. Song et al. have shown that applying selenium can significantly increase the total sterol content of *Camellia* oil [39]. Pinto et al. found that sterol-related enzymes were significantly altered only in the presence of a selenium deficiency [40]. Under the treatment of boron deficiency, the downregulation of two sterol methyltransferases decreased the sterol content [41]. This study shows that high concentrations of selenium can promote the squalene and total flavonoid content in *Camellia* oil, while boron and zinc have no direct impact. When the selenium concentration was 200 mg/L, the squalene and total flavonoid content was at its highest. Ma et al. have shown that the squalene content of *Camellia* oil is positively correlated with the selenium content, in accordance with the current study [42]. Spraying zinc fertilizer can significantly induce the expression of phenolic synthesis-related genes such as *VvLDOX* and *VvMYBF1* during fruit development, further promoting phenolic synthesis [42]. The study's results indicate that the application of selenium at a concentration of 200 mg/L can lead to a significant increase in the polyphenol content of *Camellia* oil, However, the effect of zinc fertilizer is not significant. This finding suggests that selenium may play a crucial role in the production of high-quality *Camellia* oil. The mechanism behind the influence of selenium, therefore, warrants further exploration. The polyphenol content of *Camellia* oil was significantly correlated with the total flavonoid and squalene content, suggesting that these three substances have the same response mechanism to selenium, boron, and zinc, and that they synergistically promote each other.

5. Conclusions

The present study is the first to apply zinc, boron, and selenium as foliar fertilizers to *C. oleifera* cultivated in nutrient-poor soil. The results show that the effects of micronutrient sprays on the phenotypic and economic traits of *C. oleifera* varied, with treatments Y4, 7, 8, and 11 all increasing the transverse, longitudinal diameter and fruit weight of *C. oleifera* fruit. Optimal concentrations of zinc and selenium are critical for promoting the growth and development of *C. oleifera* fruit. The transverse diameter of the fruit, the single fruit weight, the number of seeds per fruit, the single fresh seed weight, the oil content in the fruit, and the oil yield per plant of other treatments can be increased by up to 3.07%, 10.57%, 23.66%, 30.23%, 7.94%, and 21.95%, respectively, at most, compared to the control group. Diluting zinc from 1000 to 1500 times and maintaining a selenium concentration from 100 to 200 mg/L has been found to be beneficial for fruit growth. While low concentrations of selenium may promote an increase in fruit transverse diameter, high concentrations of selenium, along with high dilutions of zinc, can have the opposite effect, leading to a reduction in fruit diameter. However, a high concentration of selenium can positively impact the number of seeds per fruit. Treatments Y1 and Y2 were the most effective in increasing the oil yield of fresh fruits and the oil production per plant. The most effective combination was found to be a selenium concentration of 0 mg/L, a boron concentration of 4 mg/L, and a zinc dilution of 1500. Interestingly, lower concentrations of selenium and boron, as well as lower dilutions of zinc, were found to increase the oil yield per plant. This suggests that a careful balance of trace elements is required to promote both fruit growth and oil content.

This study's findings suggest that high concentrations of selenium and boron can play a crucial role in promoting the content of functional active compounds in *Camellia* oil. The total sterol, squalene, total flavonoid, and polyphenol content of other treatments can be increased by up to 28.81%, 32.07%, 188.04%, and 92.61%, respectively, at most, compared to the control group. Specifically, the results reveal that treatments Y2, 6, 9, and 11 exhibited better performance in terms of increasing the total sterol content of the oil, with selenium and boron, showing a highly significant positive correlation with this compound. Moreover, the application of high selenium concentrations was found to have a significant Impact on the squalene and total flavonoid contents of *Camellia* oil, with treatments Y2, 11, 12, and 13 showing better performance in increasing the squalene content of the oil, while treatments Y2, 4, and 7 showed better performance in increasing the total flavonoid content. Furthermore, the study shows that spraying selenium at a concentration of 200 mg/L can lead to a significant increase in the polyphenol content of *Camellia* oil, with treatments Y2, 13, and 14 performing better in this regard.

Overall, the study provides a theoretical basis for improving the quality and yield of the *C. oleifera* industry by selecting suitable fertilizer combinations to increase the fresh kernel oil rate, the oil yield per plant, and the content of functional active components such as total sterols, squalene, total flavonoids, and polyphenols in *Camellia* oil. These findings emphasize the interrelationship between trace elements and the functional active components of *Camellia* oil and offer guidance for enhancing the growth of *C. oleifera* forests in nutrient-deficient soils.

Author Contributions: Conceptualization, Z.D. and Q.D.; methodology, Z.D. and Q.D.; software, Z.D. and Q.D.; validation, L.P., Y.Y. and Y.H.; formal analysis, Z.D. and Q.D.; investigation, L.P. and Q.D.; resources, L.P.; data curation, L.N. and Q.D.; writing—original draft preparation, Z.D. and Q.D.; writing—review and editing, Z.D.; visualization, L.N. and Q.D.; supervision, Y.H. and J.H.; project administration, Y.H. and J.H.; funding acquisition, J.H. All authors have read and agreed to the published version of the manuscript.

Funding: This research was supported by the Forestry Science and Technology Innovation Project of Guangdong Province (2017KJCX005), and the Forestry Science and Technology Innovation Project of Guangdong Province (2023KJCX003).

Data Availability Statement: All data relevant to the study are included in the article.

Conflicts of Interest: The authors declare no potential conflict of interest.

References

1. Dong, B.; Wu, B.; Hong, W.; Li, X.; Li, Z.; Xue, L.; Huang, Y. Transcriptome analysis of the tea oil camellia (*Camellia oleifera*) reveals candidate drought stress genes. *PLoS ONE* **2017**, *12*, e0181835. [CrossRef] [PubMed]
2. Cao, Y.; Xie, Y.; Ren, H. Fatty acid composition and tocopherol, sitosterol, squalene components of *Camellia reticulata* oil. *J. Consum. Prot. Food Saf.* **2018**, *13*, 403–406. [CrossRef]
3. Yu, J.; Yan, H.; Wu, Y.; Wang, Y.; Xia, P. Quality Evaluation of the Oil of *Camellia* spp. *Foods* **2022**, *11*, 2221. [CrossRef] [PubMed]
4. Xu, Z.G.; Cao, Z.R.; Yao, H.; Li, C.; Zhao, Y.; Yuan, D.; Yang, G. The physicochemical properties and fatty acid composition of two new woody oil resources: Camellia hainanica seed oil and Camellia sinensis seed oil. *CyTA J. Food* **2021**, *19*, 208–211.
5. Xiao, X.M.; He, L.; Chen, Y.Y.; Wu, L.; Wang, L.; Liu, Z.P. Anti-inflammatory and antioxidative effects of *Camellia oleifera* Abel components. *Future Med. Chem.* **2017**, *9*, 2069–2079. [CrossRef] [PubMed]
6. Li, H.; Zhou, G.Y.; Zhang, H.; Liu, J.A. Research progress on the health function of tea oil. *J. Med. Plants Res.* **2011**, *5*, 485–489.
7. Xiang, Z.Y.; Xia, C.; Feng, S.; Chen, T.; Zhou, L.; Liu, L.; Kong, Q.; Yang, H.; Ding, C. Assessment of free and bound phenolics in the flowers and floral organs of two *Camellia* species flower and their antioxidant activities. *Food Biosci.* **2022**, *49*, 101905. [CrossRef]

8. Zhang, L.X.; Wang, S.J.; Yang, R.; Mao, J.; Jiang, J.; Wang, X.; Zhang, W.; Zhang, Q.; Li, P.W. Simultaneous determination of tocopherols, carotenoids and phytosterols in edible vegetable oil by ultrasound-assisted saponification, LLE and LC-MS/MS. *Food Chem.* **2019**, *289*, 313–319. [CrossRef]
9. Tang, G.X.; Tang, S.B.; Li, Y. Study on soil characters of *Camellia oleifera* land. *Jiangxi For. Sci. Technol.* **2002**, *1*, 20–24.
10. Mukhopadhyay, M.; Mondal, T.K. Effect of Zinc and Boron on Growth and Water Relations of *Camellia sinensis* (L.) O. Kuntze cv. T-78. *Natl. Acad. Sci. Lett.-India* **2015**, *38*, 283–286. [CrossRef]
11. Ma, J.; Zhang, R.; Ye, H.; Wang, J.; Li, X.; Wang, Q.; Rui, Y. Determination of Trace Element Nutrition and Risk Assessment of Heavy Metals in Se-Enriched *Camellia oleifera* by ICP-MS. *Asian J. Chem.* **2013**, *25*, 8833–8834. [CrossRef]
12. Mukhopadhyay, M.; Das, A.; Subba, P.; Bantawa, P.; Sarkar, B.; Ghosh, P.; Mondal, T.K. Structural, physiological, and biochemical profiling of tea plantlets under zinc stress. *Biol. Plant.* **2013**, *57*, 474–480. [CrossRef]
13. Upadhyaya, H.; Dutta, B.K.; Panda, S.K. Zinc Modulates Drought-Induced Biochemical Damages in Tea *Camellia sinensis* (L) O Kuntze. *J. Agric. Food Chem.* **2013**, *61*, 6660–6670. [CrossRef]
14. Mei, X.C.; Du, Y.X. Application Effect Test of Boron on *Camellia oleifera*'s Production. *Jiangxi For. Sci. Technol.* **2014**, *42*, 22–23.
15. Wang, Y.H.; Liu, J.; Ma, X.L.; Jiang, J.M.; Wu, L.; Liu, F. Effects of different selenium–rich methods on quality characteristics and functional components of *Camellia oleifera* oil. *Sci. Technol. Food Ind.* **2017**, *38*, 54–59.
16. Xu, W. Enrichment of selenium in *Camellia oleifera* and selenium fertilizer effect on the qualiy of *Camellia* oil. Master's Thesis, Anhui Agricultural University, Hefei, China, 2018.
17. Zhao, J.P.; Wu, L.C.; Chen, Y.Z.; Li, F.C.; Qin, J. Effects of active selenium on soil chemical properties in *Camellia oleifera* forest. *J. Cent. South Univ. For. Technol.* **2011**, *31*, 75–81.
18. National Meteorological Science Data Center of China. Available online: http://data.cma.cn/analysis/yearbooks.html (accessed on 20 March 2023).
19. Wen, Y.; Zhang, Y.; Su, S.; Yang, S.; Ma, L.; Zhang, L.; Wang, X. Effects of Tree Shape on the Microclimate and Fruit Quality Parameters of *Camellia oleifera* Abel. *Forests* **2019**, *10*, 563. [CrossRef]
20. *GB/T 25223-2010*; Animal and Vegetable Fats and Oils-Determination of Individual and Total Sterols Contents-Gas Chormatographic Method. ISO: Nanjing, China, 2010.
21. *LS/T 6120-2017*; Inspection of Grain and Oil—Determination of Squalene in Vegetable Oil—Gas Chromatographic. ISO: Beijing, China, 2017.
22. Yue, C.; Yu, N.; Li, H.; She, J.; Zhou, W.; Li, Z. Research on fatty acids and active components in *Camellia oleifera* Abel. seed oil of Hunan province. *J. Food Saf. Qual.* **2021**, *12*, 1972–1977.
23. *LS/T 6119-2017*; Inspection of Grain and Oils—Determination of Polyphenols in Vegetable Oil—Spectrometric Method. ISO: Beijing, China, 2017.
24. Zhou, B.; Chen, Y.Y.; Zeng, L.T.; Cui, Y.Y.; Li, J.L.; Tang, H.; Liu, J.Y.; Tang, J.C. Soil nutrient deficiency decreases the postharvest quality-related metabolite contents of tea (*Camellia sinensis* (L.) Kuntze) leaves. *Food Chem.* **2022**, *377*, 132003. [CrossRef]
25. Zhu, X.L.; Tang, J.M.; Qin, H.Z.; Bai, K.D.; Chen, Z.Y.; Zou, R.; Liu, S.Y.; Yang, Q.G.; Xiao, W.; Chai, S.F. Contrasting Adaptation Mechanisms of Golden Camellia Species to Different Soil Habitats Revealed by Nutrient Characteristics. *Agronomy* **2022**, *12*, 1511. [CrossRef]
26. Xiang, J.; Rao, S.; Chen, Q.; Zhang, W.; Cheng, S.; Cong, X.; Zhang, Y.; Yang, X.; Xu, F. Research Progress on the Effects of Selenium on the Growth and Quality of Tea Plants. *Plants* **2022**, *11*, 2491. [CrossRef]
27. Hajiboland, R.; Bahrami-Rad, S.; Bastani, S. Phenolics Metabolism in Boron-Deficient Tea *Camellia sinensis* (L.) O. Kuntze Plants. *Acta Biol. Hung.* **2013**, *64*, 196–206. [CrossRef]
28. Huang, G.W.; Zhang, W.J.; Deng, S.G. Effects of Foliar Fertilizer on the Growth and Physiological Characteristics of *Camellia oleifera* Seedlings. *J. West China For. Sci.* **2021**, *50*, 56–63.
29. Jiang, J.Y.; Yang, H.Q.; Ouyang, W.; An, M.; Yang, X.; Long, X.; Hu, Y. Effects of Combined Application of Organic Fertilizer and NPK on Growth and Economic Characters of *Camellia oleifera*. *J. Sichuan Agric. Univ.* **2022**, *40*, 73–82.
30. Yan, C.; Yao, X.; Yin, H.; Wang, K.; Yin, R.; Teng, J. Fruit Development Dynamics and Lignin Accumulation Law of Oil Tea. *Acta Agric. Univ. Jiangxiensis* **2020**, *42*, 788–801.
31. Yan, C. Molecular Regulation of Lignin Biosynthesis in Fruit Development of Oil-Camellia. Ph.D. Thesis, China Academy of Forestry Sciences, Beijing, China, 2020.
32. Li, Z.J.; Hua, J.Q.; Zeng, Y.R. Oil content of *Camellia oleifera* fruit trees. *J. Zhejiang A F Univ.* **2010**, *27*, 935–940.
33. Liang, W.J.; Xiao, P.; Cui, M.; Fu, Y.; Luo, L. The growth and development dynamics of *Camellia oleifera* Abel. fruits and seeds. *J. Nanchang Univ.* **2019**, *43*, 46–52.
34. HU, Y.L.; Long, X.Y.; Yang, H.; Yang, X.; Yang, S.; Pan, Z. Effects of different foliar fertilization concentrations on the chlorophyll content and productivity of oil-tea *Camellia*. *J. For. Environ.* **2021**, *41*, 527–535.
35. Zan, Y.L.; Wang, L. Effect of exogenous selenium on seed oil and fatty acid content in *Helianthus annuus*. *Guizhou Agric. Sci.* **2016**, *44*, 19–22.
36. He, P.; Zhang, J.D.; Bai, X.Q. Effects of Some Trace Elements on the Flower and Fruit Drop and Oil Content of *Camellia oleifera*. *Jiangxi For. Sci. Technol.* **1984**, *5*, 5–8.
37. Lin, P.; Wang, K.; Wang, Y.; Hu, Z.; Yan, C.; Huang, H.; Ma, X.; Cao, Y.; Long, W.; Liu, W.; et al. The genome of oil-Camellia and population genomics analysis provide insights into seed oil domestication. *Genome Biol.* **2022**, *23*, 14. [CrossRef] [PubMed]

38. Song, Y.R.; Jiang, X.G.; Peng, S.F.; Li, F.; Liu, J.; Chen, D.; Wu, L.; Liu, F. Effect of Selenium Content on the Quality and Functional Components of Selenium-riched *Camellia oleifera* Oil. *J. Chin. Inst. Food Sci. Technol.* **2015**, *15*, 142–149.
39. Pinto, A.; Juniper, D.T.; Sanil, M.; Morgan, L.; Clark, L.; Sies, H.; Rayman, M.P.; Steinbrenner, H. Supranutritional selenium induces alterations in molecular targets related to energy metabolism in skeletal muscle and visceral adipose tissue of pigs. *J. Inorg. Biochem.* **2012**, *114*, 47–54. [CrossRef]
40. Su, W.L.; Liu, N.; Mei, L.; Luo, J.; Zhu, Y.J.; Liang, Z. Global Transcriptomic Profile Analysis of Genes Involved in Lignin Biosynthesis and Accumulation Induced by Boron Deficiency in Poplar Roots. *Biomolecules* **2019**, *9*, 156. [CrossRef] [PubMed]
41. Ma, X.L.; Tang, W.L.; Liu, J.; Liu, J.; Wu, L.C.; Liu, F. Effects of selenium-enriched fertilizer on the main chemical characteristics and functional components of *Camellia* oil. *Food Sci. Technol.* **2015**, *40*, 148–154.
42. Song, C.Z.; Liu, M.Y.; Meng, J.F.; Chi, M.; Xi, Z.M.; Zhang, Z.W. Promoting Effect of Foliage Sprayed Zinc Sulfate on Accumulation of Sugar and Phenolics in Berries of Vitis vinifera cv. Merlot Growing on Zinc Deficient Soil. *Molecules* **2015**, *20*, 2536–2554. [CrossRef] [PubMed]

forests

MDPI

Article

Responses of Fine Root Traits and Soil Nitrogen to Fertilization Methods and Nitrogen Application Amounts in a Poplar Plantation

Xiaoli Yan [1], Tengfei Dai [2], Yuan Gao [2], Nan Di [2] and Liming Jia [2,*]

[1] College of Forestry, Fujian Agriculture and Forestry University, Fuzhou 350002, China
[2] Ministry of Education Key Laboratory of Silviculture and Conservation, Beijing Forestry University, Beijing 100083, China
[*] Correspondence: jlm@bjfu.edu.cn; Tel.: +86-010-62337055

Abstract: Inappropriate fertilization management practices have led to low timber production in intensive plantation systems in China. Thus, optimized conventional or advanced fertilization management practices are needed. We aimed to quantify whether optimized furrow fertilization (FF) is comparable to advanced drip fertigation (DF) and to make recommendations regarding fertilization management strategies for poplar plantations. A completely randomized block design experiment with two fertilization methods (DF and FF) and four N application amounts (F_0: 0, F_1: 68, F_2: 113, and F_3: 158 kg N·ha^{-1}·yr^{-1}) was carried out on a *Populus* × *euramericana* cv. 'Guariento' plantation. Fine root biomass density (FRBD), fine root length density (FRLD), specific root length (SRL), soil total nitrogen (STN), soil inorganic nitrogen (SIN), soil ammonium (NH_4^+-N) and nitrate nitrogen (NO_3^--N) were measured. The productivity increment was calculated based on tree surveys. The results showed that FRBD and FRLD decreased with the soil depth, and more than 86% was distributed within the 40 cm soil depth. FRBD, FRLD, productivity increment and soil N increased with an increasing amount of N application. DF treatments achieved 117%, 94% and 10% higher FRBD, FRLD and productivity increments, respectively, than did FF treatments. The averages of STN, SIN, NH_4^+-N and NO_3^--N under FF were higher than those under DF, leading to higher concentrations of residual NO_3^--N in deep soil. Beneficial management practices for fine root growth were evaluated in the following order: water coupled with N > only N $\geq$ only water > control. FRBD was positively correlated with the productivity increment. Therefore, fine root extension to increase soil resource absorption yields greater productivity under DF treatments. Drip fertilization is recommended as a better fertilization method to greatly promote the growth of fine roots, as well as productivity and residual lower soil N for poplar plantations.

Keywords: drip fertigation; optimized furrow fertilization; soil nitrogen; fine root traits; poplar plantation

Citation: Yan, X.; Dai, T.; Gao, Y.; Di, N.; Jia, L. Responses of Fine Root Traits and Soil Nitrogen to Fertilization Methods and Nitrogen Application Amounts in a Poplar Plantation. *Forests* **2023**, *14*, 282. https://doi.org/10.3390/f14020282

Academic Editor: Benjamin L. Turner

Received: 9 December 2022
Revised: 27 January 2023
Accepted: 30 January 2023
Published: 1 February 2023

1. Introduction

Poplar is one of the most important species for timber use and is widely planted in North America, Asia and Europe [1]. In China, there exist 8.54 million ha of poplar plantations, constituting 18.14% of the national plantation area, the largest in the world [2,3]. Due to the higher nitrogen (N) requirements for high productivity in poplar than in most other plantation species, N fertilization is widely used in poplar plantations [4,5]. However, the average productivity under existing silvicultural practices is approximately 15 m^3·ha^{-1}·yr^{-1}, which is far lower than the highest international level of productivity (53 m^3·ha^{-1}·yr^{-1}), and outdated fertilization methods and inappropriate N application rates are the main reasons for the low average productivity in China [6,7]. Therefore, there is a need for optimized conventional or advanced fertilization methods, as well as appropriate N application rates, to address this problem.

Conventional fertilization methods, such as furrow, hole, annular and radiation fertilization are currently the major approaches to fertilization management practices in plantations or orchards [6]. To increase productivity or yield, overuse of N fertilization is a common practice in China. In this mode of fertilization management, the N concentration in the root zone soil may be in excess of that required by plants for growth during the first period of fertilization. After this period, the N concentration diminishes gradually until deficit levels are reached, which inhibits plant growth [8]. Although optimized conventional fertilization methods [6], such as adjusting fertilizer application rates as well as the frequency, application distance and furrow depth, have been applied to solve the aforementioned problems, there is currently a lack of information on whether optimized conventional fertilization methods can efficiently achieve high productivity with low environmental risk.

In recent years, many advanced fertilization methods have been studied regarding their potential to improve yield production efficiently and without high environmental risk, such as sprinkling fertilization, exponential fertilization, layered fertilization and drip fertigation (DF) [9,10]. DF is an advanced fertilization technique, and its advantages include supplying fertilizer directly to the root zone, increasing resource-use efficiency, reducing N loss, reducing the need for fertilizer applications, improving soil fertility and reducing labor costs [9,11–13]. To date, DF has been increasingly applied for the cultivation of plantations worldwide [3,4,14], but few studies have investigated its intensive use in timber-producing plantations.

The selection of the optimal fertilization method can be evaluated by collectively considering the availability and distribution of soil nutrients, plant root growth and yield [15]. Among these parameters, the morphological characteristics (such as root length, specific root length, root diameter and root volume) and spatial configuration of fine roots can significantly affect soil water content, nutrient transformation and plant nutrient-use efficiency [16]. In addition, specific root length can be used as an indicator of environmental changes [17]. Moreover, fine root growth can be affected by spatiotemporal variations in water and nutrients [15]. To date, the effects of soil N availability on fine root biomass and morphology have been extensively studied, but the results are inconsistent due to variation in the diversity of tree species, the timing of sampling, specific soil properties and site conditions among the studies. Some studies have shown that root length or biomass increase with increased soil N availability [18–20], whereas other studies have shown no such relationship [21] or even the opposite relationship [22]. Because fertilization management practices can directly affect soil N availability, it is important to measure the N distribution in the soil. Most tree roots mainly take up N from soil in inorganic forms (NH_4^+-N and NO_3^--N), and chemical N fertilizers are often used to supply sufficient inorganic N in order to achieve high yields. Previous research has revealed that the overuse of N fertilizer leads to excessive N concentrations in the soil profile and is associated with NO_3^--N leaching and N_2O emission to the environment. In addition, it causes underground water pollution and atmospheric pollution [23]. Therefore, understanding the spatial distribution of soil N and the characteristics of fine root traits, as well as their interactions to increase nutrient availability in the rhizosphere, will be beneficial for both meeting the N demands of plants and making decisions involving fertilization management strategies in poplar plantations [24,25].

The species *Populus* × *euramericana* cv. 'Guariento' has received much attention in northern China because of its high growth rate and timber output [3,10,26]. Previous studies on the effects of N applications on this species have concentrated mainly on yield responses [3,4], nutrient uptake [27], and aboveground biomass [28]. DF and optimized furrow fertilization (FF) may promote NO_3^--N leaching, necessitating the careful calculation of fertilizer rates to minimize that risk. Hence, a two-year field experiment involving DF and optimized FF was conducted on a *Populus* × *euramericana* cv. 'Guariento' plantation in the North China Plain. This study was undertaken to achieve two objectives: (1) to quantitatively evaluate the fine root traits, productivity increment and residual soil N responses to different fertilization methods and N application rates; and (2) to make recommendations regarding fertilization management strategies for poplar plantations.

2. Materials and Methods

2.1. Study Site

A field experiment was carried out on former agricultural land in the district of Shunyi, a northern suburb of Beijing, China (40°05′48.7″ N and 116°49′35.6″ W), in a flat, low alluvial region of the Chaobai River. The elevation at the site was approximately 28 m above sea level [3,10]. The region was classified as having a warm-temperate, semi-humid, continental, monsoon climate (dry and windy in spring, hot and rainy in summer) and has an average annual temperature of 11.5 °C. The highest monthly temperature reached 40.5 °C in July and August, and the lowest was −19.1 °C in January. The mean annual precipitation is approximately 560 mm, of which 11% occurs between March and May, 70% occurs between June and August and 16% occurs between September and November. There are approximately 195 frost-free days per annum, and the monthly average underground water ranges from 6 m to 13 m in depth. These meteorological data were based on 4 years (2011–2014) of data from Shunyi District meteorological stations. Table 1 shows the basic physical and chemical properties of the soil [10]. The soil texture at the study site is sandy loam, according to the USDA classification system [3]. There are no stones larger than 2 mm in the soil. The soil data represent the average values from five random soil profiles from the experimental site in 2012, with each soil profile having three layers (0–20 cm, 20–40 cm and 40–60 cm).

Table 1. Physical and chemical properties of the soil at the experimental site.

Depth cm	Sand %	Silt %	Clay %	Soil Texture (USDA Classification)	Bulk Density $g \cdot cm^{-3}$	Organic Matter $g \cdot kg^{-1}$	Total N $g \cdot kg^{-1}$	NH_4^+-N $mg \cdot kg^{-1}$	NO_3^--N $mg \cdot kg^{-1}$	Available P $mg \cdot kg^{-1}$
0–20	69.96 ± 0.62	29.52 ± 0.64	0.52 ± 0.02	Sandy loam	1.68 ± 0.02	10.73 ± 0.20	0.58 ± 0.05	5.12 ± 0.05	3.88 ± 0.43	4.91 ± 0.97
20–40	67.19 ± 0.79	32.28 ± 0.77	0.53 ± 0.02	Sandy loam	1.64 ± 0.01	6.65 ± 0.29	0.49 ± 0.09	5.34 ± 0.44	2.56 ± 0.30	4.74 ± 0.69
40–60	63.52 ± 0.98	35.92 ± 0.96	0.56 ± 0.02	Sandy loam	1.62 ± 0.01	5.78 ± 0.23	0.44 ± 0.01	4.22 ± 0.39	6.63 ± 0.32	5.02 ± 0.68

Note: Soil texture is according to the USDA classification system; total N represents soil total nitrogen (SIN); NH_4^+-N and NO_3^--N represent soil ammonium and nitrate nitrogen; available P represents available phosphorous. The data are the average values from five random soil profiles in the experimental site in 2012, with each soil profile having three layers. Data are mean ± SE (n = 5).

2.2. Plant Material

A 4 ha experimental plantation was established with 3-year-old poplar clonal trees of *Populus* × *euramericana* cv. 'Guariento' in the spring of 2011. *Populus* × *euramericana* cv. 'Guariento' is a hybrid between the American black poplar (*Populus deltoides* cl. '8/67') and the European black poplar (*Populus* nigra) [3]. In northern China, the growing season is from March to October and leaf shedding occurs in November. The trees were planted with an alternate narrow- (6 m) and wide-row (12 m) spacing scheme with a within-row spacing of 4 m, resulting in a planting density of 300 trees·ha^{-1}. The average stem height and base diameter of the trees were 7.0 m and 5.0 cm, respectively. A surface drip irrigation tube system was installed in the plantation in the spring of 2012. The system was established such that one drip pipe was laid along each tree row. The emitter interval spacing was 1.0 m, and the flow rate was approximately 2 L·h^{-1}. After 1 year, no significant differences in tree diameter at breast height (DBH = 10.35 ± 0.35 cm, tree diameter at 1.3 m, $p > 0.05$) or height (H = 10.22 ± 0.45 m, $p > 0.05$) were observed from the treatments at the beginning of the experiment (March of 2013). The fertilization experiment was conducted from 2013 to 2014.

2.3. Experimental Design

The experiment was arranged in a completely randomized block design with 3 replicate blocks. Each block had 8 plots (72 m × 18 m, 38 trees·plot^{-1}). There was a buffer of at least a 16 m between adjacent plots. Therefore, 22 trees in each plot were included in the experiment, and the buffer reduced the likelihood that N fertilizer applied to one plot was transported to another plot. The experiment consisted of two fertilization methods and four N application rates, eight treatments in total. The fertilization methods included furrow fertilization (FF) and drip fertigation (DF). For the FF, a trench was dug underneath the emitter, which was away from the tree (approximately 1.0 m), and the length, width and depth of the trench were 0.8 m, 0.2 m and 0.15 m, respectively. The application frequency was set to 6 times greater than that of conventional FF (1–2 applications in the spring and autumn) [10,29]. The 4 N application rates were as follows: 0 (F_0), 68 (F_1), 113 (F_2) and 158 (F_3) kg $N \cdot ha^{-1}$. N fertilizer was applied as a urea solution (46% N, approximately 95 g $N \cdot L^{-1}$) which was injected directly into the main line of the drip system via a hydraulic-driven injector (Mix Rite Model 2504, Tefen, Israel). Treatment of FF_0 represents the conventional management practice in the north of China, which involves no irrigation or fertilization during the growing season. DF_0 represents only drip irrigation, DF_2 represents drip fertigation with a moderate N application rate and FF_2 represents furrow fertilization with a moderate N application rate. Before the leaf expansion period, furrow irrigation (approximately 3800 $m^3 \cdot ha^{-1} \cdot yr^{-1}$) was applied to all the treatments to promote leaf expansion and tree growth. Weeds were controlled during the experimental period using herbicides. The specific applications of water and N fertilizer in each treatment are summarized in Table 2.

Table 2. Experimental design and implementation overview at the study site.

Treatment	Spring Irrigation ($m^3 \cdot ha^{-1} \cdot yr^{-1}$)	Irrigation Amount ($m^3 \cdot ha^{-1} \cdot yr^{-1}$)	Fertilizer Rate (kg $N \cdot ha^{-1} \cdot yr^{-1}$)	Fertilization Frequency (times yr^{-1})
DF_1	3800	1395	68	6
DF_2	3800	1485	113	6
DF_3	3800	1580	158	6
DF_0	3800	1485	0	0
FF_1	3800	0	68	6
FF_2	3800	0	113	6
FF_3	3800	0	158	6
FF_0	3800	0	0	0

Note: The six N applications each year occurred on 28 April, 20 May, 13 June, 30 June, 26 July and 17 August for 2013, and 21 April, 12 May, 7 June, 7 July, 2 August and 28 August for 2014.

2.4. Sample Collection and Analysis

The diameter at breast height (DBH at 1.3 m, cm) and height (H, m) of each tree in the field was measured at the beginning of the experiment (March) and at the end of growing season (November). Standard trees (average trees) were marked in each plot. Points for soil and root sampling were randomly selected under the standard trees in each plot in November of 2013 and 2014. Each soil core was collected underneath the emitter at approximately 1.0 m away from the nearest standard tree. The soil cores were extracted using a cylindrical soil core (5 cm inner diameter and 20 cm length for soil sampling, 10 cm inner diameter and 20 cm length for root sampling) at 3 depth intervals: 0–20 cm, 20–40 cm and 40–60 cm (because most of the fine poplar roots were distributed above 40 cm [30], soil sampling down to 60 cm was appropriate for the research purposes of this study to examine the spatial distribution of soil N). A total of 72 soil samples and 72 root samples were collected in each year.

After removing roots and plant residues, each soil sample was thoroughly mixed, sieved (2 mm), placed in plastic bags and transported to the lab. The samples were later partitioned to measure soil water content (via the drying method), and the concentrations of NH_4^+-N (via indophenol blue colorimetry) and NO_3^--N (via dual wave-length colorimetry) [31,32] were measured using a continuous flow analysis instrument (AA3, Bran and Luebbe, Norderstedt, Germany). The remaining soil samples were air-dried, ground and sieved (0.15 mm) in preparation for STN using H_2SO_4-H_2O_2 digestion, and then they were measured using a Kjeldahl nitrogen meter (Hangzhou, Zhejiang, China) [32]. Inorganic N was calculated as the sum of ammonium and nitrate.

Each root sample was gently rinsed with fresh water to separate the roots from the soil particles and organic materials. Two sieves (0.8 and 0.125 mm mesh) were used to avoid losing fine roots. After washing, the roots were placed in fresh water, and all living roots < 2 mm in diameter were manually collected from the residual soil particles and organic materials using forceps and filters. Living roots were distinguished from dead roots by their lighter color and greater resilience; yellow-brown or brown fine roots were living, while black roots were dead [33]. All living fine roots in each layer were digitally scanned using a flatbed scanner at 400 dpi, and the files were saved in tif format. The root images were analyzed using image analysis software (WinRHIZO Pro 2008a, Regent Instruments Inc., Canada) for root length. The roots were then placed into labeled envelopes and oven-dried at 65–70 °C to a constant mass (for at least 48 h), and then weighed to estimate the fine root biomass [29,34].

Tree growth data were used to calculate the productivity increment (m^3 ha^{-1}) using the following formula:

$$P = \left[g_{1.3} \times (H_{end} + 3) \times f - g_{1.3} \times \left(H_{begin} + 3\right) \times f\right] \times N \quad (1)$$

where P is the productivity increment, $g_{1.3}$ is the cross-sectional area at breast height, H_{begin} is tree height at the beginning of growing season, H_{end} is the tree height at the end of the growing season, f is the experimental form factor and N is the tree density in the field (300 tree ha^{-1}). The factor f was determined by comparing several approaches to stem volume calculation based on destructive sampling of average tree stems from 2012–2014, and a form factor of 0.41 was obtained from the Newton approximation method as the relatively accurate value for this plantation [3].

2.5. Data Analysis

The fine root trait variables which were calculated included fine root biomass density (FRBD, g·m^{-3} soil) = fine root biomass/volume of soil block; fine root length density (FRLD, m·m^{-3} soil) = fine root length/volume of soil block and specific root length (SRL, m·g^{-1}) = fine root length/fine root biomass. All data were expressed as the means ± standard errors.

The Kolmogorov–Smirnov test and the Levene test were used to verify the assumptions of normality and homogeneity of the variance of the data for each variable before further analysis. The variables that did not conform to these assumptions were mathematically transformed using logarithms or reciprocal functions. Multivariate-way ANOVAs were carried out to determine the effects of the year of experiment (Y), fertilization methods (M), fertilizer amount (A) and soil layer (S) on fine root traits and soil N ($p < 0.05$). One-way ANOVA was used to test the differences between N addition treatments (Duncan's test, $P = 0.05$). An independent-sample test was used to test for differences between two fertilization methods. Pearson correlations were used to measure the degree of association between the fine root traits and soil N. Linear relationships were assessed to examine the relationships between FRBD and productivity increments. All statistical analyses were performed using SPSS, version 20.0 (SPSS Inc., Chicago, IL, USA). All figures and tables were produced using Excel 2016.

3. Results

3.1. FRBD, FRLD and SRL

The significance levels were analyzed for the effects of the year of experiment, fertilization method, fertilizer amount, soil layers and their interactions on FRBD, FRLD and SRL (Table 3). The effects of the year of experiment (Y), fertilization method (M), fertilizer amount (A), soil layers (S) and Y × S, M × S and Y × M × S interactions on FRBD, FRLD and SRL were all statistically significant ($p < 0.001$ or $p < 0.05$), but the effects of Y × A and Y × M × A interactions were not significant ($p > 0.05$). (Table 3).

Table 3. p values of repeated measures ANOVA for the year of experiment (Y), fertilization method (M), fertilizer amount (A), soil layers (S) and their interactions on fine root biomass density (FRBD), fine root length density (FRLD) and specific root length (SRL) of fine roots in a poplar plantation.

Source of Variation	df	FRBD	FRLD	SRL
Year of experiment (Y)	1	<0.001	<0.001	<0.001
Fertilization method (M)	1	<0.001	<0.001	<0.001
Fertilizer amount (A)	4	<0.001	<0.001	0.013
Soil layers (S)	2	<0.001	<0.001	<0.001
Y × M	1	0.435	0.064	<0.001
Y × A	1	0.135	0.545	0.719
Y × S	2	<0.001	<0.001	0.002
M × S	2	<0.001	<0.001	<0.001
A × S	8	<0.001	<0.001	0.472
M × A	3	<0.001	<0.001	0.707
Y × M × S	2	0.012	<0.001	<0.001
Y × M × A	1	0.298	0.130	0.370
M × S × A	6	<0.001	<0.001	0.946
Y × S × A	2	0.180	0.025	0.419
Y × M × S × A	2	0.203	0.004	0.824

FRBD and FRLD decreased with increasing soil layer depth (Figure 1a–d), and SRL tended to be greater at a depth of 40–60 cm under the FF method (Figure 1e,f). FRBD and FRLD were increased with the amount of N applied (0–180 kg·ha^{-1}) under treatments of DF_{0-3} and FF_{0-3}, but showed a tendency to increase initially and then decrease with the N application amount under the FF method in 2014 (Figure 1a–d).

Most of the fine roots were concentrated at a depth of 0–40 cm, with FRBD and FRLD at this depth accounting for 81%–89% and 86%–90%, respectively, of the total under each treatment. In the first 20 cm depth, both FRBD and FRLD were significantly higher in the DF- than the FF-treated plots for almost all the N treatments tested (Figure 1a–d). At 40–60 cm soil depth, the average FRBD and FRLD with DF were 117% and 94% higher than with FF, respectively (Figure 1).

Comparing the two fertilization methods, the average FRBD and FRLD were higher with the DF method than with the FF method. They were significantly enhanced, by 101%–86% and 50%–66%, compared to the FF method in 2013 and 2014, respectively. Comparing the four N application levels, the average FRBD and FRLD with treatments of DF_{0-3} were significantly higher than with treatments of FF_{0-3}, from 0–180 kg ·ha^{-1}, in the years of 2013 and 2014. The values of SRL with the FF method were higher than with the DF method in 2013 and 2014. SRL tended to be higher at a depth of 40–60 cm, especially for the FF method in 2013. SRL was higher in 2013 than in 2014 (Figure 1e,f).

3.2. NH_4^+-N, NO_3^--N, SIN and STN

The significance levels were analyzed for the effects of the year of experiment, fertilization method, fertilizer amount, soil layers and their interactions on the soil NH_4^+-N, NO_3^--N, SIN and STN (Table 4). The effects of the fertilization method (M), fertilizer amount (A), soil layers (S) and Y × M, Y × S, M × S, A × S, M × A, Y × M × A and M × S × A interactions on the soil NH_4^+-N, NO_3^--N, SIN and STN were all statistically significant ($p < 0.001$ or $p < 0.05$), but the effects of Y, Y × A, Y × S × A and Y × M × S × A interactions on NO_3^--N were not significant ($p > 0.05$) (Table 4).

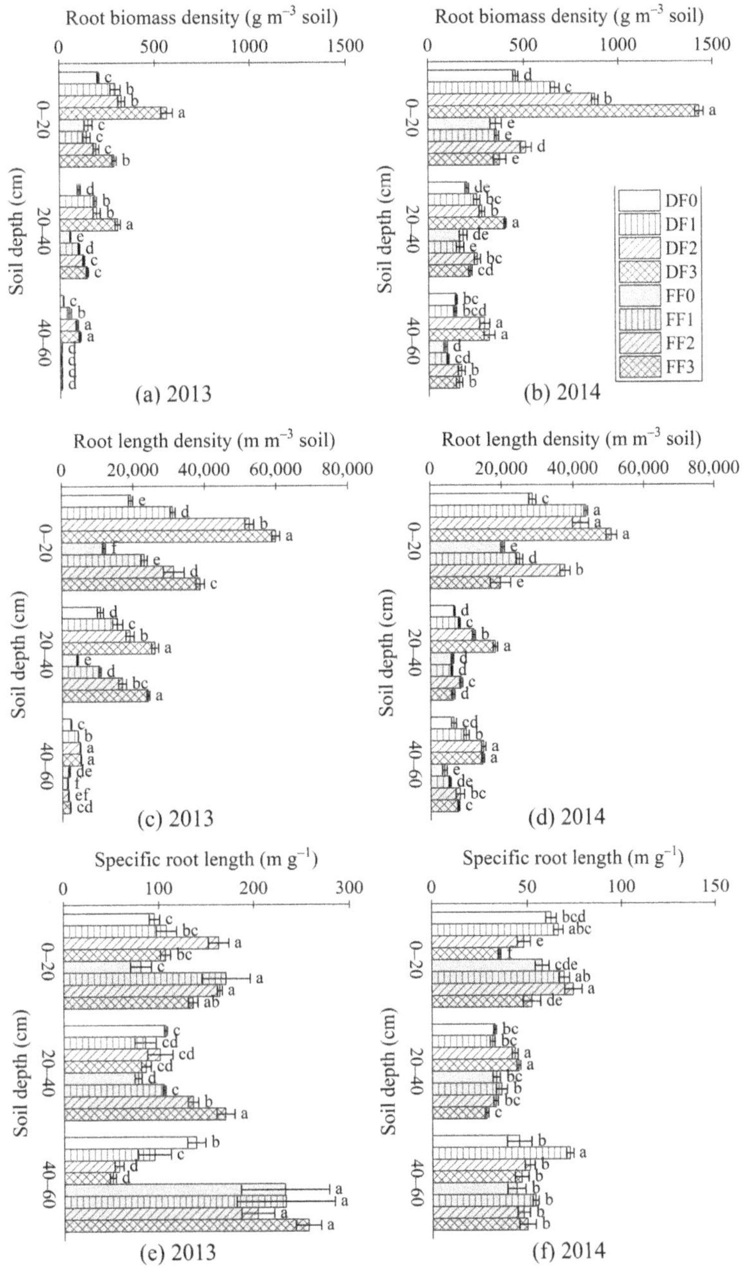

Figure 1. Values of fine root biomass density (FRBD, (**a**): 2103, (**b**): 2014), fine root length density (FRLD, (**c**): 2013, (**d**): 2014) and specific root length (SRL, (**e**): 2013, (**f**): 2014) of each soil core at different soil depths (0–20 cm, 20–40 cm and 40–60 cm, respectively) in a poplar plantation. The data are the means ± SE (n = 3). Different lowercase letters within the same soil layer indicate significant differences among eight N addition treatments.

Table 4. p values of repeated measures ANOVA for the year of experiment (Y), fertilization method (M), fertilizer amount (A), soil layers (S) and their interactions on the soil NH_4^+-N, NO_3^--N, SIN and STN in a poplar plantation.

Source of Variation	df	NH_4^+-N	NO_3^--N	SIN	STN
Year of experiment (Y)	1	<0.001	0.105	<0.001	0.090
Fertilization method (M)	1	<0.001	<0.001	<0.001	<0.001
Fertilizer amount (A)	4	<0.001	<0.001	<0.001	<0.001
Soil layers (S)	2	<0.001	<0.001	<0.001	<0.001
Y × M	1	<0.001	<0.001	<0.001	0.002
Y × A	1	<0.001	0.069	<0.001	0.289
Y × S	2	<0.001	<0.001	<0.001	<0.001
M × S	2	<0.001	<0.001	<0.001	<0.001
A × S	8	<0.001	<0.001	<0.001	<0.001
M × A	3	<0.001	<0.001	<0.001	<0.001
Y × M × S	2	<0.001	<0.001	0.626	<0.001
Y × M × A	1	<0.001	<0.001	<0.001	<0.001
M × S × A	6	<0.001	<0.001	<0.001	<0.001
Y × S × A	2	<0.001	0.202	<0.001	0.264
Y × M × S × A	2	<0.001	0.465	0.011	0.002

STN decreased with increasing soil layer depth under both DF and FF (Figure 2a,b). SIN, NH_4^+-N and NO_3^--N decreased with increasing soil depth under the FF method, whereas they accumulated within the deep soil under the DF method (Figures 2c,d and 3a–d). The average NO_3^--N in the 40–60 cm soil layer under the FF method was 53% higher than that under DF ($p < 0.05$). Particularly, the concentrations of all N forms tested were clearly higher with the FF than the DF treatments at lower depths (Figure 3c,d).

Treatments of $DF_{1\text{-}3}$ and $FF_{1\text{-}3}$ were significantly higher than the control treatments (DF_0 and FF_0). The average values of STN, SIN, NH_4^+-N and NO_3^--N concentration with the FF method were significantly higher than with the DF method, which improved by 67%, 85%, 8% and 241% in 2013 and by 47%, 280%, 89% and 576% in 2014, respectively. The average concentration of STN, SIN, NH_4^+-N and NO_3^--N were increased with N application amounts under the two fertilization methods.

3.3. Relationships between Fine Root Traits and Soil N

Table 5 shows that FRBD and FRLD, under both the DF and FF methods, were positively correlated with the STN. FRBD and FRLD were not correlated with soil NH_4^+-N, NO_3^--N or SIN under the DF method, but were negatively correlated with SRL. Meanwhile, FRBD and FRLD were positively correlated with soil NH_4^+-N, NO_3^--N and SIN, but negatively correlated with SRL, under the FF method ($p < 0.05$).

3.4. Relationships among N Application Amount, FRBD and Productivity Increment under Two Fertilization Methods

Productivity increment and fine root biomass density increased with increasing N application under DF (Figure 4a,b); 158 kg·ha^{-1} yielded the highest value. However, in FF, these factors showed a trend of first increasing and then decreasing with increasing N application (Figure 4a,b); 113 kg·ha^{-1} yielded the highest FRBD and productivity increment (Figure 4). DF treatments achieved significantly higher productivity increments (by 1%–20%) than did FF treatments ($p < 0.05$). Figure 4c,d show that FRBD was positively correlated with the productivity increment. The linear regression relationship was more significant with DF treatment than with FF treatment.

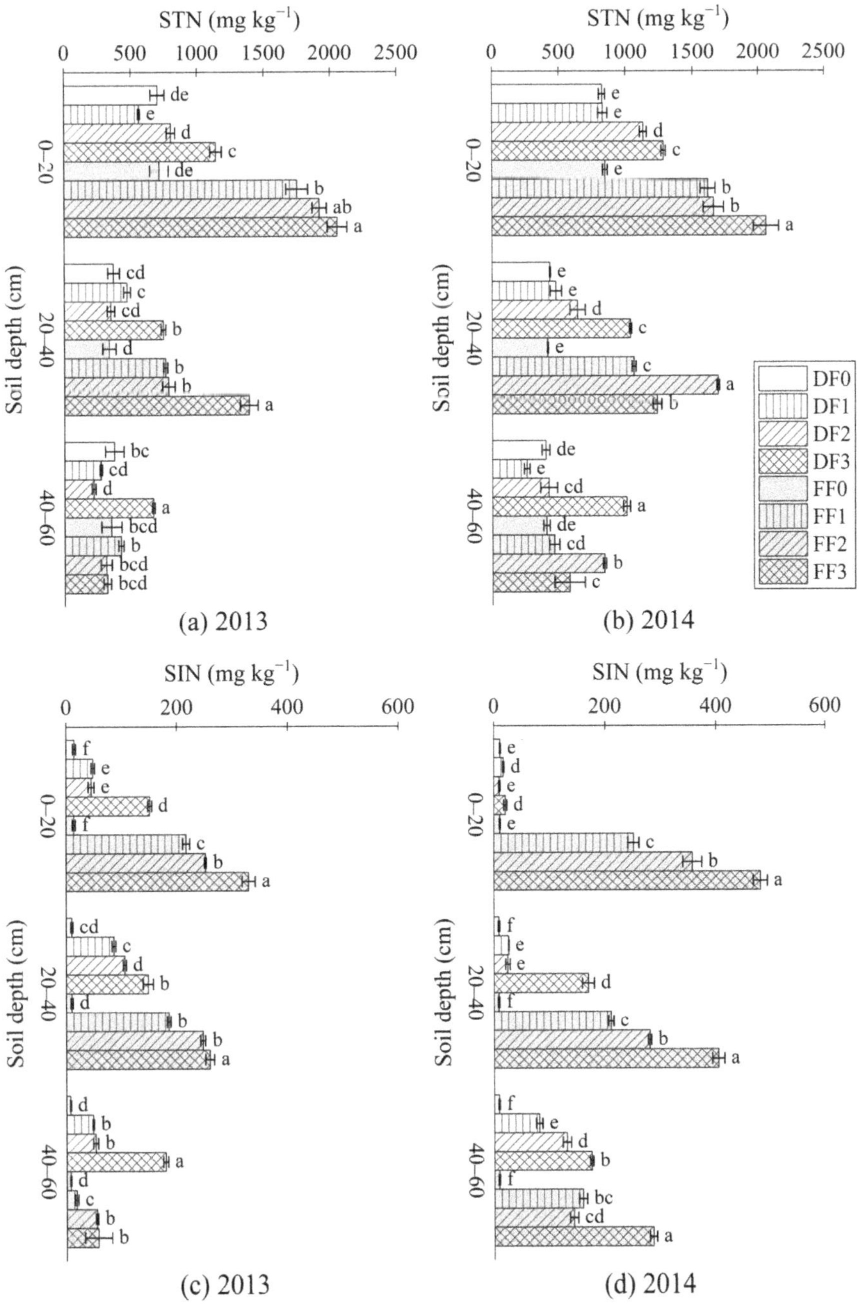

Figure 2. Values of total soil nitrogen (STN, (**a**): 2013, (**b**): 2014) and total soil inorganic nitrogen (SIN, (**c**): 2013, (**d**): 2014) at different soil depths (0–20 cm, 20–40 cm and 40–60 cm) in a poplar plantation. The data are the means ± SE (n = 3). Different lowercase letters within the same soil layer indicate significant differences among eight N addition treatments.

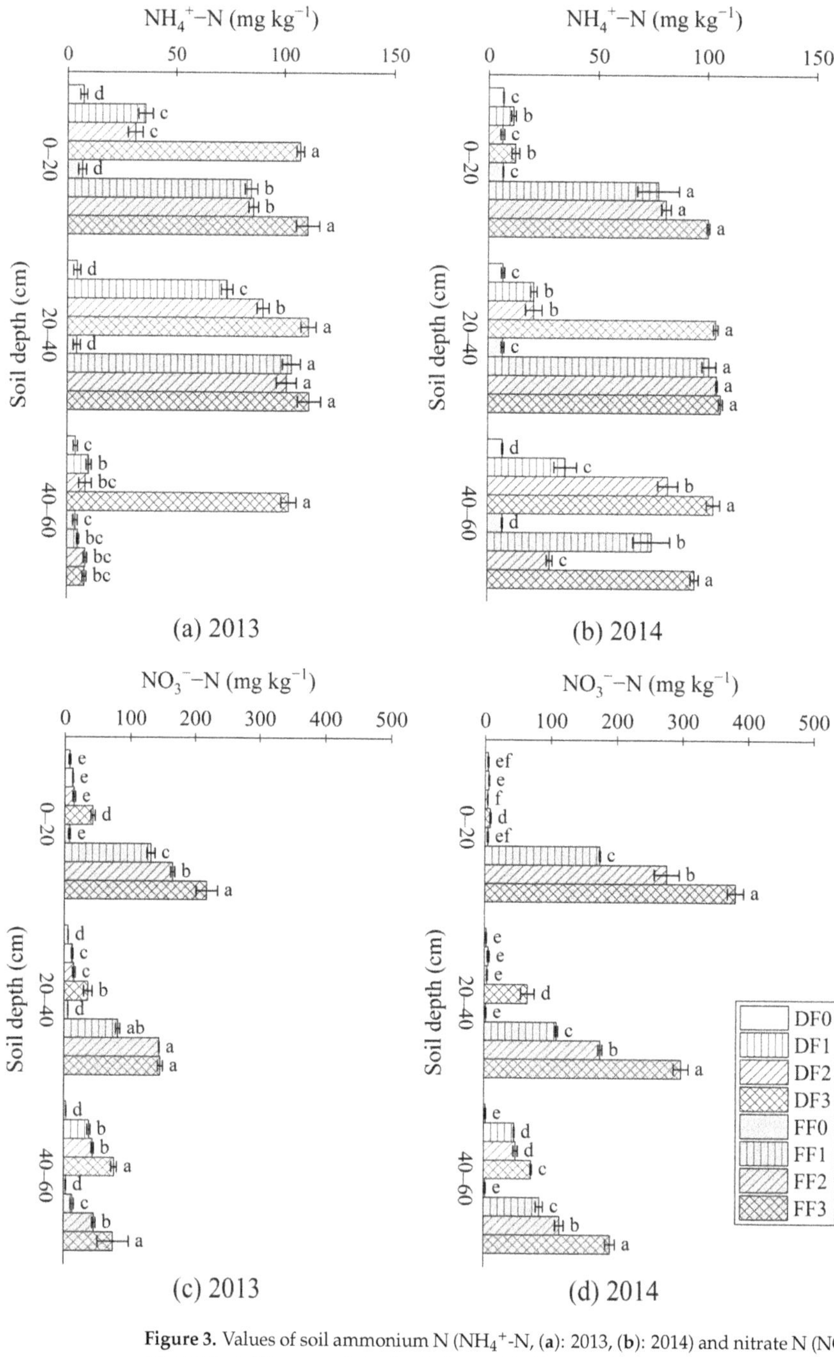

Figure 3. Values of soil ammonium N (NH_4^+-N, (**a**): 2013, (**b**): 2014) and nitrate N (NO_3^--N, (**c**): 2013, (**d**): 2014) at different soil depths (0–20 cm, 20–40 cm and 40–60 cm) in a poplar plantation. The data are the means ± SE (n = 3). Different lowercase letters within the same soil layer indicate significant differences among eight N addition treatments.

Table 5. Pearson correlation coefficients between fine root traits and soil N under the drip fertigation (DF) and optimized furrow fertilization (FF) methods in a poplar plantation. Fine root traits are fine root biomass density (FRBD), fine root length density (FRLD) and specific root length (SRL); soil N variables are ammonium N (NH_4^+-N), nitrate N (NO_3^--N), total soil inorganic nitrogen (SIN) and total soil nitrogen (STN). $p < 0.05$ *, $p < 0.01$ **.

Methods	Traits	NH_4^+-N	NO_3^--N	SIN	STN
DF	FRBD	0.15	−0.09	0.06	0.86 **
	FRLD	0.03	−0.19	−0.07	0.82 **
	SRL	−0.59 **	−0.55 **	−0.61 **	0.06
FF	FRBD	0.50 **	0.66 **	0.63 **	0.88 **
	FRLD	0.55 **	0.74 **	0.71 **	0.93 **
	SRL	−0.37 *	−0.24	−0.29	−0.47 **

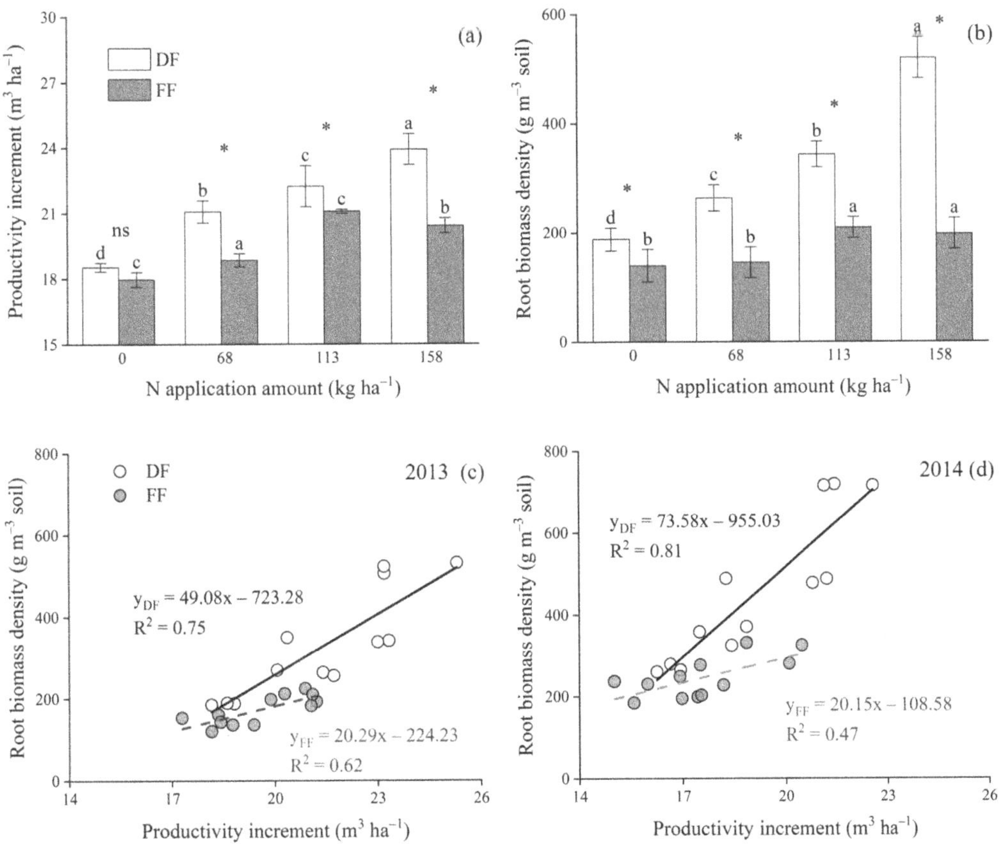

Figure 4. Relationships between productivity increment and N application amount (**a**), fine root biomass density and N application amount (**b**) and productivity increment and fine root biomass density ((**c**): 2013, (**d**): 2014) under DF and FF methods. Solid black circles represent DF; hollow circles represent FF. Different small letters labeling the bars indicate significant differences among N application rates at the 0.05 level, and * indicates a significant difference between the two fertilization methods at the 0.05 level.

4. Discussion

4.1. Influence of Fertilization Methods and Application Rates on Fine Root Traits

Fine roots are vital organs that provide water and mineral nutrients to plants. They not only determine soil's utilization effectiveness and potential, but also reflect the distribution pattern of water and mineral nutrients in soils [35,36]. In this study, fine root biomass density and fine root length density decreased with soil depth in all of the treatments, and more than 86% and 89% of the total FRBD and FRLD was distributed within the 40 cm soil depth. Similarly, the fine root distribution within the surface soil layer has been reported in plantations of other species under different site conditions [37–40]. This phenomenon can be attributed to the role of the surface soil layer as the main provider of soil N and water for plant roots [41]. Soil N was mainly concentrated within 40 cm of soil depth in all treatments, which may have favored the proliferation of fine roots. Nonetheless, fine root development should be affected by both nutrients and water along the soil profile. It is certain that the soil's water distribution along the profile would change according to the irrigation method. The affecting mechanism should include water application and absorption efficiency along different soil layers, and this may be investigated in our future works. Furthermore, the average FRBD and FRLD in the 40–60 cm soil layer were 117% and 94% higher with DF than with FF, implying that DF could promote extension of fine roots to greater soil depths. We speculate that the fine root distribution pattern observed in our study was the result of the integrated effects of both intrinsic tree characteristics and fertilization methods [37,39,42].

In this study, FRBD and FRLD were increased with increasing N application more significantly under DF than under FF (Figure 1). High-frequency DF is conducive to maintaining a stable water and N status in the root zone, as water and urea liquid are simultaneously applied in this method [42,43]. A potential reason for the greater average FRBD and FRLD under the DF method than under the FF method is the greater proportion of photosynthate allocation to fine roots [41]. In practice, FF methods usually require that an artificial gully be dug that reaches the trees, resulting in damage of the fine roots to some extent. This might disrupt root growth balance and decrease the amount of roots. As the roots start to recover and regrow, the system becomes a sink of C and N [44]. Additionally, greater amounts of resource inputs are consumed during the recovery and regrowth processes of fine roots. Accordingly, by consuming the same amount of resources, DF achieved higher fine root biomass and fine root length than FF.

To date, the results regarding the effects of soil N availability on fine root production are still inconclusive. Some studies have found that fertilizer application variably affects soil N mineralization and increases plant uptake of soil N via plant-mediated mechanisms, such as increases in fine root biomass and length [44]. In the present study, FRBD and FRLD increased with N application rates under the DF method. Poplar fine roots are considered highly plastic, and some studies have reported that poplar roots proliferate in the presence of increased N [45]. Artacho and Bonomelli [20] proposed that higher N concentrations in roots could increase the speed of fine root turnover or increase the N concentrations in leaves, which could then increase the availability of photosynthates and promote greater root growth. However, under the FF method, N application had a relatively small effect on FRBD (Figure 1). Similarly, Price et al. [46] studied the fine root production dynamics in an intensively managed sweetgum (*Liquidambar styraciflua* L.) coppice and reported no differences between the two studied fertilizer treatments. Other studies have also reported reductions in fine root growth with N application. Jia et al. [22] studied the responses of belowground biological processes to soil N availability in *Larix gmelinii* and *Fraxinus mandshurica* plantations and reported that, compared with the control treatment, the N fertilization treatment induced 52% and 25% decreases in fine root biomass in larch and ash, respectively. Rytter et al. [47] considered that, in absolute terms, a lower standing fine root biomass develops where N availability is high due to the simultaneous increase in turnover rate and decrease in standing fine root biomass. In summary of the abovementioned results, the relationship between available soil N and fine root growth is complex, and multiple factors, such as tree species diversity and site conditions, as mentioned above, should be considered when designing experiments. We found that, compared with the four fertilization practices, the different management practices benefited fine root growth (FRBD) in the following order: water coupled with N (DF_2) > N application alone (FF_2) ≥ water application alone (DF_0) > control (FF_0) (Figure 1). This result confirmed that soil resource availability affected fine root growth [15].

4.2. Influence of Fertilization Methods and Application Rates on Soil N

The spatial and temporal distribution of N in soil are heterogeneous. Thus, understanding the distribution of soil N under different fertilization methods is helpful for developing appropriate management strategies for poplar plantations. In this study, we found that soil N decreased with soil depth under the FF method (Figures 2 and 3), which corroborates the results of previous works [43,48]. However, the vertical distributions of SIN, NH_4^+-N and NO_3^--N under DF differed. These inorganic forms of N increasingly accumulated within the deep soil layer with increasing N application (Figures 2 and 3). This result is similar to Li's findings concerning subsurface DF [49]. The results also showed that the average STN, SIN, NH_4^+-N and NO_3^--N values under FF were higher than those under DF (Figures 2 and 3), which is consistent with Qi and Huang's findings [43,50]. More time was required for available N (NH_4^+-N and NO_3^--N) to be hydrolyzed and transformed from urea under the FF method than under the DF method; the former released more available N into the soil in the short term than the latter [10]. As a result of the movement of available N with water flow and its redistribution, available N accumulated in the deeper soil layers.

Irrigation and fertilization are linked, and inappropriate fertilization methods and fertilization applications can greatly increase nutrient losses [51]. In the present study, inappropriate fertilization methods significantly enhanced the concentration of NO_3^--N in the soil (Figures 2–4). The average concentration of NO_3^--N in the 40–60 cm soil layer was 53% higher under the FF method than under the DF method. This result demonstrated that the FF method led to higher concentrations of inorganic N in the soil and caused the deep migration of large concentrations of NO_3^--N, which may increase the risk of N leaching.

4.3. Relationships among Soil N, Fine Root Traits and Productivity Increment

Fine root distribution is largely influenced by soil resource availability [52]. In this study, under the DF method, both FRBD and FRLD were positively correlated with STN, but not with SIN, NH_4^+-N or NO_3^--N. In contrast, under the FF method, FRBD and FRLD were positively correlated with STN, SIN, NH_4^+-N and NO_3^--N (Table 5). These differences between treatments may reflect the different fertilization methods that were used. FRBD and FRLD, as well as STN, decreased with increasing soil depth under all treatments, indicating a greater proportion of fine roots in the surface soil layer than in deeper layers in response to ample moisture and nitrogen availability [41,52,53]. A larger portion of the SIN was efficiently absorbed by fine roots under the DF method, which led to lower available soil N, ultimately reflecting the observed weaker relationship between SIN and fine root traits. However, negative relationships between FRBD and nutrient supplies in beech (*Fagus sylvatica* L.) forests and tropical rainforests have been reported [54,55].

Our results show that productivity increments and FRBD increased with increasing N application under DF (Figure 4a,b). This indicates that drip fertigation is an efficient silvicultural practice to increase the fine root and tree growth on poplar plantations, which is similar to the findings of previous studies. For example, Yan et al. reported a significant cumulative effect of N fertigation on productivity in poplar plantation, with 180 kg·ha^{-1} N application achieving the highest increments [3]. However, through the results of the DF method, we were unable to achieve the optimal N application rate for the poplar plantation with the existing gradient, which indicated that a further improvement in productivity could be achieved with increased fertilization. Xi et al. recommended an optimal N application rate of 192 kg·ha^{-1} for a 3–5-year-old poplar plantation under drip fertigation [28]. Thus, we speculated that drip fertigation positively promoted fine root growth and productivity, and that the highest level of 180 kg N·ha^{-1} yr^{-1} would be within the range of an optimal N application rate. However, with FF, they showed trends of first increasing and then decreasing with the increasing N application rate. This result implies that fine root growth and productivity did not increase with the fertilization rate when it was above 158 kg N ·ha^{-1} under FF, which has also been observed in other studies. For example, Wang reported that the use of an N application rate above 115 kg N·ha^{-1} yr^{-1} did not result in growth nor N uptake benefits [4]. Hochmuth summarized the findings of more than 15 fertilization trials under drip irrigation and concluded that tomato yield did not increase with the fertilization rate when it was above 227 kg N·ha^{-1} yr^{-1} [56].

The positive correlation between fine root biomass density and productivity increment were observed under the DF and FF methods (Figure 4c,d). The results were in accordance with previous research [20,57] which has suggested that fine root extension to increase soil resource absorption yields greater productivity. However, the opposite results were obtained for pine seedlings; the effects of the plasticity of fine root growth on stem growth resulted in increased allocation of biomass to foliage and decreased allocation to fine roots [58]. In our study, ample availability of N and water contributed to higher FRBD and productivity increments under the DF method.

5. Conclusions

In summary, drip fertilization achieved higher fine root biomass density, fine root length density and productivity increments than optimized furrow fertilization, whereas optimized furrow fertilization contributed to higher concentrations of STN, SIN, NH_4^+-N and NO_3^--N than drip fertigation. Therefore, drip fertigation is recommended as a superior fertilization method, as it greatly promoted the growth of fine roots as well as productivity and residual lower soil N on the poplar plantation. It is suggested that fine roots adjust their growth and morphology in response to N availability, which varies along with the soil profile and the fertilization method. It can be deduced that, in fields with similar soil types to that of our experimental site, the practice of drip fertilization management can be generalized as an effective management regime in fast-growing forest plantations.

Author Contributions: Conceptualization, L.J., T.D. and X.Y.; methodology, T.D. and X.Y.; formal analysis, T.D., X.Y. and N.D.; investigation, X.Y. and T.D.; writing—original draft preparation, T.D. and X.Y.; writing—review and editing, X.Y., Y.G. and N.D.; funding acquisition, L.J. and X.Y. All authors have read and agreed to the published version of the manuscript.

Funding: This research was funded by the National Natural Science Foundation of China (31670625) and the Distinguished Young Scientific Research Talent Program of Fujian Agriculture and Forestry University (KXJQ20012).

Institutional Review Board Statement: Not applicable.

Informed Consent Statement: Not applicable.

Data Availability Statement: Not applicable.

Conflicts of Interest: The authors declare no conflict of interest.

References

1. Dickmann, D.I.; Nguyen, P.V.; Pregitzer, K.S. Effects of irrigation and coppicing on above-ground growth, physiology, and fine-root dynamics of two field-grown hybrid poplar clones. *For. Ecol. Manag.* **1996**, *80*, 163–174. [CrossRef]
2. Wang, N.N.; Huang, J.; Ding, C.J.; Zhang, H.; Shen, Y.B.; Su, X.H. The relationship of shade tolerance of poplar and biomass production under different plantation density. *For. Res.* **2015**, *28*, 691–700.
3. Yan, X.L.; Dai, T.F.; Jia, L.M. Evaluation of the cumulative effect of drip irrigation and fertigation on productivity in a poplar plantation. *Ann. For. Sci.* **2018**, *75*, 5. [CrossRef]
4. Wang, Y.; Xi, B.Y.; Bloomberg, M.; Moltchanova, E.; Li, G.D.; Jia, L.M. Response of diameter growth, biomass allocation and n uptake to n fertigation in a triploid *populus tomentosa* plantation in the North China Plain: Ontogenetic shift does not exclude plasticity. *Eur. J. For. Res.* **2015**, *134*, 889–898. [CrossRef]
5. Rennenberg, H.; Wildhagen, H.; Ehlting, B. Nitrogen nutrition of poplar trees. *Plant Biol.* **2010**, *12*, 275–291. [CrossRef] [PubMed]
6. Zheng, S.K. *High Yield Cultivation of Poplar*; Golden Shield: Beijing, China, 2006. (In Chinese)

7. Xi, B.Y.; Wang, Y.; Jia, L.M.; Bloomberg, M.; Li, G.D.; Di, N. Characteristics of fine root system and water uptake in a triploid *Populus tomentosa* plantation in the North China Plain: Implications for irrigation water management. *Agric. Water Manag.* **2013**, *117*, 83–92. [CrossRef]
8. Matsuo, N.; Mochizuki, T. Assessment of three water-saving cultivations and different growth responses among six rice cultivars. *Plant Prod. Sci.* **2009**, *12*, 514–525. [CrossRef]
9. Tarkalson, D.D.; Van Donk, S.J.; Petersen, J.L. Effect of nitrogen application timing on corn production using subsurface drip irrigation. *Soil Sci.* **2009**, *174*, 174–179. [CrossRef]
10. Dai, T.F.; Xi, B.Y.; Yan, X.L.; Jia, L.M. Effects of fertilization method and nitrogen application rate on soil nitrogen vertical migration in a *Populus × euramericana* cv. 'Guariento' plantation. *Chin. J. Appl. Ecol.* **2015**, *26*, 1641–1648, (In Chinese with English Abstract).
11. Bhat, R.; Sujatha, S. Cost-benefit analysis of ferti-drip irrigation in arecanut (*Areca catechu* L.). *J. Plant. Crops* **2006**, *34*, 263.
12. Bhat, R.; Sujatha, S. Soil fertility and nutrient uptake by arecanut (*Areca catechu* L.) as affected by level and frequency of fertigation in a laterite soil. *Agr. Water Manag.* **2009**, *96*, 445–456. [CrossRef]
13. Valin, M.I.; Cameira, M.R.; Teodoro, P.R.; Pereira, L.S. Depivot: A model for center-pivot design and evaluation. *Comput. Electron. Agric.* **2012**, *87*, 159–170. [CrossRef]
14. O'Neill, M.K.; Allen, S.C.; Heyduck, R.F.; Lombard, K.A.; Smeal, D.; Arnold, R.N. Hybrid poplar (*Populus* spp.) adaptation to a semi-arid region: Results from northwest new mexico (2002–2011). *Agrofor. Syst.* **2014**, *88*, 387–396. [CrossRef]
15. Liang, X.S.; Gao, Y.A.; Zhang, X.Y.; Tian, Y.Q.; Zhang, Z.X.; Gao, L.H. Effect of optimal daily fertigation on migration of water and salt in soil, root growth and fruit yield of cucumber (*Cucumis sativus* L.) in solar-greenhouse. *PLoS ONE* **2014**, *9*, e86975. [CrossRef] [PubMed]
16. Lynch, J.P.; Brown, K.M. New roots for agriculture: Exploiting the root phenome. *Philos. T. R. Soc. B* **2012**, *367*, 1598–1604. [CrossRef] [PubMed]
17. Ostonen, I.; Puttsepp, U.; Biel, C.; Alberton, O.; Bakker, M.R.; Lohmus, K.; Majdi, H.; Metcalfe, D.; Olsthoorn, A.F.M.; Pronk, A.; et al. Specific root length as an indicator of environmental change. *Plant Biosyst.* **2007**, *141*, 426–442. [CrossRef]
18. King, J.S.; Albaugh, T.J.; Allen, H.L.; Buford, M.; Strain, B.R.; Dougherty, P. Below-ground carbon input to soil is controlled by nutrient availability and fine root dynamics in loblolly pine. *New Phytol.* **2002**, *154*, 389–398. [CrossRef]
19. Noguchi, K.; Nagakura, J.; Konopka, B.; Sakata, T.; Kaneko, S.; Takahashi, M. Fine-root dynamics in sugi (*Cryptomeria japonica*) under manipulated soil nitrogen conditions. *Plant Soil* **2013**, *364*, 159–169. [CrossRef]
20. Artacho, P.; Bonomelli, C. Net primary productivity and allocation to fine-root production in field-grown sweet cherry trees under different soil nitrogen regimes. *Sci. Hortic.* **2017**, *219*, 207–215. [CrossRef]
21. Phillips, D.L.; Johnson, M.G.; Tingey, D.T.; Storm, M.J.; Ball, J.T.; Johnson, D.W. CO_2 and n-fertilization effects on fine-root length, production, and mortality: A 4-year ponderosa pine study. *Oecologia* **2006**, *148*, 517–525. [CrossRef]
22. Jia, S.X.; Wang, Z.Q.; Li, X.P.; Sun, Y.; Zhang, X.P.; Liang, A.Z. N fertilization affects on soil respiration, microbial biomass and root respiration in *Larix gmelinii* and *Fraxinus mandshurica* plantations in China. *Plant Soil* **2010**, *333*, 325–336. [CrossRef]
23. Xie, B.H.; Gu, J.X.; Yu, J.B.; Han, G.X.; Zheng, X.H.; Xu, Y.; Lin, H.T. Effects of n fertilizer application on soil N_2O emissions and CH_4 uptake: A two-year study in an apple orchard in eastern China. *Atmosphere* **2017**, *8*, 181. [CrossRef]
24. Coleman, M. Spatial and temporal patterns of root distribution in developing stands of four woody crop species grown with drip irrigation and fertilization. *Plant Soil* **2007**, *299*, 195–213. [CrossRef]
25. Xue, X.R.; Mai, W.X.; Zhao, Z.Y.; Zhang, K.; Tian, C.Y. Optimized nitrogen nitrogen fertilizer application enhances absorption of soil nitrogen and yield of castor with drip irrigation under mulch film. *Ind. Crop. Prod.* **2017**, *95*, 156–162. [CrossRef]
26. Jia, L.M.; Liu, S.Q.; Zhu, L.H.; Hu, J.J.; Wang, X.P. Carbon storage and density of poplars in China. *J. Nanjing For. Univ.* **2013**, *37*, 1–7, (In Chinese with English Abstract).
27. He, Y.; Lan, Z.P.; Sun, S.W.; Fu, J.P.; Liu, J.Q. Study on n, p, and k uptake and fertilization of young '107' poplar clone with drip irrigation. *For. Res.* **2015**, *25*, 426–430, (In Chinese with English Abstract).
28. Xi, B.; Wang, Y.; Jia, L. Effects of nitrogen application rate and frequency on biomass accumulation and nitrogen uptake of *Populus tomentosa* under drip fertigation. *Sci. Silvae Sin.* **2017**, *53*, 63–73, (In Chinese with English Abstract).
29. Fang, C.H.; Zhou, K.Y.; Zhang, Y.W.; Li, B.Z.; Han, M.Y. Effect of root pruning and nitrogen fertilization on growth of young 'fuji' apple (*Malus domestica* borkh.) trees. *J. Plant Nutr.* **2017**, *40*, 1538–1546. [CrossRef]
30. Yan, X.; Dai, T.; Xing, C.; Jia, L.; Zhang, L. Coupling effect of water and nitrogen on the morphology and distribution of fine root in surface soil layer of young *Populus × euramericana* plantation. *Acta Ecol. Sin.* **2015**, *35*, 3692–3701, (In Chinese with English Abstract).
31. Chen, J.; Xiao, G.L.; Kuzyakov, Y.K.; Jenerette, G.D.; Ma, Y.; Liu, W.; Wang, Z.F.; Shen, W.J. Soil nitrogen transformation responses to seasonal precipitation changes are regulated by changes in functional microbial abundance in a subtropical forest. *Biogeosciences* **2017**, *14*, 2513–2525. [CrossRef]
32. Bao, S.D. *Soil Agro-Chemistrical Analysis*; China Agriculture Press: Beijing, China, 2007.
33. Shen, Y.F.; Wang, N.; Cheng, R.M.; Xiao, W.F.; Yang, S.; Guo, Y.; Lei, L.; Zeng, L.X.; Wang, X.R. Characteristics of fine roots of *Pinus massoniana* in the three gorges reservoir area, China. *Forests* **2017**, *8*, 183. [CrossRef]
34. Bo, H.; Wen, C.; Song, L.; Yue, Y.; Nie, L. Fine-root responses of *Populus tomentosa* forests to stand density. *Forests* **2018**, *9*, 562. [CrossRef]

35. Schenk, H.J.; Jackson, R.B. Mapping the global distribution of deep roots in relation to climate and soil characteristics. *Geoderma* **2005**, *126*, 129–140. [CrossRef]
36. Zhang, Y.Q.; Ke, Z.Q.; Shi, Q.I.; Yan, Z.; Mei, W.D. Root system distribution characteristics of plants on the terrace banks and their impact on soil moisture. *Acta Ecol. Sin.* **2005**, *25*, 500–506, (In Chinese with English Abstract).
37. Singh, B.; Tripathi, K.P.; Jain, R.K.; Behl, H.M. Fine root biomass and tree species effects on potential n mineralization in afforested sodic soils. *Plant Soil* **2000**, *219*, 81–89. [CrossRef]
38. Chang, R.Y.; Fu, B.J.; Liu, G.H.; Yao, X.L.; Wang, S. Effects of soil physicochemical properties and stand age on fine root biomass and vertical distribution of plantation forests in the loess plateau of China. *Ecol. Res.* **2012**, *27*, 827–836. [CrossRef]
39. Yan, H.; Su, Y.Q.; Zhu, Y.Y.; Zhang, J.Q. Distribution characters of fine root of poplar plantation and its relation to properties of soil in the northern slope of Qinling Mountain. *J. Nanjing For. Univ.* **2009**, *33*, 85–89, (In Chinese with English Abstract).
40. Verma, K.S.; Kohli, S.; Kaushal, R.; Chaturvedi, O.P. Root structure, distribution and biomass in five multipurpose tree species of Western Himalayas. *J. Mt. Sci. Engl.* **2014**, *11*, 519–525. [CrossRef]
41. Chen, L.L.; Mu, X.M.; Yuan, Z.Y.; Deng, Q.; Chen, Y.L.; Yuan, L.Y.; Ryan, L.T.; Kallenbach, R.L. Soil nutrients and water affect the age-related fine root biomass but not production in two plantation forests on the loess plateau, China. *J. Arid Environ.* **2016**, *135*, 173–180. [CrossRef]
42. Yan, X.L.; Dai, T.F.; Jia, L.M.; Dai, L.L.; Xin, F.M. Responses of the fine root morphology and vertical distribution of *Populus × euramericana* 'guariento' to the coupled effect of water and nitrogen. *Chin. J. Plant Ecol.* **2015**, *39*, 825–837, (In Chinese with English Abstract).
43. Qi, D.; Hu, T.; Niu, X. Responses of root growth and distribution of maize to nitrogen application patterns under partial root-zone irrigation. *Int. J. Plant Prod.* **2017**, *11*, 209–224.
44. Liu, X.J.A.; van Groenigen, K.J.; Dijkstra, P.; Hungate, B.A. Increased plant uptake of native soil nitrogen following fertilizer addition—Not a priming effect? *Appl. Soil Ecol.* **2017**, *114*, 105–110. [CrossRef]
45. Block, R.M.A.; Rees, K.C.J.; Knight, J.D. A review of fine root dynamics in populus plantations. *Agroforest Syst.* **2006**, *67*, 73–84. [CrossRef]
46. Price, J.S.; Hendrick, R.L. Fine root length production, mortality and standing root crop dynamics in an intensively managed sweetgum (*Liquidambar styraciflua* L.) coppice. *Plant Soil* **1998**, *205*, 193–201. [CrossRef]
47. Rytter, R.M. The effect of limited availability of n or water on c allocation to fine roots and annual fine root turnover in *Alnus incana* and *Salix viminalis*. *Tree Physiol.* **2013**, *33*, 924–939. [CrossRef]
48. Hu, J.L.; Lin, X.G.; Wang, J.H.; Dai, J.; Chen, R.R.; Zhang, J.B.; Wong, M.H. Microbial functional diversity, metabolic quotient, and invertase activity of a sandy loam soil as affected by long-term application of organic amendment and mineral fertilizer. *J. Soil Sediment* **2011**, *11*, 271–280. [CrossRef]
49. Li, J.S.; Liu, Y.C. Water and nitrate distributions as affected by layered-textural soil and buried dripline depth under subsurface drip fertigation. *Irrig. Sci.* **2011**, *29*, 469–478. [CrossRef]
50. Huang, J. Effects of bagging fertilization on plant growth and soil nutrient in cunninghamia lanceolata plantation. *J. For. Environ.* **2017**, *37*, 163–168, (In Chinese with English Abstract).
51. Fan, Z.B.; Lin, S.; Zhang, X.M.; Jiang, Z.M.; Yang, K.C.; Jian, D.D.; Chen, Y.Z.; Li, J.L.; Chen, Q.; Wang, J.G. Conventional flooding irrigation causes an overuse of nitrogen fertilizer and low nitrogen use efficiency in intensively used solar greenhouse vegetable production. *Agr. Water Manag.* **2014**, *144*, 11–19. [CrossRef]
52. Imada, S.; Taniguchi, T.; Acharya, K.; Yamanaka, N. Vertical distribution of fine roots of *Tamarix ramosissima* in an arid region of southern nevada. *J. Arid Environ.* **2013**, *92*, 46–52. [CrossRef]
53. February, E.C.; Higgins, S.I. The distribution of tree and grass roots in savannas in relation to soil nitrogen and water. *S. Afr. J. Bot.* **2010**, *76*, 517–523. [CrossRef]
54. Leuschner, C.; Hertel, D.; Schmid, I.; Koch, O.; Muhs, A.; Holscher, D. Stand fine root biomass and fine root morphology in old-growth beech forests as a function of precipitation and soil fertility. *Plant Soil* **2004**, *258*, 43–56. [CrossRef]
55. Okada, K.; Aiba, S.; Kitayama, K. Influence of temperature and soil nitrogen and phosphorus availabilities on fine-root productivity in tropical rainforests on Mount Kinabalu, Borneo. *Ecol. Res.* **2017**, *32*, 145–156. [CrossRef]
56. Hochmuth, G.; Hanlon, E. *A Summary of N, P and K Research with Tomato in Florida*; University of Florida Institute of Food and Agricultural Sciences: Gainesville, FL, USA, 2020; Available online: http://edis.ifas.ufl.edu/cv236 (accessed on 8 December 2022).
57. Yamashita, T.; Kasuya, N.; Nishimura, S.; Takeda, H. Comparison of two coniferous plantations in Central Japan with respect to forest productivity, growth phenology and soil nitrogen dynamics. *For. Ecol. Manag.* **2004**, *200*, 215–226. [CrossRef]
58. Grissom, J.E.; Wu, R.; McKeand, S.E.; O'Malley, D.M. Phenotypic plasticity of fine root growth increases plant productivity in pine seedlings. *Ann. Occup. Environ. Med.* **2004**, *4*, 14.

forests

MDPI

Article

Genome-Wide Identification and Expression Analyses of the *PP2C* Gene Family in *Paulownia fortunei*

Zhenli Zhao [1,2], Peiyuan Zhang [1], Minjie Deng [1,2], Yabing Cao [1] and Guoqiang Fan [1,2,*]

[1] College of Forestry, Henan Agricultural University, Zhengzhou 450002, China
[2] Institute of *Paulownia*, Henan Agricultural University, Zhengzhou 450002, China
* Correspondence: zlxx64@henau.edu.cn; Tel.: +86-371-6355-8605

Abstract: We explored the composition and roles of the protein phosphatase 2C (*PP2C*) family in *Paulownia fortunei*. The genome *P. fortunei* harbored 91 *PfPP2C* genes, encoding proteins with 120–1107 amino acids (molecular weight range, 13.51–124.81 kDa). The 91 *PfPP2Cs* were distributed in 12 subfamilies, with 1–15 *PfPP2Cs* per subfamily. The number and types of conserved structure domains differed among *PP2Cs*, but the distribution of conserved motifs within each subfamily was similar, with the main motif structure being motifs 3, 16, 13, 10, 2, 6, 12, 4, 14, 1, 18, and 8. The *PfPP2C* genes had 2 to 20 exons. There were ABA-response elements in the promoters of 42 *PfPP2C* genes, response elements to phytohormones, and stress in the promoters of other *PfPP2C* genes. A covariance analysis revealed that gene fragment duplication has played an important role in the evolution of the *PfPP2C* family. There were significant differences in the transcript levels of some *PfPP2C* genes in *P. fortunei* affected by witches' broom (PaWB) and after treatment with rifampicin and methyl methanesulfonate. *PfPP2C02*, *PfPP2C12*, *PfPP2C19*, and *PfPP2C80* were strongly related to PaWB. These findings provide a foundation for further studies on the roles of *PP2Cs* in PaWB.

Keywords: *Paulownia fortunei*; *PP2C* family; identification; hormone treatment; expression analysis

Citation: Zhao, Z.; Zhang, P.; Deng, M.; Cao, Y.; Fan, G. Genome-Wide Identification and Expression Analyses of the *PP2C* Gene Family in *Paulownia fortunei*. *Forests* **2023**, *14*, 207. https://doi.org/10.3390/f14020207

Academic Editor: Tadeusz Malewski

Received: 9 December 2022
Revised: 18 January 2023
Accepted: 18 January 2023
Published: 21 January 2023

1. Introduction

Protein phosphorylation and dephosphorylation are important for the functional expression of proteins and are essential regulatory mechanisms in living organisms [1]. Protein phosphatases catalyze the dephosphorylation of phosphorylated proteins, thereby playing important roles in plants' responses to abiotic stresses and hormones [2]. Protein phosphatases can be classified according to their substrates into protein tyrosine phosphatases (PTPs), serine/threonine phosphatases (STPs), and dual-substrate PTPs (DSPTPs) [3–5]. The STPs can be further divided into phosphoprotein phosphatases (PPP) and protein phosphatase metal-dependent (PPM) proteins according to their crystal structure [6]. The protein phosphatase 2C (*PP2C*) and phosphopyruvate dehydrogenase phosphatase (PDP) families are the two main PPM families [6]. *PP2C*s are Mg^{2+}- or Mn^{2+}-dependent monomeric enzymes that are widely found in Archaea, bacteria, fungi, animals, and plants [7]. Compared with other organisms, plants tend to have more *PP2C* proteins [8]. Plant *PP2C* proteins have a specific structural pattern; most have a conserved catalytic region at the C-terminus, while the N-terminus is less conserved, with a variable-length extension region that contains sequences related to intracellular signaling, such as transmembrane and kinase interaction sequences [9]. *PP2C*s are widely involved in physiological processes, such as abscisic acid (ABA) signaling and trauma signaling, plant growth and development, and plant disease resistance [10]. *Arabidopsis thaliana* has 80 *PP2C* gene family members in 12 subfamilies [11]. The *PP2C* proteins in subfamily A negatively regulate ABA signaling by binding to the ABA receptor proteins PYR/PYL/RCAR, thereby causing physiological responses, such as inhibition of germination and stomatal closure [12,13]. The *AtPP2C* proteins in subfamily B play a regulatory role in the mitogen-activated protein kinase (MAPK) pathway [14].

Members of subfamily C are involved in physiological processes such as plant flower organ development [15]. Members of subfamily D in *A. thaliana* are involved in regulating seed germination in the dark, seed growth, and the ABA signaling pathway by mediating the activity of the plasma membrane H^+-ATPase in cells [16,17]. Less is known about the other subfamilies. However, various studies have shown that *PP2Cs* are also involved in biological responses under abiotic stresses; for example, *AtPP2C31* and *AtPP2CG1* negatively regulate the response to high salt and low temperatures, respectively, in *A. thaliana* [18,19].

Previous studies have shown that members of the *PP2C* family play important regulatory roles in responses to abiotic stresses, such as low temperature, high temperature, and drought, as well as in hormonal regulation. For example, gene microarray expression profiling and real-time fluorescence quantification analyses revealed that members of the subclades C, E, and G of the *PP2C* family in *Vitis vinifera* were up-regulated under stress conditions, while members of subclades A, D, F, H, and K were down-regulated [20]. In *Broussonetia papyrifera*, four members of the 18 *BpPP2Cs* tested were found to be up-regulated under low temperature (4 °C) [21]. Furthermore, two proteins showed increased phosphorylation levels at 6 h of the low-temperature treatment, demonstrating that *PP2Cs* are involved in the cold stress response in plants [21]. In *Brachypodium distachyum*, almost all *BdPP2Cs* were up-regulated under low-temperature stress (4 °C) [22]. In *Oryza sativa*, *OsPP2C09* mediated ABA desensitization, which contributed to root elongation, under drought stress [23]. Under low-temperature, high-temperature, and drought conditions, 10, 8, and 9 *PP2C* genes respectively, were found to be continuously up-regulated in *Poncirus trifoliata* [24]. Six *Phyllostachys heterocycla PP2C* genes, including *PH02Gene33357.t1* and *PH02Gene38274.t1*, were up-regulated under high-salt conditions (200 mmol·L^{-1} NaCl) and by ABA (100 μmol·L^{-1}) [25]. In *Dendrobium catenatum* treated with 20% PEG-6000, 200 mmol·L^{-1} NaCl, 100 μmol·L^{-1} ABA, and 200 μmol·L^{-1} salicylic acid, *DcPP2C5, DcPP2C5, DcPP2C5*, and *DcPP2C5* were up-regulated under drought and salt stress, and *DcPP2C20, DcPP2C38*, and *DcPP2C56* were up-regulated in the roots under ABA and SA treatment [26]. These findings indicated that *D. catenatum PP2Cs* are not only involved in responses to abiotic stresses, but also in responses to hormones [26].

Paulownia fortunei is a deciduous tree in the genus *Paulownia* (family *Scrophulariaceae*) [27]. It is a source of timber and is planted for farmland protection in China [27]. It has fast growth, produces high-quality wood, and shows strong adaptability and resistance [27]. It contributes to alleviating timber shortages, improving the ecological environment, ensuring food security, and improving people's living standards [27]. However, there are some serious problems in its production, such as the occurrence of witches' broom disease (PaWB), which increases tree mortality, slows tree growth, and seriously affects the development of the *Paulownia* industry. Although the genome of *P. fortunei* has been sequenced [28], the members of the *PfPP2C* gene family in this species have not yet been reported. In this study, using the *PP2C* gene sequences from *Arabidopsis thaliana* as search queries, members of the *PfPP2C* gene family in *P. fortunei* were screened and identified using homologous alignment analyses. The genes and their encoded proteins were analyzed using a series of bioinformatic tools. Differences in gene expression between diseased and healthy *P. fortunei* seedlings were determined by analyses of RNA-Seq data. A preliminary investigation of the expression patterns of *PfPP2Cs* in *P. fortunei* under various stress conditions and in PaWB-affected plants reveal potential functions of the *PfPP2C* family, and provide a theoretical basis for exploring their roles in the development of PaWB.

2. Materials and Methods

2.1. Identification, Physicochemical Properties, and Prediction of Subcellular Localization of PP2C Family Members in P. fortunei

The sequences of *Arabidopsis thaliana PP2C* proteins were obtained from the TAIR database (https://www.arabidopsis.org/) (accessed on 24 May 2022). The *P. fortunei* genome database was searched for homologous protein sequences with high structural similarity to the *Arabidopsis thaliana PP2C* family using BlastP. The hidden Markov model

(PF00481) file of the *PP2C* protein structural domain was downloaded from the Pfam database (http://pfam.xfam.org/) (accessed on 25 May 2022), and then used in Biolinux to search the genome of *P. fortunei* using hmmersearch. Candidate protein sequences were those with an e-value of $\leq 10^{-2}$. The candidate protein sequences of the *PP2C* family in *P. fortunei* were those that were detected in both the BlastP and hmmer analyses. The candidate protein sequences were verified by Pfam, and protein structural domains were identified using SMART (http://smart.embl.de/smart/batch.pl) (accessed on 26 May 2022) and CDD (https://www.ncbi.nlm.nih.gov/cdd) (accessed on 26 May 2022). The protein sequences without *PP2C* structural domains were removed to obtain the final set of *PP2C* family members in *P. fortunei*. The number of amino acids, isoelectric point, and molecular weight of putative *PP2C* proteins of *P. fortunei* were predicted using Expasy (https://web.expasy.org/protparam/) (accessed on 22 June 2022). Subcellular localization of the *PP2C* gene family members by using the Cell-PLoc 2.0 online tool (http://www.csbio.sjtu.edu.cn/bioinf/Cell-PLoc-2/) (accessed on 14 January 2023).

2.2. Chromosomal Localization and Phylogenetic Analysis of PP2C Genes in P. fortunei

The genome annotation files were downloaded from the *P. fortunei* genome database. Information about chromosome length and the location of *PP2C* genes on chromosomes was extracted from the *P. fortunei* genome annotation files using TBtools software. The Map MG2C online tool (http://mg2c.iask.in/mg2c_v2.0/) (accessed on 10 July 2022) was used to map the distribution of *Paulownia PP2C* genes on chromosomes. The *PP2C* protein sequences were downloaded from the *A. thaliana* genome website and the *Poncirus trifoliata* protein sequence file was downloaded from the citrus genome database (http://citrus.hzau.edu.cn/) (accessed on 19 July 2022). The amino acid multiple sequence alignment analysis of *PP2C* proteins from *A. thaliana*, *P. trifoliata*, and *P. fortunei* was performed using MEGA-X software, and the phylogenetic tree was constructed using NJ in MEGA-X software, with the bootstrap value set to 1000 and other parameters set to default values.

2.3. Conserved Structural Domains, Conserved Motifs, and Gene Structure Analysis of PP2C Family Members in P. fortunei

The online tool Pfam search (http://pfam.xfam.org/search#tabview=tab1) (accessed on 17 August 2022) was used to identify conserved structural domains in *P. fortunei PP2C* proteins, and the results were visualized using TBtools software. The conserved motifs of *PP2C* protein sequences were analyzed using the online tool MEME (https://meme-suite.org/meme/tools/meme) (accessed on 7 August 2022), with the number of motifs set to 20, and the results were visualized using TBtools software. The structure of each member of the *PP2C* gene family, based on its coding sequence, was analyzed online by GSDS (http://gsds.gaolab.org/) (accessed on 12 July 2022).

2.4. Analysis of Promoter Cis-Acting Elements and Covariance in Members of the PP2C Family in P. fortunei

TBtools software was used to extract the 2000-bp upstream sequence of the start codon of each *PP2C* gene as the promoter sequence. The *cis*-acting elements in the promoters of *PP2C* genes were detected using PlantCARE online software (http://bioinformatics.psb.ugent.be/webtools/plantcare/html/) (accessed on 27 August 2022), and the results were visualized using TBtools software [29]. Fasta Satas, File Merge For MCScanX, Text Block Extract, and Filter tools in TBtools were used to obtain files with information about chromosome length and associations between gene family members of *P. fortunei*. Table Row Extract or the Filter tool was used to obtain gene ID display files. The Advanced Circos tool was used to conduct the *P. fortunei PP2C* gene family covariance analysis.

2.5. Expression Analysis of PP2C Genes in P. fortunei

The materials used to generate RNA-Seq data were healthy *P. fortunei* (PF) and infected *P. fortunei* (PFI) grown in the intelligent greenhouse of the *Paulownia* Institute of Henan Agricultural University. At the age of 3 months, seedlings with the same growth status were selected and treated with rifampicin (Rif) at 30 mg/L or with methyl methanesulfonate (MMS) at 20 mg/L. Leaves of five seedlings were randomly taken after 5, 10, 15, and 30 d of treatment, snap-frozen in liquid nitrogen, and stored at −80 °C. This experiment was conducted with three biological replicates, using healthy seedlings in normal culture and seedlings with PaWB as controls. Affymetrix GeneChip 16K gene IDs with identical sequences were retrieved using the *PP2C* nucleic acid sequence of *P. fortunei* as a probe, and then the RNASeq data for the *PP2C* genes of P. fortunei (PF and PFI) treated or not with Rif or MMS were extracted, log2-transformed using Excel, and used to generate a heat-map using TBtools software.

3. Results

3.1. Identification and Physicochemical Properties of the PP2C Family in P. fortunei

A total of 91 *PP2C* genes were identified in the *P. fortunei* genome. The protein sequences were extracted using TBtools software. Redundant sequences were manually deleted, and the remaining sequences were further validated using tools at the Pfam, SMART, and CDD databases. The 91 *PP2C* genes of *P. fortunei* were finally identified and named *PfPP2C1* through *PfPP2C91* (Table 1). The lengths of the putative proteins encoded by the *PfPP2C* genes ranged from 120 aa (*PfPP2C61*) to 1107 aa, with an average length of 432.39 aa; the predicted molecular weight ranged from 13.51 kDa to 124.81 kDa (average, 47.62 kDa); and the theoretical isoelectric point ranged from 4.60 to 9.51. In total, 20 of the *PfPP2C* proteins were predicted to be basic (including *PfPP2C01* and *PfPP2C90*) and the remaining 71 were predicted to be acidic (including *PfPP2C03* and *PfPP2C47*). Among the 91 *PfPP2C* proteins, 32.97% were predicted to be stable proteins, but the majority (67.03%) were predicted to be unstable. The theoretical instability index ranged from 42.32 to 94.39. The GRAVY values were all less than 0, ranging from −0.586 to −0.101, indicating that all 91 *PP2C* proteins were hydrophilic.

Table 1. Physicochemical properties of *PP2C* family members in *P. Fortunei*.

Gene Name	Gene ID	Number of Amino Acids	Molecular Weight	Theoretical PI	Instability Index	Aliphatic Index	GRAVY	Predicted Location(s)
PfPP2C01	Pfo01g001610.1	294	32,296.66	8.24	40.85	83.54	−0.408	Nucleus
PfPP2C02	Pfo01g002750.1	473	52,519.65	5.22	48.98	91.46	−0.268	Nucleus
PfPP2C03	Pfo01g006380.1	392	43,064.51	4.81	40.50	81.84	−0.283	Nucleus
PfPP2C04	Pfo01g009870.1	655	72,156.19	5.58	31.28	90.37	−0.137	Nucleus
PfPP2C05	Pfo02g010590.1	369	40,963.87	6.89	32.03	86.40	−0.213	Chloroplast/Nucleus
PfPP2C06	Pfo02g010660.1	426	46,080.74	5.87	37.47	93.31	−0.133	Chloroplast
PfPP2C07	Pfo02g014240.1	279	30,485.43	7.12	42.24	83.23	−0.368	Nucleus
PfPP2C08	Pfo02g016010.1	386	43,014.95	8.48	48.79	87.10	−0.317	Nucleus
PfPP2C09	Pfo02g019750.1	397	44,228.38	8.66	45.22	87.63	−0.258	Nucleus
PfPP2C10	Pfo03g000530.1	379	42,820.76	6.44	49.31	89.74	−0.339	Nucleus
PfPP2C11	Pfo03g006870.1	1081	119,853.48	5.03	39.87	89.16	−0.207	Cell membrane/Nucleus
PfPP2C12	Pfo03g008490.1	553	60,968.45	5.33	53.04	93.24	−0.203	Nucleus
PfPP2C13	Pfo03g009450.1	294	32,627.04	7.67	47.43	79.25	−0.473	Nucleus
PfPP2C14	Pfo03g013130.1	377	42,129.12	9.51	46.15	90.69	−0.359	Nucleus
PfPP2C15	Pfo03g013680.1	631	70,004.71	5.50	38.11	79.41	−0.380	Chloroplast/Nucleus
PfPP2C16	Pfo03g015100.1	433	48,321.81	5.24	37.09	82.66	−0.379	Chloroplast/Mitochondrion
PfPP2C17	Pfo04g000480.1	397	44,173.32	8.72	44.84	87.88	−0.247	Nucleus
PfPP2C18	Pfo04g003840.1	280	30,717.54	6.76	35.40	80.79	−0.444	Nucleus
PfPP2C19	Pfo04g006590.1	429	46,293.98	7.49	39.12	87.48	−0.190	Chloroplast
PfPP2C20	Pfo04g006660.1	372	41,415.26	6.42	33.56	85.43	−0.252	Nucleus
PfPP2C21	Pfo05g000400.1	349	38,472.54	4.73	54.69	91.35	−0.101	Nucleus
PfPP2C22	Pfo05g003700.1	397	43,555.96	5.25	44.41	83.73	−0.188	Nucleus
PfPP2C23	Pfo05g003720.1	1107	124,807.73	5.56	46.73	82.18	−0.366	Nucleus
PfPP2C24	Pfo05g010690.1	405	44,340.19	5.25	64.32	78.49	−0.347	Nucleus
PfPP2C25	Pfo05g011250.1	669	74,601.59	5.15	42.67	78.73	−0.456	Chloroplast/Nucleus
PfPP2C26	Pfo06g004460.1	270	30,030.46	6.76	51.43	89.59	−0.260	Nucleus
PfPP2C27	Pfo06g004710.1	196	22,216.76	8.92	53.29	94.39	−0.143	Nucleus
PfPP2C28	Pfo07g001190.1	449	48,913.54	7.16	45.84	76.88	−0.407	Nucleus
PfPP2C29	Pfo07g005140.1	422	45,245.83	8.34	28.48	88.06	−0.159	Nucleus
PfPP2C30	Pfo07g009030.1	801	88,560.94	5.27	46.37	74.59	−0.481	Chloroplast
PfPP2C31	Pfo07g014460.1	293	31,684.04	4.93	36.38	79.52	−0.355	Nucleus
PfPP2C32	Pfo07g014670.1	353	39,022.99	5.20	34.35	75.47	−0.392	Nucleus
PfPP2C33	Pfo07g015080.1	282	30,638.32	5.50	49.48	75.46	−0.321	Nucleus
PfPP2C34	Pfo08g002420.1	348	38,000.29	5.69	49.59	88.76	−0.228	Nucleus
PfPP2C35	Pfo08g010120.1	343	37,581.19	5.17	39.89	70.52	−0.551	Nucleus

Table 1. *Cont.*

Gene Name	Gene ID	Number of Amino Acids	Molecular Weight	Theoretical PI	Instability Index	Aliphatic Index	GRAVY	Predicted Location(s)
PfPP2C36	Pfo08g013880.1	555	60,319.53	4.97	40.94	91.98	−0.166	Nucleus
PfPP2C37	Pfo09g009830.1	526	57,753.16	5.26	41.75	79.94	−0.358	Nucleus
PfPP2C38	Pfo09g014660.1	284	31,332.79	6.84	40.26	90.63	−0.348	Nucleus
PfPP2C39	Pfo09g017020.1	283	31,287.57	6.14	32.44	88.55	−0.332	Chloroplast/Cytoplasm
PfPP2C40	Pfo10g007720.1	358	38,635.14	6.27	54.49	84.41	−0.159	Chloroplast/Nucleus
PfPP2C41	Pfo10g009060.1	425	46,090.32	6.83	59.09	72.68	−0.363	Nucleus
PfPP2C42	Pfo10g009070.1	425	46,090.32	6.83	59.09	72.68	−0.363	Nucleus
PfPP2C43	Pfo10g012720.1	394	44,063.26	6.55	43.71	91.75	−0.293	Nucleus
PfPP2C44	Pfo10g013980.1	491	54,041.38	5.23	41.96	72.06	−0.542	Chloroplast/Nucleus
PfPP2C45	Pfo11g000240.1	379	42,063.30	5.08	63.53	83.43	−0.275	Nucleus
PfPP2C46	Pfo11g001200.1	505	54,945.55	4.75	45.45	88.22	−0.202	Nucleus
PfPP2C47	Pfo11g001760.1	349	38,194.48	4.60	34.90	80.37	−0.296	Nucleus
PfPP2C48	Pfo11g002190.1	601	65,992.20	6.34	31.58	91.70	0.155	Chloroplast
PfPP2C49	Pfo11g008040.1	388	42,005.64	5.32	56.68	83.69	−0.206	Chloroplast/Cytoplasm
PfPP2C50	Pfo11g011060.1	438	47,647.39	5.22	37.03	83.70	−0.217	Nucleus
PfPP2C51	Pfo11g013940.1	313	34,056.08	4.93	46.41	81.02	−0.337	Nucleus
PfPP2C52	Pfo11g015860.1	403	44,718.89	6.54	44.29	90.77	−0.345	Nucleus
PfPP2C53	Pfo12g008400.1	731	81,194.22	5.62	39.63	77.36	0.479	Chloroplast/Nucleus
PfPP2C54	Pfo14g000920.1	557	60,286.22	4.86	39.82	89.57	−0.218	Nucleus
PfPP2C55	Pfo14g006170.1	265	28,851.94	8.59	34.98	93.89	−0.114	Nucleus
PfPP2C56	Pfo14g009710.1	348	37,801.87	5.31	51.37	89.60	−0.255	Chloroplast/Nucleus
PfPP2C57	Pfo15g011470.1	461	50,501.81	5.87	41.40	82.47	−0.206	Chloroplast/Nucleus
PfPP2C58	Pfo15g012450.1	390	42,239.89	5.11	53.71	82.54	−0.201	Chloroplast
PfPP2C59	Pfo16g002080.1	471	52,221.33	5.27	52.56	69.92	−0.493	Nucleus
PfPP2C60	Pfo16g008470.1	526	57,751.38	5.05	38.31	80.68	−0.319	Chloroplast/Nucleus
PfPP2C61	Pfo16g013780.1	120	13,508.30	4.68	36.38	89.33	−0.458	Chloroplast/Cytoplasm
PfPP2C62	Pfo17g001200.1	489	53,137.05	8.85	34.27	87.36	−0.150	Chloroplast
PfPP2C63	Pfo17g006640.1	439	47,981.43	6.65	44.24	76.79	−0.424	Chloroplast
PfPP2C64	Pfo18g001250.1	548	59,083.53	4.61	40.50	88.03	−0.126	Chloroplast
PfPP2C65	Pfo18g001790.1	346	37,658.61	6.55	40.99	85.69	−0.249	Nucleus
PfPP2C66	Pfo18g003260.1	387	42,305.59	5.10	61.40	82.17	−0.318	Nucleus
PfPP2C67	Pfo18g003900.1	360	39,673.58	5.00	35.39	73.44	−0.414	Nucleus
PfPP2C68	Pfo18g005750.1	461	49,990.82	8.62	57.48	74.49	−0.279	Chloroplast
PfPP2C69	Pfo18g006540.1	373	40,537.83	6.49	59.32	72.92	−0.463	Nucleus
PfPP2C70	Pfo19g000370.1	397	43,812.26	4.92	61.73	84.08	−0.222	Nucleus
PfPP2C71	Pfo19g001430.1	375	41,296.74	7.55	35.74	80.13	−0.324	Nucleus
PfPP2C72	Pfo19g002000.1	506	55,328.18	4.94	49.31	87.11	−0.198	Nucleus
PfPP2C73	Pfo19g002890.1	378	42,036.04	4.83	35.01	79.89	−0.337	Nucleus
PfPP2C74	Pfo19g003480.1	600	66,319.14	6.20	47.53	88.42	−0.215	Chloroplast
PfPP2C75	Pfo19g005950.1	233	26,049.77	7.05	52.31	72.79	−0.389	Nucleus
PfPP2C76	Pfo20g000210.1	538	59,481.42	6.99	45.87	78.79	−0.353	Chloroplast
PfPP2C77	Pfo20g001120.1	394	43,765.70	6.12	39.81	90.10	−0.275	Nucleus
PfPP2C78	Pfo20g004520.1	427	46,141.06	5.53	63.43	71.66	−0.367	Nucleus
PfPP2C79	Pfo20g005770.1	363	39,132.58	7.02	44.44	80.55	−0.199	Cell membrane/Nucleus
PfPP2C80	Pfo20g008850.1	387	42,878.69	5.11	45.02	67.05	−0.586	Nucleus
PfPP2C81	Pfo20g009100.1	389	42,573.53	7.55	44.69	92.06	−0.119	Nucleus
PfPP2C82	Pfoxxg008780.1	385	42,539.37	9.11	43.32	89.82	−0.331	Nucleus
PfPP2C83	Pfoxxg011050.1	432	48,285.76	5.78	38.35	77.64	−0.458	Nucleus
PfPP2C84	Pfoxxg015210.1	380	42,806.71	5.93	49.34	92.63	−0.236	Nucleus
PfPP2C85	Pfoxxg021750.1	631	69,947.47	5.56	40.22	79.10	−0.394	Nucleus
PfPP2C86	Pfoxxg021830.1	433	48,153.28	5.54	40.15	73.93	−0.310	Nucleus
PfPP2C87	Pfoxxg022160.1	293	31,764.12	4.87	36.59	78.19	−0.369	Nucleus
PfPP2C88	Pfoxxg025250.1	432	48,177.56	5.58	34.23	77.87	−0.419	Nucleus
PfPP2C89	Pfoxxg026240.1	632	70,046.61	5.63	39.96	79.13	−0.392	Nucleus
PfPP2C90	Pfoxxg026500.1	385	42,563.43	9.11	42.32	42.32	−0.318	Nucleus
PfPP2C91	Pfoxxg028980.1	380	42,792.68	5.93	49.56	92.37	−0.237	Nucleus

3.2. *Prediction of the Subcellular Localization of PP2C Family Members of P. fortunei*

Proteins are distributed throughout the cell to participate in physiological activities, and subcellular localization prediction can clearly show the respective protein prediction sites of *PfPP2C* gene family members (Table 1), thus inferring the functions of related genes. The predicted subcellular cellular localization results showed that the predicted sites of *P. fortunei PP2C* protein were in the nucleus, chloroplast, cytoplasm, mitochondria, and cell membrane (Figure 1). The protein prediction sites had 83.5% (76) of their members in the nucleus, followed by the chloroplasts (24) and the least number of members in the mitochondria (1). Members of the protein prediction site in the nucleus are distributed except for the K subfamily, and the distribution of members in the C, D, I, J, and L subfamilies reaches 100%. Members of protein prediction sites in chloroplasts are distributed in all but the D, I, J, and L subfamilies, with the K subfamily having a 100% distribution. Based on the results of the study, it can be inferred that most of the action sites of the *PP2C* gene family members of *P. fortunei* are in the nucleus and chloroplasts.

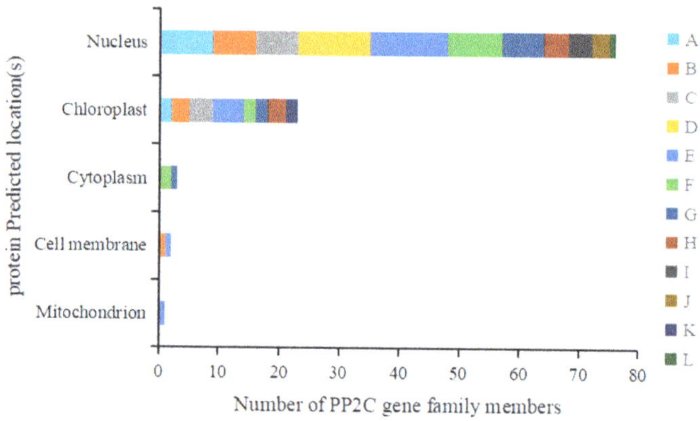

Figure 1. Prediction of subcellular localization.

3.3. *Chromosomal Localization of PP2C Genes in P. fortunei*

The location of *PP2C* genes was mapped onto the chromosomes of *P. fortunei* (Figure 2). The *PP2C* genes were unevenly distributed on 19 chromosomes. Chromosome 11 had the highest number of *PP2C* genes (8), followed by chromosome Chr03 (7), and then Chr7, Chr18, Ch19, and Chr20 (6 on each). The lowest number of *PP2C* genes (1 gene) was on Chr12. The distribution of *PP2C* genes within the same chromosome was also uneven, with two or more genes forming gene clusters. The results of sequence and chromosomal localization analyses revealed that *PfPP2C41* and *PfPP2C42* encoded the same amino acid sequence but were located at different chromosomal positions. In general, there was no positive correlation between the length of a chromosome and the number of *PP2C* genes it contained. Most genes on the same chromosome did not belong to the same subclade in the evolutionary tree. These results suggested that different genes on the same chromosome may encode proteins with different functions.

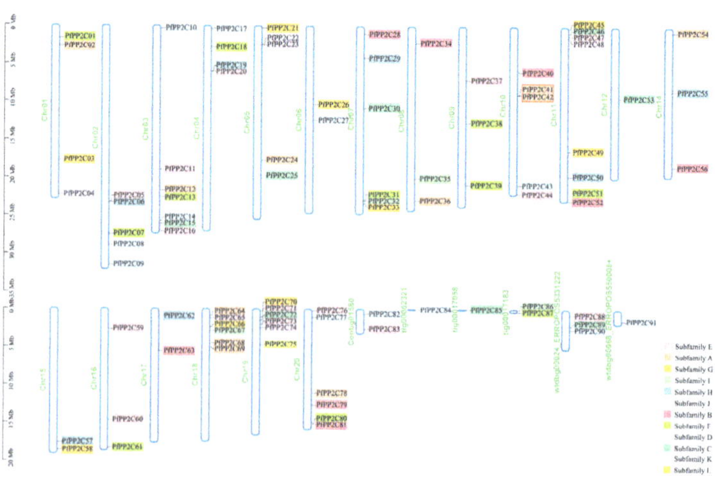

Figure 2. Chromosome mapping of *PP2C* family members in *P. fortunei*.

3.4. Phylogenetic Analysis of Members of the PP2C Family in P. fortunei

Sequence analyses showed that the *PP2C* family genes in *P. fortunei* were poorly conserved at the N-terminal end, but strongly conserved at the C-terminal end, which contained common structural subdomains that are presumed to be functionally similar. To clarify the evolutionary relationships among *PP2C* family members in *P. fortunei*, 80 protein sequences of the *A. thaliana PP2C* family, 53 protein sequences of the *P. trifoliata PP2C* family, and 91 protein sequences of the *P. fortunei PP2C* family were used to construct a phylogenetic evolutionary tree using the neighbor-joining method with MEGA-X software. In the tree (Figure 3), the *PP2C* sequences from the three species were divided into 12 subfamilies, with those of *Paulownia* distributed among the 12 subfamilies. Subfamily E had the highest number of *Paulownia PP2C* genes (15 genes), followed by subfamily D (12 genes), while subfamily L had the lowest number of *Paulownia PP2C* genes (1 gene). The number of *Paulownia PP2C* genes in the other families ranged from 2 to 11. In general, all subfamilies A–L contained *PP2C* genes from *A. thaliana*, *P. trifoliata*, and *P. fortunei*. All three species showed similar distribution ratios of *PP2C* genes in the subfamilies, indicative of relatively consistent evolutionary relationships among *A. thaliana*, *P. trifoliata*, and *P. fortunei*.

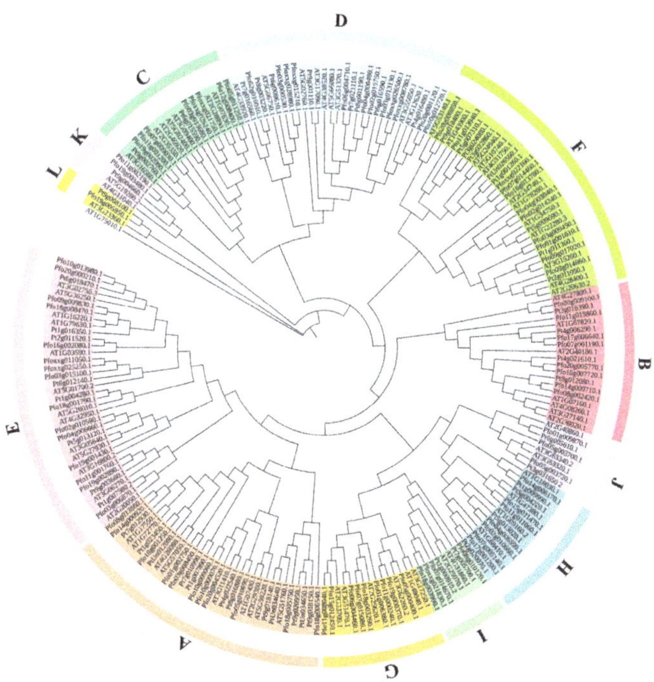

Figure 3. Phylogenetic tree of the *PP2C* gene family in *A. thaliana* (*At*), *P. trifoliata* (*Pt*), and *P. fortunei* (*Pf*).

3.5. Analysis of Conserved Structural Domains and Conserved Motifs of PP2C Family Members of P. fortunei

Analysis of the conserved structural domains revealed that all 91 *PP2C* protein sequences of *P. fortunei* contained conserved *PP2C* structural domains (Figure 4b). Among them, eight Paulownia *PP2C* proteins (*PfPP2C16*, *PfPP2C45*, *PfPP2C52*, *PfPP2C69*, *PfPP2C74*, *PfPP2C81*, *PfPP2C83*, and *PfPP2C88*) contained *PP2C*-2 structural domains; *PfPP2C11* contained the most diverse conserved structural domains (*PP2C*, Pkinase, cNMP_binding, and PK_Tyr_Ser-Thr domains); and *PfPP2C04* contained three conserved structural domains (*PP2C*, PK_Tyr_Ser-Thr, and Pkinase_fungal). The conserved structural domains of Yop-YscD_cpl and FHA were only present in *PfPP2C48* and *PfPP2C74*, respectively.

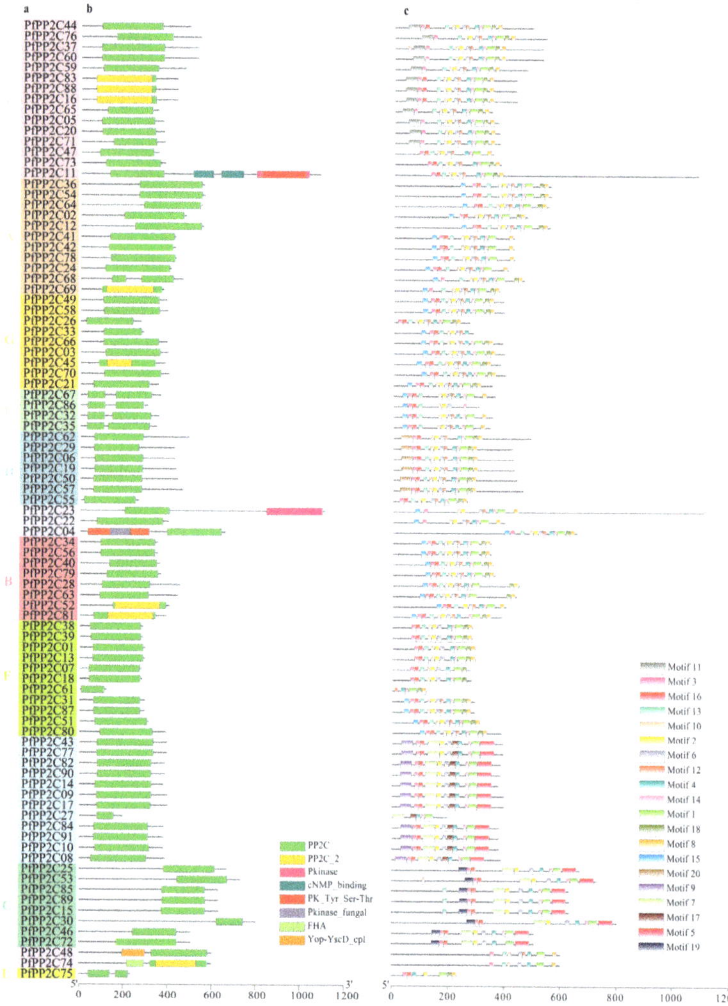

Figure 4. Subfamily grouping (**a**), conserved structural domain analysis (**b**) and conserved motif analysis (**c**) of *PP2C* gene family of *P. fortunei*.

We detected 20 conserved motifs in members of the *PP2C* family in *P. fortune* (Table 2), but the distribution of motifs differed significantly among subfamilies (Figure 4c). Among the 20 conserved motifs, motif 5 was only present in subfamilies C and D; motif 11 was only present in subfamily E; motif 20 was only present in subfamily H; motifs 9 and 17 were only present in subfamily D; and motif 19 was only present in subfamily C. This situation may be indicative of different functions of proteins in the different subfamilies. While each subfamily had unique motifs, subfamilies E, A, G, I, H, J, B, and F all contained the following motif structure: motifs 3, 16, 13, 10, 2, 6, 12, 4, 14, 1, 18, and 8; and the common motif structure in members of subfamilies C and D was motifs 3, 16, 7, 2, 6, 4, 14, 1, and 5. These findings indicated that many *PfPP2C* proteins share a high degree of similarity in the composition of their conserved motifs. *PfPP2C27* and *PfPP2C61* had a number of motifs missing compared with other members of the same subfamily, which may have resulted from sequence losses during tandem duplication of genes.

Table 2. Conserved motif information of the *PP2C* gene family in *P. fortunei*.

Motif	Length	Amino Acid Sequence Information
motif 1	29	LTPEDEFLILASDGLWDVLSNZEAVDJVR
motif 2	16	DLYVANVGDSRAVLCR
motif 3	15	TFFGVFDGHGGPGAA
motif 4	15	GGLAVSRAIGDRYLK
motif 5	50	NPRGGPARRLVKAALFRAAKKREMRYSELKKIDQGVRRHYHDDITVIVIF
motif 6	15	AIQLTVDHKPNREDE
motif 7	41	DVJKKAFSATEEEFLSLVDRQWMIKPZJASVGSCCLVGVIC
motif 8	15	RGSKDBITVIVVDFK
motif 9	41	GRVDGLLWYKDLGHHVNGEFSMAVVQANNLLEDQSQLESGP
motif 10	11	SGTTAVTALVI
motif 11	41	TPGRVFLNGSSKYASLFTQQGKKGVNQDAMIVWENFGGQED
motif 12	11	RERIEAAGGRV
motif 13	15	KKAJKKAFLKTDKEL
motif 14	11	PYLIAEPEVTV
motif 15	18	GRRREMEDAVAAIPDLCG
motif 16	15	FVKDNLFENVLKELK
motif 17	21	RSLHPDDSQIVVLKHKVWRVK
motif 18	15	PDPEAAAKRLVEEAL
motif 19	29	SLGSQNLQWAQGKAGEDRVHVVVSEEHGW
motif 20	41	NEKIEKPTVK YGQAAQSKKGEDYFLIKTDCQRVPGBPSTSF

3.6. Structure of PP2C Genes in P. fortunei

To understand the structure of *PfPP2C* genes, their intron and exon composition was determined (Figure 5b). All the *PfPP2C* genes contained introns and exons in their sequences, with the number of exons ranging from 2 (in *PfPP2C47*, *PfPP2C73*, *PfPP2C26*, *PfPP2C45*, and *PfPP2C70*) to 20 (in *PfPP2C23*) and the number of introns ranging from 1 to 19. There were 15 exons 14 introns in *PfPP2C11*. In total, 36 *PfPP2C* genes (39.5%) contained four exons and 23 contained five exons. In total, 36 *PfPP2C* genes (39.5%) contained three introns and 22 contained four introns. These results indicated that the gene structure of *PfPP2Cs* is relatively well conserved. Apart from genes in subfamilies E and J, those in the other subfamilies contained similar numbers of exons and introns, with a difference of no more than three. For example, all twelve members of subfamily D had four exons and three introns except for *PfPP2C27*, which had five exons and four introns, and all four members of subfamily I had ten exons and nine introns. In addition, the exon distribution and sequence lengths of *PfPP2C* genes belonging to the same subfamily in the phylogenetic tree were not very different and somewhat conserved, suggesting that the genes within these subfamilies have similar functions.

3.7. Analysis of Cis-Acting Elements in Promoters of PP2C Genes in P. fortunei

We identified 20 *cis*-acting elements in the promoter regions of *PP2C* genes in *P. fortunei* (Figure 5c), including hormone-responsive elements, light-responsive elements, and stress-responsive elements. Among all the *PfPP2C* genes, 62.6% (57), 48.4% (44), 47.3% (43), 35.2% (32), and 33.0% (30) had methyl jasmonate (MeJA)-responsive, gibberellin (GA)-responsive, abscisic acid (ABA)-responsive, indole acetic acid (IAA)-responsive, and salicylic acid (SA)-responsive elements, respectively, in their promoter regions; moreover, 96.7% (88), 70.3% (64), 40.7% (37), and 56.0% (51) had response elements related to light regulation, anaerobic induction, meristem expression and abiotic stress/defense, respectively, in their promoter regions. A small number of *PfPP2C* gene promoters also contained specific response elements related to circadian rhythm regulation, cell cycle regulation, endosperm expression, wound response, zein metabolism, tissue growth, and development related to palisade mesophyll cell differentiation. These results suggested that members of the *PP2C* family of *P. fortunei* play important regulatory roles in responses to hormone induction, light regulation, and stress under adverse conditions.

Figure 5. Subfamily grouping (a), analysis of the gene structure (b) and *cis*-element (c) of the *PP2C* gene family in *P. fortunei*.

3.8. Covariance Analysis of Members of the PP2C Family of P. fortunei

To explore the evolution of the *PP2C* family in *P. fortunei*, an intraspecific covariance analysis was conducted (Figure 6a). The results showed that 67 of the 91 *PP2Cs* in *P. fortunei* (74% of all *PP2Cs* in *P. fortunei*) were involved in 56 pairs of gene covariation events, suggesting that gene fragment duplication has played an important role in the evolution of the *PP2C* family in *P. fortunei*. The largest number of *PFPP2C* family members involved in covariance events was on chromosome Chr11 (six genes), followed by Chr3 and Chr19 (five genes each). To further elucidate the evolutionary relationships of *PP2C* genes between different species, a covariance analysis was performed on *P. fortunei* and *A. thaliana* (Figure 6b). We detected 98 pairs of gene covariation events between the two species, of which 54 *AtPP2C* genes (67.5% of all *AtPP2Cs*) and 67 *PfPP2C* genes (73.6% of all *PfPP2Cs*) were involved in gene covariation events. These results indicate a high degree of homology and similar evolutionary relationships between the *PP2C* family members of *P. fortunei* and *A. thaliana*, which have been highly conserved during evolution.

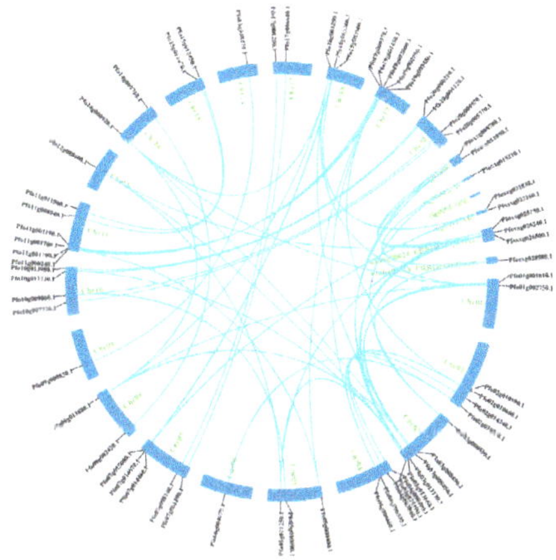

(a)

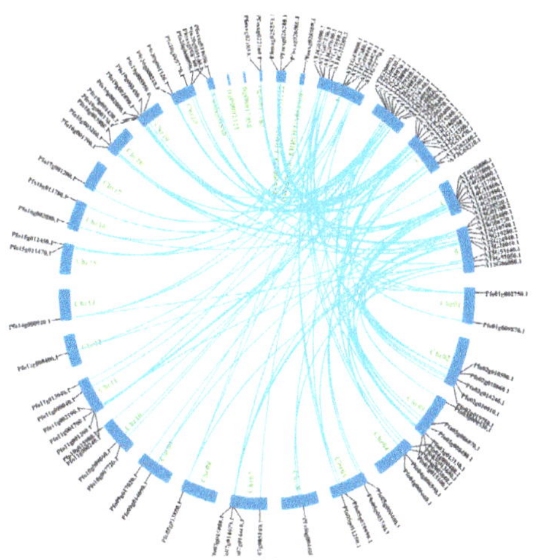

(b)

Figure 6. Collinearity analysis of *PP2C* gene families in different species: (**a**) Collinearity analysis of the *PP2C* gene family in *P. fortunei*; (**b**) Collinearity analysis of *PP2C* gene families in *P. fortune* and *A. thaliana*.

3.9. Expression Analysis of PP2C Family Members in P. fortunei

Analyses of RNAseq data from *P. fortune* infected (PFI) or uninfected (PF) with the mycoplasma that causes PaWB revealed that 80 of the 91 *PfPP2C* genes were expressed, and 11 were not (Figure 7a). The transcript levels of *PP2C19*, *PfPP2C57*, *PfPP2C59*, *PfPP2C80*, and *PfPP2C81* significantly increased after the development of PaWB. A total of 10 *PfPP2C* genes showed decreased transcript levels after the development of PaWB in *P. fortunei*, among which *PfPP2C02*, *PfPP2C12*, and *PfPP2C24* showed the most obvious decreases. The transcript levels of the remaining 66 *PfPP2C* genes, including *PfPP2C38* and *PfPP2C55*, did not change significantly. After Rif treatment, 80 *PfPP2C* genes were expressed, while 11 *PfPP2C* genes were not (Figure 7b). As the Rif treatment time extended, the transcript levels of 5 *PfPP2C* genes, including *PfPP2C02*, *PfPP2C12*, and *PfPP2C20*, significantly continued increase, while the transcript levels of 19 *PfPP2C* genes, including *PfPP2C19*, *PfPP2C52*, *PfPP2C69*, and *PfPP2C80*, significantly continued to decrease. The transcript levels of 23 *PfPP2C* genes, including *PfPP2C57* and

PfPP2C68, showed no significant monotonous trend, and the transcript levels of the remaining 19 and 15 genes increased and then decreased and decreased and then increased, respectively. After MMS treatment, 78 *PfPP2C* genes had detectable transcript levels and 13 *PfPP2C* genes had no detectable expression (Figure 7c). Among them, eight *PfPP2C* genes, including *PfPP2C02*, *PfPP2C12*, and *PfPP2C47*, were up-regulated over time under MMS treatment. The transcript levels of *PfPP2C19*, *PfPP2C52*, *PfPP2C69*, *PfPP2C80*, and 14 other genes showed an overall decreasing trend, compared with their respective levels in the control. Another 10 genes showed an increasing and then decreasing transcript levels, and 25 genes showed a decreasing and then increasing transcript levels, while 21 genes did not show significant changes in transcript levels under MMS treatment.

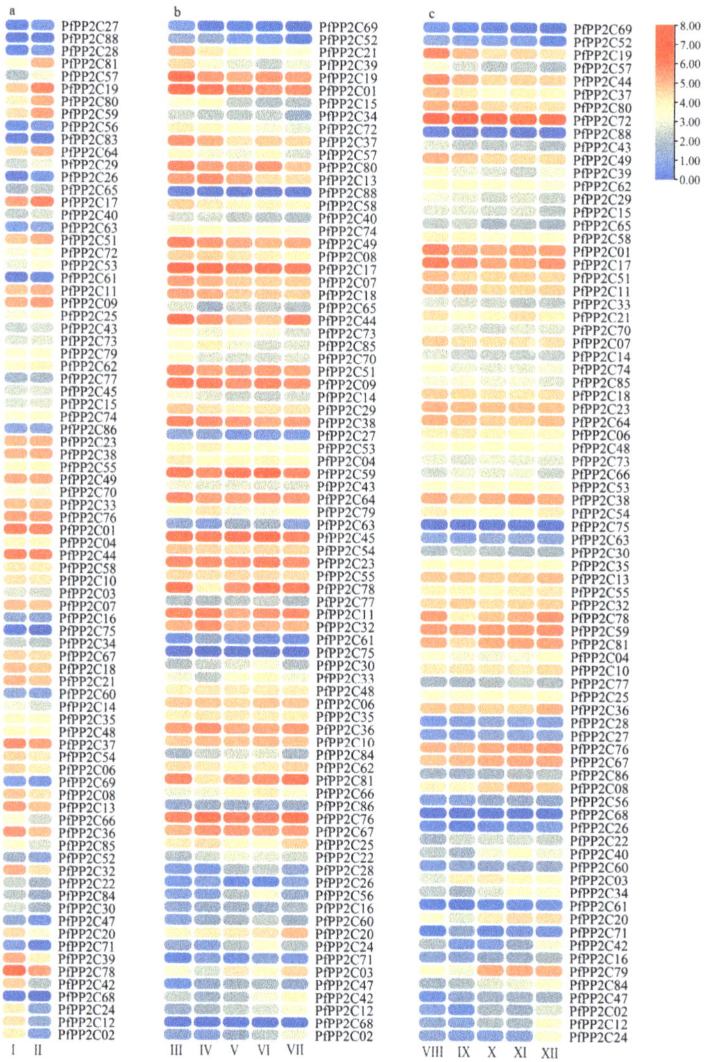

Figure 7. Expression analysis of the *PP2C* gene of *P. fortunei* during the development of witches' broom: (**a**) I is the *P. fortunei* seedling, II is the PaWB seedling; (**b**) III is the PaWB seedling, IV, V, VI, and VII are the PaWB seedlings treated with Rif (30 mg/L) for 5, 10, 15, and 30 days, respectively; (**c**) VIII is the PaWB seedling, IX, X, XI, and XII are the PaWB seedlings treated with MMS (20 mg/L) for 5, 10, 15, and 30 days, respectively.

Summarizing the above results, *PfPP2C19* and *PfPP2C80* were up-regulated in *P. fortunei* affected by PaWB, but down-regulated by Rif and MMS treatments. *PfPP2C02* and *PfPP2C12* were down-regulated in in *P. fortunei* affected by PaWB and up-regulated by Rif and MMS treatments. These findings indicated that these genes play an important regulatory role in the development of PaWB. Further in-depth analyses of their roles will provide further insights into the molecular mechanism of PaWB.

4. Discussion

In plants, *PP2Cs* are an important class of protein phosphatases that regulate plant metabolism by catalyzing the dephosphorylation of phosphorylated proteins [30]. The number of *PP2C* family members varies widely among different plant species. For example, there are 80 *PP2C* genes in *A. thaliana* [11], 27 in *V. vinifera* [20], 86 in *B. distachyum* [22], 53 in *P. trifoliata* [24], 125 in *P. heterocycla* [25], 67 in *Dendrobium catenatum* [26], 90 in *O. sativa* [31], 122 in *Vigna radiata* [32], and 81 in *Fagopyrum tataricum* [33]. This suggests that the number of *PP2C* gene family members may be related to the genome size of the species, or may have changed during the course of evolution. In this study, we identified 91 *PP2C* family members in *P. fortunei*. The amino acid length, isoelectric point, and relative molecular weight varied widely among the putative *PfPP2C* proteins, and such variations may be related to their functional diversity. The chromosomal localization analyses revealed that the *PP2C* genes in *P. fortunei* are unevenly distributed on 19 chromosomes, with one to eight *PP2C* genes per chromosome. The distribution of *PP2C* genes was also uneven within the same chromosome, with two or more genes arranged in gene clusters. These patterns of chromosomal localization are similar to those of *PP2C* genes in *P. trifoliata* and *V. radiata* [24,32].

In the phylogenetic evolutionary analysis of *PP2C* genes in *A. thaliana*, *P. trifoliata*, and *P. fortunei*, the *PP2C* genes were divided into 12 subfamilies, each of which harbored *P. fortunei PP2C* genes. Consistent with this, the *PP2C* genes of *A. thaliana* and *P. heterocycla* are also distributed among 12 subfamilies [11,24]. Our results indicate that the *PP2C* genes of *A. thaliana*, *P. trifoliata*, and *P. fortunei* are similarly distributed among the 12 subfamilies, indicative of relatively consistent evolutionary relationships among these three species. Thus, the *PP2C* gene family has been conserved during evolution. We detected clear differences in the number and type of conserved structural domains among *PfPP2C* family members. Our results show that the cNMP_binding and Pkinase_fungal domains are conserved domains unique to *PfPP2C11* and *PfPP2C04*, respectively. We also detected some variability in the distribution of conserved motifs among different subfamilies, and some similarities in the distribution of conserved motifs within each subfamily. Nearly 3/4 of *PfPP2C* proteins have the following motif structure: motifs 3, 16, 13, 10, 2, 6, 12, 4, 14, 1, 18, and 8. The distribution of conserved motifs in members of the A subclade of *PPfPP2C* is similar to that in members of the A subclade in *A. thaliana*. This high degree of affinity suggests that A subclade members in *P. fortunei* participate in the regulation of ABA signaling, similar to their counterparts in *A. thaliana*.

In terms of gene structure, *PfPP2C* genes have between 2 and 20 exons, similar to the members of the *PP2C* families in *P. trifoliata* and *S. italica* [24,34]. The type and number of *cis*-acting elements in the promoter region affect differential gene expression [35]. One of the most important roles of the *PP2C* gene family is in the regulation of ABA signaling [36]. In plants, subfamily A *PP2Cs* regulate early events in the ABA signaling pathway [37]. The involvement of *PP2C* proteins in this pathway varies among species, and among different organs and tissues of the same species [37]. For example, in *A. thaliana*, subfamily A *PP2C* proteins negatively regulate the ABA pathway, while *AtPP2C-G1* positively regulates the ABA pathway and the response to high salt stress [38], and *AtPP2C2* can significantly increase the response to ABA when overexpressed [39]. In this study, we detected *cis*-acting elements responsive to various phytohormones, such as MeJA, GA, ABA, IAA, and SA in the promoter regions of *PfPP2C* genes. In total, 7 of the 11 members of subfamily A contained ABA-responsive elements in their promoter regions, suggesting that subfamily

A *PfPP2Cs* are involved in the regulation of the ABA signaling pathway. These results are consistent with those reported in studies on *Zea mays* and *B. papyrifera* [21,40]. Overall, 42 of the *PfPP2C* family members contain ABA-responsive elements in their promoter regions, and half of the *PfPP2C* genes have stress-responsive elements in their promoters. Therefore, we speculate that *PfPP2Cs* may regulate various physiological activities in plants via the ABA signaling pathway.

Previous studies have demonstrated that *PP2Cs* also participate in the disease resistance signaling pathway. When plants are attacked by fungi, bacteria, and viruses, various signaling molecules such as SA and JA are produced, and they trigger the expression of genes encoding components of the disease resistance response [30]. It has been suggested that *PP2Cs* may play a role in plant resistance to biological stresses, such as rust and powdery mildew [41]. In this study, we found that the expression of some *PP2C* genes in *P. fortunei* were significantly affected by PaWB and treatment with Rif and MMS. The expression of *PfPP2C19* and *PfPP2C80* increased during the formation of PaWB in *P. fortunei*, but decreased under Rif and MMS treatments, while *PfPP2C02* and *PfPP2C12* were down-regulated during the formation of PaWB, but up-regulated in response to Rif and MMS treatments. These findings suggest that these genes play an important regulatory role in the development of PaWB, but may have a complex regulatory network. Further research is needed to explore their roles in the development of PaWB.

5. Conclusions

At present, research on *P. fortunei* and its molecular biology lags behind that on other plants, and much less is known about the function of its *PP2C* proteins than about those of model plants, such as *A. thaliana*, *O. sativa*, and *Glycine max*. In this study, we analyzed *P. fortunei PP2C* family members to determine the chromosomal distribution, evolutionary relationships, and gene structure, and the conserved structural domains and conserved motifs in their encoded proteins. We also determined their transcript profiles before and after the development of PaWB, and identified four genes closely related to PaWB development (*PfPP2C02*, *PfPP2C12*, *PfPP2C19*, and *PfPP2C80*) among the 91 *PP2C* genes in *P. fortunei*. The results of this study provide a reference for future studies on the structure, function, and regulatory roles of the *PP2C* gene family in *Paulownia*, and provide clues about the *PP2C* proteins that may participate in the formation of PaWB.

Author Contributions: G.F. conceived and designed the experiments; Z.Z. and P.Z. performed the experiments and wrote the paper; M.D. and Y.C. contributed reagents and analyzed the data. All authors have read and agreed to the published version of the manuscript.

Funding: This research was funded by the Academic Scientist Fund for Zhongyuan Scholars of Henan Province (grant 2018 [99]) and the National Key Research and Development Program (Grant No. 2016YFD0600106).

Data Availability Statement: Not applicable.

Acknowledgments: We thank Jennifer Smith for editing the English text of a draft of this manuscript.

Conflicts of Interest: The authors declare no conflict of interest.

References

1. Hunter, T. Protein kinases and phosphatases: The Yin and Yang of protein phosphorylation and signaling. *Cell* **1995**, *80*, 225–236. [CrossRef] [PubMed]
2. Luan, S. Protein phosphatases and signaling cascades in higher plants. *Trends Plant Sci.* **1998**, *3*, 4–10. [CrossRef]
3. Hunter, T. Protein-tyrosine phosphatases: The other side of the coin. *Cell* **1989**, *58*, 1013–1016. [CrossRef] [PubMed]
4. Mumby, M.C.; Walter, G. Protein serine/threonine phosphatases: Structure, regulation, and functions in cell growth. *Physiol. Rev.* **1993**, *73*, 673–699. [CrossRef]
5. Luan, S. Protein Phosphatases in Plants. *Annu. Rev. Plant Biol.* **2003**, *54*, 63–92. [CrossRef] [PubMed]
6. Cohen, P. The structure and regulation of protein phosphatases. *Annu. Rev. Biochem.* **1989**, *58*, 453–508. [CrossRef]
7. Schweighofer, A.; Hirt, H.; Meskiene, I. Plant PP2C phosphatases: Emerging functions in stress signaling. *Trends Plant Sci.* **2004**, *9*, 236–243. [CrossRef]

8. Kerk, D.; Bulgrien, J.; Smith, D.W.; Barsam, B.; Veretnik, S.; Gribskov, M. The complement of protein phosphatase catalytic subunits encoded in the genome of Arabidopsi. *Plant Physiol.* **2002**, *129*, 908–925. [CrossRef]
9. Ruan, H.H.; Xu, L.L. Progress in the structure and function of PP2C-type protein in phosphatases. *J. Nanjing Agric. Univ.* **2007**, *30*, 136–141.
10. Hu, X.B.; Song, F.M.; Zheng, Z. Structure and function of protein phosphatase 2C in higher plants. *Chin. J. Cell Biol.* **2005**, *27*, 29–34.
11. Fuchs, S.; Grill, E.; Meskiene, I.; Schweighofer, A. Type 2C protein phosphatases in plants. *FEBS J.* **2013**, *280*, 681–693. [CrossRef] [PubMed]
12. Raghavendra, A.S.; Gonugunta, V.K.; Christmann, A.; Grill, E. ABA perception and signalling. *Trends Plant Sci.* **2010**, *15*, 395–401. [CrossRef] [PubMed]
13. Chu, M.L.; Che, P.W.; Meng, S.F.; Xu, P.; Lan, W.Z. The Arabidopsis phosphatase PP2C49 negatively regulates salt tolerance through inhibition of At HKT1;1. *J. Integr. Plant Biol.* **2021**, *63*, 528–542. [CrossRef] [PubMed]
14. Schweighofer, A.; Kazanaviciute, V.; Scheikl, E.; Teige, M.; Doczi, R.; Hirt, H.; Schwanninger, M.; Kant, M.; Schuurink, R.; Mauch, F.; et al. PP2C-Type phosphatase AP2C1, which negatively regu-lates MPK4 and MPK6, modulates innate immunity, jasmonic acid, and ethylene levels in Arabidopsis. *Plant Cell* **2007**, *19*, 2213–2224. [CrossRef]
15. Gagne, J.M.; Clark, S.E. The Arabidopsis stem cell factor POLTERGEIST is membrane localized and phospholipid stimulated. *Plant Cell* **2010**, *22*, 729–743. [CrossRef]
16. Xue, T.; Wang, D.; Zhang, S.; Ehlting, J.; Ni, F.; Jakab, S.; Zheng, C.; Zhong, Y. Genome-wide and expression analysis of protein phosphatase 2C in rice and Arabidopsis. *BMC Genom.* **2008**, *9*, 550. [CrossRef] [PubMed]
17. Akiyama, M.; Sugimoto, H.; Inoue, S.I.; Takahashi, Y.; Hayashi, M.; Hayashi, Y.; Mizutani, M.; Ogawa, T.; Kinoshita, D.; Ando, E.; et al. Type 2C protein phosphatase clade D family members dephosphorylate guard cell plasma membrane H+-ATPase. *Plant Physiol.* **2022**, *188*, 2228–2240. [CrossRef] [PubMed]
18. Zhang, C.; Yu, J.L.; Chu, M.L.; Zhang, B.M.; Lan, W.Z. Functional analysis of protein phosphatase PP2C31 in salt reponse in Arabidopsis thaliana. *China Sci.* **2018**, *13*, 2070–2075.
19. Lv, J.; Liu, J.; Ming, Y.; Shi, Y.; Song, C.; Gong, Z.; Yang, S.; Ding, Y. Reciprocal regulation between the negative regulator PP2CG1 phosphatase and the positive regulator OST1 kinase confers cold response in *Arabidopsis*. *J. Integr. Plant Biol.* **2021**, *63*, 1568–1587. [CrossRef]
20. He, H.H.; Lu, Z.H.; Ma, Z.H.; Liang, G.P.; Ma, L.J.; Wan, P.; Mao, J. Genome-Wide Identification and Expression Analysis of the PP2C Gene Family in Vitis vinifera. *Acta Hortic. Sin.* **2018**, *45*, 1237–1250. [CrossRef]
21. Zhang, B.; Chen, N.; Peng, X.; Shen, S. Identification of the PP2C gene family in paper mulberry (*Broussonetia papyrifera*) and its roles in the regulation mechanism of the response to cold stress. *Biotechnol. Lett.* **2021**, *43*, 1089–1102. [CrossRef] [PubMed]
22. Cao, J.M.; Jiang, M.; Li, P.; Chu, Z.Q. Genome-wide identification and evolutionary analyses of the PP2C gene family with their expression profiling in response to multiple stresses in Brachypodium distachyon. *BMC Genom.* **2016**, *17*, 175. [CrossRef] [PubMed]
23. Miao, J.; Li, X.; Li, X.; Tan, W.; You, A.; Wu, S.; Tao, Y.; Chen, C.; Wang, J.; Zhang, D.; et al. Os*PP2C09*, a negative regulatory factor in abscisic acid signalling, plays an essential role in balancing plant growth and drought tolerance in rice. *New Phytol.* **2020**, *227*, 1417–1433. [CrossRef] [PubMed]
24. Yang, J.; Chen, R.; Hu, W.J.; Wu, Q.L.; Tong, X.N.; Li, X.T. Identification and expression analysis of PP2C gene family in *Poncirus trifoliata*. *J. Fruit Sci.* **2022**, *39*, 532–547. [CrossRef]
25. Hu, Q.T.; Hou, D.; Zhao, Z.Y.; Wei, H.T.; Lin, X.C. Identification and Expression Analysis of PP2C Gene Family in Phyllostachys edulis. *J. Agric. Biotechnol.* **2020**, *28*, 1776–1786. [CrossRef]
26. Zhang, T.T.; Li, Y.X.; Zhang, D.Y.; Kang, Y.Q.; Wang, J.; Song, X.Q.; Zhou, Y. Genome-wide Identification and Expression Analyses of PP2C Gene Family in Dendrobium catenatum. *Acta Hortic. Sin.* **2021**, *48*, 2458–2470. [CrossRef]
27. Zhai, X.Q.; Wang, Z.Q.; Fan, G.Q. Direct plantlet regeneration via organogenesis of Paulownia plants. *Acta Agric. Nucleatae Sin.* **2004**, *18*, 357–360.
28. Cao, Y.; Sun, G.; Zhai, X.; Xu, P.; Ma, L.; Deng, M.; Zhao, Z.; Yang, H.; Dong, Y.; Shang, Z.; et al. Genomic insights into the fast growth of paulownias and the formation of Paulownia witches' broom. *Mol. Plant* **2021**, *14*, 1668–1682. [CrossRef] [PubMed]
29. Tao, Y.T.; Chen, L.X.; Jin, J.; Du, Z.K.; Li, J.M. Genome-wide identification and analysis of bZIP gene family reveal their roles during development and drought stress in Wheel Wingnut (*Cyclocarya paliurus*). BMC *Genom.* **2022**, *23*, 743. [CrossRef]
30. Hu, X.L.; Li, D.Q. Protein Phosphatase 2C in Plants and Its Functions of Signal Transduction. *Plant Physiol. J.* **2007**, *43*, 407–412. [CrossRef]
31. Singh, A.; Giri, J.; Kapoor, S.; Tyagi, A.K.; Pandey, G.K. Protein phosphatase complement in rice: Genome-wide identification and transcriptional analysis under abiotic stress conditions and reproductive development. *BMC Genom.* **2010**, *11*, 435. [CrossRef] [PubMed]
32. Chen, H.L.; Hu, L.L.; Wang, L.X.; Wang, S.H.; Cheng, X.Z. Genome-Wide Identification and Bioinformatics Analysis of bHLH Transcription Factor Family in Mung Bean(*Vigna radiata* L.). *J. Plant Genet. Resour.* **2017**, *18*, 1159–1167. [CrossRef]
33. Liu, Y.D.; Xiao, S.Y.; Wang, A.H.; Liu, Y.; Fang, Y.; Li, X.Y.; Liu, Z.B.; Li, X.F.; Wang, J.M.; Yang, Y. Genome-wide identification and expression analysis of protein phosphatase 2C family in Tartary buckwheat. *J. Sichuan Univ.* **2021**, *58*, 163–171. [CrossRef]

34. Min, D.H.; Xue, F.Y.; Ma, Y.N.; Chen, M.; Xu, Z.S.; Li, L.C.; Diao, X.M.; Jia, G.Q.; Ma, Y.Z. Characteristics of PP2C Gene Family in Foxtail Millet (*Setaria italica*). *Acta Agron. Sin.* **2013**, *39*, 2135–2144. [CrossRef]
35. Jin, H.; Xing, M.; Cai, C.; Li, S. B-box Proteins in Arachis duranensis: Genome-Wide Characterization and Expression Profiles Analysis. *Agronomy* **2020**, *10*, 23. [CrossRef]
36. Lorenzo, O.; Nicolás, C.; Nicolás, G.; Rodríguez, D. Molecular cloning of a functional protein phosphatase 2C (*FsPP2C2*) with unusual features and synergistically up-regulated by ABA and calcium in dormant seeds of Fagus sylvatica. *Physiol. Plant* **2002**, *114*, 482–490. [CrossRef]
37. Zhang, J.H.; Tao, N.G. Research progress on regulation mechanism of plant PP2C protein phosphatase ABA signal transduction and abiotic stress. *Plants Guangxi* **2015**, *35*, 935–941.
38. Liu, X.; Zhu, Y.; Zhai, H.; Cai, H.; Ji, W.; Luo, X.; Li, J.; Bai, X. *AtPP2CG1*, a protein phosphatase 2C, positively regulates salt tolerance of Arabidopsis in abscisic acid-dependent manner. *Biochem. Biophys. Res. Commun.* **2012**, *422*, 710–715. [CrossRef]
39. Reyes, D.; Rodríguez, D.; González-García, M.P.; Lorenzo, O.; Nicolás, G.; García-Martínez, J.L.; Nicolás, C. Overexpression of a protein phosphatase 2C from beech seeds in Arabidopsis shows phenotypes related to abscisic acid responses and gibberellin biosynthesis. *Plant Physiol.* **2006**, *141*, 1414–1424. [CrossRef]
40. He, Z.; Wu, J.; Sun, X.; Dai, M. The Maize Clade A PP2C Phosphatases Play Critical Roles in Multiple Abiotic Stress Responses. *Int. J. Mol. Sci.* **2019**, *20*, 3573. [CrossRef]
41. Yu, X.; Han, J.; Wang, E.; Xiao, J.; Hu, R.; Yang, G.; He, G. Genome-Wide Identification and Homoeologous Expression Analysis of PP2C Genes in Wheat (*Triticum aestivum* L.). *Front Genet.* **2019**, *10*, 561. [CrossRef] [PubMed]

forests

MDPI

Article

Overexpression of the Poplar *WRKY51* Transcription Factor Enhances Salt Tolerance in *Arabidopsis thaliana*

Yangyan Zhou [1], Qing Li [2] and Yue Zhang [3,*]

1 School of Advanced Agricultural Sciences, Peking University, Beijing 100871, China
2 Key Research Institute of Yellow River Civilization and Sustainable Development & Collaborative Innovation Center on Yellow River Civilization Jointly Built By Henan Province and Ministry of Education, Henan University, Kaifeng 475001, China
3 Grassland Agri-Husbandry Research Center, College of Grassland Science, Qingdao Agricultural University, Qingdao 266109, China
* Correspondence: zhang_yue@qau.edu.cn; Tel.: +86-0532-5895-7222

Abstract: Salt is a severe environmental stressor that affects growth and development in plants. It is significant to enhance the salt tolerance in plants. In this study, a salt-responsive WRKY transcription factor *PtrWRKY51* was isolated from *Populus trichocarpa* (clone 'Nisqually-1'). *PtrWRKY51* was highly expressed in mature leaves and root and induced by salt stress. The *PtrWRKY51* was overexpressed in *Arabidopsis* to investigate its biological functions. Compared with Col-0 lines, Overexpressed lines had an increase in germination rate of seed, root length, higher photosynthetic rate, instantaneous leaf WUE, chlorophyll content to improve salt tolerance under salt stress conditions. In contrast, compared to overexpressed and Col-0 lines, the mutant *wrky51* was more sensitive to salt stress with lower photosynthetic rate and WUE. Additionally, it was found that the complementary lines (*wrky51/PtrWRKY51*) had almost the same salt response as Col-0. In conclusion, *PtrWRKY51* is a potential target in the enhancement of poplar tolerance by genetic engineering strategies.

Keywords: poplar; transcription factor; *PtrWRKY51*; salt tolerance; photosynthetic rate; water-use efficiency

Citation: Zhou, Y.; Li, Q.; Zhang, Y. Overexpression of the Poplar *WRKY51* Transcription Factor Enhances Salt Tolerance in *Arabidopsis thaliana*. *Forests* **2023**, *14*, 191. https://doi.org/10.3390/f14020191

Academic Editors: Jie Luo and Wentao Hu

Received: 6 December 2022
Revised: 12 January 2023
Accepted: 12 January 2023
Published: 18 January 2023

1. Introduction

As one of the most widely distributed and adaptable forest species around the world, poplar is one of the most promising tree species for traditional afforestation and dealing with wood shortages. With the gradual aggravation of soil salinization, salt stress has become an essential factor restricting tree growth. Salt is one of the key environmental stress factors, which affects plant growth and development [1]. It is reported that more than 800 million hectares of land are affected by salinization in world, which across all continents including Africa, Asia, Australasia, Americas, and China, occupying an estimated 37.4 million hectares of salinized soil [2,3]. Moreover, soil salinization is becoming more serious due to environmental degradation, improper irrigation, climate change, growing population and industrial pollution [4,5]. Excessive salinity in plants will lead to excessive accumulation of intracellular metal ions, destroy ion balance, damage plant cell membrane structure, and thus affect the physiological and biochemical metabolic processes in plants [6,7]. Many studies have shown that the water potential of the soil in the saline-alkali soil was decreased, and the absorption of water by seed was inhibited, which seriously affected the germination rate, vigor index and germination index of seeds, which have a reduction in seed emergence rate and strong seedling rate [8,9]. Photosynthesis is the sum of a series of complex metabolic reactions which are essential in plant growth. Photosynthesis provides material and energy to maintain normal growth and development in plants, and chloroplasts in leaves are the place in which photosynthesis is carried out in higher plants. Salt stress affects plant absorb water and inorganic ions, inhibit the normal synthesis of chlorophyll, and reduce the content of chlorophyll, which affect the normal function of the pigment protein complexes, reduce the rate of plant convert light energy into chemical energy, which eventually make plant to a serious shortage of energy supply, inhibit the growth and development in plants [10,11]. At the same time, salt stress can also cause damage to chloroplast structure, and then affect its photosynthetic performance. It has been reported that under salt stress, the net photosynthetic rate, chlorophyll fluorescence and stomatal conductance of leaves are greatly reduced in citrus, which inhibit growth [12]. Similarly, it has been confirmed that salt stress seriously inhibits the photosynthetic rate and electron transport rate in photosynthesis, which affected plant growth in mustard [13]. Therefore, it is important to improve the salt tolerance in plants. During evolution, plants have developed various strategies to adapt to salt stress [14]. Signal sensing and signal transduction are key ways for plants to cope with challenges in environment, which involve many regulatory genes and proteins. Among these, the transcription factors, such as WRKY, NAC, bZIP, MYB, HSFs, AP2/EREBP have been identified as key regulators in controlling intrinsic development processes and regulating stress resistance by stimulating the expression of downstream targets in improving salt resistance in plants [15–19].

WRKY transcription factor is a key regulator in plants, and significant study has been made in WRKY transcription factor over the past 20 years [20]. *SPF1* is the first WRKY gene, cloned from sweet potato [21]. Subsequently, WRKY genes were cloned and reported in *Arabidopsis thaliana*, wild oats and parsley, which were found that the proteins were encoded by WRKY genes that could specifically bind to DNA sequence element (T) (T) TGAC (C/T) [22]. The specific DNA sequence element (T) (T) TGAC (C/T) is also known as the W-box [22–24]. It is well known that WRKY transcription factor contain a highly conserved WRKY domains with WRKYGQK sequence and a zinc-finger-like motif CX4–7-CX23–28-HX1–2-(H/C) (C2H2 or C2HC) at N and C termini [25]. In general, WRKY proteins can be categorized into three groups based on the number of WRKY domains and the pattern of the zinc finger motif [25]. Group I WRKY transcription contains two WDs and C2H2 zinc finger, and group II and III WRKY transcription factors contain one WD and C2H2 zinc finger motif [25].

Many studies have found that WRKY transcription factors is a key role in plant response to abiotic stresses such as drought and salt stress. Northern blot hybridization showed that 10 WRKY genes in rice had different responses to salt, osmotic stress, cold and heat treatments [26]. In wheat, eight of the 15 WRKY genes also showed response to PEG, low temperature, high temperature and NaCl stress treatments [27]. In *Arabidopsis*, overexpression of *OSWRKY45* and *AtWRKY46* can regulate stomatal movement in response to drought and salt stresses [28,29]. Heterologous expression of *GmWRKY54* in *Arabidopsis* increased drought and salt tolerance [30]. In addition, WRKY genes such as *TaWRKY2* and *TaWRKY19*, *TaWRKY146*, *TaWRKY1* and *TaWRKY33*, *VaWRKY*14, *ZmWRKY17* and *ZmWRKY40*, and *PeWRKY83* were overexpressed in *Arabidopsis* can improve plant tolerance to drought and/or salt stresses, respectively [31–37]. Previous study has shown that *PtrWRKY51* responds to salt stress [38], but functional characterization of *PtrWRKY51* in salt stress response remains unclear.

In this study, *PtrWRKY51* was cloned from the *Populus trichocarpa* to investigate its function in salt stress in *Arabidopsis*. We analyzed its expression pattern in root, stem and leaf tissues and in response to salt stress. Moreover, the functional characterization of *PtrWRKY51* in salt response was studied in *Arabidopsis*. Our results demonstrate that overexpression of *PtrWRKY51* enhanced salt tolerance in *Arabidopsis*.

2. Materials and Methods

2.1. Plant Materials and Growth Conditions

The solid Lloyd and McCown's Woody Plant Basal Salts (WPM) medium was used for cultivation of the *Populus trichocarpa* in vitro [39]. After plantlet regeneration, they were individually transplanted and grown in pots (20.5 × 17 × 14.5 cm; top diameter/height/bottom diameter) containing a mixture of soil and vermiculite (2:1) and incubated in the greenhouse at 22 °C under a 16 h/8 h (light/dark) photoperiod (150 μmol m^{-2} s^{-1}) and 70% relative humidity.

Arabidopsis Col-0 was selected as the wild-type control, and the mutant *wrky51* (stock name SALK_022198.56.00.x) were obtained from Arabidopsis Biological Resource Center (ABRC; https://abrc.osu.edu/). *Arabidopsis* seeds were washed two times with water, then with 75% alcohol for 1 min followed by 1% NaClO within 10 min and five washes in distilled water. Seeds were sown on 1/2MS medium plates containing 30 g/L sucrose and 6 g/L agar. The seeded plates were vernalized at 4 °C in the dark for 2 days and then moved to 22 °C under 16/8 light/dark cycle. When *Arabidopsis* sprouted, it was transplanted at a density of four plants per pot (7 × 7 × 6.5 cm; top diameter/height/bottom diameter) with a mixture of soil and vermiculite (2:1) at 22 °C under 16/8 light/dark cycle and 70% relative humidity.

2.2. *PtrWRKY51 Cloning and Transformation*

Total RNA was extracted from the young leaves of *Populus trichocarpa* by using the RN38 EASYspin Plus Plant RNA Kit (Aidlab Biotech, Beijing, China). The first strand cDNA synthesis was reverse-transcribed using the Tiangen FastQuant RT Kit (Tiangen) according to the manufacturer's instructions. The resultant cDNA was used as a template, and the *PtrWRKY51* (Accession number: *Potri.005G085200*) sequence was amplified by PCR with gene-specific primers (Supplementary Table S1). The gene-specific primers were designed by using Primer Premier 5 software.

For overexpression of *PtrWRKY51* and complementation of the *Arabidopsis wrky51* mutant with *PtrWRKY51*, the *PtrWRKY51* cDNA was cloned into the pCAMBIA-1301 binary vector under the control of the Cauliflower mosaic virus (CaMV) 35S promoter, and the plasmids were introduced into *Agrobacterium tumefaciens* strain GV3101 by heat shock. Then, the 35S: PtrWRKY51: GFP plasmid transformed into the *Arabidopsis* Col-0 and *wrky51* mutant lines respectively by the floral dip method [40]. The transgenic lines were identified using half-strength MS plates containing 100 mg l^{-1} hygromycin. The third (T_3) generation of seeds were used in the experiments to make sure hereditary stability in transgenic lines.

2.3. *Molecular Verification of Transgenic Plants*

Genomic DNA of Col-0 and transgenic *Arabidopsis* lines was extracted from leaves by using the CTAB method [41]. Transformants were identified by PCR amplification using the gene-specific primers PtrWRKY51-F and GFP-R (Supplementary Table S1). The transcript levels of *PtrWRKY51* were quantified by Quantitative real-time PCR (RT-qPCR) in transgenic lines.

2.4. *Quantitative Real-Time PCR Analysis*

The ABI StepOnePlus™ Real-Time PCR System (Applied Biosystems, Inc., Carlsbad, CA, USA) is used to perform Quantitative real-time PCR (RT-qPCR) experiments according to the manufacturer's instructions. The $2^{-\Delta\Delta CT}$ method [42] was used to calculate the relative expression level of *PtrWRKY51*, with poplar *Actin* employed as the internal control (Supplementary Table S1).

2.5. *Physiological Experiments*

To determine germination rates of overexpressed *PtrWRKY51*(*oePtrWRKY51*), Col-0, *wrky51* and *wrky51*/*PtrWRKY51* seeds, 100 seeds of different genotypes were seeded separately on the same 1/2MS medium with or without 200 mM NaCl stress and germination rates were recorded daily. When the seeds germinated, 10 plants of each genotype were selected and placed vertically on the medium with or without 200 mM NaCl stress to grow for 10 days, and the root length was recorded.

Net photosynthetic rate and transpiration rate were measured of *oePtrWRKY51*, Col-0, *wrky51* and *wrky51*/*PtrWRKY51* plants by using an infrared gas analysis system

(LI-COR 6400, Lincoln, NE, USA) as previously described [43,44]. Instantaneous leaf WUE was defined as the ratio of the rate of photosynthetic rate/transpiration rate [45].

Leaves of *oePtrWRKY51*, Col-0, *wrky51* and *wrky51/ PtrWRKY51* plants grown for 4 weeks were used for chlorophyll content determination with 80% acetone. A UV/visible spectrophotometer (YHB-061; GE Healthcare, Little Chalfont, Buckinghamshire, UK) was used to measure absorbance at 663 for chlorophyll a and 645 nm for chlorophyll b, and then the chlorophyll contents were calculated according to Lichtenthaler's method [46].

2.6. Salt Experiments

For the salt treatments of *Populus trichocarpa*, 4-week-old seedlings were removed from the soil carefully and then placed in a salt solution containing 200 mM NaCl for 24 h. Subsequently, leaves were harvested at 0 h, 1 h, 3 h, 6 h, 8 h, 12 h and 24 h for RT-qPCR analysis.

Arabidopsis seeds of *oePtrWRKY51*, Col-0, *wrky51* and *wrky51/PtrWRKY51* were seeded on 1/2MS medium and transplanted at a density of four plants per pot (7 × 7 × 6.5 cm; top diameter/height/bottom diameter) with a mixture of soil and vermiculite (2:1) at 22 °C under 16/8 light/dark cycle and 70% relative humidity after two true leaves had grown. It was cultured in the greenhouse with normal watered for two weeks, and then it was watered with 200 mM NaCl solution every 3 days for 15 days to observe its phenotype. The photosynthetic rates and instantaneous leaf WUE were measured at each stage as previously.

3. Results

3.1. The Analysis of Expression Pattern in PtrWRKY51

The tissue-specific expression of *PtrWRKY51* transcripts in young leaf, mature leaf, senescent leaf, stem and root of *P. trichocarpa* was detected by RT-qPCR. The result showed that *PtrWRKY51* transcript levels in mature leaves and root was higher than that in young leaf, senescent leaf and stem (Figure 1A). To study the response of *PtrWRKY51* to salt stress, *P. trichocarpa* plants were subjected to salt stress and the *PtrWRKY51* transcript levels were quantified by RT-qPCR. The results showed that the *PtrWRKY51* transcript level gradually increased and peaked at 6 h of the NaCl treatments, and thereafter decreased (Figure 1B). These results indicated that *PtrWRKY51* was mainly expressed in mature leaves and roots and up-regulated in response to salt stress.

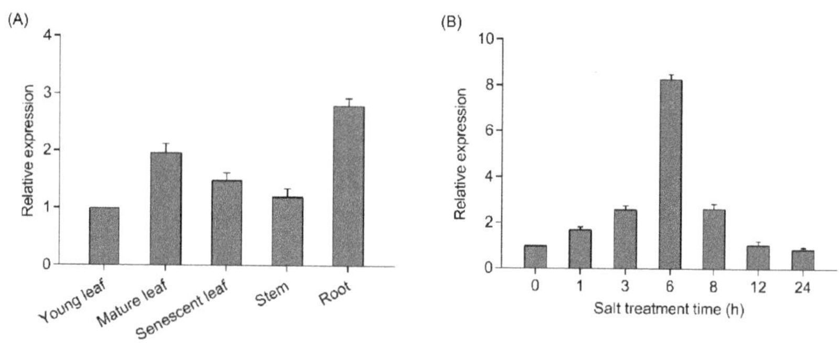

Figure 1. Expression patterns of *PtrWRKY51*. (**A**)Transcript level of *PtrWRKY51* in various organs including young leaf, mature leaf, senescent leaf, stem and root of *P. trichocarpa*. (**B**) RT-qPCR analysis of *PtrWRKY51* transcript levels under salt conditions. Data are shown as mean ± SE (*n* = 6).

3.2. Identification of PtrWRKY51-Overexpressing Transgenic Plant

To further explore the function of *PtrWRKY51* in salt stress conditions, the 35S: PtrWRKY51: GFP is transformed into Col-0 *Arabidopsis*. Eight transgenic lines (*oePtrWRKY51 #1-7*) were obtained and confirmed by PCR analyses using the combination of the gene-specific primers PtrWRKY51-F and GFP-R. The expected band was amplified in all transgenic plants but not in Col-0 plants (Figure 2A). In addition, RT-qPCR analysis of the above transgenic plants showed that *PtrWRKY51* was overexpressed in all selected transgenic lines (Figure 2B). Then three of *PtrWRKY51*-overexpressing lines (*oePtrWRKY51#1*, *#2*, and #3) with higher expression level than other lines were selected for further analysis in salt tolerance.

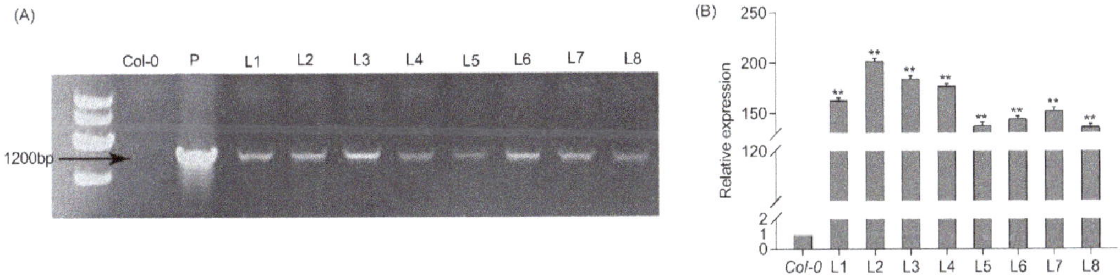

Figure 2. Analysis of overexpressing *PtrWRKY51* in transgenic plants. (**A**) Eight putatively transgenic plants of *PtrWRKY51* were identified by PCR. (**B**) RT-qPCR was used to detect the transcription level of *PtrWRKY51* in different lines. The data are represented as means ± SE (n = 6). The asterisks (**) represent a significant differences (** $p < 0.01$) compared to the control.

3.3. Plant Phenotype Analysis under Well-Watered Conditions

To explore the function of *PtrWRKY51*, the growth phenotype of *oePtrWRKY51*, Col-0, *wrky51* and *wrky51/ PtrWRKY51* plants were observed under normal conditions. The result showed that the primary root length of overexpressed plants was significantly longer than that of Col-0 (1.17-fold) and *wrky51* (1.61-fold) after 10 days of vertical growth on 1/2 MS medium (Figure 3A,B). These results indicated that overexpression of *PtrWRKY51* could promote root growth.

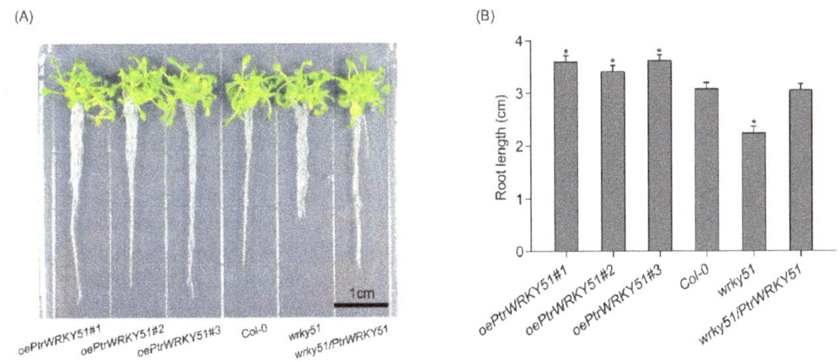

Figure 3. Root length analysis of overexpressing *PtrWRKY51* in plants under normal conditions. (**A**) Phenotypic analysis of the primary root length for 10 days on 1/2MS medium. Bars, 1 cm. (**B**) Difference in the primary root length after 10 days of vertical growth on 1/2MS medium. The data are represented as means ± SE (n = 12). The asterisks (*) represent significant differences (* $p < 0.05$) compared to the control.

3.4. Overexpression of PtrWRKY51 Enhanced WUE in Arabidopsis

Photosynthetic physiological analysis that the net photosynthetic rates of *oePtrWRKY51* lines was higher than that of Col-0 and *wrky51* under normal conditions, and the net photosynthetic rate of *wrky51/ PtrWRKY51* lines was equal to that of Col-0 (Figure 4A). In addition, the transpiration rate of each genotype was found that the *oePtrWRKY51* lines showed slower rates than the Col-0 and *wrky51* mutant (Figure 4B). As a result, based on the higher photosynthetic capability and lower transpiration level, the instantaneous WUE values of the *oePtrWRKY51* lines were observably higher than that of Col-0 and *wrky51* mutant (Figure 4C). In general, overexpression of *PtrWRKY51* enhanced the WUE in *Arabidopsis*.

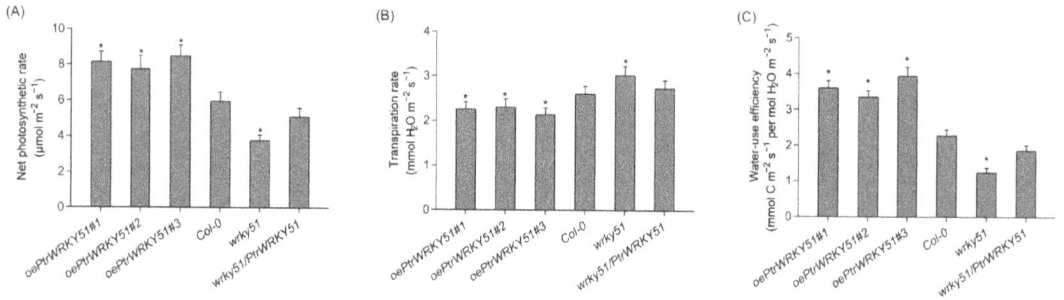

Figure 4. Gas exchange analysis of overexpressing *PtrWRKY51* lines showing higher instantaneous WUE in *Arabidopsis*. The Net photosynthesis rate (**A**), transpiration rate (**B**), instantaneous leaf WUE (**C**) of 3-week-old seedlings. The data are represented as means ± SE (n = 12). The asterisks (*) represent significant differences (* $p < 0.05$) compared to the control.

3.5. Overexpression of PtrWRKY51 Increases Salt Tolerance under Salt Stress Conditions

A series of experiments were conducted to explore the differences in salt tolerance of *oePtrWRKY51*, Col-0, *wrky51* and *wrky51*/ *PtrWRKY51* plants. The seeds of *oePtrWRKY51*, Col-0, *wrky51* and *wrky51*/ *PtrWRKY51* were sown on 1/2 MS culture containing 200 mM NaCl to observe germination rates. After 4 days, the *oePtrWRKY51* plants had a higher germination rate (73.4%) than the Col-0 (42.8%) and mutant (20.2%) plants (Figure 5A). In addition, the primary root length of the seedlings was also different after 10 days of vertical growth under salt stress conditions. The primary root length of *oePtrWRKY51* plants was significantly longer (1.48- and 1.85- fold, respectively) than that of Col-0 and mutant (Figure 5B,C). After the seedlings were transplanted to soil, the salt stress was imposed for 15 days. As expected, the Col-0 and *wrky51* mutant plants showed more severe wilting than over-expressing plants, especially the *wrky51* mutants, whereas those of the *oePtrWRKY51* lines continued to show development and growth (Figure 5D).

The net photosynthesis rate and WUE of *oePtrWRKY51*, Col-0, *wrky51* and *wrky51*/ *PtrWRKY51* plants were measured and calculated under the salt treatments conditions (Figure 6A,B). The result showed that compared with the Col-0 and *wrky51* mutant plants, the *oePtrWRKY51* lines could maintain a higher photosynthetic rate and have a higher instantaneous leaf WUE under salt stress conditions.

To further assess the potential biological functions of *PtrWRKY51* in regulation of salt tolerance, the chlorophyll *a*, chlorophyll *b*, and total chlorophyll contents of *oePtrWRKY51*, Col-0, *wrky51* and *wrky51*/ *PtrWRKY51* plants were also measured under normal and salt stress conditions. The result showed that there was no significant difference in chlorophyll content between the *oePtrWRKY51* and the Col-0 plants, but there was a difference between them and the *wrky51* mutant under normal conditions. However, under salt stress conditions, the chlorophyll content of *oePtrWRKY51* plants was markedly higher than that of Col-0 and *wrky51* mutant (Figure 7), indicating that the *oePtrWRKY51* plants showed a better ability to absorb light energy compared with Col-0 and *wrky51* mutant, thus the *oePtrWRKY51* plants maintained a higher photosynthetic rate under salt stress conditions. Therefore, the expression of *PtrWRKY51* was beneficial for plant growth under salt stress conditions.

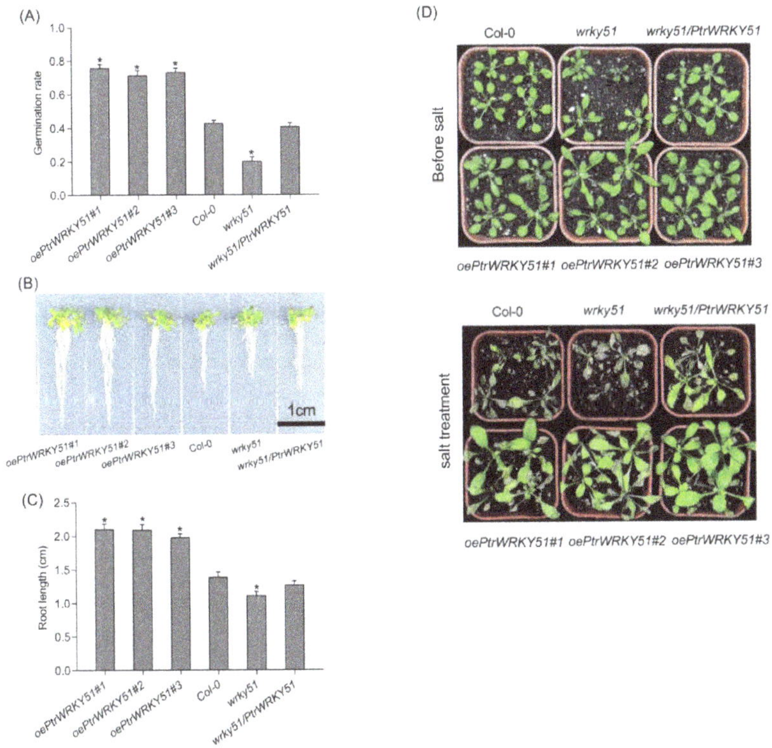

Figure 5. Overexpression of *PtWRKY51* confers salt tolerance in *Arabidopsis*. (**A**) Difference in germination ratio among oe*PtWRKY51*, Col-0, *wrky51*, and *wrky51/ PtWRKY51* plants grown on 1/2 MS medium with 200 mM NaCl. (**B**) Morphological comparisons of the primary root length for 10 d under 1/2 MS medium with 200 mM NaCl conditions. (**C**) Difference in the primary root length after 10 days of vertical growth on 1/2 MS medium with 200 mM NaCl conditions. (**D**) Morphological differences in salt treatments experiments. The data are represented as means ± SE (n = 12). The asterisks (*) represent significant differences (* $p < 0.05$) compared to the control.

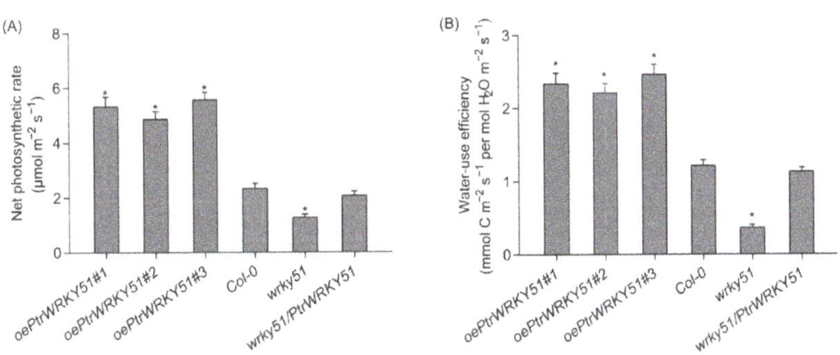

Figure 6. Physiological analysis of over-expressing *PtWRKY51* lines under salt stress conditions. Difference in net photosynthetic rate (**A**) and instantaneous leaf WUE (**B**) among oe*PtWRKY51*, Col-0, *wrky51*, and *wrky51/ PtWRKY51* plants under salt stress conditions. The data are represented as means ± SE (n = 12). The asterisks (*) represent significant differences (* $p < 0.05$) compared to the control.

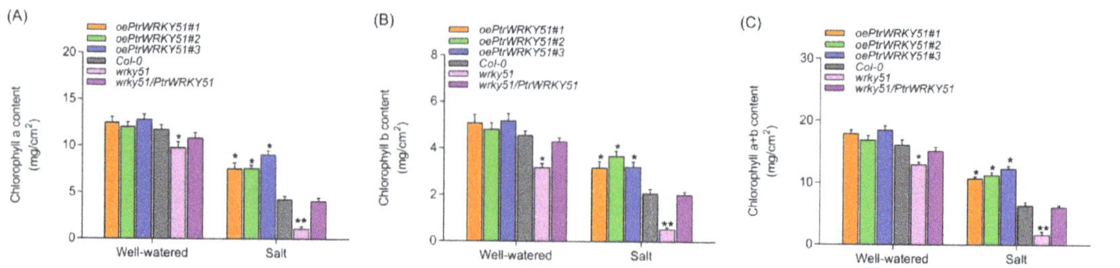

Figure 7. The chlorophyll content analysis of over-expressing *PtWRKY51* plants under salt treatment conditions. Determine of the content of chlorophyll a (**A**), chlorophyll b (**B**) and chlorophyll a + b (**C**) among oe*PtWRKY51*, Col-0, *wrky51*, and *wrky51*/ *PtWRKY51* plants under salt stress conditions. The data are represented as means ± SE (n = 12). The asterisks (* and **) represent significant differences (* $p < 0.05$ and ** $p < 0.01$) compared to the control.

4. Discussion

Salt stress has seriously affected and inhibited plant growth and development. Stress-induced transcription factors (TFs) are important pivotal regulators in response to stress. They can bind membrane protein acting elements and promote related genes expression in response to stress to reduce damage of the stress. WRKY is a plant-specific zinc finger type transcription factor. It is involved in a wide variety of biotic and abiotic stress responses [20,38,47–51]. Studies have shown that *OsWRKY51* and *OsWRKY71* regulate GA and ABA signal transduction in seed germination [52]. In addition, *AtWRKY51* was involved in JA-induced defense response [53]. Nevertheless, little was known about the function of *WRKY51* in abiotic, especially in salt stress. Previous studies have shown that *PtrWRKY51* responds to salt [38]. However, the function of *PtrWRKY51* in salt stress remains unclear. In the study, *PtrWRKY51* was cloned from poplar and genetically modified in *Arabidopsis* to determine its potential role resistance to salt stress.

To explore the role of genes in response to biological or abiotic stress, tissue-specific gene expression patterns play a crucial role in evaluating the potential function of genes [54–56]. Therefore, a novel idea was provided to investigate the potential functions of *PtrWRKY51* by studying the tissue-specific expression. Analysis of transcription patterns showed that *PtrWRKY51* was predominantly expressed in mature leaves and root (Figure 1A). In addition, the result showed that the primary root length of *oePtrWRKY51* plants was significantly longer than that of Col-0 and *wrky51* on 1/2 MS medium (Figure 3A,B). Moreover, *PtrWRKY51* were up-regulated by salt stress (Figure 1B), and *oePtrWRKY51* plants have a prominently longer root that of Col-0 and *wrky51* on 1/2 MS medium with 200 mM NaCl (Figure 5B,C). These findings suggested that *PtrWRKY51* plays a key role in the regulation of growth and salt tolerance in plants.

In generally, the growth status and external morphology of a plant can reflect the stress degree in the plant. Seed germination is a critically and extremely sensitive period in plant growth, and most plants in salt environment have a negative impact on their growth rate. In the germination stage, dry seed germination is promoted through water expansion, soil water potential is decreased in salt soil, and the absorption of water by seeds is inhibited, which seriously affects the germination rate, vigor index and germination index of seeds, which leading to the reduction of seed emergence rate [8,9]. The result showed that the *oePtrWRKY51* plants had a higher germination rate than the Col-0 and mutant plants under salt treatments (Figure 5A). Therefore, overexpression of *PtrWRKY51* is beneficial to plant in growth under salt stress conditions. Salt stress not only inhibited seed germination, but also significantly inhibited the growth of seedlings. Under salt stress conditions, the root structure of the plant was significantly damaged, and the root length was dramatically reduced, which affected the transport of water and nutrients from the root system to the aboveground part, which led to the obstruction of organic synthesis, and ultimately affected the growth and development of plants [57]. In this study, the primary root length of *oePtrWRKY51* lines was significantly longer than that of Col-0 and *wrky51* mutant under salt stress conditions, which was conducive to plant growth and development (Figure 5B, C). In summary, these results further supported the conclusion that overexpression of PtrWRKY51 enhances salt tolerance in salt stress conditions.

Photosynthesis is one of the most sensitive physiological processes in plant response to stress [58]. Under normal growth conditions, the photosynthesizing organic matter of plants is enough to supply their needs, but under some special stress, the photosynthesis of plants will be affected, and they cannot synthesize enough organic matter to ensure the growth and breeding in plants. These stresses include cold, drought, high temperature, salt and so on. Among them, the damage area of salt stress is wider, and the damage degree of salt stress is deeper. The net photosynthetic rate analysis showed that photosynthesis was higher in the *oePtrWRKY51* lines compared to Col-0 and *wrky51* mutant under normal conditions, while the transpiration rate was lower than that of Col-0 and *wrky51* (Figure 4A,B). Therefore, the instantaneous WUE of the *oePtrWRKY51* lines were significantly higher than those of Col-0 and *wrky51* mutant lines (Figure 4C). WUE has been considered as one of the key factors which affect plant growth, and high WUE can promote plant growth [36]. WUE is also an important physiological index to assess plant stress resistance [59,60]. Under salt stress, the net photosynthetic rate and WUE of plants decreased, but *oePtrWRKY51* lines still had a higher net photosynthetic rate and WUE (Figure 6). Therefore, overexpression of *PtrWRKY51* improve plant growth and salt tolerance.

The chlorophyll content of leaves is closely related to photosynthesis in plants [61]. Salt stress can damage the ultrastructure of chloroplasts and affect the normal function of cells, which affecting the growth and development of plants [62]. We observed that under normal growth conditions, there was no significant difference in chlorophyll content between *oePtrWRKY51* and Col-0 plants, but both were higher than that of *wrky51* mutant plants. Under salt stress, chlorophyll content of *oePtrWRKY51* plants was significantly higher than that of Col-0 and *wrky51* mutant, which suggested *oePtrWRKY51* lines have a growth advantage under salt stress conditions. Taken together, these data indicate that *PtrWRKY51* is a promising gene target in increasing the tolerance of plants under salt stress conditions.

5. Conclusions

In this study, we illustrate the functional characterization of the poplar WRKY transcription factor PtrWRKY51 in the salt response. A PtrWRKY51 was isolated from *Populus trichocarpa*. RT-qPCR analysis revealed that PtrWRKY51 was mainly expressed in mature leaves and root. In addition, PtrWRKY51 is induced by salt stress. Overexpression of PtrWRKY51 in *Arabidopsis* improved salt tolerance. Consistently, overexpression of PtrWRKY51 exhibit an increase in seed germination rate, root length, photosynthetic rate, instantaneous leaf WUE, chlorophyll content under salt stress conditions. Taken together, our data indicate that PtrWRKY51 is a potential candidate gene in the improvement of salt tolerance in poplar by biotechnological strategies.

Supplementary Materials: The following supporting information can be downloaded at: https://www.mdpi.com/article/10.3390/f14020191/s1, Table S1: Primer sequences used for cloning of PtrWRKY51 cDNA and RT-PCR.

Author Contributions: Conceptualization, Y.Z. (Yangyan Zhou), Q.L. and Y.Z. (Yue Zhang); Formal analysis, Y.Z. (Yangyan Zhou) and Q.L.; Funding acquisition, Y.Z. (Yangyan Zhou); Investigation, Y.Z. (Yangyan Zhou) and Q.L.; Methodology, Y.Z. (Yangyan Zhou); Supervision, Y.Z. (Yangyan Zhou), Q.L. and Y.Z. (Yue Zhang); Writing—original draft, Y.Z. (Yangyan Zhou); Writing—review & editing, Y.Z. (Yue Zhang). All authors have read and agreed to the published version of the manuscript.

Funding: This research was financially supported by China Postdoctoral Science Foundation (2022 M710215) and the Natural Science Foundation of Shandong Province (ZR2022QC162).

Data Availability Statement: The data set used in this study can be made available from the authors upon reasonable request.

Conflicts of Interest: The authors declare no conflict of interest.

References

1. Yang, Y.Q.; Guo, Y. Unraveling salt stress signaling in plants. *J. Integr. Plant Biol.* **2018**, *60*, 796–804. [CrossRef] [PubMed]
2. Chen, F.; Fasoli, M.; Tornielli, G.B.; Dal Santo, S.; Pezzotti, M.; Zhang, L.S.; Cai, B.; Cheng, Z.M. The evolutionary history and diverse physiological roles of the grapevine calcium-dependent protein kinase gene family. *PLoS ONE* **2013**, *8*, e80818. [CrossRef] [PubMed]
3. Martinez-Beltran, J.; Manzur, C.L. An overview of the word's salinity issues and FAO's strategy to address this issue. In Proceedings of the International Salinity Forum Proceedings, Riverside, CA, USA, 25–27 April 2005; Volume 5, pp. 311–313.
4. Park, H.J.; Kim, W.Y.; Yun, D.J. A New Insight of Salt Stress Signaling in Plant. *Mol. Cells* **2016**, *39*, 447–459. [CrossRef] [PubMed]
5. Ouhibi, C.; Attia, H.; Rebah, F.; Msilini, N.; Chebbi, M.; Aarrouf, J.; Urban, L.; Lachaal, M. Salt stress mitigation by seed priming with UV-C in lettuce plants: Growth, antioxidant activity and phenolic compounds. *Plant Physiol. Biochem.* **2014**, *83*, 126–133. [CrossRef] [PubMed]
6. Bo, C.; Chen, H.W.; Luo, G.W.; Li, W.; Zhang, X.G.; Ma, Q.; Cheng, B.J.; Cai, R.H. Maize WRKY114 gene negatively regulates salt-stress tolerance in transgenic rice. *Plant Cell Rep.* **2020**, *39*, 135–148. [CrossRef] [PubMed]
7. Keyster, M.; Klein, A.; Ludidi, N. Caspase-like enzymatic activity and the ascorbate-glutathione cycle participate in salt stress tolerance of maize conferred by exogenously applied nitric oxide. *Plant Signal Behav.* **2012**, *7*, 349–360. [CrossRef]
8. Hu, H.; Liu, H.; Liu, F. Seed germination of hemp (*Cannabis sativa* L.) cultivars responds differently to the stress of salt type and concentration. *Ind. Crops Prod.* **2018**, *123*, 254–261. [CrossRef]
9. Wang, W.B.; Kim, Y.H.; Lee, H.S.; Kim, K.Y.; Deng, X.P.; Kwak, S.S. Analysis of antioxidant enzyme activity during germination of alfalfa under salt and drought stresses. *Plant Physiol. Biochem.* **2009**, *47*, 570–577. [CrossRef] [PubMed]
10. İbrahimova, U.; Kumari, P.; Yadav, S.; Rastogi, A.; Antala, M.; Suleymanova, Z.; Zivcak, M.; Tahjib-Ul-Arif, M.; Hussain, S.; Abdelhamid, M.; et al. Progress in understanding salt stress response in plants using biotechnological tools. *J. Biotechnol.* **2021**, *329*, 180–191. [CrossRef]
11. Chen, Y.; Hoehenwarter, W. Changes in the Phosphoproteome and Metabolome Link Early Signaling Events to Rearrangement of Photosynthesis and Central Metabolism in Salinity and Oxidative Stress Response in Arabidopsis. *Plant Physiol.* **2015**, *169*, 3021–3033. [CrossRef]
12. López-Climent, M.F.; Arbona, V.; Pérez-Clemente, R.M.; Gómez-Cadenas, A. Relationship between salt tolerance and photosynthetic machinery performance in citrus. *Environ. Exp. Bot.* **2008**, *62*, 176–184. [CrossRef]
13. Mittal, S.; Kumari, N.; Sharma, V. Differential response of salt stress on *Brassica juncea*: Photosynthetic performance, pigment, proline, D1 and antioxidant enzymes. *Plant Physiol. Biochem.* **2012**, *54*, 17–26. [CrossRef] [PubMed]
14. He, L.; Wu, Y.H.; Zhao, Q.; Wang, B.; Liu, Q.L.; Zhang, L. Chrysanthemum *DgWRKY2* Gene Enhances Tolerance to Salt Stress in Transgenic Chrysanthemum. *Int. J. Mol. Sci.* **2018**, *19*, 2062. [CrossRef] [PubMed]
15. Jakoby, M.; Weisshaar, B.; Dröge-Laser, W.; Vicente-Carbajosa, J.; Tiedemann, J.; Kroj, T.; Parcy, F. bZIP Research Group. bZIP transcription factors in *Arabidopsis*. *Trends Plant Sci.* **2002**, *7*, 106–111. [CrossRef]
16. Abe, H.; Urao, T.; Ito, T.; Seki, M.; Shinozaki, K.; Yamaguchi-Shinozaki, K. Arabidopsis AtMYC2 (bHLH) and AtMYB2 (MYB) function as transcriptional activators in abscisic acid signaling. *Plant Cell* **2003**, *15*, 63–78. [CrossRef] [PubMed]
17. Marè, C.; Mazzucotelli, E.; Crosatti, C.; Francia, E.; Stanca, A.M.; Cattivelli, L. Hv-WRKY38: A new transcription factor involved in cold- and drought-response in barley. *Plant Mol. Biol.* **2004**, *55*, 399–416. [CrossRef] [PubMed]
18. Aharoni, A.; Dixit, S.; Jetter, R.; Thoenes, E.; van Arkel, G.; Pereira, A. The SHINE clade of AP2 domain transcription factors activates wax biosynthesis, alters cuticle properties, and confers drought tolerance when overexpressed in Arabidopsis. *Plant Cell* **2004**, *16*, 2463–2480. [CrossRef]
19. Hu, H.; You, J.; Fang, Y.; Zhu, X.; Qi, Z.; Xiong, L. Characterization of transcription factor gene SNAC2 conferring cold and salt tolerance in rice. *Plant Mol. Biol.* **2008**, *67*, 169–181. [CrossRef]
20. Rushton, P.J.; Somssich, I.E.; Ringler, P.; Shen, Q.J. WRKY transcription factors. *Trends Plant Sci.* **2010**, *15*, 247–258. [CrossRef]
21. Ishiguro, S.; Nakamura, K. Characterization of a cDNA encoding a novel DNA-binding protein, SPF1, that recognizes SP8 sequences in the 5′ upstream regions of genes coding for sporamin and beta-amylase from sweet potato. *Mol. Gen. Genet.* **1994**, *244*, 563–571. [CrossRef]
22. Rushton, P.J.; Macdonald, H.; Huttly, A.K.; Lazarus, C.M.; Hooley, R. Members of a new family of DNA binding proteins bind to a conserved cis-element in the promoters of a-Amy2 genes. *Plant Mol. Biol.* **1995**, *29*, 691–702. [CrossRef]
23. Rushton, P.J.; Torres, J.T.; Parniske, M.; Wernert, P.; Hahlbrock, K.; Somssich, I.E. Interaction of elicitor-induced DNA binding proteins with elicitor response elements in the promoters of parsley PR1 genes. *EMBO J.* **1996**, *15*, 5690–5700. [CrossRef] [PubMed]
24. de Pater, S.; Greco, V.; Pham, K.; Memelink, J.; Kijne, J. Characterization of a zinc-dependent transcriptional activator from Arabidopsis. *Nucleic Acids Res.* **1996**, *24*, 4624–4631. [CrossRef] [PubMed]
25. Eulgem, T.; Rushton, P.J.; Robatzek, S.; Somssich, I.E. The WRKY superfamily of plant transcription factors. *Trends Plant Sci.* **2000**, *5*, 199–206. [CrossRef] [PubMed]
26. Qiu, Y.P.; Jing, S.J.; Fu, J.; Li, L.; Yu, D.Q. Cloning and analysis of expression profile of 13 WRKY genes in rice. *Chin. Sci. Bull.* **2004**, *49*, 2159–2168. [CrossRef]

27. Wu, H.L.; Ni, Z.F.; Yao, Y.Y.; Guo, G.G.; Sun, Q.X. Cloning and expression profiles of 15 genes encoding WRKY transcription factor in wheat (*Triticum aestivem L.*). *Prog. Nat. Sci.* **2008**, *18*, 697–705. [CrossRef]
28. Qiu, Y.P.; Yu, D.Q. Over-expression of the stress-induced *OsWRKY45* enhances disease resistance and drought tolerance in *Arabidopsis*. *Environ. Exp. Bot.* **2008**, *65*, 35–47. [CrossRef]
29. Ding, Z.J.; Yan, J.Y.; Xu, X.Y.; Yu, D.Q.; Li, G.X.; Zhang, S.Q.; Zheng, S.J. Transcription factor WRKY46 regulates osmotic stress responses and stomatal movement independently in *Arabidopsis*. *Plant J.* **2014**, *79*, 13–27. [CrossRef]
30. Zhou, Q.Y.; Tian, A.G.; Zou, H.F.; Xie, Z.M.; Lei, G.; Huang, J.; Wang, C.M.; Wang, H.W.; Zhang, J.S.; Chen, S.Y. Soybean WRKY-type transcription factor genes, GmWRKY13, GmWRKY21, and GmWRKY54, confer differential tolerance to abiotic stresses in transgenic Arabidopsis plants. *Plant Biotechnol. J.* **2008**, *6*, 486–503. [CrossRef]
31. Niu, C.F.; Wei, W.; Zhou, Q.Y.; Tian, A.G.; Hao, Y.J.; Zhang, W.K.; Ma, B.; Lin, Q.; Zhang, Z.B.; Zhang, J.S.; et al. Wheat WRKY genes TaWRKY2 and TaWRKY19 regulate abiotic stress tolerance in transgenic Arabidopsis plants. *Plant Cell Environ.* **2012**, *35*, 1156–1170. [CrossRef]
32. He, G.H.; Xu, J.Y.; Wang, Y.X.; Liu, J.M.; Li, P.S.; Chen, M.; Ma, Y.Z.; Xu, Z.S. Drought-responsive WRKY transcription factor genes TaWRKY1 and TaWRKY33 from wheat confer drought and/or heat resistance in Arabidopsis. *BMC Plant Biol.* **2016**, *16*, 116. [CrossRef]
33. Cai, R.; Dai, W.; Zhang, C.; Wang, Y.; Wu, M.; Zhao, Y.; Ma, Q.; Xiang, Y.; Cheng, B. The maize WRKY transcription factor ZmWRKY17 negatively regulates salt stress tolerance in transgenic Arabidopsis plants. *Planta* **2017**, *246*, 1215–1231. [CrossRef]
34. Ma, J.; Gao, X.; Liu, Q.; Shao, Y.; Zhang, D.; Jiang, L.; Li, C. Overexpression of *TaWRKY146* Increases Drought Tolerance through Inducing Stomatal Closure in *Arabidopsis thaliana*. *Front. Plant Sci.* **2017**, *8*, 2036. [CrossRef] [PubMed]
35. Wu, M.; Liu, H.; Han, G.; Cai, R.; Pan, F.; Xiang, Y. A moso bamboo WRKY gene PeWRKY83 confers salinity tolerance in transgenic Arabidopsis plants. *Sci. Rep.* **2017**, *7*, 11721. [CrossRef] [PubMed]
36. Wang, C.T.; Ru, J.N.; Liu, Y.W.; Yang, J.F.; Li, M.; Xu, Z.S.; Fu, J.D. The Maize WRKY Transcription Factor ZmWRKY40 Confers Drought Resistance in Transgenic *Arabidopsis*. *Int. J. Mol. Sci.* **2018**, *19*, 2580. [CrossRef] [PubMed]
37. Zhang, L.; Cheng, J.; Sun, X.; Zhao, T.; Li, M.; Wang, Q.; Li, S.; Xin, H. Overexpression of VaWRKY14 increases drought tolerance in Arabidopsis by modulating the expression of stress-related genes. *Plant Cell Rep.* **2018**, *37*, 1159–1172. [CrossRef] [PubMed]
38. Jiang, Y.Z.; Duan, Y.J.; Yin, J.; Ye, S.L.; Zhu, J.R.; Zhang, F.Q.; Lu, W.X.; Fan, D.; Luo, K.M. Genome-wide identification and characterization of the *Populus* WRKY transcription factor family and analysis of their expression in response to biotic and abiotic stresses. *J. Exp. Bot.* **2014**, *65*, 6629–6644. [CrossRef] [PubMed]
39. Song, J.; Lu, S.; Chen, Z.Z.; Lourenco, R.; Chiang, V.L. Genetic transformation of Populus trichocarpa genotype Nisqually-1: A functional genomic tool for woody plants. *Plant Cell Physiol.* **2006**, *47*, 1582–1589. [CrossRef]
40. Clough, S.J.; Bent, A.F. Floral dip: A simplified method for Agrobacterium-mediated transformation of *Arabidopsis thaliana*. *Plant J.* **1998**, *16*, 735–743. [CrossRef]
41. Porebski, S.; Bailey, L.G.; Baum, B.R. Modification of ctab dna extraction protocol for plants containing high polysaccharide and polyphenol components. *Plant Mol. Biol. Rep.* **1997**, *15*, 8–15. [CrossRef]
42. Schmittgen, T.D.; Livak, K.J. Analyzing real-time PCR data by the comparative C(T) method. *Nat. Protoc.* **2008**, *3*, 1101–1108. [CrossRef]
43. Wang, C.; Liu, S.; Dong, Y.; Zhao, Y.; Geng, A.; Xia, X.; Yin, W. PdEPF1 regulates water-use efficiency and drought tolerance by modulating stomatal density in poplar. *Plant Biotechnol. J.* **2016**, *14*, 849–860. [CrossRef] [PubMed]
44. Han, X.; Tang, S.; An, Y.; Zheng, D.; Xia, X.; Yin, W. Overexpression of the poplar NF-YB7 transcription factor confers drought tolerance and improves water-use efficiency in Arabidopsis. *J. Exp. Bot.* **2013**, *64*, 4589–4601. [CrossRef] [PubMed]
45. Wong, S.C.; Cowan, I.R.; Farquhar, G.D. Leaf Conductance in Relation to Assimilation in Eucalyptus pauciflora Sieb. ex Spreng: Influence of Irradiance and Partial Pressure of Carbon Dioxide. *Plant Physiol.* **1978**, *62*, 670–674. [CrossRef] [PubMed]
46. Lichtenthaler, H.K. Chlorophylls and carotenoids-pigments of photosynthetic biomembranes. *Methods Enzymol.* **1987**, *148*, 350–382.
47. Parinita, A.; Reddy, M.P.; Jitendra, C. WRKY: Its structure, evolutionary relationship, DNA-binding selectivity, role in stress tolerance and development of plants. *Mol. Biol. Rep.* **2011**, *38*, 3883–3896.
48. Chen, L.; Song, Y.; Li, S.; Zhang, L.; Zou, C.; Yu, D. The role of WRKY transcription factors in plant abiotic stresses. *Biochim. Et Biophys. Acta* **2012**, *1819*, 120–128. [CrossRef] [PubMed]
49. Wu, X.L.; Shiroto, Y.; Kishitani, S.; Ito, Y.; Toriyama, K. Enhanced heat and drought tolerance in transgenic rice seedlings overexpressing *OsWRKY11* under the control of HSP101 promoter. *Plant Cell Rep.* **2009**, *28*, 21–30. [CrossRef] [PubMed]
50. Shen, H.S.; Liu, C.T.; Zhang, Y.; Meng, X.P.; Zhou, X.; Chu, C.C.; Wang, X.P. *OsWRKY30* is activated by MAP kinases to confer drought tolerance in rice. *Plant Mol. Biol.* **2012**, *80*, 241–253. [CrossRef]
51. Zhu, H.; Zhou, Y.Y.; Zhai, H.; He, S.Z.; Zhao, N.; Liu, Q.C. A Novel Sweetpotato WRKY transcription factor, IbWRKY2, positively regulates drought and salt tolerance in transgenic *Arabidopsis*. *Biomolecules* **2020**, *10*, 506. [CrossRef]
52. Xie, Z.; Zhang, Z.L.; Zou, X.; Yang, G.; Komatsu, S.; Shen, Q.J. Interactions of two abscisic-acid induced WRKY genes in repressing gibberellin signaling in aleurone cells. *Plant J.* **2006**, *46*, 231–242. [CrossRef]
53. Gao, Q.M.; Venugopal, S.; Navarre, D.; Kachroo, A. Low oleic acid-derived repression of jasmonic acid-inducible defense responses requires the WRKY50 and WRKY51 proteins. *Plant Physiol.* **2011**, *155*, 464–476. [CrossRef] [PubMed]

54. Siefers, N.; Dang, K.K.; Kumimoto, R.W.; Bynum, W.E.; Tayrose, G.; Holt, B.F. Tissue-specific expression patterns of Arabidopsis NF-Y transcription factors suggest potential for extensive combinatorial complexity. *Plant Physiol.* **2009**, *149*, 625–641. [CrossRef] [PubMed]
55. Jeong, J.S.; Kim, Y.S.; Redillas, M.C.; Jang, G.; Jung, H.; Bang, S.W.; Choi, Y.D.; Ha, S.H.; Reuzeau, C.; Kim, J.K. OsNAC5 overexpression enlarges root diameter in rice plants leading to enhanced drought tolerance and increased grain yield in the field. *Plant Biotechnol. J.* **2013**, *11*, 101–114. [CrossRef]
56. Horváth, B.M.; Magyar, Z.; Zhang, Y.; Hamburger, A.W.; Bakó, L.; Visser, R.G.; Bachem, C.W.; Bögre, L. EBP1 regulates organ size through cell growth and proliferation in plants. *EMBO J.* **2006**, *25*, 4909–4920. [CrossRef]
57. Sun, W.; Xu, X.; Zhu, H.; Liu, A.; Liu, L.; Li, J.; Hua, X. Comparative transcriptomic profiling of a salt-tolerant wild tomato species and a salt-sensitive tomato cultivar. *Plant Cell Physiol.* **2010**, *51*, 997–1006. [CrossRef]
58. Rivero, R.M.; Mestre, T.C.; Mittler, R.; Rubio, F.; Garcia-Sanchez, F.; Martinez, V. The combined effect of salinity and heat reveals a specific physiological, biochemical and molecular response in tomato plants. *Plant Cell Environ.* **2014**, *37*, 1059–1073. [CrossRef]
59. Zhou, Y.; Zhang, Y.; Wang, X.; Han, X.; An, Y.; Lin, S.; Shen, C.; Wen, J.; Liu, C.; Yin, W.; et al. Root-specific NF-Y family transcription factor, PdNF-YB21, positively regulates root growth and drought resistance by abscisic acid-mediated indoylacetic acid transport in Populus. *New Phytol.* **2020**, *227*, 407–426. [CrossRef]
60. Karaba, A.; Dixit, S.; Greco, R.; Aharoni, A.; Trijatmiko, K.R.; Martinez, N.M.; Krishnan, A.; Nataraja, K.N.; Udayakumar, M.; Pereira, A. Improvement of water use efficiency in rice by expression of HARDY, an Arabidopsis drought and salt tolerance gene. *Proc. Natl. Acad. Sci. USA* **2007**, *104*, 15270–15275. [CrossRef]
61. An, Y.; Han, X.; Tang, S.; Xia, X.; Yin, W. Poplar GATA transcription factor PdGNC is capable of regulating chloroplast ultrastructure, photosynthesis, and vegetative growth in Arabidopsis under varying nitrogen levels. *Plant Cell Tissue Organ Cult.* **2014**, *119*, 313–327. [CrossRef]
62. Peharec Štefanić, P.; Koffler, T.; Adler, G.; Bar-Zvi, D. Chloroplasts of salt-grown Arabidopsis seedlings are impaired in structure, genome copy number and transcript levels. *PLoS ONE* **2013**, *8*, e82548. [CrossRef] [PubMed]

forests

MDPI

Article

Root-Growth-Related MaTCP Transcription Factors Responsive to Drought Stress in Mulberry

Wuqi Wei, Jinzhi He, Yiwei Luo, Zhen Yang, Xiaoyu Xia, Yuanxiang Han and Ningjia He *

State Key Laboratory of Silkworm Genome Biology, Southwest University, Chongqing 400715, China
* Correspondence: hejia@swu.edu.cn

Abstract: Root growth regulation plays a crucial role in the acclimatization of plants to their surroundings, but the molecular mechanisms underlying this process remain largely uncertain. Teosinte branched1/cycloidea/proliferating cell factor (TCP) transcription factors are crucial elements linking together plant growth and development, phytohormone signaling, and stress response. In this study, 15 TCP transcription factors were identified in the mulberry (*Morus alba*) genome. Gene structure, conserved motif, and phylogenetic analyses revealed the conservation and divergence of these MaTCPs, thus providing insights into their functions. A promoter analysis uncovered distinct numbers and compositions of *cis*-elements in *MaTCP* gene promoter regions that may be connected to reproductive growth and phytohormone and stress responses. An expression pattern analysis of the 15 *MaTCP* genes in mulberry roots indicated that transcriptional levels of *MaTCP2*, *MaTCP4-1*, *MaTCP8*, *MaTCP9-1*, and *MaTCP20-2* are correlated with root development. As revealed by changes in their expressions after drought treatment, these five *MaTCP* genes are involved in root growth and may increase mulberry tolerance to drought. Our findings lay the foundation for future functional studies of these genes.

Keywords: mulberry; MaTCP transcription factor; drought tolerance; root development

Citation: Wei, W.; He, J.; Luo, Y.; Yang, Z.; Xia, X.; Han, Y.; He, N. Root-Growth-Related MaTCP Transcription Factors Responsive to Drought Stress in Mulberry. *Forests* **2023**, *14*, 143. https://doi.org/10.3390/f14010143

Academic Editor: Stéphane Maury

Received: 4 December 2022
Revised: 5 January 2023
Accepted: 11 January 2023
Published: 12 January 2023

1. Introduction

Given their sessile nature, plants are constantly subjected to a range of adverse environmental conditions, including drought, heat, cold, and light. Future climate change will most likely intensify these environmental pressures. For example, many climate models anticipate that drought will increasingly threaten crop growth and productivity [1], and a deeper comprehension of the factors influencing drought tolerance is thus required. Plants have evolved a series of complex mechanisms to perceive and respond to environmental stimuli and can withstand drought using a variety of physiological, morphological, and biochemical processes [2,3]. Reduced transpiration, the development of a deep and robust root system, increased water intake, and maintenance of tissue water potential are the key methods used by plants to resist drought [4]. Nevertheless, the underlying physiological and molecular mechanisms are not fully understood.

Recent studies in diverse plant species have shown that teosinte branched1/cycloidea/proliferating cell factor (TCP) transcription factors are crucial for the development of plant roots [5–7]. Investigation into TCP in plant stress resistance has gained a lot of attention since it links together plant development, stress response, and hormone signaling [8–10]. TCP transcription factors constitute a plant-specific protein family with a conserved TCP domain containing a 59-amino-acid non-canonical basic helix–loop–helix (bHLH) structure that enables DNA binding and protein–protein interactions [11–14]. TCP proteins are categorized into two classes according to the characteristics of their TCP domains: class I (PCF or TCP-P class) and class II (TCP-C class) [12]. Class II TCP members are subdivided in turn into CIN and CYC/TB1 subclasses [15]. Genes encoding TCPs have been identified and analyzed in many plant species, and accumulating evidence indicates that the TCP family plays important regulatory roles in plant growth and development, hormone signaling, and stress response [8,9,13,16]. AtTCP14 and AtTCP15 have been proposed to regulate cell proliferation and organ growth and promote gibberellin-induced seed germination in Arabidopsis [17,18]. In Arabidopsis embryonic root apical meristem, growth-repressor DELLA proteins bind and inhibit AtTCP14 and AtTCP15 activities during cell proliferation in the presence of low levels of gibberellic acid [7]. Moreover, *GbTCP*, the cotton homolog of *AtTCP15*, positively regulates jasmonic acid (JA) biosynthesis and response as well as other pathways. Silencing of *GbTCP* results in plants with lower JA levels and reduced cotton fiber elongation. When overexpressed in Arabidopsis, *GbTCP* also promotes the initiation and extension of root hair development [19]. In rice, OsTCP19 synthesis is triggered by exposure to water deficit and salt stress conditions. OsTCP19 induces the expression of ABI4, which encodes a transcription factor involved in ABA signal transduction, and directly interacts with the ABI4 protein to positively regulate its activity. Increased stress resistance in *OsTCP19*-overexpressing Arabidopsis is accompanied by decreased water loss, decreased production of reactive oxygen species, and lipid droplet hyperaccumulation [20]. In addition, *TCP* genes in Arabidopsis and tomato have been revealed to be targets of the microRNA319 (miR319) family [21,22], which is involved in plant stress response. Transgenic creeping bentgrass (*Agrostis stolonifera*) overexpressing *Osa-miR319a* exhibits improved drought and salt tolerance along with higher leaf wax content and increased water retention [23]. Taken together, these results demonstrate that the TCP family is essential for plant development and stress response.

Mulberry (*Morus* L.), an ecologically and economically important deciduous tree [24,25], has been used for thousands of years, both as a food source for domesticated silkworms and as a raw material for the production of juice, jam, wine, and traditional Chinese medicines. Because of its widespread distribution, rapid maturation cycle, well-developed root system, and capacity to resist a variety of environmental stresses, mulberry can be grown under many conditions that are unsuitable for other cash crops [26]. Mulberry is even a prospective species for the ecological rehabilitation of the drawdown zone of the Three Gorges Reservoir in China [27]. An understanding of the ability of mulberry to tolerate harsh settings is thus crucial for the selective breeding and cultivation of this species. Such knowledge would benefit the advancement of sericulture and the pharmaceutical industry and help preserve the ecological environment.

In the present study, we systematically identified and analyzed mulberry TCP transcription factors. We performed analyses of gene structures, expression patterns, subcellular localizations, and responses to drought stress, thereby generating new data for elucidating the roles of mulberry TCP transcription factors in future functional studies. This would aid in the selection and breeding of mulberry cultivars that are drought-tolerance.

2. Materials and Methods

2.1. Plant Materials and Culture Conditions

Seeds of mulberry (*Morus atropurpurea* 'Guisangyou12', abbreviated as GY12) were soaked in aseptic water for 48 h at 4 °C and then transferred to a 25 °C climate chamber under a 16-h light/8-h dark cycle. After germination, the mulberry seedlings were transplanted and raised on sterile soil under well-watered conditions. In subsequent experiments, 21- to 49-day-old mulberry seedlings were used to study root growth; the root lengths of 10 seedlings at each development stage were measured with a ruler with stamped millimeter graduation, while 1-month-old seedlings were used for drought treatment.

2.2. Phylogenetic and Comparative Sequence Analyses of the TCP Family

Annotated sequences of *Arabidopsis thaliana* and *Populus euphratica* [28] TCP transcription factors and CDS regions were downloaded from the PlantTFDB (http://planttfdb.gao-lab.org/, accessed on 12 July 2022) and NCBI (https://www.ncbi.nlm.nih.gov/, accessed on 12 July 2022) databases (Table S1). After performing tblastn searches (E-value < 10^{-10}) to identify candidate TCP members in the *M. alba* genome [29], we applied Interpro (https://www.ebi.ac.uk/interpro/, accessed on 25 July 2022) to confirm the presence of the conserved TCP domain in each candidate [30]. A phylogenetic tree based on full-length amino acid sequences from *A. thaliana*, *P. euphratica*, and *M. alba* was then constructed in MEGA11 by the neighbor-joining method with 1000 bootstrap repetitions, with all other parameters kept at default settings [31]. Multiple alignments of the MaTCP proteins were performed by using CLUSTALW (https://www.genome.jp/tools-bin/clustalw, accessed on 6 August 2022). Conserved motifs of MaTCP proteins were determined using MEME (https://meme-suite.org/meme/, accessed on 5 August 2022).

2.3. Promoter Element Analyses

The PlantCARE database (https://bioinformatics.psb.ugent.be/webtools/plantcare/html/, accessed on 10 November 2022) was used to determine the number and composition of stress- and development-related elements in the 2-kb promoter of *MaTCP* genes. The data analysis and mapping were carried out with TBtools [32].

2.4. RNA Extraction, Gene Cloning, and Quantitative Real-Time PCR (RT-qPCR) Analysis

RNA was extracted from GY12 roots using an RNAprep Pure Plant Plus kit (Tiangen, Beijing, China). Next, 1 µg of RNA extracted from GY12 was synthesized into cDNA by using a PrimeScript RT Reagent kit with gDNA Eraser (Takara, Beijing, China). We used the NCBI Primer-BLAST online tool to design RT-qPCR primer pairs specific to the *MaTCP* genes shown in Table S2. *RPL15* was used as an internal reference gene [33]. RT-qPCR amplifications were conducted using SuperReal PreMix Plus (Tiangen, Beijing, China) on Applied Biosystems StepOne and StepOnePlus Real-Time PCR systems (Thermo Fisher Scientific, Waltham, MA, USA). Each experiment was performed with three technical replicates.

2.5. Subcellular Localization

The subcellular location of each MaTCP protein was predicted using Cell-PLoc 2.0 (http://www.csbio.sjtu.edu.cn/bioinf/Cell-PLoc-2/, accessed on 14 August 2022). For the subcellular localization experiment, the ORFs of *MaTCP2*, *MaTCP4-1*, *MaTCP8*, *MaTCP9-1*, and *MaTCP20-2* were inserted into a pZYGC plant expression vector containing a GFP reporter gene to generate *35S:MaTCP:GFP* fusion expression vectors. Transient transformation of onion epidermal cells was performed by the *Agrobacterium*-mediated method. Firstly, 1 cm^2 of the onions' inner epidermis was soaked in Agrobacterium solution for 20 min, then transferred to MS solid medium and incubated at 25 °C in dark for 8 h, followed by another 24 h under a photoperiod of 16 h light/8 h dark. The transformed cells were placed on a glass slide, stained with DAPI, and observed under an Olympus IX73 inverted fluorescent microscope (Olympus, Tokyo, Japan).

2.6. Stress Treatment

For the stress treatment, 1-month-old GY12 seedlings were treated with 10% PEG6000 solution for 0, 1, 3, 5, and 7 days. The initial day was determined to be day 0 of treatment. The 0-day plants were the control. The roots of the control and the treated plants were then sampled, frozen in liquid nitrogen, and stored at −80 °C for RNA extraction.

2.7. Data Processing and Statistical Analysis

Relative levels of *MaTCP* transcripts were determined by the $2^{-\Delta\Delta CT}$ method [34]. A time series analysis was performed online using BioLadder (https://www.bioladder.cn/web/, accessed on 8 October 2022). Correlation coefficients between GY12 root lengths and *MaTCP* gene expressions were calculated using Office 2010. Statistical comparisons of samples were performed by Student's t test for one-way ANOVA followed by the Student–Newman–Keuls (*SNK*) post-hoc test for multiple comparisons (p-values < 0.05).

3. Results

3.1. Characteristics of MaTCP Transcription Factors in Mulberry

A total of 15 *MaTCPs* were identified in the mulberry genome. As shown in Dataset S1 and Table 1, the highly conserved TCP domain (Interpro accession number PF03634) was present in each MaTCP protein. Eight and seven of the identified mulberry TCPs belonged to classes I and II, respectively. Seven class II MaTCPs were further subdivided into one CYC/TB1 MaTCP12 and six CIN-type MaTCPs.

Table 1. Information on MaTCP proteins identified in this study.

TFID	Common Name	Type	Conserved Domain (aa)	Chromosome Location	CDS Length (bp)	Protein Length (aa)	Predicted Protein Localization
M.alba_G0010963	MaTCP2	CIN	78–236	Chr04: 10504440..10506373	1491	496	Nucleus
M.alba_G0019145	MaTCP4-1	CIN	55–170	Chr09: 2738157..2740500	1332	443	Nucleus
M.alba_G0002273	MaTCP4-2	CIN	24–117	Chr10: 10539871..10541516	1077	358	Nucleus
M.alba_G0018610	MaTCP5	CIN	54–147	Chr08: 11690488..11691912	1083	360	Nucleus
M.alba_G0008688	MaTCP7	PCF	51–128	Chr02: 11919887..11922133	831	276	Nucleus
M.alba_G0015857	MaTCP8	PCF	172–336	Chr06: 21186236..21188526	1665	554	Nucleus
M.alba_G0005407	MaTCP9-1	PCF	113–190	Chr12: 15460670..15462253	1215	404	Nucleus
M.alba_G0015368	MaTCP9-2	PCF	17–144	Chr06: 17004501..17004995	495	164	Nucleus
M.alba_G0010736	MaTCP10	CIN	135–219	Chr04: 6280140..6281898	1083	360	Nucleus
M.alba_G0007510	MaTCP12	CYC/TB1	126–257	Chr14: 7692279..7693628	1350	449	Nucleus
M.alba_G0001832	MaTCP13	CIN	61–207	Chr10: 4961787..4963406	1125	374	Nucleus
M.alba_G0018434	MaTCP14	PCF	107–282	Chr08: 7338007..7339840	1308	435	Nucleus
M.alba_G0003268	MaTCP19	PCF	91–157	Chr11: 3914056..3915332	1102	366	Nucleus
M.alba_G0006782	MaTCP20-1	PCF	37–125	Chr14: 728387..729849	810	269	Nucleus
M.alba_G0001316	MaTCP20-2	PCF	69–156	Chr10: 593920..595287	1005	334	Nucleus

To investigate the relationships of MaTCPs to other TCPs and to obtain insights into their potential functions, we constructed a phylogenetic tree based on full-length amino acid sequences of TCPs from *M. alba*, *A. thaliana*, and *P. euphratica*. As shown in Figure 1A, all of the sequences were classified into two clades. Clade I was named the PCF clade and contained 39 TCP proteins. Clade II was subdivided into 10 TCPs with CYC/TB1 and 23 TCPs with CIN. MaTCPs were more closely related to the TCPs in *Populus* than to those in Arabidopsis.

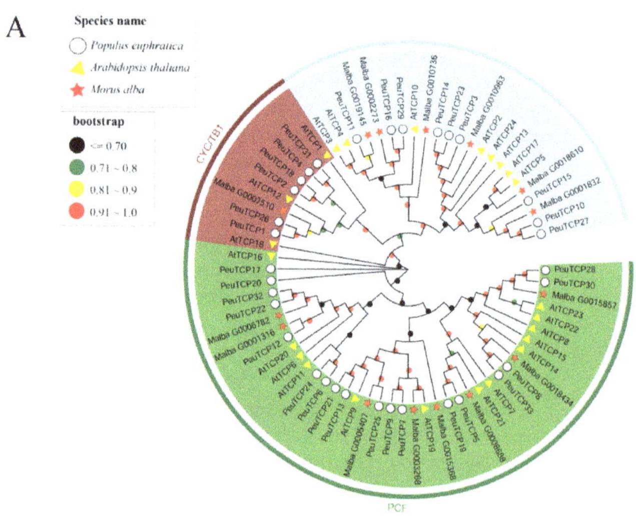

Figure 1. *Cont.*

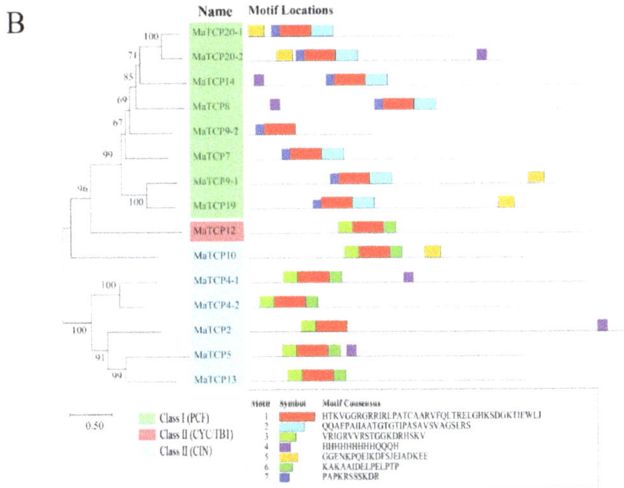

Figure 1. The phylogenetic and protein structure analyses of MaTCP family members. (**A**) Neighbor-joining phylogenetic tree of TCP transcription factors in *M. alba*, *P. euphratica*, and *A. thaliana*. Three different colors were designated as three subclades of PCF, CYC/TB1, and CIN. (**B**) Conserved motifs of MaTCP proteins. Different motifs were shown by different colors numbered 1–7.

To further examine the sequence features of MaTCP proteins, we looked for conserved motifs in the aligned set of TCP amino acid sequences (Figure S1). Seven conserved motifs were identified among MaTCP proteins. The N-terminal conserved core region (marked as motif 1) was present in all 15 MaTCP proteins. Within a given MaTCP subclass, the composition of motifs was similar, whereas compositions differed greatly between the groups. For instance, the majority of MaTCP proteins in the TCP-P class contained motif 2, whereas most TCP-C class MaTCP proteins possessed motifs 3 and 6 (Figure 1B).

3.2. The Promoter Analyses of MaTCP Genes

Cis-acting elements in the 2-kb upstream region of the *MaTCP* gene translational start site were also analyzed and classified into three functional categories: hormone responsiveness, abiotic stress responsiveness, and tissue-specific expression (Figure 2). Moreover, the *MaTCP2* gene promoter region contained three abiotic stress elements related to low-temperature responsiveness and drought inducibility and seven hormone elements involved in GA, MeJA, and SA responsiveness. *MaTCP4-1* possessed an abiotic stress element associated with defense and stress responsiveness and nine hormone elements related to ABA, GA, MeJA, and SA responsiveness. *MaTCP8* included two abiotic stress elements concerned with drought inducibility and defense and stress responsiveness and six hormone elements involved in ABA and MeJA responsiveness. *MaTCP9-1* harbored two abiotic stress elements associated with drought inducibility and 15 hormone elements with a function in ABA, GA, and MeJA responsiveness. *MaTCP20-2* contained a tissue-specific expression element related to meristem expression and six hormone elements involved in ABA, GA, SA, and MeJA responsiveness.

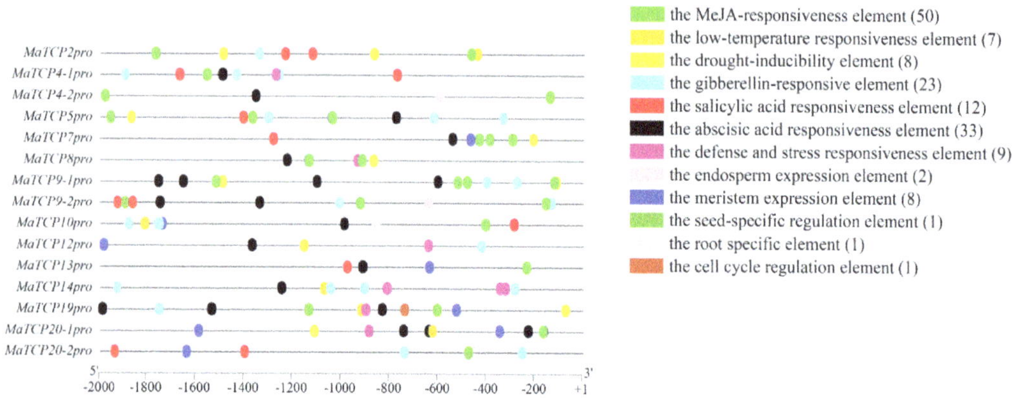

Figure 2. The promoter analyses of *MaTCP* genes. The number in parentheses indicated the number of *cis*-acting elements.

3.3. Expression Patterns of MaTCP Genes during Mulberry Seedling Growth

To analyze *MaTCP* expression patterns during mulberry seedling growth, transcript levels of 15 *MaTCP* genes were measured in GY12 seedlings at five developmental stages. Meanwhile, the roots of ten individuals for each stage were collected and their lengths were measured (Figure 3A,B). As shown in Figure 3C,D, the 15 *MaTCP* genes were grouped according to their expression profiles into three clusters. Clusters 1 and 2 contained four and five *MaTCP* genes, respectively, whereas six *MaTCP* genes were grouped into cluster 3. The expressions of *MaTCP* genes in cluster 3 gradually increased over time, with the highest transcript levels observed in 49-day-old GY seedlings (Figure 3C); this trend was the same as the increases in root length detailed in Table S3. A correlation analysis was also performed between the expressions of the 15 *MaTCP* genes and root lengths at five developmental stages. As shown in Table 2, the expressions of *MaTCP2*, *MaTCP4-1*, *MaTCP8*, *MaTCP9-1*, *MaTCP14*, *MaTCP19*, and *MaTCP20-2* were significantly correlated with root length, with all of them having correlation coefficients higher than 0.8.

3.4. Association of Five MaTCP Genes with Root Development

The expression trend of genes in cluster 3 was similar to that of root growth (Figure 3), and the relative expressions of *MaTCP2*, *MaTCP4-1*, *MaTCP8*, *MaTCP9-1*, and *MaTCP20-2* genes in cluster 3 were significantly correlated with root length (Table 2). These five genes were considered to be highly associated with root development and were thus subjected to further analysis (Figure 4).

3.5. Subcellular Localization Analyses of Five MaTCP Proteins

To confirm the predicted nuclear locations of the five abovementioned MaTCP proteins, we carried out a subcellular localization analysis. As shown in Figure 5, green fluorescence was only detected in the nuclei of the onion epidermal cells transformed with *35S:MaTCP:GFP*, compared with those transformed with *35S:GFP*. MaTCP2, MaTCP4-1, MaTCP8, MaTCP9-1, and MaTCP20-2 proteins were thus localized to the nucleus.

Table 2. Correlation coefficients between the relative expression of MaTCP genes and root length.

	Root Length	*MaTCP2*	*MaTCP4-1*	*MaTCP4-2*	*MaTCP5*	*MaTCP7*	*MaTCP8*	*MaTCP9-1*	*MaTCP9-2*	*MaTCP10*	*MaTCP12*	*MaTCP13*	*MaTCP14*	*MaTCP19*	*MaTCP20-1*	*MaTCP20-2*
Root Length	1.0000															
MaTCP2	**0.8743**	1.0000														
MaTCP4-1	**0.8234**	0.9902	1.0000													
MaTCP4-2	0.6699	0.3515	0.2348	1.0000												
MaTCP5	−0.4257	−0.4922	−0.3966	−0.6194	1.0000											
MaTCP7	0.6955	0.8982	0.9492	0.0031	−0.1111	1.0000										
MaTCP8	**0.8761**	0.9808	0.9472	0.4683	−0.6480	0.7989	1.0000									
MaTCP9-1	**0.8481**	0.9968	0.9875	0.3488	−0.5323	0.8911	0.9839	1.0000								
MaTCP9-2	−0.2840	−0.3748	−0.2780	−0.4582	0.9480	0.0143	−0.5367	−0.4068	1.0000							
MaTCP10	0.5372	0.1679	0.0528	0.9789	−0.5054	−0.1547	0.2840	0.1647	−0.3416	1.0000						
MaTCP12	−0.6220	−0.6888	−0.6220	−0.6790	0.8925	−0.4070	−0.7914	−0.7299	0.7122	−0.5611	1.0000					
MaTCP13	−0.3205	−0.6062	−0.5719	−0.0533	0.7356	−0.3836	−0.6815	−0.6290	0.8311	0.1258	0.5628	1.0000				
MaTCP14	**0.8765**	0.7762	0.6946	0.8480	−0.7511	0.4828	0.8541	0.7840	−0.5664	0.7320	−0.8993	−0.4104	1.0000			
MaTCP19	**0.9249**	0.8014	0.7117	0.8032	−0.7134	0.4834	0.8760	0.7908	−0.6072	0.6673	−0.7799	−0.5066	0.9523	1.0000		
MaTCP20-1	0.7055	0.5086	0.4407	0.8764	−0.5261	0.3010	0.5691	0.5223	−0.2542	0.8447	−0.7812	0.0261	0.8811	0.7290	1.0000	
MaTCP20-2	**0.9031**	0.9684	0.9312	0.5514	−0.6460	0.7850	0.9895	0.9734	−0.4947	0.3818	−0.8315	−0.5890	0.9052	0.8945	0.6782	1.0000

Note: Significant positive correlations ($r > 0.8$) are indicated in bold.

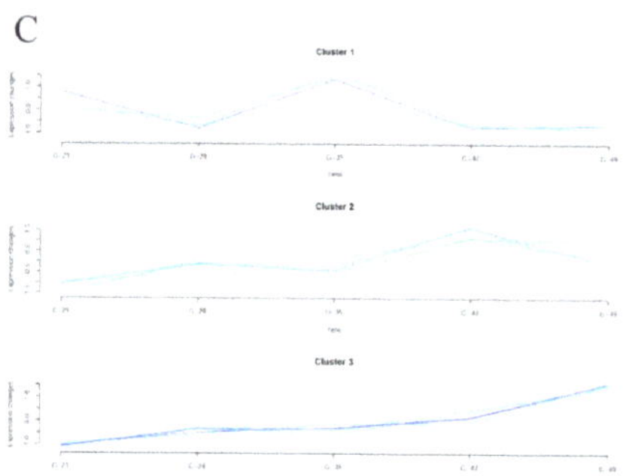

D

Class	Gene Name
Cluster 1	*MaTCP5*
Cluster 1	*MaTCP9-2*
Cluster 1	*MaTCP12*
Cluster 1	*MaTCP13*
Cluster 2	*MaTCP4-2*
Cluster 2	*MaTCP10*
Cluster 2	*MaTCP14*
Cluster 2	*MaTCP19*
Cluster 2	*MaTCP20-1*
Cluster 3	*MaTCP2*
Cluster 3	*MaTCP4-1*
Cluster 3	*MaTCP7*
Cluster 3	*MaTCP8*
Cluster 3	*MaTCP9-1*
Cluster 3	*MaTCP20-2*

Figure 3. Expressions of *MaTCP* genes in GY12 seedlings during root development. (**A**) Images of 21-, 28-, 35-, 42- and 49-day-old GY12 seedlings. Scale bars, 10 cm. (**B**) Corresponding root lengths. Each *black triangle* represents a GY12 seedling. (**C**) RT-qPCR-based time series analysis of the expressions of 15 *MaTCP* genes in GY12 seedlings at five developmental time points. Roots were collected from 10 individuals at each time point (21, 28, 35, 42, and 49 days after germination), and their lengths were measured. (**D**) Clustering of *MaTCP* genes based on the results of the time series analysis.

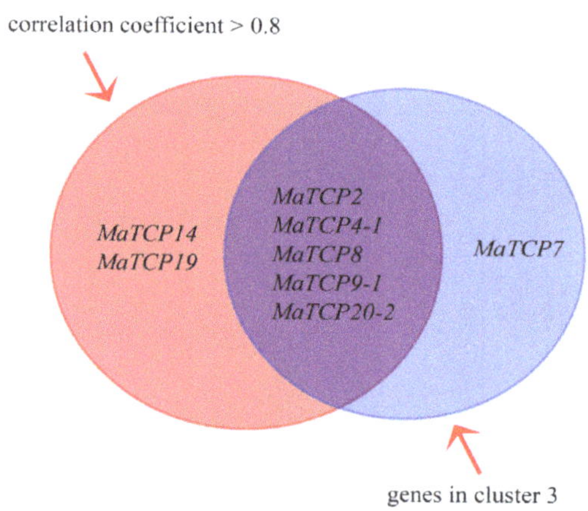

Figure 4. Venn diagram of *MaTCP* genes significantly associated with root length in a correlation analysis (red circle) and those exhibiting a similar trend in regard to root development in a time series analysis (blue circle).

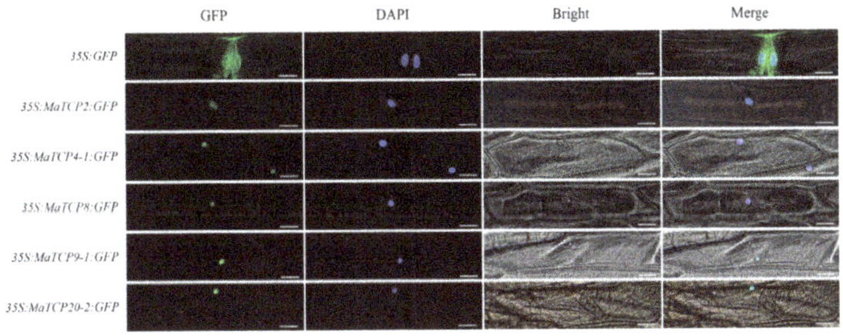

Figure 5. Subcellular localization of MaTCP2, MaTCP4-1, MaTCP8, MaTCP9-1, and MaTCP20-2 transcription factors. *35S:GFP* is an empty vector control, and DAPI was used for nucleic acid staining. Scale bars, 50 µm.

3.6. Expression of Five MaTCP Genes under Drought Treatment

To investigate whether the five *MaTCP* genes are involved in plant responses to drought stress, we examined the expression levels of these genes in GY12 seedlings treated with a 10% PEG6000 solution. RT-qPCR analysis revealed a rapid decrease in *MaTCP2*, *MaTCP4-1*, *MaTCP8*, and *MaTCP9-1* transcripts following 1 day of drought stress. Transcript levels of *MaTCP2*, *MaTCP8*, and *MaTCP9-1* gradually recovered as the treatment time was extended, and *MaTCP20-2* expression significantly increased after 7 days of treatment (Figure 6).

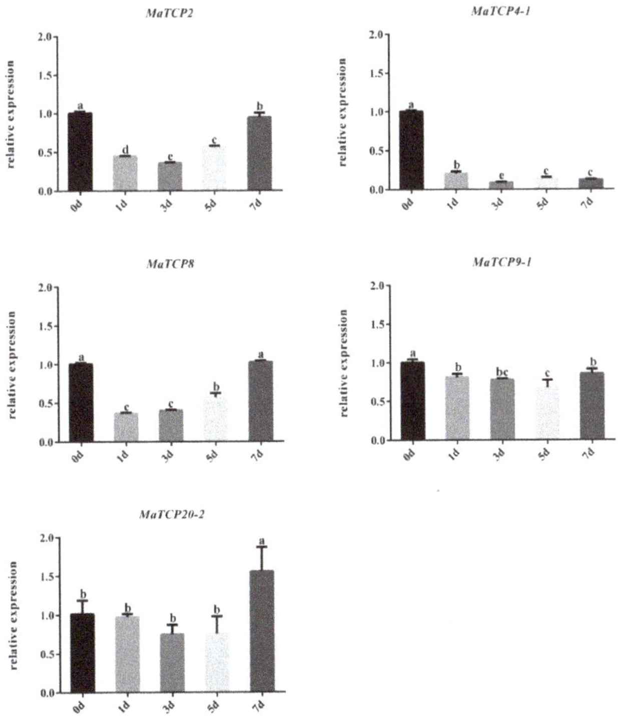

Figure 6. Expressions of *MaTCP* genes during drought stress. The plants were treated with 10% PEG6000 solution for 0, 1, 3, 5, and 7 days. The 0-day plants were the control. Error bars indicate SE. Different lowercase letters indicate statistically significant differences among treatments ($p < 0.05$).

4. Discussion

As the only food source for cultivated silkworms, mulberry trees are a crucial component of sericulture. In light of global warming and increasing water scarcity, the cultivation of mulberry species adapted to various agroclimatic conditions is vital for sustainable sericulture. Consequently, a top research objective for sericulture is the expansion and exploitation of the natural capacity of mulberry to withstand stress and adversity. Mulberry trees are characterized by their extensive root systems, fast growth, and high biomass, and are quite tolerant to their surroundings [35]. According to a previous study, mulberry seedlings can retain high root activities under long-term drought stress by increasing their root absorptive area and improving their capacity for water retention [36]. Moreover, drought tolerance can be improved by drought hardening during the seedling stage [37].

Although attention has focused on mulberry as a plant suitable for ecological restoration, little is known about the molecular mechanisms underlying its tolerance to drought. In particular, studies on the root system and drought tolerance of mulberry seedlings are lacking. Sequencing of the mulberry genome has recently been completed [24,29], and transcriptional data have been obtained from the roots of several mulberry species [38]. A foundation has thus been laid for the investigation of the mulberry root system.

In the present study, we identified 15 MaTCP transcription factors. A protein multiple sequence analysis revealed that MaTCP proteins of the same subclass share a similar motif makeup, but considerable differences exist between subclasses. For instance, the majority of TCP-P MaTCP-class proteins contain motif 2, whereas most TCP-C MaTCP-class proteins possess motifs 3 and 6 (Figure 1B). The TCP family is important for plant growth and development, hormone signaling, and stress response. Prolonged moderate drought also promotes root growth, which boosts a plant's drought resilience. We therefore hypothesized that increased expression of a previously undiscovered *MaTCP* gene cluster enhances root development in mulberry and thereby improves plant drought tolerance. In a time series analysis, *MaTCP2, MaTCP4-1, MaTCP7, MaTCP8, MaTCP9-1,* and *MaTCP20-2* gene expression patterns mirrored those of GY12 root development (Figure 3). In addition, the correlation analysis indicated that *MaTCP2, MaTCP4-1, MaTCP8, MaTCP9-1, MaTCP14, MaTCP19,* and *MaTCP20-2* expressions were significantly correlated with root length (Table 2). Taking into account the results of the two experiments, we thus identified *MaTCP2, MaTCP4-1, MaTCP8, MaTCP9-1,* and *MaTCP20-2* as the key candidate genes involved in GY12 root development. A promoter analysis revealed that *cis*-acting elements involved in MeJA, ABA, SA, GA, and abiotic stress responsiveness are abundant in the promoter regions of the five *MaTCP* genes. Finally, a gene expression analysis indicated that the five *MaTCP* genes are involved in the response of mulberry to drought stress.

All of these results suggest that *MaTCP*s play an essential role linking together mulberry growth, drought response, and phytohormone signaling. Further research on mulberry to elucidate the detailed functions and regulatory mechanisms of *MaTCP* genes would thus be valuable. The resulting findings should contribute to the selection and cultivation of mulberry varieties resistant to harsh settings, the advancement of sericulture, and the preservation of the ecological environment.

5. Conclusions

In conclusion, MaTCP2, MaTCP4-1, MaTCP8, MaTCP9-1, and MaTCP20-2 act as transcription factors and localize to the nucleus. The promoter regions of these five *MaTCP* genes contain a large number of *cis*-acting elements involved in MeJA, ABA, SA, GA, and abiotic stress responsiveness. Their transcriptional levels are highly correlated with root development and may increase mulberry tolerance to drought. Our results have increased the understanding of the function of the mulberry TCP protein and have provided candidate genes for the selective breeding and cultivation of mulberry.

Supplementary Materials: The following supporting information can be downloaded at https://www.mdpi.com/article/10.3390/f14010143/s1. Figure S1. Multiple alignments of the amino acid sequences of the MaTCP proteins. Table S1. TF IDs and common names of the TCP transcription factors used in phylogenetic analyses. Table S2. List of primers used in the RT-qPCR analyses of MaTCP genes. Table S3. The length of mulberry seedling roots at five developmental periods. Dataset S1. FASTA files of MaTCP proteins in mulberry.

Author Contributions: W.W., Z.Y., J.H., Y.L., X.X. and Y.H. performed the experiment. N.H., W.W. and J.H. analyzed the data and were the major contributors to writing the manuscript. W.W. wrote the manuscript and N.H. revised the manuscript. All authors have read and agreed to the published version of the manuscript.

Funding: This work was funded by the Chongqing Research Program of Basic Research and Frontier Technology (No. cstc2021yszx-jcyj0004) and the National Natural Science Foundation of China (No. 32001328).

Data Availability Statement: Most of the data presented in this study are contained within the article and in the Supplementary Materials. Data not shown in the article are available on request from the corresponding author.

Conflicts of Interest: The authors declare no conflict of interest.

References

1. Mega, R.; Abe, F.; Kim, J.S.; Tsuboi, Y.; Tanaka, K.; Kobayashi, H.; Sakata, Y.; Hanada, K.; Tsujimoto, H.; Kikuchi, J.; et al. Tuning water-use efficiency and drought tolerance in wheat using abscisic acid receptors. *Nat. Plants* **2019**, *5*, 153–159. [CrossRef] [PubMed]
2. Yue, B.; Xue, W.Y.; Xiong, L.Z.; Yu, X.Q.; Luo, L.J.; Cui, K.H.; Jin, D.M.; Xing, Y.Z.; Zhang, Q.F. Genetic basis of drought resistance at reproductive stage in rice: Separation of drought tolerance from drought avoidance. *Genetics* **2006**, *172*, 1213–1228. [CrossRef] [PubMed]
3. Fang, Y.J.; Xiong, L.Z. General mechanisms of drought response and their application in drought resistance improvement in plants. *Cell. Mol. Life Sci.* **2015**, *72*, 673–689. [CrossRef] [PubMed]
4. Shavrukov, Y.; Kurishbayev, A.; Jatayev, S.; Shvidchenko, V.; Zotova, L.; Koekemoer, F.; de Groot, S.; Soole, K.; Langridge, P. Early flowering as a drought escape mechanism in plants: How can it aid wheat production? *Front. Plant Sci.* **2017**, *8*, 1950. [CrossRef]
5. Wang, Y.; Yu, Y.H.; Chen, Q.J.; Bai, G.H.; Gao, W.W.; Qu, Y.Y.; Ni, Z.Y. Heterologous expression of GbTCP4, a class II TCP transcription factor, regulates trichome formation and root hair development in *Arabidopsis*. *Genes* **2019**, *10*, 726. [CrossRef] [PubMed]
6. Wang, Y.; Yu, Y.H.; Wang, J.D.; Chen, Q.J.; Ni, Z.Y. Heterologous overexpression of the *GbTCP5* gene increased root hair length, root hair and stem trichome density, and lignin content in transgenic *Arabidopsis*. *Gene* **2020**, *758*, 144954. [CrossRef]
7. Resentini, F.; Felipo-Benavent, A.; Colombo, L.; Blazquez, M.A.; Alabadi, D.; Masiero, S. TCP14 and TCP15 mediate the promotion of seed germination by gibberellins in *Arabidopsis thaliana*. *Mol. Plant.* **2015**, *8*, 482–485. [CrossRef]
8. Lopez, J.A.; Sun, Y.; Blair, P.B.; Mukhtar, M.S. TCP three-way handshake: Linking developmental processes with plant immunity. *Trends Plant Sci.* **2015**, *20*, 238–245. [CrossRef]
9. Danisman, S. TCP transcription factors at the interface between environmental challenges and the plant's growth responses. *Front. Plant Sci.* **2016**, *7*, 1930. [CrossRef]
10. Fang, Y.J.; Zheng, Y.Q.; Lu, W.; Li, J.; Duan, Y.J.; Zhang, S.; Wang, Y.P. Roles of miR319-regulated TCPs in plant development and response to abiotic stress. *Crop J.* **2021**, *9*, 17–28. [CrossRef]
11. Cubas, P.; Lauter, N.; Doebley, J.; Coen, E. The TCP domain: A motif found in proteins regulating plant growth and development. *Plant J.* **1999**, *18*, 215–222. [CrossRef]
12. Martin-Trillo, M.; Cubas, P. *TCP* genes: A family snapshot ten years later. *Trends Plant Sci.* **2010**, *15*, 31–39. [CrossRef] [PubMed]
13. Manassero, N.G.U.; Viola, I.L.; Welchen, E.; Gonzalez, D.H. TCP transcription factors: Architectures of plant form. *Biomol. Concepts* **2013**, *4*, 111–127. [CrossRef] [PubMed]
14. Aggarwal, P.; Das Gupta, M.; Joseph, A.P.; Chatterjee, N.; Srinivasan, N.; Nath, U. Identification of specific DNA binding residues in the TCP family of transcription factors in *Arabidopsis*. *Plant Cell* **2010**, *22*, 1174–1189. [CrossRef]
15. Horn, S.; Pabon-Mora, N.; Theuss, V.S.; Busch, A.; Zachgo, S. Analysis of the CYC/TB1 class of TCP transcription factors in basal angiosperms and magnoliids. *Plant J.* **2015**, *81*, 559–571. [CrossRef] [PubMed]
16. Nicolas, M.; Cubas, P. TCP factors: New kids on the signaling block. *Curr. Opin. Plant Biol.* **2016**, *33*, 33–41. [CrossRef]
17. Kieffer, M.; Master, V.; Waites, R.; Davies, B. TCP14 and TCP15 affect internode length and leaf shape in *Arabidopsis*. *Plant J.* **2011**, *68*, 147–158. [CrossRef]

18. Peng, Y.C.; Chen, L.L.; Lu, Y.R.; Wu, Y.B.; Dumenil, J.; Zhu, Z.G.; Bevan, M.W.; Li, Y.H. The ubiquitin receptors DA1, DAR1, and DAR2 redundantly regulate endoreduplication by modulating the stability of TCP14/15 in *Arabidopsis*. *Plant Cell* **2015**, *27*, 649–662. [CrossRef]
19. Hao, J.; Tu, L.L.; Hu, H.Y.; Tan, J.F.; Deng, F.L.; Tang, W.X.; Nie, Y.C.; Zhang, X.L. GbTCP, a cotton TCP transcription factor, confers fibre elongation and root hair development by a complex regulating system. *J. Exp. Bot.* **2012**, *63*, 6267–6281. [CrossRef]
20. Mukhopadhyay, P.; Tyagi, A.K. *OsTCP19* influences developmental and abiotic stress signaling by modulating ABI4-mediated pathways. *Sci. Rep.* **2015**, *5*, 9998. [CrossRef]
21. Ori, N.; Cohen, A.R.; Etzioni, A.; Brand, A.; Yanai, O.; Shleizer, S.; Menda, N.; Amsellem, Z.; Efroni, I.; Pekker, I.; et al. Regulation of *LANCEOLATE* by *miR319* is required for compound-leaf development in tomato. *Nat. Genet.* **2007**, *39*, 787–791. [CrossRef]
22. Nag, A.; King, S.; Jack, T. miR319a targeting of *TCP4* is critical for petal growth and development in *Arabidopsis*. *Proc. Natl. Acad. Sci. USA* **2009**, *106*, 22534–22539. [CrossRef] [PubMed]
23. Zhou, M.; Li, D.Y.; Li, Z.G.; Hu, Q.; Yang, C.H.; Zhu, L.H.; Luo, H. Constitutive expression of a *miR319* gene alters plant development and enhances salt and drought tolerance in transgenic creeping bentgrass. *Plant Physiol.* **2013**, *161*, 1375–1391. [CrossRef] [PubMed]
24. He, N.J.; Zhang, C.; Qi, X.W.; Zhao, S.C.; Tao, Y.; Yang, G.J.; Lee, T.H.; Wang, X.Y.; Cai, Q.L.; Li, D.; et al. Draft genome sequence of the mulberry tree *Morus notabilis*. *Nat. Commun.* **2013**, *4*, 2445. [CrossRef] [PubMed]
25. Huang, H.-P.; Ou, T.-T.; Wang, C.-J. Mulberry (sang shen zǐ) and its bioactive compounds, the chemoprevention effects and molecular mechanisms in vitro and in vivo. *J. Tradit. Complement. Med.* **2013**, *3*, 7–15. [CrossRef]
26. Rohela, G.K.; Shukla, P.; Muttanna; Kumar, R.; Chowdhury, S.R. Mulberry (*Morus* spp.): An ideal plant for sustainable development. *Trees For. People* **2020**, *2*, 100011. [CrossRef]
27. Liu, Y.; Willison, J.H.M. Prospects for cultivating white mulberry (*Morus alba*) in the drawdown zone of the Three Gorges Reservoir, China. *Environ. Sci. Pollut. Res.* **2013**, *20*, 7142–7151. [CrossRef]
28. Ma, X.D.; Ma, J.C.; Fan, D.; Li, C.F.; Jiang, Y.Z.; Luo, K.M. Genome-wide identification of TCP family transcription factors from *Populus euphratica* and their involvement in leaf shape regulation. *Sci. Rep.* **2016**, *6*, 32795. [CrossRef]
29. Jiao, F.; Luo, R.S.; Dai, X.L.; Liu, H.; Yu, G.; Han, S.H.; Lu, X.; Su, C.; Chen, Q.; Song, Q.X.; et al. Chromosome-level reference genome and population genomic analysis provide insights into the evolution and improvement of domesticated mulberry (*Morus alba*). *Mol. Plant.* **2020**, *13*, 1001–1012. [CrossRef]
30. Mulder, N.J.; Apweiler, R.; Attwood, T.K.; Bairoch, A.; Bateman, A.; Binns, D.; Bradley, P.; Bork, P.; Bucher, P.; Cerutti, L.; et al. InterPro, progress and status in 2005. *Nucleic Acids Res.* **2005**, *33*, D201–D205. [CrossRef]
31. Tamura, K.; Stecher, G.; Kumar, S. MEGA11: Molecular evolutionary genetics analysis version 11. *Mol. Biol. Evol.* **2021**, *38*, 3022–3027. [CrossRef] [PubMed]
32. Chen, C.J.; Chen, H.; Zhang, Y.; Thomas, H.R.; Frank, M.H.; He, Y.H.; Xia, R. TBtools: An integrative toolkit developed for interactive analyses of big biological data. *Mol. Plant.* **2020**, *13*, 1194–1202. [CrossRef] [PubMed]
33. Qi, X.W.; Shuai, Q.; Chen, H.; Fan, L.; Zeng, Q.W.; He, N.J. Cloning and expression analyses of the anthocyanin biosynthetic genes in mulberry plants. *Mol. Genet. Genom.* **2014**, *289*, 783–793. [CrossRef] [PubMed]
34. Livak, K.J.; Schmittgen, T.D. Analysis of relative gene expression data using real-time quantitative PCR and the $2^{-\Delta\Delta CT}$ method. *Methods* **2001**, *25*, 402–408. [CrossRef]
35. Huang, R.Z.; Jiang, Y.B.; Jia, C.H.; Jiang, S.M.; Yan, X.P. Subcellular distribution and chemical forms of cadmium in *Morus alba* L. *Int. J. Phytoremediat.* **2018**, *20*, 448–453. [CrossRef]
36. Feng, D.L.; Huang, X.H.; Liu, Y.; Willison, J.H.M. Growth and changes of endogenous hormones of mulberry roots in a simulated rocky desertification area. *Environ. Sci. Pollut. Res.* **2016**, *23*, 11171–11180. [CrossRef]
37. Huang, X.H.; Liu, Y.; Li, J.X.; Xiong, X.Z.; Chen, Y.H.; Yin, X.H.; Feng, D.L. The response of mulberry trees after seedling hardening to summer drought in the hydro-fluctuation belt of Three Gorges Reservoir Areas. *Environ. Sci. Pollut. Res.* **2013**, *20*, 7103–7111. [CrossRef]
38. Checker, V.G.; Saeed, B.; Khurana, P. Analysis of expressed sequence tags from mulberry (*Morus indica*) roots and implications for comparative transcriptomics and marker identification. *Tree Genet. Genomes* **2012**, *8*, 1437–1450. [CrossRef]

forests

MDPI

Article

Grafting with Different Rootstocks Induced DNA Methylation Alterations in Pecan [*Carya illinoinensis* (Wangenh.) K. Koch]

Zhuangzhuang Liu [1,2], Pengpeng Tan [1], Youwang Liang [1], Yangjuan Shang [3], Kaikai Zhu [1], Fangren Peng [1,*] and Yongrong Li [4]

1 College of Forestry, Nanjing Forestry University, Nanjing 210037, China
2 Jiangsu Key Laboratory for the Research and Utilization of Plant Resources, Institute of Botany, Jiangsu Province and Chinese Academy of Sciences (Nanjing Botanical Garden Mem. Sun Yat-Sen), Nanjing 210014, China
3 College of Life and Environment Sciences, Huangshan University, Huangshan 245021, China
4 Green Universe Pecan Science and Technology Co., Ltd., Nanjing 210007, China
* Correspondence: frpeng@njfu.edu.cn; Tel.: +86-025-8542-7995

Abstract: Rootstocks are well known to have important effects on scion growth performance. However, the involved mechanisms remain unclear. Recent studies provided some clues on the potential involvement of DNA methylation in grafting, which open up new horizons for exploring how rootstocks induce the growth changes. To better understand the involvement of DNA methylation in rootstock-induced growth alterations, whole-genome bisulfite sequencing (WGBS) was used to evaluate the methylation profiles of two sets of pecan grafts with different growth performances on different sizes of rootstocks. The results showed that methylated cytosines accounted for 24.52%–25.60% of all cytosines in pecan. Methylation levels in CG were the highest, with the lowest levels being in CHH (C= cytosine; G= guanine; H = adenine, thymine, or cytosine). Rootstocks induced extensive methylation alterations in scions with 934, 2864, and 15,789 differentially methylated regions (DMRs) determined in CG, CHG, and CHH contexts, respectively. DMR-related genes (DMGs) were found to participate in various processes associated with plant growth, among which 17 DMGs were found, most likely related to hormone response, that may play particularly important roles in graft growth regulation. This study revealed DNA methylomes throughout the pecan genome for the first time, and obtained abundant genes with methylation alterations that were potentially involved in rootstock-induced growth changes in pecan scions, which lays a good basis for further epigenetic studies on pecan and deeper understanding of grafting mechanisms in pecan grafts.

Keywords: pecan; scion growth; regulation; epigenetics; whole-genome bisulfite sequencing (WGBS)

Citation: Liu, Z.; Tan, P.; Liang, Y.; Shang, Y.; Zhu, K.; Peng, F.; Li, Y. Grafting with Different Rootstocks Induced DNA Methylation Alterations in Pecan [*Carya illinoinensis* (Wangenh.) K. Koch]. *Forests* **2023**, *14*, 4. https://doi.org/10.3390/f14010004

Academic Editors: Jie Luo and Wentao Hu

Received: 27 October 2022
Revised: 16 December 2022
Accepted: 16 December 2022
Published: 20 December 2022

1. Introduction

Grafting is the process of connecting two plant parts with the root piece known as "rootstock" and the shoot part called "scion" [1]. Plants with desirable characteristics can be produced by combining the favorable traits of these two graft segments. As an ancient, clonal propagation technique, grafting is widely used for reproducing horticultural crops, such as fruit trees, vegetables, and flowers. For crops harvested from aboveground parts, attention is usually paid to the effects of rootstocks on their scions. Various studies have revealed that rootstocks can regulate scion growth vigor, tree architecture, mineral element composition, fruit quality and yield, and stress tolerance [2–4].

Although grafting has been applied for a long time, the mechanisms on how rootstocks influence the growth vigor of scions remain elusive. Previous studies on the effects of dwarfing rootstocks have put forth various hypotheses from physiological aspects to explain rootstock-conferred differences in vigor. In summary, the rootstocks can regulate scion growth, through their effects on anatomy of graft union, water and solutes supply to the scion, and synthesis and transportation of hormones [5,6]. Compared with much attention on the physiological mechanisms involved in rootstocks-induced growth differences, there are few studies to reveal the molecular mechanisms.

Previous studies have revealed the movement of mRNA, microRNA and siRNAs (small interfering RNAs) in grafts [7]. As a type of non-coding RNAs, siRNAs are known to direct transcriptional gene silencing (TGS) by RNA-directed DNA methylation (RdDM) [8]. Recent studies in *A. thaliana* and tobacco (*Nicotiana tabacum* L.) suggested the small interfering RNAs (siRNAs), a type of non-coding RNAs, can move from across the graft union and direct DNA methylation in the genome of scion recipient cells [9,10]. On the basis that DNA methylation is an important epigenetic modification and plays critical roles in regulation of plant development [11], it is possible that DNA methylation may be involved in the regulatory roles of rootstocks in scions. Several studies using methylation-sensitive amplified polymorphism (MSAP) technique suggested that rootstocks can induce different extent of DNA methylation changes in amplified sites of scions [12–14], and the sites with different methylation patterns were sequenced to analyze their potential functions [12,14]. By grafting eggplant (*Solanum melongena* L.) on *Solanum torvum* Swartz rootstock or tomato F1 commercial hybrid Emperador RZ rootstock, Cerruti et al. investigated the epigenetic bases of the grafting-induced vigor and found that CHH (C= cytosine; H = adenine, thymine, or cytosine) methylation levels decreased significantly in eggplant scions on the two rootstocks compared to that of self-grafted plants, which was associated with enhanced vigor in hetero-grafted eggplant scions [15]. Although these studies provided some clues on the involvement of DNA methylation in grafting process [12–15], the mechanisms on mediation of DNA methylation in rootstock-induced vigor changes in scions remain largely unknown, and systematic studies are urgently needed regarding this issue.

Pecan [*Carya illinoinensis* (Wangenh.) K. Koch] is an important nut tree native to North America [16]. Besides delicious and nutritious nuts, it can also produce fine timber or can be planted for afforestation. With significant economic, social, and ecological benefits, it has been planted widely throughout the world. As the main clonal propagation method of pecan, grafting plays important roles in the maintenance of excellent characteristics of cultivars, early flowering and fruit-bearing, and improvement of stress resistance. With the development of pecan industry, the directional cultivation of fine pecan grafts is required to suit multiple purposes. For example, in addition to the requirement of dwarfing trees for fruits in areas where land resources are scarce, there is also an increasing demand for the timber forests with fast vegetative growth. As an important part of the graft, rootstocks are known to have important effects on performance of scions. Thus, studies on the cultivation of rootstock resources and rootstocks-conferred effects on scions are critical to the directional cultivation of fine pecan grafts. In production, no clonal rootstock of pecan is available, and the rootstocks for grafting are mainly originated from the open-pollinated seeds of pecan trees. The limited breeding works on pecan rootstock mainly focus on the selection of the seedling rootstocks of different pecan cultivars, and their resistances to pests and diseases are regarded as important characteristics to test [17,18]. Currently, there is a still lack of rootstock resources to meet the requirements of pecan graft production, and little is known on the effects of rootstocks on scion performance and the involved mechanisms in pecan. In our previous study, we observed that rootstocks with small height can significantly reduce growth vigor in pecan scions. Using deep sequencing technology, we identified 24 significantly differentially expressed miRNAs between the two groups of pecan grafts with different growth vigor and further revealed their potential roles in growth regulation [19]. However, nothing is currently known regarding how DNA methylation changes in response to grafting with different rootstocks in this fruit tree, nor its involvement in regulation of pecan graft. Studies on the DNA methylation in grafts will help us fully understand the roles of epigenetic factors in grafting process.

Different methods have been developed for detecting DNA methylation, such as high-performance liquid chromatography (HPLC), MSAP method, immunoprecipitation technology, and whole-genome bisulfite sequencing (WGBS). Among these, MSAP has been more widely used because of its low cost and lack of need for genomic information.

We previously used the MSAP method to successfully detect methylation in different pecan cultivars and tissues at different stages [20,21]. However, unlike MSAP technology, which covers preselected CCGG sites using a limited number of primer pairs (G= guanine), WGBS allows DNA methylation to be detected on a genome-wide scale with single-base resolution and is considered the "gold standard" [22]. Recently, the pecan genome was revealed [23], and it is now feasible for the first time to reveal DNA methylomes throughout the pecan genome using WGBS, which can provide comprehensive methylation information for epigenetic studies in pecan. In the current study, WGBS was performed to detect the DNA methylation profiles in pecan grafts with different growth vigor on tall and short rootstocks. Differentially methylated regions (DMRs) and DMR-related genes (DMGs) in the two sets of grafts were then identified. Finally, the DMGs were subjected to Gene Ontology (GO) and Kyoto Encyclopedia of Genes and Genomes (KEGG) analyses to reveal gene functions.

2. Materials and Methods

2.1. Plant Materials

At the end of 2015, 180 tall and 180 short 1-year-old pecan seedlings were selected from Nanjing Green Universe Pecan Science and Technology Co., Nanjing, China. In January of 2016, they were transferred and cultivated in the test site of Nanjing Forestry University, Zhenjiang, Jiangsu Province (height data collected after seedling transplantation: mean height of 68.3 cm for tall seedlings, 16.7 cm for short seedlings). In September of 2016, the seedlings were reselected, and short seedlings of 30.0 cm height and tall seedlings of 94.6 cm height were obtained. Then, Pawnee and Shaoxing scions from the same trees of respective cultivars were grafted on the two types of seedling rootstocks (tall rootstocks: TR; short rootstocks: SR) using patch budding. After observation and selection of the grafts in 2017 and 2018, three grafts with significantly strong growth vigor (SV) on TR and three grafts with poor growth vigor (PV) on SR were obtained for each cultivar. The graft height and stem diameter of SV grafts were significantly higher than those of PV grafts, respectively ($p < 0.05$; the growth indexes of 'Pawnee' grafts are shown in Figure S1). Phloem samples from selected 'Pawnee' grafts were subjected to WGBS. The experimental plants were arranged in three blocks in a split-plot design with rootstock types set as main plots and cultivars as secondary plots. The detailed experimental design and selection method of the rootstocks and experimental grafts were previously described [19].

2.2. WGBS Library Construction and Sequencing

Genomic DNA was extracted from each pecan sample using the cetyltrimethylammonium bromide (CTAB) method and the DNA quality was measured using 1% agarose gel electrophoresis, a K5500 micro-spectrophotometer (Beijing Kaiao Technology Development Co., Ltd., Beijing, China), and a Qubit fluorometer 2.0 (Invitrogen, Carlsbad, CA, USA). The qualified DNA was segmented firstly by ultrasound, and the DNA fragments were purified, end-repaired, adenylated at the 3′ end, and ligated with methylated adapters. The targeted fragments were obtained from 2% agarose gel electrophoresis, treated with bisulfite, and amplified by PCR to generate the WGBS library. The prepared library was sequenced on an Illumina HiSeq PE125/PE150 platform (Biomarker Technologies, Beijing, China).

2.3. Sequence Data Processing and Analysis

Raw image files were produced by high-throughput sequencing and converted into sequenced reads, known as raw reads. Clean reads were obtained from the raw reads for subsequent analysis by removing the reads with adapters and excluding reads with more than 10% N content or more than 50% low quality bases (quality value < 10). Clean reads need to be aligned with a reference genome to conduct the analysis of DNA methylation. We used Bismark software to blast align the clean reads with the pecan genome to determine the uniquely mapped reads and calculate mapping efficiency (the number of uniquely mapped reads per number of total clean reads). Under bisulfite treatment and PCR amplification of the treated fragments, unmethylated cytosines were converted into thymine (T), while methylated cytosines (^{m}C) remained unchanged. Therefore, the blast information on cytosine throughout the genome was extracted according to the alignment of clean reads with the pecan genome, and the reads supporting methylated and unmethylated cytosines were counted. Detailed steps of alignment analysis using Bismark are shown on the following website: https://github.com/FelixKrueger/Bismark/blob/master/Docs/README.md (accessed on 15 December 2022). To determine if each individual cytosine site (C site) was methylated, a binomial distribution test was conducted. With a coverage depth $\geq 4\times$ and false discovery rate (FDR) < 0.05, methylation status of the C sites was confirmed.

In the current study, methylation level of a single C site was calculated using the following formula:

$$\text{Methylation level of C site} = Ci/(Ci + Ti) \quad (1)$$

The methylation level in regions was counted as follows:

$$\text{Weighted methylation level} = \sum_{i=1}^{n} Ci/\sum_{i=1}^{n}(Ci + Ti) \quad (2)$$

where C represented the number of reads supporting methylated cytosine, T represented the number of reads supporting unmethylated cytosine, i was the position of cytosine, and n was the summation of cytosine positions [24].

Model-based analysis of bisulfite sequencing data (MOABS) [25] was used to determine DMRs in SV and PV grafts in which the coverage depth was no less than $10\times$. There were at least three different methylation sites, the minimum difference in methylation levels was 0.2 (0.3 for CG type), and p-value from Fisher's exact test was less than 0.05. Annotation of the DMGs was obtained through the comparison of DMGs with functional databases of GO and KEGG using BLAST to analyze gene functions. Fisher's exact test with $p < 0.05$ was used to determine the significantly enriched GO annotations and KEGG pathways.

3. Results

3.1. Pecan DNA Methylomes

Using WGBS, a total of 66,502,614 to 84,008,102 clean reads and 19,950,784,200 to 25,202,430,600 clean bases were generated from the pecan grafts, including the grafts with strong growth vigor on TR (SV1, SV2, and SV3) and grafts with poor growth vigor on SR (PV1, PV2, and PV3). Guanine and cytosine accounted for 21.53%–22.42% of all the bases. Through mapping analysis, 75.50%–77.95% of the clean reads could be uniquely mapped to the pecan genome for subsequent analysis, and all of the bisulfite conversion rates for the six samples were over 99% (Table 1).

Table 1. Whole-genome bisulfite sequencing data in two sets of pecan grafts with different growth performances. SV: grafts with strong growth vigor on tall rootstocks; PV: grafts with poor growth vigor on short rootstocks.

Sample	Clean Bases	GC (%)	Clean Reads	Unique Reads	Mapped (%)	Conversion Rate (%)
SV1	25,202,430,600	22.42	84,008,102	65,482,398	77.95	99.32
SV2	19,950,784,200	21.94	66,502,614	51,044,261	76.76	99.27
SV3	22,336,860,600	21.57	74,456,202	56,215,238	75.50	99.21
PV1	22,141,198,200	21.53	73,803,994	56,631,166	76.73	99.29
PV2	21,764,759,400	21.79	72,549,198	55,861,160	77.00	99.18
PV3	21,762,984,300	21.78	72,543,281	54,898,021	75.68	99.24

Note: GC (%): The number of guanines (G) and cytosines (C)/the number of all bases. Mapped (%): The number of unique reads/number of total clean reads. Conversion rate (%): Bisulfite conversion rate, the number of clean reads mapped to lambda DNA that support methylated cytosines/the number of total clean reads mapped to lambda DNA.

Methylation in pecan was shown to occur in all cytosine sequence contexts, CG, CHG, and CHH. Based on the methylation status of each individual C site and statistics of ^{m}C, total methylated cytosines (^{m}CG + ^{m}CHG + ^{m}CHH) accounted for 24.52%–25.60% of all cytosines (total methylation levels) in the six pecan samples (Table S1). The methylation levels were highest in CG (66.77%–67.85%), with methylation levels of 54.16%–55.31% in CHG and 14.21%–15.37% in CHH (Methylation level in C context was determined based the number of ^{m}C of each context/the number of all C sites of the same C context) (Figure 1a). Among the methylated cytosine sites, ^{m}CHH contexts accounted for 44.85%–46.41% (relative methylation levels), which was the highest percentages, followed by 27.98%–28.70% ^{m}CHG contexts and 25.61%–26.45% ^{m}CG context (Figure 1b). Through student's t-tests, there were no significant differences in the methylation levels (including total methylation levels, methylation levels in different C contexts, and relative methylation levels) between VG and PG grafts ($p > 0.05$).

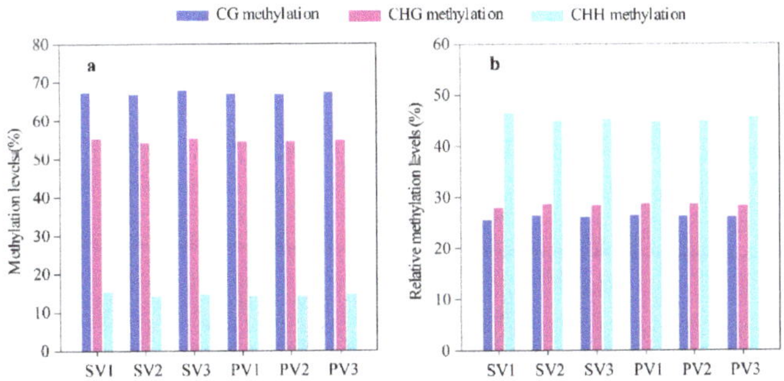

Figure 1. The methylation levels in pecan grafts. SV1, SV2, SV3: grafts with strong growth vigor; PV1, PV2, PV3: grafts with poor growth vigor. (**a**) Methylation level was determined based the number of methylated cytosines (^{m}C) of each context/the number of all C sites of the same C context; (**b**) Relative methylation level represented the percentage of methylated CG, CHG or CHH among all methylated cytosines (H = A, T, or C).

Violin graphs demonstrate the distribution of methylation levels at single C sites (Figure 2). Overall, ^{m}CG and ^{m}CHG, with methylation levels more than 75%, had higher density in pecan, while ^{m}CHH with levels lower than 25% possessed higher density. Analysis was then conducted on the characteristics of 9-base pair (bp) sequences in which ^{m}C was in the fifth position. Based on the results of all pecan samples, the frequencies of adenine (A) and T were higher than those of C in the sixth position of the CHG sequences. In addition, A and T were also found to occur at higher frequencies than C in the sixth and seventh positions of the CHH sequences (Figures 3 and S2).

Chromosome methylation maps were plotted to visually demonstrate the distribution of ^{m}C throughout pecan chromosomes (Figures 4a and S3). The maps showed that methylation levels were highest in CG in the whole, with the lowest methylation levels detected in the CHH context. We also analyzed methylation in different gene regions of pecan genome, including the upstream regions [2 kilobase (kb)], first intron, inner intron, first exon, inner exon, last exon, and downstream regions (2 kb). Methylation levels in the three contexts, especially CG, exhibited an obvious drop between the gene bodies and flanking regions near the transcription start site (TSS) and transcription termination site (TTS). In the CG context, highest methylation levels were found in the inner introns compared to those in other regions, while methylation levels were highest in the upstream and downstream regions for the CHG and CHH contexts (Figures 4b and S4). In the gene bodies, methylation levels in the introns appeared higher than those in the exons. We also investigated methylation levels in different regions of repeats and found that methylation levels in repeats were higher than those in gene regions, and repeat bodies had higher methylation levels than those in the upstream (1.5 kb) and downstream (1.5 kb) regions of repeats (Figures 4c and S5).

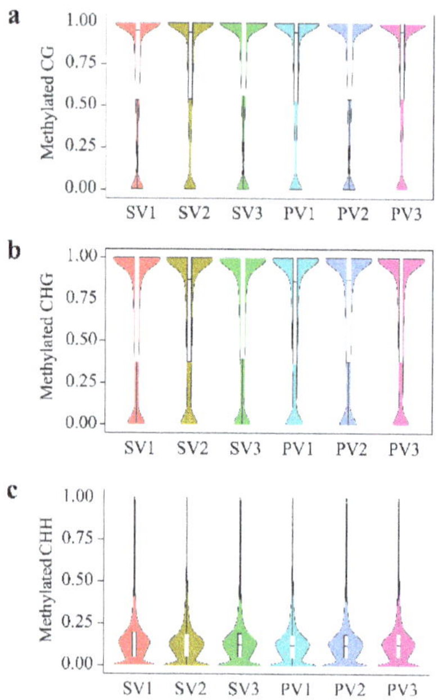

Figure 2. Distribution of methylation levels at single cytosine (C) sites of CG (**a**), CHG (**b**) and CHH (**c**) contexts in pecan grafts (H = A, T, or C). SV1, SV2, SV3: grafts with strong growth vigor; PV1, PV2, PV3: grafts with poor growth vigor. The ordinate represents the methylation level of C sites, and the width of each violin graph represents the density of C sites at that methylation level.

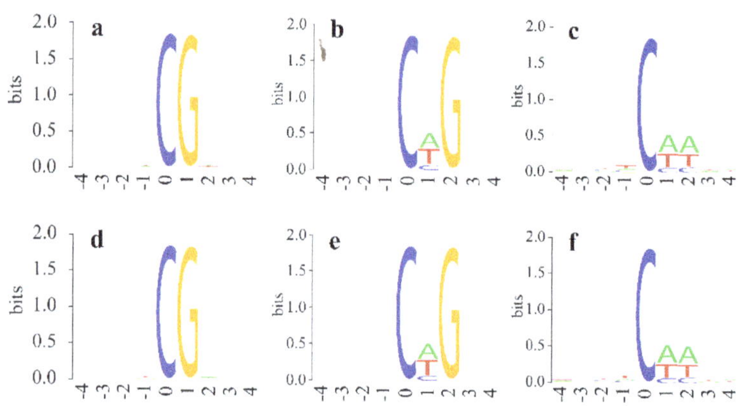

Figure 3. Characteristics of 9-base pair sequences with methylated CG, CHG and CHH (H = A, T, or C) in pecan. (**a**–**c**) represent the sequence characteristics in the Watson strand, and (**d**–**f**) represent that in Crick strand. The abscissa refers to the position of bases at the methylation sites and the height of base signal at each position refers to the relative frequency of the bases at that position. The graphs were obtained based on the methylation data of SV1 pecan graft (One of the three pecan grafts with strong growth vigor).

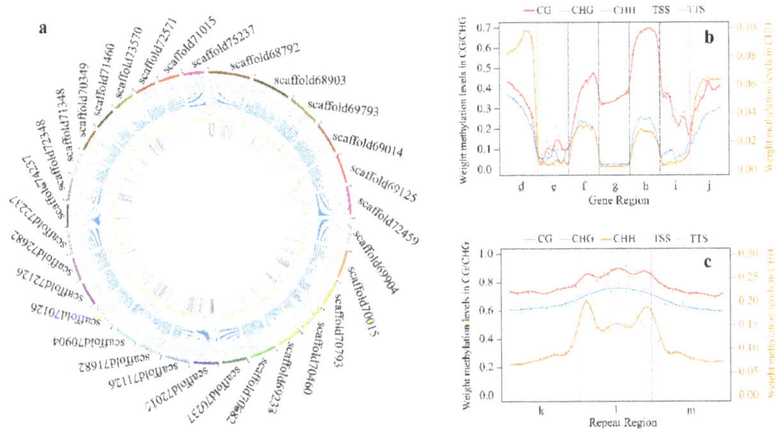

Figure 4. DNA methylation distribution in pecan genome. (**a**) Circle plot of methylation distribution in pecan chromosomes. The outermost circle refers to the chromosomes of the pecan. Light blue, blue and green bars represent the methylation levels in CG, CHG, and CHH contexts, respectively (H = A, T, or C). The innermost circle with color of gray represent the number of genes, and the darker the color is, the more genes there are. (**b**) Methylation levels in different gene regions (d: upstream; e: first exon; f: first intron; g: inner exon, h: inner intron; i: last exon; j: downstream). (**c**) Methylation levels in different repeat regions (k: upstream; l: repeat body; m: downstream). TSS: Transcription start site; TTS: Transcription termination site. The graphs were plotted based on the methylation data of SV1 pecan graft (one of the three pecan grafts with strong growth vigor).

Hypermethylation CG island (CGI) regions were defined as CGI regions with methylation levels greater than 70%. We also excluded those in which the rates of C sites with coverage depth > 5× were less than 0.1, and annotated the remaining hypermethylation CGI regions. The results showed that distal intergenic regions had the highest proportions of hypermethylation CGI regions, ranging from 81.35% to 84.26% in the six pecan samples. Hypermethylation CGI regions in $\leq$ 1 kb promoter, 1–2 kb promoter and 2–3 kb promoter accounted for 3.03%–4.52%, 5.74%–6.61% and 3.65%–4.45%, respectively. The lowest percentages of hypermethylation CGI were found in untranslated regions (UTRs) and exons ($\leq$ 0.3%; Figure 5).

3.2. Analysis of DMRs in Pecan Grafts with Different Growth Performances

DMRs were detected in the SV and PV grafts and the results showed that a total of 934 DMRs were identified in CG contexts (CG-DMRs), 2864 in CHG contexts (CHG-DMRs), and 15,789 in CHH contexts (CHH-DMRs) in the two types of grafts. Compared with the methylation status in SV grafts, 452 CG-DMRs, 1303 CHG-DMRs, and 7758 CHH-DMRs were hypermethylated in PV grafts, with 482 CG-DMRs, 1561 CHG-DMRs, and 8031 CHH-DMRs showing hypomethylation. The identified DMRs were annotated according to their position in the genome and the annotation information of the genome. Most DMRs were located in the distal intergenic regions with fewer DMRs in promoters, introns, and exons and very few DMRs in the 3′UTR and 5′UTR. In the distal intergenic regions and promoters, the number of hypomethylated DMRs for each C context was uniformly found to be more than that of hypermethylated DMRs. In addition, the number of hypermethylated or hypomethylated DMRs in the CHH context was the highest, followed by those of CHG-DMRs and CG-DMRs in the distal intergenic regions, promoters, and introns (Figure 6; Table S2).

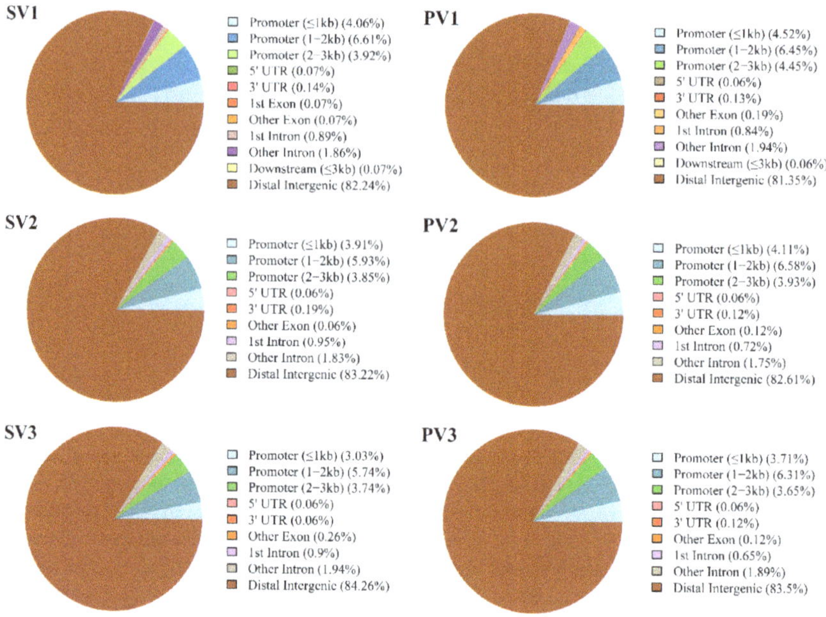

Figure 5. The percentages of hypermethylation CG island (CGI) regions distributed in different gene regions of pecan genome. SV1, SV2, SV3: grafts with strong growth vigor; PV1, PV2, PV3: grafts with poor growth vigor. UTR: untranslated region.

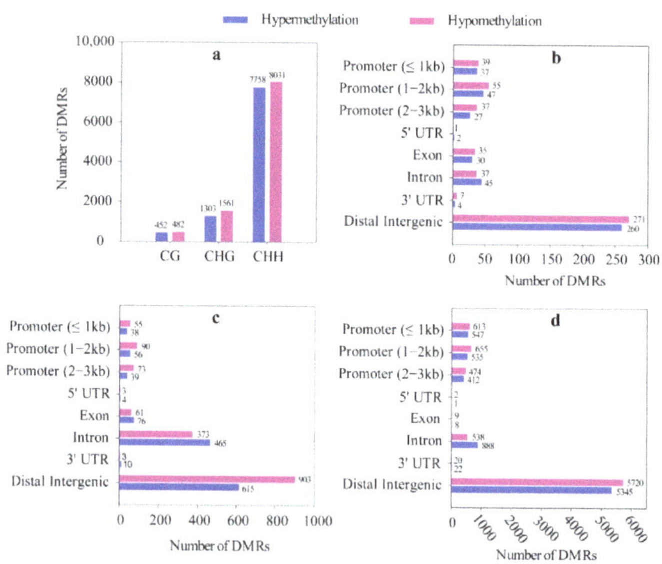

Figure 6. Statistics of differentially methylated regions (DMRs) in SV and PV grafts (SV: grafts with strong growth vigor; PV: grafts with poor growth vigor). (**a**) The number of DMRs in CG, CHG and CHH contexts. The number of DMRs in different gene regions in the CG (**b**), CHG (**c**) and CHH (**d**) contexts (H = A, T, or C). Hypermethylation and hypomethylation in DMRs represent the methylation changes in PV grafts relative to the methylation status in SV grafts. UTR: untranslated region.

3.3. GO and KEGG Analysis of DMGs

In the pecan grafts, a total of 849, 2114, and 6333 DMGs were detected for the CG, CHG, and CHH contexts, respectively (Table S2). GO and KEGG analyses were performed to analyze the gene functions of these DMGs. The GO annotation results were consistently found based on the analysis of CG-DMGs, CHG-DMGs, and CHH-DMGs contexts (Figures 7 and S6). In terms of biological processes, the largest number of DMGs were annotated with terms of metabolic process, cellular process, and single-organism process. In addition, the DMGs were involved in regulation activities with most genes enriched at three cellular components, cell, cell part, and membrane. In terms of molecular function, catalytic activity and binding enriched the most DMGs. Based on enrichment analysis using Fisher's exact test, DMRs were significantly enriched in the biological processes that potentially affect graft growth ($p < 0.05$; Table S3). For example, the CG-DMGs were enriched in regulation of the ethylene-activated signaling pathway (GO:0010104) and jasmonic acid biosynthetic process (GO:0009695), the CHG-DMGs were enriched in defense response to virus (GO:0051607), and the CHH-DMGs were enriched in response to auxin (GO:0009733), abscisic acid (GO:0009737), cell wall macromolecule catabolic process (GO:0016998), cell differentiation (GO:0030154), asymmetric cell division (GO:0008356), and defense response (GO:0006952).

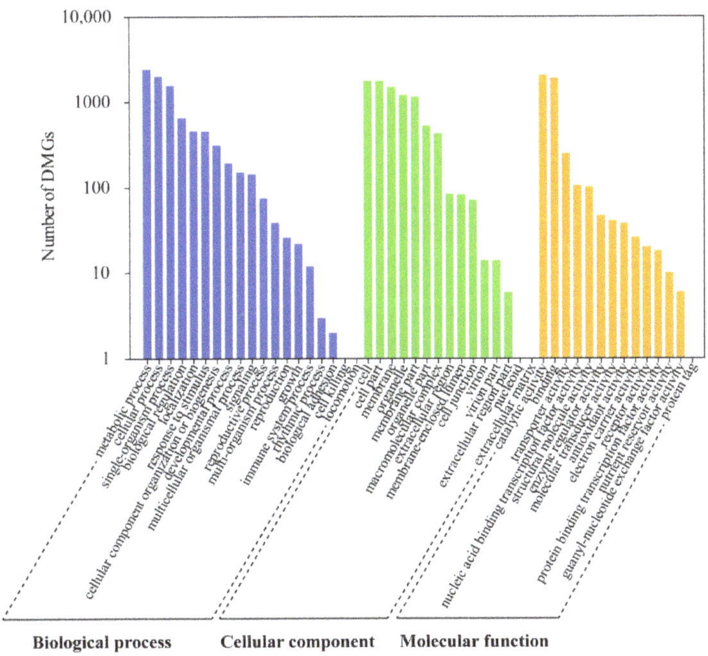

Figure 7. Gene Ontology (GO) annotation of differentially methylated region (DMR)-related genes (DMGs) in CHH type (H =A, T, or C) in SV and PV grafts (SV: grafts with strong growth vigor; PV: grafts with poor growth vigor).

According to enrichment analysis of KEGG pathways, the CG-DMGs were significantly involved in phenylpropanoid biosynthesis, steroid biosynthesis, arginine biosynthesis, and alanine, aspartate and glutamate metabolism ($p < 0.05$). The CHG-DMGs were found to significantly correlate with the regulation of valine, leucine, and isoleucine biosynthesis, other glycan degradation, and glycosaminoglycan degradation ($p < 0.05$). For the CHH-DMGs, most (91 genes) were significantly assigned to plant hormone signal transduction ($p < 0.05$; Figures 8 and Figure S7).

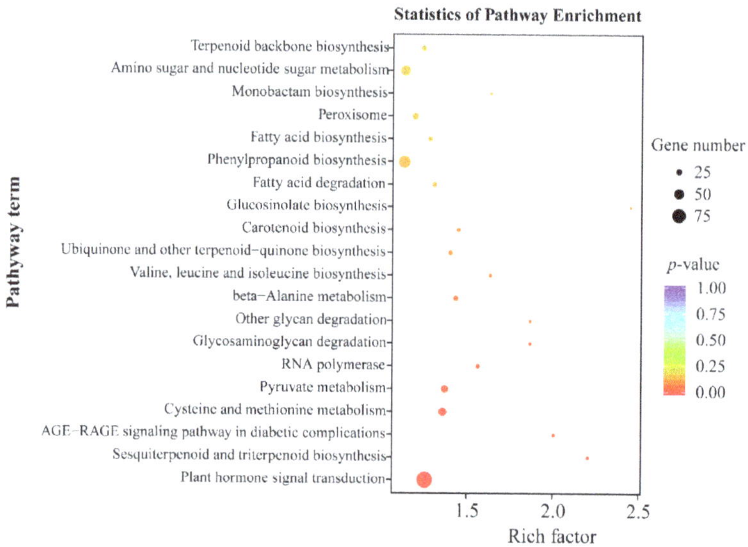

Figure 8. Kyoto Encyclopedia of Genes and Genomes (KEGG) analysis of differentially methylated region (DMR)-related genes (DMGs) in CHH type (H = A, T, or C) in SV and PV grafts (SV: grafts with strong growth vigor; PV: grafts with poor growth vigor). The circle size represents the gene number. The coloring of circles represents *p*-value, indicating the enrichment significance of DMGs in KEGG pathways.

Based on analysis of GO and KEGG, a total of 17 genes were identified that were significantly enriched in the GO terms of response to hormones (15 for GO terms of response to auxin and 2 for the terms of response to abscisic acid) and the KEGG pathway of plant hormone signal transduction. The DMGs contained 31 DMRs, including 14 hypermethylated DMRs and 17 hypomethylated DMRs, all of which were located in the promoter and distal intergenic regions. Additional information regarding the selected genes is provided in Table S4.

4. Discussion

As an important research topic in horticultural crops, the mechanisms driving rootstock-induced changes in scion vigor remains poorly understood. DNA methylation is an important regulator of gene expression and plant developmental processes. Recent studies suggested that DNA methylation may be involved in the grafting process, but little is known about the involved mechanisms. As a reliable technology to detect DNA methylation on a genome-wide scale at single-base resolution, WGBS has been applied to some plants, such as *A. thaliana*, rice (*Oryza sativa* L.), soybean (*Glycine max* L. Merrill Cv. Jack), poplar (*Populus trichocarpa* Torr. and Gray), and white birch (*Betula platyphylla* Suk.) [26–30]. To date, there have been no reports on methylation sequencing in pecan. In this study, WGBS was conducted to detect DNA methylation in two sets of pecan grafts with different growth vigor (SV and PV) on tall and short rootstocks, revealing the DNA methylomes in pecan for the first time. Based on analysis of the methylation profiles, DMRs and DMGs were identified between SV and PV grafts. Furthermore, the DMGs were subjected to functional analysis to explore the potential involvement of DNA methylation in rootstock-induced growth alterations.

4.1. Pecan Methylation Patterns

Currently, we have obtained more knowledge on the genetic characteristics of pecan based on previous studies, such as the application of molecular markers [31,32] and recent revelation of pecan genome [23]. However, little is known on pecan epigenome. In present study, WGBS was applied for the first time to reveal methylome of pecan on a genome-wide scale, which is of great significant to full understanding of pecan epigenome and further epigenetic research on pecan.

According to detection of methylation status in the cytosines, methylated cytosines accounted for 24.52%–25.60% of all cytosines in the six pecan samples, which were similar to the cytosine percentage in rice (24.3%) [28], but were about four times higher than that in wild-type *A. thaliana* [26]. It was found that the methylation levels were highest in the CG context (66.77%–67.85%), followed by those in the CHG (54.16%–55.31%) and CHH (14.21%–15.37%) contexts, which was a common trend consistent with other plants [27,29]. However, obvious differences in methylation levels can be observed. For instance, methylation levels in poplar were reported to be 41.9%, 20.9%, and 3.25% in CG, CHG, and CHH, respectively [27], which were lower than those we observed in pecan. In all methylated cytosines in pecan, the percentages of ^{m}CG were found to be the lowest, while the largest percentages were found to be of ^{m}CHH. The opposite results were reported for rice and wild-type *A. thaliana* with the highest ratios being for ^{m}CG and the lowest ratios for ^{m}CHH [26,28]. In addition, a methylation study on soybean seed development suggested that methylation distribution levels can also change in the same species during different developmental stages, with the largest percentages being observed for ^{m}CHH during the middle and late seed maturation stages, but with the largest percentages for ^{m}CG being during early seed maturation stages [30]. It is known that the establishment, maintenance, and removal of methylation in CG, CHG, and CHH contexts are catalyzed by various enzymes, which results in a specific methylation state [33]. Therefore, in response to the different intrinsic environments of different species or developmental processes, the regulation activities of the enzymes related to methylation may be affected, which may induce alterations in methylation. In addition, a previous study suggested the higher methylation ratios for each context in *Brassica rapa* L. compared with those in *A. thaliana* may result from higher levels of repeat sequences [34]. Thus, the distinct genome composition in a species may also affect the methylation levels.

Genes contain different functional elements, and the distribution of methylation in different gene regions always attracts our attention for investigation. In this study, it was found that methylation levels in CG context tend to show a peak in the gene bodies and presented an obvious drop between the gene bodies and flanking regions. The similar characteristics in methylation distribution was also found in other plants, such as *A. thaliana*, soybean, and white birch [27,29,30], which may be attributed to their common evolutionary origin. To further explore gene body methylation, the methylation levels across introns and exons were compared. It was obtained herein that methylation levels in the introns were higher than those in the exons, which was consistent with the results from white birch [29]. However, as Feng et al. reported, the eight diverse plant and animals commonly showed the reverse trend, with the lower CG methylation levels in the introns relative to those in the exons [27]. It has been shown that nucleosome positioning affected the targeting of DNA methyltransferases in different DNA regions [35]. Therefore, there may be differences in nucleosome enrichment across exons and introns among diverse species, due to the distinct evolutional degrees where they are, which possibly influences the DNA methylation levels in exons and introns. Certainly, the reasons for the different distribution characteristics of methylation across exons and introns in diverse species are complicated, and need to be further explored. In addition, this study suggested that methylation levels in the repeats were higher than those in the gene regions, and the repeat bodies had higher methylation levels than those in the upstream and downstream of repeats. This was in agreement with the results from *A. thaliana*, rice, poplar and white birch [27,29]. Notably, methylation levels in CHG were found higher in the repeats of poplar and pecan than those of *A. thaliana* and rice. From this, it can be seen that there were certain methylation differences in the repeats between herbaceous and woody plants.

4.2. Methylation Alterations Induced by Grating with Different Rootstocks

The finding that siRNAs can move through grafts to direct DNA methylation in scions [9,10] prompted us to start being concerned about the involvement of methylation changes in grafting process. One hetero-grafting study demonstrated that cucumber (*Cucumis sativus* L.) and melon (*Cucumis melo* L.) scions grafted on pumpkin [*Cucurbita moschata* (Lam.) Poir.] presented significantly increased methylation levels compared to that of their respective seed-plant control [13]. The study on rubber [*Hevea brasiliensis* (Willd. ex A, Juss.) Muell. Arg.] trees obtained methylation levels ranging from 16.42% to 19.80% in the scions on different rootstocks [14]. Hetero-grafting experiments on eggplant also revealed that CHH methylation levels significantly decreased by 3.37% and 2.58% in eggplant scions grafted on *S. torvum* and 'Emperador RZ' rootstocks, respectively, compared to those of the self-grafted plants [15]. The detailed data provided in these studies reflected the overall methylation changes induced by rootstocks. Wu et al. found that the methylation levels were largely not affected by rootstocks, and mainly focused on the methylation changes in the amplified sites [12]. They revealed that more than 10% of the detected sites exhibited methylation changes in tomato scions grafted on eggplant, and eggplant scions on tomato, compared with their respective seed-plants. In the present study, by analysis of the methylation profiles in a genome-wide scale, no significant differences were found in the methylation levels between SV and PV grafts, while a total of 934, 2864, and 15,789 DMRs were identified in the CG, CHG, and CHH contexts, respectively. In the cytosine sites of SV and PV grafts, the reverse variation direction of methylation patterns may be one of the reasons for no significant differences in methylation levels. Furthermore, at the genome-wide level, the methylation variation in part of the sites may not significantly affect the overall methylation levels. The DMRs reflect more specifically the methylation changes, and the identification of so many DMRs in the three C contexts fully explained the extensive methylation alterations in pecan scions induced by different rootstocks. However, the exact mechanism is unknown on driving the methylation differences, even though it has been reported that the mobile siRNAs can direct DNA methylation [9,10]. Rootstocks are known to be able to change the physiological environments in scions, such as water potentials, concentrations of mineral elements, and hormones [36–38]. In view of the fact that DNA methylation can be alterated in response to various environment stresses [39–41], we speculate that the methylation alterations in scions may be responsive to the environmental changes induced by different rootstocks, in addition to being possibly directed by the siRNA transported from rootstocks.

According to the obtained results in the present study, although the methylation levels in CG were highest throughout genome, the methylation changes in CHG and CHH were more frequent. The similar results were also reported previously [15,30,42]. Apart from the significant decrease in CHH methylation found in grafted eggplants [15], CHH methylation was reported to change significantly and the CHH-DMRs account for most of the DMRs during the developmental processes of soybean seeds [30]. Domb et al. also found that specific elimination of CG methylation did not dysregulate genes or transposons, and, in contrast, exclusive removal of non-CG methylation massively up-regulated genes and transposons [42]. In the light of these results, it is speculated that CHH or CHG methylation may be more sensitive to plant growth signal and more closely related to growth regulation. This study mainly revealed the methylation alterations in pecan grafts induced by different rootstocks, and, on this basis, further research on their transcriptional roles could help elucidate the grafting mechanisms.

In this study, we also localized the identified DMRs and found that most DMRs were in the distal intergenic regions and promoters, with very few DMRs in the 3′UTR or 5′UTR. According to the existing reports, methylation of the promoter in the upstream region of a gene generally repressed gene expression, while different effects of methylation in the gene body have been revealed in previous studies [28,43]. However, little is known regarding the regulatory roles of methylation in distal intergenic regions. As is well known, the intergenic regions account for a large proportion of the genome in plants. These regions were previously thought to lack biological functions and were considered as "junk DNA"; however, their biological functions have started to be revealed in recent years. For example, a recent study on maize (*Zea mays* L.) demonstrated that *KERNELROW NUMBER4* (*KRN4*), an intergenic quantitative trait locus (QTL) for kernel row number, can regulate the expression of *UNBRANCHED3* (*UB3*) as a distal enhancer and mediate inflorescence development [44]. Therefore, it is necessary to further study the effects of methylation on the regulatory roles of intergenic regions based on the substantial methylation differences in distal intergenic regions.

4.3. Involvement of Methylation in Graft Growth Regulation

Given the many methylation alterations between SV and PV grafts and the important roles of methylation in biological processes, it was reasonable to believe that methylation changes may produce functional consequences. Functional analysis of the DMGs could enable us to understand the potential involvement of methylation in graft growth. In the current study, 849 CG-DMGs, 2114 CHG-DMGs, and 6333 CHH-DMGs were detected. According to GO enrichment analysis of the genes, the DMGs were found to be involved in various biological processes that potentially affect plant growth, such as cell activities, plant hormone synthesis and signal regulation, and defense response. It can be presumed that methylation may have regulated theses growth-related genes to influence growth vigor of the pecan grafts. There have been previous studies revealing gene expression changes in scions with different growth vigor induced by rootstocks [45–47]. However, little is known regarding the genes that are potentially regulated by methylation alterations induced by grafting. Through WGBS, this study provided abundant information on methylation variation and the related genes for a deep revelation regarding the molecular regulatory mechanisms of grafting-induced growth vigor.

Notably, this study identified 17 DMGs that were significantly enriched in both GO terms (15 genes enriched in response to auxin and two genes enriched in response to abscisic acid) and a KEGG pathway (plant hormone signal transduction). This implies that these genes may be closely associated with response to auxin and abscisic acid (ABA), which are generally regarded as plant growth promoter and inhibitor, respectively [48,49]. It has been known that hormone-responsive genes are activated or repressed under control of hormones, and play important roles in physiological effects through mediation of hormone signal transduction [50–52]. Therefore, we speculate that under the stimulus of hormones in grafts, the methylation alterations may be the important factor controlling the gene expression to influence the growth of pecan grafts.

This study located genes with methylation changes in SV and PV grafts that had different growth vigor, which lays a good foundation for revealing the molecular mechanisms driving grafting-induced growth alterations. To better understand the molecular mechanisms associated with grafting, the effects of methylation on gene expression need to be explored in the future, even though it seems difficult to reveal potential joint regulatory roles of methylation in the distinct C contexts of different gene regions and possible interference of other regulation factors, such as miRNAs. Further studies are also needed to validate the roles of DMGs in graft growth regulation.

5. Conclusions

In the current study, we used WGBS to develop a DNA methylation study on two sets of pecan grafts with different growth vigor (SV and PV) on tall and short rootstocks. The DNA methylomes throughout the pecan genome were revealed for the first time. Based on pecan methylation profiles, large amount of DMRs were identified in SV and PV grafts, reflecting extensive differences in methylation induced by different rootstocks. The functional analysis of DMGs showed that they were involved in the biological processes that likely affect graft growth, such as cell activities, plant hormone synthesis and signal regulation, and defense response. In particular, we identified 17 DMGs that were most likely related to response to auxin and ABA, which may have especially important roles in the regulation of graft growth. This study demonstrated the potential involvement of DNA methylation in rootstocks-induced growth changes in pecan scions.

Supplementary Materials: The following supporting information can be downloaded at: https://www.mdpi.com/article/10.3390/f14010004/s1. Figure S1: Significantly different growth vigor in two groups of pecan grafts (SV and PV), Figure S2: Characteristics of 9-base pair sequences with methylated CG, CHG and CHH (SV2, SV3, PV1, PV2, and PV3), Figure S3: Circle plot of methylation distribution in pecan chromosomes (SV2, SV3, PV1, PV2, and PV3), Figure S4: Methylation distribution in different gene regions (SV2, SV3, PV1, PV2, and PV3), Figure S5: Methylation distribution in repeat regions (SV2, SV3, PV1, PV2, and PV3), Figure S6: GO annotation of DMGs in CG and CHG contexts, Figure S7: KEGG enrichment analysis of DMGs in CG and CHG contexts, Table S1: Statistics of methylated cytosines, Table S2: DMRs and DMGs, Table S3: GO enrichment analysis of DMGs, Table S4: The DMGs in CHH context potentially related with hormone regulation.

Author Contributions: F.P. and Z.L. conceived and designed the study. Z.L. collected experimental data, analyzed, and wrote the manuscript. P.T., Y.L. (Youwang Liang) and K.Z. provided help in data analysis and improving manuscript. Y.S. participated in collection of samples. Y.L. (Yongrong Li) provided the rootstocks. All authors have read and agreed to the published version of the manuscript.

Funding: This research was funded by a grant from the National Key R&D Program of China (2021YFD1000403), the Promotion and Demonstration Project of Forestry Science and Technology of Central Government ([2022]TG04), the Priority Academic Program Development of Jiangsu Higher Education Institutions (PAPD) and the Doctorate Fellowship Foundation of Nanjing Forestry University.

Data Availability Statement: Not applicable.

Conflicts of Interest: The authors declare no conflict of interest.

References

1. Goldschmidt, E.E. Plant grafting: New mechanisms, evolutionary implications. *Front. Plant Sci.* **2014**, *5*, 727. [CrossRef] [PubMed]
2. Seleznyova, A.; Thorp, T.; White, M.; Tustin, S.; Costes, E. Application of architectural analysis and AMAPmod methodology to study dwarfing phenomenon: The branch structure of 'Royal Gala'apple grafted on dwarfing and non-dwarfing rootstock/interstock combinations. *Ann. Bot.* **2003**, *91*, 665–672. [CrossRef] [PubMed]
3. Wallis, C.M.; Wallingford, A.K.; Chen, J. Grapevine rootstock effects on scion sap phenolic levels, resistance to *Xylella fastidiosa* infection, and progression of Pierce's disease. *Front. Plant Sci.* **2013**, *4*, 502. [CrossRef] [PubMed]
4. Dubey, A.; Sharma, R. Effect of rootstocks on tree growth, yield, quality and leaf mineral composition of lemon (*Citrus limon* (L.) Burm.). *Sci. Hortic.* **2016**, *200*, 131–136. [CrossRef]
5. Webster, A. Vigour mechanisms in dwarfing rootstocks for temperate fruit trees. In Proceedings of the Ist International Symposium on Rootstocks for Deciduous Fruit Tree Species, Zaragoza, Spain, 11–14 June 2002; Sanchez, M.A., Webster, A.D., Eds.; International Society Horticultural Science: Leuven, Belgium, 2004; pp. 29–41.
6. Koepke, T.; Dhingra, A. Rootstock scion somatogenetic interactions in perennial composite plants. *Plant Cell Rep.* **2013**, *32*, 1321–1337. [CrossRef]
7. Harada, T. Grafting and RNA transport via phloem tissue in horticultural plants. *Sci. Hortic.* **2010**, *125*, 545–550. [CrossRef]
8. Movahedi, A.; Sun, W.; Zhang, J.; Wu, X.; Mousavi, M.; Mohammadi, K.; Yin, T.; Zhuge, Q. RNA-directed DNA methylation in plants. *Plant Cell Rep.* **2015**, *34*, 1857–1862. [CrossRef]
9. Molnar, A.; Melnyk, C.W.; Bassett, A.; Hardcastle, T.J.; Dunn, R.; Baulcombe, D.C. Small silencing RNAs in plants are mobile and direct epigenetic modification in recipient cells. *Science* **2010**, *328*, 872–875. [CrossRef]
10. Bai, S.; Kasai, A.; Yamada, K.; Li, T.; Harada, T. A mobile signal transported over a long distance induces systemic transcriptional gene silencing in a grafted partner. *J. Exp. Bot.* **2011**, *62*, 4561–4570. [CrossRef]
11. Finnegan, E.J.; Genger, R.K.; Peacock, W.J.; Dennis, E.S. DNA methylation in plants. *Annu. Rev. Plant Physiol. Plant Mol. Biol.* **1998**, *49*, 223–247. [CrossRef]
12. Wu, R.; Wang, X.; Lin, Y.; Ma, Y.; Liu, G.; Yu, X.; Zhong, S.; Liu, B. Inter-species grafting caused extensive and heritable alterations of DNA methylation in *Solanaceae* plants. *PLoS ONE* **2013**, *8*, e61995. [CrossRef]
13. Avramidou, E.; Kapazoglou, A.; Aravanopoulos, F.A.; Xanthopoulou, A.; Ganopoulos, I.; Tsaballa, A.; Madesis, P.; Doulis, A.G.; Tsaftaris, A. Global DNA methylation changes in *Cucurbitaceae* inter-species grafting. *Crop. Breed. Appl. Biotechnol.* **2015**, *15*, 112–116. [CrossRef]
14. Uthup, T.K.; Karumamkandathil, R.; Ravindran, M.; Saha, T. Heterografting induced DNA methylation polymorphisms in *Hevea brasiliensis*. *Planta* **2018**, *248*, 579–589. [CrossRef]
15. Cerruti, E.; Gisbert, C.; Drost, H.; Valentino, D.; Portis, E.; Barchi, L.; Prohens, J.; Lanteri, S.; Comino, C.; Catoni, M. Epigenetic bases of grafting-induced vigour in eggplant. *bioRxiv* **2019**, 831719.

16. Zhang, R.; Peng, F.; Li, Y. Pecan production in China. *Sci. Hortic.* **2015**, *197*, 719–727. [CrossRef]
17. Grauke, L.J.; Starr, J.L. Phenotypic screening of pecan seedling rootstocks in search of nematode resistance. *Trees* **2014**, *28*, 1333–1341. [CrossRef]
18. Sanderlin, R.S. Susceptibility of some common pecan rootstocks to infection by *Xylella fastidiosa*. *Hortscience* **2015**, *50*, 1183–1186. [CrossRef]
19. Liu, Z.; Li, F.; Peng, F.; Tan, P.; Zhu, K.; Feng, G.; Mo, Z.; Li, Y. Identification of grafting-responsive microRNAs associated with growth regulation in pecan [*Carya illinoinensis* (Wangenh.) K. Koch]. *Forests* **2020**, *11*, 196. [CrossRef]
20. Liu, Z.Z.; Chen, T.; Peng, F.R.; Liang, Y.W.; Tan, P.P.; Mo, Z.H.; Cao, F.; Shang, Y.J.; Zhang, R.; Li, Y.R. Variation in cytosine methylation among pecan cultivars at different developmental stages. *J. Am. Soc. Hortic. Sci.* **2018**, *143*, 173–183. [CrossRef]
21. Liu, Z.; Zhou, F.; Shang, J.; Peng, F.; Mo, Z.; Li, Y. Changes of cytosine methylation in pecan tissues of different stages by quantitative methylation-sensitive amplified polymorphism. *Biol. Plant.* **2020**, *64*, 473–484. [CrossRef]
22. Li, Q.; Hermanson, P.J.; Springer, N.M. Detection of DNA methylation by whole-genome bisulfite sequencing. *Methods Mol. Biol.* **2018**, 185–196.
23. Huang, Y.; Xiao, L.; Zhang, Z.; Zhang, R.; Wang, Z.; Huang, C.; Huang, R.; Luan, Y.; Fan, T.; Wang, J. The genomes of pecan and Chinese hickory provide insights into *Carya* evolution and nut nutrition. *GigaScience* **2019**, *8*, giz036. [CrossRef] [PubMed]
24. Schultz, M.D.; Schmitz, R.J.; Ecker, J.R. 'Leveling' the playing field for analyses of single-base resolution DNA methylomes. *Trends Genet.* **2012**, *28*, 583–585. [CrossRef] [PubMed]
25. Sun, D.; Xi, Y.; Rodriguez, B.; Park, H.J.; Tong, P.; Meong, M.; Goodell, M.A.; Li, W. MOABS: Model based analysis of bisulfite sequencing data. *Genome Biol.* **2014**, *15*, R38. [CrossRef] [PubMed]
26. Lister, R.; Omalley, R.C.; Tontifilippini, J.; Gregory, B.D.; Berry, C.C.; Millar, A.H.; Ecker, J.R. Highly integrated single-base resolution maps of the epigenome in *Arabidopsis*. *Cell* **2008**, *133*, 523–536. [CrossRef]
27. Feng, S.; Cokus, S.J.; Zhang, X.; Chen, P.; Bostick, M.; Goll, M.G.; Hetzel, J.; Jain, J.; Strauss, S.H.; Halpern, M.E. Conservation and divergence of methylation patterning in plants and animals. *Proc. Natl. Acad. Sci. USA* **2010**, *107*, 8689–8694. [CrossRef]
28. Li, X.; Zhu, J.; Hu, F.; Ge, S.; Ye, M.; Xiang, H.; Zhang, G.; Zheng, X.; Zhang, H.; Zhang, S. Single-base resolution maps of cultivated and wild rice methylomes and regulatory roles of DNA methylation in plant gene expression. *BMC Genom.* **2012**, *13*, 300. [CrossRef]
29. Su, C.; Wang, C.; He, L.; Yang, C.; Wang, Y. Shotgun bisulfite sequencing of the *Betula platyphylla* genome reveals the tree's DNA methylation patterning. *Int. J. Mol. Sci.* **2014**, *15*, 22874–22886. [CrossRef]
30. An, Y.C.; Goettel, W.; Han, Q.; Bartels, A.; Liu, Z.; Xiao, W. Dynamic changes of genome-wide DNA methylation during soybean seed development. *Sci. Rep.* **2017**, *7*, 12263. [CrossRef]
31. Jia, X.D.; Wang, T.; Zhai, M.; Li, Y.R.; Guo, Z.R. Genetic diversity and identification of Chinese-grown pecan using ISSR and SSR markers. *Molecules* **2011**, *16*, 10078–10092. [CrossRef]
32. Conner, P.J.; Wood, B.W. Identification of pecan cultivars and their genetic relatedness as determined by randomly amplified polymorphic DNA analysis. *J. Am. Soc. Hortic. Sci.* **2001**, *126*, 474–480. [CrossRef]
33. Zhang, H.; Lang, Z.; Zhu, J.K. Dynamics and function of DNA methylation in plants. *Nat. Rev. Mol. Cell Biol.* **2018**, *19*, 489–506. [CrossRef]
34. Chen, X.; Ge, X.; Wang, J.; Tan, C.; King, G.J.; Liu, K. Genome-wide DNA methylation profiling by modified reduced representation bisulfite sequencing in *Brassica rapa* suggests that epigenetic modifications play a key role in polyploid genome evolution. *Front. Plant Sci.* **2015**, *6*, 836. [CrossRef]
35. Chodavarapu, R.K.; Feng, S.; Bernatavichute, Y.V.; Chen, P.Y.; Stroud, H.; Yu, Y.; Hetzel, J.A.; Kuo, F.; Kim, J.; Cokus, S.J.; et al. Relationship between nucleosome positioning and DNA methylation. *Nature* **2010**, *466*, 388–392. [CrossRef]
36. Khankahdani, H.H.; Rastegar, S.; Golein, B.; Golmohammadi, M.; Jahromi, A.A. Effect of rootstock on vegetative growth and mineral elements in scion of different Persian lime (*Citrus latifolia* Tanaka) genotypes. *Sci. Hortic.* **2019**, *246*, 136–145. [CrossRef]
37. Lordan, J.; Fazio, G.; Francescatto, P.; Robinson, T. Effects of apple (*Malus* × *domestica*) rootstocks on scion performance and hormone concentration. *Sci. Hortic.* **2017**, *225*, 96–105. [CrossRef]
38. Olien, W.C.; Lakso, A.N. Effect of rootstock on apple (*Malus domestica*) tree water relations. *Physiol. Plant.* **1986**, *67*, 421–430. [CrossRef]
39. Kou, H.P.; Li, Y.; Song, X.X.; Ou, X.F.; Xing, S.C.; Ma, J.; Wettstein, D.V.; Liu, B. Heritable alteration in DNA methylation induced by nitrogen-deficiency stress accompanies enhaced tolerance by progenies to the stress in rice (*Oryza sativa* L.). *J. Plant Physiol.* **2011**, *168*, 1685–1693. [CrossRef]
40. Peng, H.; Zhang, J. Plant genomic DNA methylation in response to stresses: Potential applications and challenges in plant breeding. *Prog. Nat. Sci.* **2009**, *19*, 1037–1045. [CrossRef]
41. Wang, W.S.; Pan, Y.J.; Zhao, X.Q.; Dwivedi, D.; Zhu, L.H.; Ali, J.; Fu, B.Y.; Li, Z.K. Drought-induced site-specific DNA methylation and its association with drought tolerance in rice (*Oryza sativa* L.). *J. Exp. Bot.* **2011**, *62*, 1951–1960. [CrossRef]
42. Domb, K.; Katz, A.; Harris, K.D.; Yaari, R.; Kaisler, E.; Nguyen, V.H.; Hong, U.V.T.; Griess, O.; Heskiau, K.G.; Ohad, N.; et al. DNA methylation mutants in *Physcomitrella patens* elucidate individual roles of CG and non-CG methylation in genome regulation. *Proc. Natl. Acad. Sci. USA* **2020**, *117*, 33700–33710. [CrossRef] [PubMed]

43. Vining, K.J.; Pomraning, K.R.; Wilhelm, L.J.; Priest, H.D.; Pellegrini, M.; Mockler, T.C.; Freitag, M.; Strauss, S.H. Dynamic DNA cytosine methylation in the *Populus trichocarpa* genome: Tissue-level variation and relationship to gene expression. *BMC Genom.* **2012**, *13*, 27. [CrossRef] [PubMed]
44. Du, Y.; Liu, L.; Peng, Y.; Li, M.; Li, Y.; Liu, D.; Li, X.; Zhang, Z. *UNBRANCHED3* expression and inflorescence development is mediated by *UNBRANCHED2* and the distal enhancer, *KRN4*, in Maize. *PLoS Genet.* **2020**, *16*, e1008764. [CrossRef] [PubMed]
45. Jensen, P.J.; Rytter, J.; Detwiler, E.A.; Travis, J.W.; McNellis, T.W. Rootstock effects on gene expression patterns in apple tree scions. *Plant Mol. Biol.* **2003**, *53*, 493–511. [CrossRef] [PubMed]
46. Prassinos, C.; Ko, J.H.; Lang, G.; Iezzoni, A.F.; Han, K.H. Rootstock-induced dwarfing in cherries is caused by differential cessation of terminal meristem growth and is triggered by rootstock-specific gene regulation. *Tree Physiol.* **2009**, *29*, 927–936. [CrossRef]
47. Cookson, S.J.; Ollat, N. Grafting with rootstocks induces extensive transcriptional re-programming in the shoot apical meristem of grapevine. *BMC Plant Biol.* **2013**, *13*, 147. [CrossRef]
48. Noda, K.; Okuda, H.; Iwagaki, I. Indole acetic acid and abscisic acid levels in new shoots and fibrous roots of citrus scion-rootstock combinations. *Sci. Hortic.* **2000**, *84*, 245–254. [CrossRef]
49. Davies, P.J. *Plant Hormones: Physiology, Biochemistry and Molecular Biology*, 2nd ed.; Kluwer Academic Publishers: London, UK, 1995.
50. Hoth, S.; Morgante, M.; Sanchez, J.; Hanafey, M.K.; Tingey, S.V.; Chua, N. Genome-wide gene expression profiling in *Arabidopsis thaliana* reveals new targets of abscisic acid and largely impaired gene regulation in the *abi1-1* mutant. *J. Cell Sci.* **2002**, *115*, 4891–4900. [CrossRef]
51. Teale, W.D.; Paponov, I.A.; Palme, K. Auxin in action: Signalling, transport and the control of plant growth and development. *Nat. Rev. Mol. Cell Biol.* **2006**, *7*, 847–859. [CrossRef]
52. Ren, H.; Gray, W.M. SAUR proteins as effectors of hormonal and environmental signals in plant growth. *Mol. Plant* **2015**, *8*, 1153–1164. [CrossRef]

forests

MDPI

Article

In Situ Rainwater Collection and Infiltration System Alleviates the Negative Effects of Drought on Plant-Available Water, Fine Root Distribution and Plant Hydraulic Conductivity

Changkun Ma [1], Haobo Meng [1], Biao Xie [2], Qian Li [2], Xiangdong Li [3], Beibei Zhou [1], Quanjiu Wang [1,*] and Yi Luo [4,*]

[1] State Key Laboratory of Eco-Hydraulics in Northwest Arid Region, Xi'an University of Technology, Xi'an 710048, China
[2] College of Horticulture, Northwest A&F University, Xianyang 712100, China
[3] School of Life Science, Yan'an University, Yan'an 716000, China
[4] Key Laboratory of Ecosystem Network Observation and Modeling, Institute of Geographic Sciences and Natural Resources Research, Chinese Academy of Sciences, Beijing 100101, China
* Correspondence: wquanjiu@163.com (Q.W.); luoyi@igsnrr.ac.cn (Y.L.)

Abstract: Soil water status and fine-root characteristics are the foundation for implementing forest water-management strategies in semiarid forest plantations, where rainwater is always the sole source of water for plant growth. Rainwater management and utilization are effective strategies to alleviate water scarcity in semiarid areas as ground water is always inaccessible there. Through the implementation of an in situ rainwater collection and infiltration system (IRCIS), we investigated the effects of IRCIS on soil water and fine-root distributions in the 0–5 m soil profile in a wet (2015, 815 mm) and a dry year (2016, 468 mm) in rainfed *Robinia pseudoacacia* forests in the Loess Plateau region of China. The results showed drought significantly decreased plant water availability and hydraulic conductivity of roots and branches, but strongly increased soil moisture deficits and fine-root (<2 mm diameter) biomass. With the implementation of IRCIS, soil profile available water and plant hydraulic conductivity can be significantly increased, but soil moisture deficits and fine-root (<2 mm diameter) biomass can be significantly decreased. Drought also significantly influenced the root distribution of *Robinia pseudoacacia*. The maximum depth of *Robinia pseudoacacia* roots in the dry year was significantly greater than in the wet year. Therefore, *Robinia pseudoacacia* can absorb shallow (0–1.5 m) soil water in wet years, while utilizing deep (>1.5 m) soil water in dry years to maintain normal growth and resist drought stress. The results of this study will contribute to the formulation of appropriate strategies for planning and managing rainwater resources in forest plantations.

Keywords: fine-root distribution; plant-available water; hydraulic characteristics; *Robinia pseudoacacia*; Loess Plateau

Citation: Ma, C.; Meng, H.; Xie, B.; Li, Q.; Li, X.; Zhou, B.; Wang, Q.; Luo, Y. In Situ Rainwater Collection and Infiltration System Alleviates the Negative Effects of Drought on Plant-Available Water, Fine Root Distribution and Plant Hydraulic Conductivity. *Forests* **2022**, *13*, 2082. https://doi.org/10.3390/f13122082

Academic Editors: Jie Luo and Wentao Hu

Received: 26 October 2022
Accepted: 5 December 2022
Published: 7 December 2022

1. Introduction

In drylands, large-scale afforestation and reforestation activities have been implemented to combat desertification, biodiversity loss and poverty, as forests are vital to preventing desertification and providing local residents with wood [1,2]. For instance, approximately 49 billion hectares of dryland were reforested between 2000 and 2010, resulting in a 0.8% increase in forest cover worldwide [3,4]. As forest plantation area increases, the amount of water consumed by these planted trees will also increase [5]. Then there may also be an increase in soil moisture deficits and a decrease in soil availability for plants. A severe soil moisture deficit will inhibit plants from growing normally (e.g., fine roots, hydraulic conductivity, etc.), and even result in the death of plants and the degradation of forest plantations [6]. Therefore, the development of these large-scale afforestation and reforestation projects will only be sustainable if adequate water resources are available. In these water-limited arid and semiarid regions, however, discrete rainfall is always the only source of soil water replenishment; therefore, regulating and utilizing rainwater rationally may be the only effective way to combat drought and alleviate forest degradation [6].

When it rains, some rainwater enters the shallow soil and is absorbed subsequently by the roots of the plants, while another part of the rainwater infiltrates into the deep soil and is stored and finally utilized by the deep roots, particularly during dry conditions in semiarid regions. Plant roots, especially fine roots, are important organs for absorbing water and nutrients from the soil [7]. Therefore, understanding the distribution of fine roots is fundamental to understanding the soil water-use pattern [8]. As a water deficiency detection sensor, fine roots are influenced by a variety of external factors, among which drought stress is the most significant. Recently, the response of fine roots of trees to drought stress has received much attention [9–13], but research on rainfed forest trees in semiarid regions is still lacking [8]. A better understanding of the distribution characteristics of rainfed plant roots, as well as their response to drought stress, is essential to understanding the water-use patterns of plants and is also important to address a number of fundamental problems in forest ecosystems, particularly in areas affected by severe droughts.

Water transfer in the roots and branches is one of the most critical aspects for ensuring plant survival. Studies have indicated that water transport within the roots and branches of plants is affected by a variety of factors, with drought stress being one of the most significant ones. When plants are subjected to drought, hydraulic failure can occur, which negatively impacts the growth and photosynthesis of plants and may lead to plant mortality as the drought persists [14,15]. In the past few decades, many studies have been conducted to determine the mechanisms causing drought-related tree mortality, and several mechanisms have been proposed [14,16], among which hydraulic failure of plants has been widely acknowledged. The hydraulic failure hypothesis states that high xylem tension induced by drought can cause air bubbles to enter the xylem and block the transport of water through the it, resulting in the plant's inability to move water and eventual desiccation [16]. Studies have been conducted to test the hydraulic failure hypothesis, and a comprehensive understanding of the water-transport characteristics of plants from the roots to the leaves has been formalized, but the roots are lagging behind compared to the branches. A better understanding of how plants adjust their water-transfer characteristics through their root systems, especially under drought conditions, will be helpful for us to adequately understand the hydraulic characteristics of plants, as well as to formulate reasonable management strategies.

In recent years, in situ rainwater collection and infiltration systems (IRCIS) have been developed and applied by farmers to harvest rainwater for sustainable plant growth, especially for orchards, in the semiarid Loess Plateau region of China [6,8]. These systems are designed to increase soil water quantity during the dry season by reducing surface runoff as well as improving rainwater-harvesting efficiency. In many studies, the effects of IRCIS have been discussed in relation to orchards [8,17]; however, very few studies have examined the effects of IRCIS on soil water content and root distribution for commonly planted trees on the Loess Plateau. However, understanding the effects of in situ rainwater collection and infiltration systems on soil moisture and vegetation growth (such as root distribution characteristics, plant hydraulic conductivity, etc.) of forest plantations under drought stress is of significant importance to develop reasonable plantation management measures and promote the healthy and sustainable development of these plantations.

The objectives of this study were (i) to determine the effects of drought on the distribution of soil moisture and fine roots (0–5 m) as well as plant hydraulic conductivity of a commonly planted tree species in the semiarid region of the Loess Plateau and (ii) to explore the effects of IRCIS on soil moisture, fine root distribution and plant hydraulic conductivity of a commonly planted tree species under drought. In order to investigate the effects of drought on soil moisture and plant growth (fine roots and plant hydraulic conductivity), a wet year with higher-than-average rainfall (2015) and a dry year with lower-than-average rainfall (2016) were chosen. The tree species *Robinia pseudoacacia*, which has been widely planted on the Loess Plateau to control soil and water erosion, prevent desertification and produce timber and fuel wood, was selected for this study.

2. Materials and Methods

2.1. Site Description

The study was conducted at Yeheshan Forestry Station (YFS; 34°33′ N, 107°54′ E, 1090 m a.s.l) in the National Natural Reserve of Fufeng County in Shaanxi Province, China. The area has a semiarid continental monsoon climate that is characterized by hot, humid summers (June–August) and cold, dry winters (December–February). According to the Fufeng meteorological station (which is located <10 km from the study area), average annual precipitation and air temperature are 580 mm and 12.7 °C, respectively. The precipitation mainly falls from May–October (~80% of annual precipitation). The main soil type in the region is silt loam, with a homogeneous texture along the 0–500 cm soil profile and mean sand, silt and clay contents of 5.8%, 73.4% and 20.9%, respectively [18]. The mean soil bulk density (BD), field capacity (FC) and permanent wilting point (PWP) of the silt loam are 1.30 g cm^{-3}, 0.304 cm^3 cm^{-3} and 0.072 cm^3 cm^{-3}, respectively [19].

YFS is a hilly-gully region with an elevation range of 449–1662 m a.s.l and area of 110 km^2. Forest cover in the region is ~90% and is dominated by *Robinia pseudoacacia* planted since the 1980s to control soil erosion. *Robinia pseudoacacia* plantation covers an area of 86.7 km^2 and has a dense understory that is a mix of *Stipa bungeana*, *Humulus scandens* and *A. codonocephala*.

2.2. Experimental Plot

Six *Robinia pseudoacacia* experimental plots (20 × 20 m^2) with an age of 13 years and density of 2000 trees/ha were selected and established on hills with slopes of 5–10° (middle slope facing south) where rainfed crops (such as maize and winter wheat) were previously cultivated. Three experimental plots (treated plots) were randomly chosen to install the in situ rainwater collection and infiltration systems (IRCIS), and another three experimental plots were used as the control plots (without IRCIS). Each plot was constructed with an aluminum composite panel ridge of 30 cm above the ground around the borders and an H-flume applied to measure surface runoff. Previous cropland (predominantly maize and winter wheat) that had been abandoned for 13 years was used as a control site for monitoring soil moisture changes.

In the treated experimental plots, IRCIS was constructed upslope of individual trees and consisted of a semi-circular ridge of radius 1.0 m and height 0.2 m. During the construction of the ridge, soil was excavated and moved in such a manner that the tree trunk formed the apex of the semicircular ridge and the soil surface within the semicircle area was relatively level. From above, the ridges formed an upslope pattern resembling a fish scale along each row in the *Robinia pseudoacacia* forest plots. Within the semicircular ridge, an 80 cm × 80 cm × 80 cm soil pit was dug. In this pit, the down-slope wall was 100 cm away from the tree trunk. This storage pit was lined with permeable geotextiles and filled with soil, weeds, branches and forest debris. The surface of this storage pit was then covered with black plastic film and a 3 cm diameter hole was drilled in the center of the plastic film. Rainwater collected from the fish-scale ridge would infiltrate into the filled material through the hole in the plastic film. Rainwater collected in the pit could then be directed laterally and downward into the surrounding soil. The design slows the flow of runoff, resulting in a reduction in sedimentation and loss of soil-pit storage capacity.

The experimental plots treated with IRCIS in 2015 and 2016 were designated, respectively, as RC2015 and RC2016. The control plots without IRCIS in 2015 and 2016 were designated as WT2015 and DT2016, respectively.

2.3. Fine-Root Measurement

The vertical distribution of fine roots in the controlled and treated *Robinia pseudoacacia* stands was investigated using the soil auger method [20]. In each plot, soil core samples were collected using a cylindrical metal corer (9 cm wide and 10 cm long) with one sharp edge. A total of 8 sample points was selected in each forest stand. Samples were taken between 10 and 20 May, 10 and 20 July, and 10 and 20 September at the beginning, middle and end of the growing season. Soil cores were collected at 20 cm intervals to a depth of 100 cm and then at 40 cm intervals to a depth of 500 cm where fine roots were not observed, and the width of the tree canopy. The collected soil samples were stored in polythene plastic bags at 4 °C for later analyses. In the laboratory, the root samples were separated from the soil by a two-stage process. In the first stage, soil samples were washed carefully over a sieve (5 mm). Then grass roots and other organic debris were discarded and tree roots were placed with tweezers into petri dishes containing water. In the second stage, the tree roots were separated by physiological status and diameter classes (d > 2 mm and d ≤ 2 mm). Then separation was performed using a microscope (10–40 × magnification). The *Robinia pseudoacacia* tree roots were identified based on color and morphology. The root samples were then digitally scanned using an Epson Perfection v700 photo scanner at 600 dpi (Seiko Epson Corporation, Naga-no-ken, Japan). Root lengths were measured using the WinRhizo image analysis software (pro 2009c, Regent Instruments Inc., Quebec City, QC, Canada). The fine-root (<2 mm) length density (FRLD, cm dm^{-3}) was calculated as:

$$FRLD = \frac{L}{Vs} \quad (1)$$

where L is fine-root length and Vs is soil volume. We also calculated the cumulative fine-root length density (CFRLD, % of total) for each experiment plot in each soil layer.

2.4. Soil Water Content and Rainfall Measurement

Volumetric soil water content was automatically measured using soil water probes. At each forest stand, 10 Hydra-Probes (Stevensons Water Monitoring Systems, Portland, OR, USA) were installed at soil depths of 5, 15, 30, 50, 80, 120, 180, 250, 350 and 500 cm below the ground surface. Soil water content was sampled every 10 min and recorded by a CR1000 datalogger (Campbell Scientific Inc., Logan, UT, USA). Before the study, all soil water probes were calibrated.

Plant-available water storage (PAMS) is the amount of water storage that can be released into the root zone at any point in time. It is defined as the difference between in situ field water content and permanent wilting point (PWP) and is calculated as:

$$PAMS_{i,j} = (SWC_{i,j} - PWP) \times BD_{i,j} \times \Delta Z_{i,j} \quad (2)$$

where $SWC_{i,j}$ is soil water content of the jth soil layer under the ith treatment (stand age); PWP is permanent wilting point; $BD_{i,j}$ is soil bulk density at depth j and sampling site i; and $\Delta Z_{i,j}$ is soil depth increment.

Soil moisture storage deficit (SMSD), relative to the values of the abandoned cropland, was used to assess the feasibility of ecological restoration [8]:

$$SMSD_{t,i} = \frac{SWS_{t,i} - SMS_{c,i}}{SMS_{c,i}} \quad (3)$$

where $SMS_{t,i}$ and $SMS_{c,i}$ are soil moisture storage in the ith soil layer in the treated experimental plot and the abandoned cropland, respectively. SMSD ≥ 0 indicates no soil moisture storage deficit, while SMSD < 0 represents soil moisture storage deficit in the specified soil layers, where the magnitude of the value is indicative of the severity of the deficit.

Rainfall was measured using both automatic and manual rain gauges. Three manual funnel-type rain gauges (30 cm orifice diameter) were installed near (<30 m away) each forest stand at a sufficiently open place. Due to financial limitations, only one T-200B weighing bucket rain gauge (Geonor, Eiksmarka, Norway) was used in this study. The automatic rain gauge was connected to a CR1000 data logger (Campbell Scientific Inc., Logan, UT, USA) and installed in a sufficiently exposed nearby place (<30 m) in the middle of the watershed. The manual rain gauges were read immediately after each rainfall event. Using the long-term yearly rainfall data (1958–2016), we constructed a rainfall frequency curve. Each year was categorized as either a wet year, a normal year or a dry year, depending on the frequency of gross rainfall (<25% are wet years; >75% are dry years; remainder are normal years) [21].

To exclude the effect of rainfall on soil water movement, seven successive days without rainfall (rain-free period) were set in this study as the investigation period. Moreover, leaves were completely unfolded and canopies remained stable during summer and therefore 6 rain-free periods (8–14 June, 25–31 July and 25–31 August in 2015 and 16–22 June, 2–8 July and 10–16 August in 2016) were ultimately used for investigation (Figure 1).

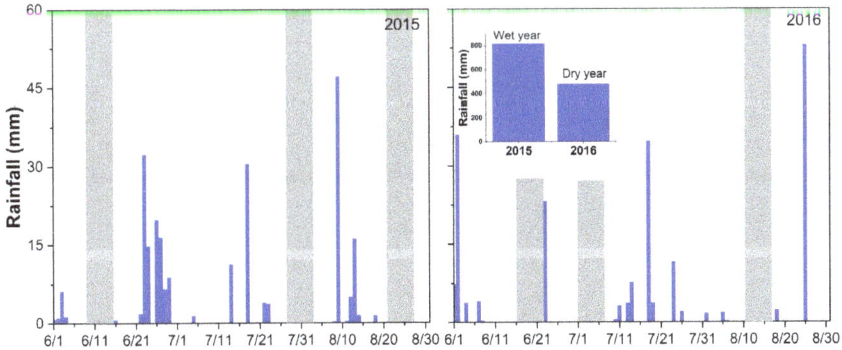

Figure 1. Gross rainfall distribution and the six selected investigation times during the experimental periods in 2015 and 2016. Gray rectangles indicate the periods during which the study was carried out.

2.5. The Percent Loss of Hydraulic Conductivity (PLC)

In this study, one-year-old branches at the middle-upper canopy on the sunny side (the south side) of the plant, as well as roots from shallow soil layers (0–100 cm), were used to determine the percent loss of hydraulic conductivity (PLC). To measure the PLC, a low-pressure flow meter was used following the approach proposed by Sperry et al. [22]. During the dawn or noon period of the fine-root collection, approximately 30–40 cm long samples of branches with a basal diameter of approximately 1.0 cm and roots in soil depths of 50 cm were collected. To avoid embolisms induced by cutting, all plant samples were cut under water. Afterwards, all the samples were wrapped in plastic bags and taken back to the lab for PLC analysis. The PLC was determined by averaging three segments (4 cm long) across three biological replicates. To determine the flow rate through the segment, the solution was collected and weighed using a balance. By measuring the flow rate of the KCl solution at a pressure differential of 4 kPa, the initial hydraulic conductivity (Ki) was determined gravimetrically. The stem segment was flushed for a period of 10 min at a pressure of 0.175 MPa in order to remove any air embolisms. Afterwards, the hydraulic conductivity of the fluid was determined again at a pressure differential of 4 kPa and was set as the maximum hydraulic conductivity (Kmax). The PLC was then calculated as PLC (%) = (1 − Ki/Kmax).

2.6. Statistical Analysis

The data were analyzed using a one-way analysis of variance (ANOVA) using SPSS Version 25.0 (Chicago, IL, USA) after verifying the assumptions of normality and homogeneity. Duncan's multiple range tests were performed at $p < 0.05$ and $p < 0.01$ for significant differences between the treatments for PAMS, SMSD, FRLD and PLC between treated and control forest plots: WT2015, RC2015, DT2016 and RC2016. Origin software 2022 (OriginLab Corporation, Hampton, MA, USA) and Excel 2022 (Microsoft Corporation, Redmond, WA, USA) were used to fit curves and plot graphs.

3. Results

3.1. Plant-Available Moisture Storage (PAMS)

Plant-available moisture storage (PAMS) in *Robinia pseudoacacia* forest stands was significantly affected by the in situ rainwater collection and infiltration system (IRCIS) in both wet and dry years (Figure 2). In general, the average PAMS profile decreased with increasing soil profile, but the decrease under the treatment of IRCIS was relatively gradual, whereas it was relatively rapid under the control treatment (Figure 2A). Furthermore, IRCIS increased the average PAMS in soil profiles (0–5 m) of *Robinia pseudoacacia* forest stands in both study years, but the increase was greater in the dry year 2016 than in the wet year 2015. IRCIS increased average PAMS in soil profiles by 4.6% in the wet year 2015 and 10.0% in the dry year 2016 (Figure 2B).

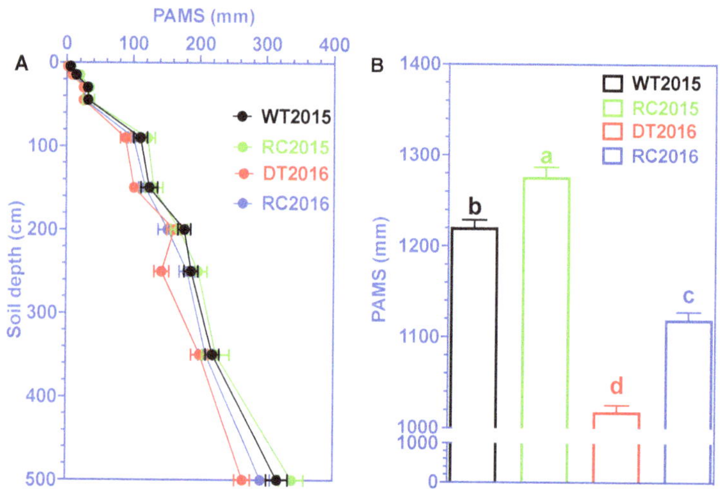

Figure 2. Vertical distribution (0–5 m, **A**) and the content (**B**) of plant-available moisture storage (PAMS) in a wet year of higher-than-average rainfall (2015) and a dry year of lower-than-average rainfall (2016) in *Robinia pseudoacacia* forest stands of different treatments. WT2015 is wet year of 2015 without in situ rainwater collection and infiltration system; RC2015 is wet year of 2015 with in situ rainwater collection and infiltration system; DT2016 is dry year of 2016 without in situ rainwater collection and infiltration system and RC2016 is wet year of 2016 with in situ rainwater collection and infiltration system. Different letters indicate a statistically significant difference at $p < 0.05$ between treatments. Error bars represent ±1 SD.

3.2. Soil Moisture Storage Deficit (SMSD)

The soil profile SMSD is shown in Figure 3. Overall, shallow soil profiles (0–1.5 m) suffered greater soil moisture deficits during the dry year of 2016, but little or no deficit during the wet year of 2015 relative to abandoned cropland (Figure 3A,B). In deep soils (>1.5 m), almost all forest treatments displayed soil moisture deficits. The vertical distribution of the profile soil moisture deficits was significantly affected by IRCIS treatment (Figure 3C). On the whole, IRCIS treatment alleviated the soil profile deficit, but the alleviation in the shallow soil (0–1.5 m) was much greater than that in the deep soil profile (>1.5 m) (Figure 3A,B). In the two study years, the amount of alleviation was similar, but slightly higher in the dry year of 2016 (decreased by 0.06) than in the wet year of 2015 (decreased by 0.05).

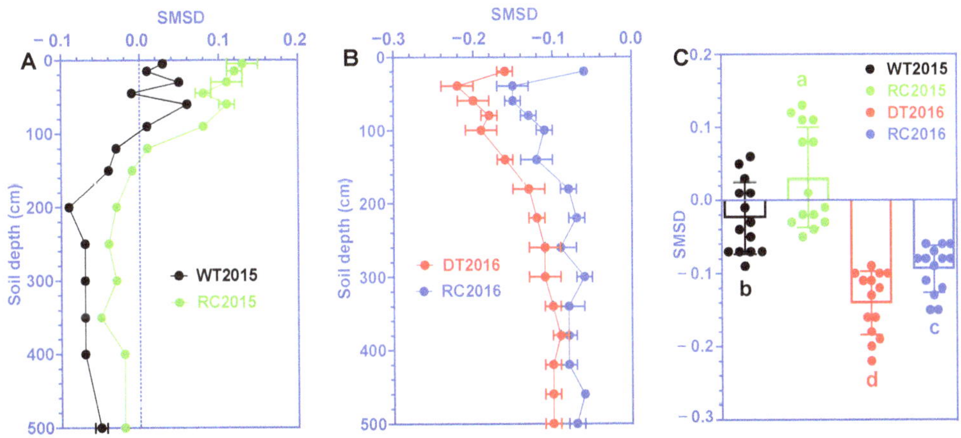

Figure 3. Vertical distribution (0–5 m, **A**,**B**) and content (**C**) of soil moisture storage deficit (SMSD) in a wet year of higher-than-average rainfall (2015, **A**) and a dry year of lower-than-average rainfall (2016, **B**) in *Robinia pseudoacacia* forest stands of different treatments. Negative and positive values represent, respectively, a negative and positive soil moisture deficit relative to the control (abandoned cropland). Different letters indicate a statistically significant difference at $p < 0.05$ between treatments. Error bars represent ±1 SD. WT2015 is wet year of 2015 without in situ rainwater collection and infiltration system; RC2015 is wet year of 2015 with in situ rainwater collection and infiltration system; DT2016 is dry year of 2016 without in situ rainwater collection and infiltration system and RC2016 is wet year of 2016 with in situ rainwater collection and infiltration system.

3.3. Fine-Root Distribution (FRLD)

The vertical distribution of fine-root length density (FRLD) is shown in Figure 4. In general, the FRLD decreases with increasing soil depth, primarily in the 0–1.5 m soil profile (Figure 4A). Fine roots were mainly concentrated in the top 0–1.5 m soil: WT2015 and RC2015 had 82.9% and 93.9% of total FRLD, respectively, in the wet year of 2015; DT2016 and RC2016 had 67.4% and 75.3% of total FRLD, respectively, in the dry year of 2016. Drought strongly increased the amount of FRLD: DT2016 and RC2016 had 31.1% and 23.8% higher total FRLD, respectively, than WT2015 and RC2015 (Figure 4B). IRCIS significantly reduced the FRLD: RC2015 and RC2016 had 7.4% and 12.7% lower total FRLD, respectively, than WT2015 and DT2016 (Figure 4B).

Over both study years, the cumulative FRLD (CFRLD) was higher in the shallow soil layer than in the deeper soil layer (Figure 5). IRCIS treatment significantly affected the distribution of cumulative fine-root length density in *Robinia pseudoacacia* forests. IRCIS tended to decrease CFRLD profiles gradually, but control treatment caused them to decline profoundly (Figure 5), especially in the 0–1.5 m soil layer. IRCIS-treated forest stands tended to have shallower D_{95} (the depth above which 95% of the root mass is present) than the control forest stands, with respective values of 1.2 m for RC2015, 1.8 m for WT2015, 2.8 m for RC2016 and 3.3 m for DT2016.

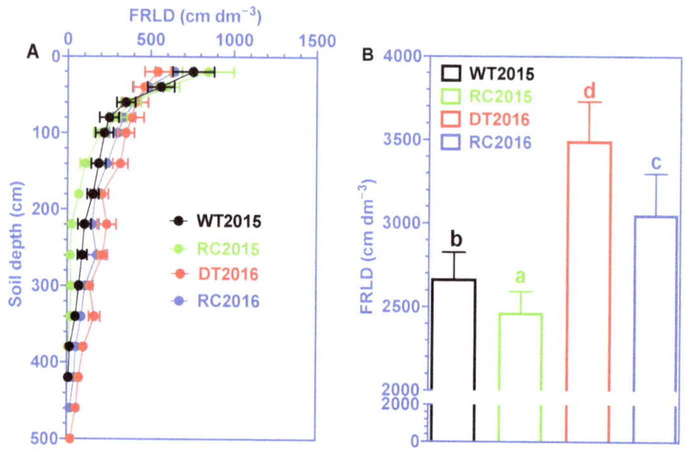

Figure 4. Vertical distribution (0–5 m, **A**) and content (**B**) of fine-root length density (FRLD, cm/dm^3) in a wet year of higher-than-average rainfall (2015) and a dry year of lower-than-average rainfall (2016) in *Robinia pseudoacacia* forest stands of different treatments. Different letters indicate a statistically significant difference at $p < 0.05$ between treatments. Error bars represent ±1 SD. WT2015 is wet year of 2015 without in situ rainwater collection and infiltration system; RC2015 is wet year of 2015 with in situ rainwater collection and infiltration system; DT2016 is dry year of 2016 without in situ rainwater collection and infiltration system and RC2016 is wet year of 2016 with in situ rainwater collection and infiltration system.

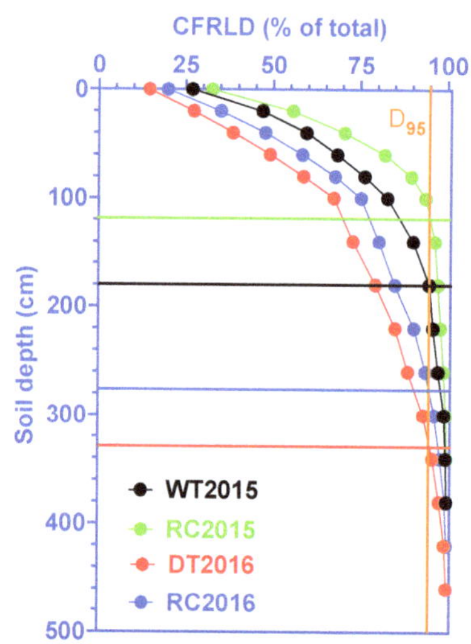

Figure 5. Comparison of cumulative fine-root length density in *Robinia pseudoacacia* forest stands of different treatments in a wet year of higher-than-average rainfall (2015) and a dry year of lower-than-average rainfall (2016). Data points represent average fine-root length densities in the indicated soil layers. The intersection of horizontal lines (with different colors) and vertical lines D_{95} indicates soil depth above which 95% of fine-root length density occur. WT2015 is wet year of 2015 without in situ rainwater collection and infiltration system; RC2015 is wet year of 2015 with in situ rainwater collection and infiltration system; DT2016 is dry year of 2016 without in situ rainwater collection and infiltration system and RC2016 is wet year of 2016 with in situ rainwater collection and infiltration system.

3.4. The Percentage Loss of Hydraulic Conductivity (PLC)

IRCIS strongly affected the percentage loss of hydraulic conductivity (PLC) in branches and roots of *Robinia pseudoacacia* forest (Figure 6). Over the two study years, IRCIS reduced the PLC for all forest stands. In branches, RC2015 and RC2016, were, respectively, 35.6% and 24.3% lower than WT2015 and DT2016 (Figure 6A). In roots, RC2015 and RC2016, were, respectively, 35.6% and 24.3% lower than WT2015 and DT2016 (Figure 6B). In general, IRCIS had a greater impact on the root PLC than on the branch PLC (Figure 6A,B). This shows that roots have a higher sensitivity to external factors, such as drought and IRCIS, than branches when it comes to hydraulic conductivity.

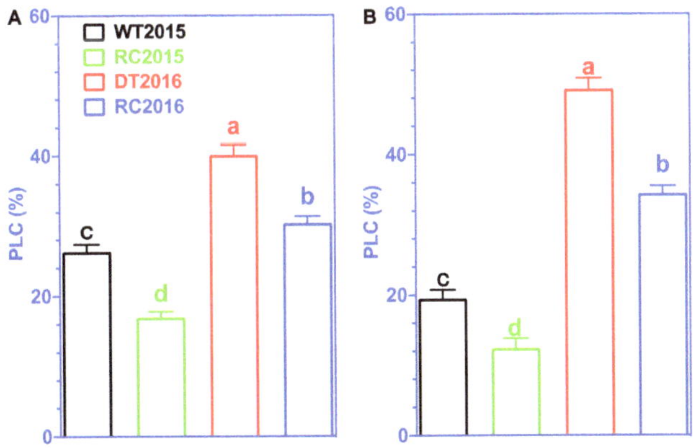

Figure 6. The percentage loss of hydraulic conductivity (PLC) in branches (**A**) and roots (**B**) of *Robinia pseudoacacia* under different treatments. Different letters indicate a statistically significant difference at $p < 0.05$ between stand age treatments. Error bars represent ±1 SD. WT2015 is wet year of 2015 without in situ rainwater collection and infiltration system; RC2015 is wet year of 2015 with in situ rainwater collection and infiltration system; DT2016 is dry year of 2016 without in situ rainwater collection and infiltration system and RC2016 is wet year of 2016 with in situ rainwater collection and infiltration system.

4. Discussion

4.1. Soil Moisture Profile

Over the last few decades, large-scale afforestation efforts have increased forest cover by 7.6% from 2001 to 2016 in the semiarid Loess Plateau region of China [23]. Due to the large amount of soil water required by forest vegetation, plant-available soil water is likely to decline dramatically, and the soil water deficit will be exacerbated, especially in drought years (Figures 2 and 3). As a result, this will negatively impact the health and stability of the forest vegetation ecosystem in the semiarid Loess Plateau region, resulting in a reduction in the normal growth of forest vegetation and even death of forest vegetation [19,24]. As the climatic and environmental conditions experienced by the forest plots are similar in our study, rainfall in different years and forest management measures (with or without IRCIS treatment) are the primary factors determining soil water balance and how soil water is distributed in the soil profile of *Robinia pseudoacacia* forests.

This study highlights the importance of IRCIS for rainfall harvesting in afforestation planting and management because it can significantly increase the availability of soil water to the plants and decrease soil water deficits (relative to the abandoned cropland). In line with our findings, Song et al. (2020) also found that IRCIS treatment can significantly increase the soil profile water content of orchard plants [8]. It has been shown that shallow (0–1.5 m) soil moisture is primarily affected by rainfall and vegetation utilization [17,25]; thus, shallow soil (0–1.5 m) has significantly wider variations in soil water content than deep soil, particularly during dry seasons or years (Figure 2). Our study also found that the effect of IRCIS on shallow soil moisture content (0–1.5 m) was greater than it was on deep soil moisture (>1.5 m), which could be attributed to the fact that the soil moisture increased by IRCIS was mainly retained and consumed by the shallow (0–1.5 m) root system of *Robinia pseudoacacia* (Figures 4 and 5), but could not be substantially replenished in the deep soils (>1.5 m). As a consequence, the shallow soil moisture content (0–1.5 m) is highly variable with the implementation of IRCIS, whereas the deep (>1.5 m) soil moisture content is relatively stable. Additionally, this suggests that deep soil profiles may serve as an important water source for forest vegetation during extremely dry years [8,26,27].

4.2. Root Distribution Pattern

Fine roots are essential to the growth and development of plants as they extract the majority of the water and nutrients from the soil. In this study, it was found that *Robinia pseudoacacia* increased its biomass of fine roots as well as its distribution depth during dry years, demonstrating that *Robinia pseudoacacia* is able to absorb and utilize large amounts of deep soil water during dry periods. These findings are similar to those of Brunner et al. 2015 [26] and Song et al., 2020 [8]. When *Robinia pseudoacacia* was treated with IRCIS, its fine-root biomass and fine-root depth decreased significantly in comparison to the control group. This may be because IRCIS can increase the water content of the shallow soil in both dry and wet years (Figure 2), making it easier for *Robinia pseudoacacia* to absorb water from the shallow soil instead of allocating more carbon for growing deeper roots to absorb deep soil water to support its growth and development. Similar results were also found by Li et al. (2022) [28] and Song et al. (2020) [8].

Additionally, this study demonstrated that the FRLD content in the surface (0–1.5 m) soil of *Robinia pseudoacacia* forests was the highest in both dry and wet years, accounting for more than 80.0% of the total fine-root biomass, among which WT2015, RC2015, DT2016 and CR2016 each had a content of 82.9%, 93.9%, 67.4% and 75.3%, respectively. The results demonstrated that this structure facilitates the efficient uptake of water from shallow soil layers, where soil water is greatly affected by rainfall and replenished rapidly [8,17,29]. Although the FRLD content in deep soil (>1.5 m) is relatively low, this part of the fine roots is critical to the growth of plants. Particularly in drought years, this part of the fine roots can absorb deep soil moisture in order to aid plants in resisting drought conditions. There is also evidence from Song et al. (2020) [8] that apple trees in semiarid regions will increase their water consumption from deep soil in drought years in order to resist drought stress [8].

A deep root system is one of the most important adaptations for plants in arid and semiarid environments. Generally, rooting depth determines how much water can be accessed by plants through transpiration from the soil [30]. In this study, it was found that, in comparison with the wet year of 2015, the growth of *Robinia pseudoacacia* fine roots (maximum depth and D_{95}) was deeper in the dry year of 2016, which may enable them to switch between shallow and deep water sources, depending on soil water availability. These findings agree well with reports by [8,25,31]. In addition, this study found that IRCIS treatment increased plant water availability and reduced soil water deficit, reducing the risk of plants suffering from drought stress. Similar results were also reported by Song et al. (2020) [8].

4.3. PLC

Xylem water transport is essential for maintaining canopy gas exchange and cell expansion and, therefore, for plant growth and survival. Studies have shown that xylem conduits with larger diameters always have greater water transport ability, but are also more vulnerable to hydraulic failure [16]. This is due to the structural characteristics of the xylem network: plants are faced with the challenge of transporting as much water as possible while minimizing the risk of drought-induced embolisms during dry periods [32]. This was also evident in the present study, where the loss of hydraulic conductivity during drought years was larger in roots (with a larger xylem diameter) than it was in branches (with a small xylem diameter). The loss of hydraulic conductivity will affect the growth processes of plants, such as photosynthesis and transpiration, resulting in slow growth and possibly even death [32]. In addition, our study found that after the application of IRCIS treatment, the PLC of the branches and roots decreased in both wet and dry years. This may indicate that IRCIS treatment plays a significant role in improving plant resistance to drought and preventing the loss of hydraulic conductivity, especially during drought conditions.

4.4. Implications for Afforest Management

In this study, we provided evidence that PAMS in shallow soil layers (0–1.5 m) was low during the rainy season of the dry year 2016, and soil water deficits were observed in most deep soil layers (>1.5 m) in both 2015 and 2016. This was mainly due to the lower rainfall and higher water consumption of *Robinia pseudoacacia* trees in dry years, which eventually reduced soil water content, increasing soil moisture deficits [19]. Severe soil moisture deficits in turn resulted in lower hydraulic conductivity, hindered *Robinia pseudoacacia* growth and even caused branch dieback and tree mortality [32]. Therefore, for the healthy development of artificial forests and sustainable ecological construction, effective water-saving management with the potential to control soil erosion without endangering further soil water availability should be implemented in the semiarid Loess Plateau region.

A rational collection, management and utilization of rainfall, the only water resource available for the growth of vegetation in this region, can result in an increase in soil water content (Figure 2), a reduction in soil water deficit (Figure 3) and an improvement in artificial forests' growth (fine roots and plant hydraulic conductivity, Figures 4–6). Diverse afforestation management strategies for the effective utilization of rainwater to increase sustainable forest development in the Loess Plateau region are well documented [6,8]. All these strategies, including engineering measures (e.g., creation of fish-scale pits and mini-catchments), level furrowing and agronomic measures (e.g., mulching with straw or stone, application of water-retaining chemicals), have all been tested and are well implemented [6]. In situ rainwater collection and infiltration systems (IRCIS) that divert rainwater and runoff to deeper soils have been introduced into orchards to optimize rainwater utilization on the Loess Plateau [8,17,33]. The results of our study also indicate that IRCIS can be used for the sustainable development of artificial forests to increase the availability of soil moisture for plants and decrease soil moisture deficits.

5. Conclusions

Afforestation is an effective measure to control soil and water erosion for sustainable ecological construction in the Loess Plateau of China. However, due to huge water requirements, forest land, such as *Robinia pseudoacacia* forests in our study, had significantly lower plant-available water than abandoned cropland, resulting in higher soil moisture deficits, especially in low rainfall years. The presence of a significant water deficit in soil will reduce the hydraulic conductivity of the roots and branches of plants, inhibiting the normal growth of plants and even resulting in their death. Our results indicated that IRCIS can increase soil water content, decrease soil moisture deficits and increase the hydraulic conductivity of plants. Moreover, our results revealed that drought significantly influenced the root distribution of *Robinia pseudoacacia*. The biomass and maximum depth of *Robinia pseudoacacia* roots in dry years were significantly greater than those in wet years, suggesting that *Robinia pseudoacacia* can absorb shallow soil water in wet years, while absorbing deep soil water in dry years to maintain normal growth and resist drought stress. The results of this study will contribute to the formulation of appropriate strategies for planning and managing rainwater resources. The use of all these strategies, such as in situ rainwater collection and infiltration systems, would help counteract the degradation of forest plantations caused by droughts, not only on the Loess Plateau, but also in other similar regions around the world.

Author Contributions: Conception and design, C.M. and Y.L.; Acquisition of data, C.M., H.M., X.L. and Q.L.; Analysis and Interpretation of data, B.X., C.M. and X.L.; Drafting the article, C.M., Y.L. and Q.L.; Conception and design, Q.W., B.Z. and Y.L.; All authors have read and agreed to the published version of the manuscript.

Funding: This research was funded by the National Natural Science Foundation of China (42107326, 41830754 and 52179042), Major Science and Technology Projects of the XPCC (2021AA003-2) and Major Science and Technology Projects of Autonomous Region (2020A01003-3).

Data Availability Statement: The datasets generated and/or analyzed during the current study are available from the corresponding author on reasonable request.

Conflicts of Interest: The authors declare no conflict of interest.

References

1. McCormack, M.L.; Dickie, I.A.; Eissenstat, D.M.; Fahey, T.J.; Fernandez, C.W.; Guo, D.; Helmisaari, H.-S.; Hobbie, E.A.; Iversen, C.M.; Jackson, R.B.; et al. Redefining fine roots improves understanding of below-ground contributions to terrestrial biosphere processes. *New Phytol.* **2015**, *207*, 505–518. [CrossRef]
2. Cheng, X.; Huang, M.; Shao, M.; Warrington, D.N. A comparison of fine root distribution and water consumption of mature Caragana korshinkii Kom grown in two soils in a semiarid region, China. *Plant Soil* **2009**, *315*, 149–161. [CrossRef]
3. Jasechko, S.; Sharp, Z.D.; Gibson, J.J.; Birks, S.J.; Yi, Y.; Fawcett, P.J. Terrestrial water fluxes dominated by transpiration. *Nature* **2013**, *496*, 347–350. [CrossRef]
4. Schenk, H.J.; Jackson, R.B. The global biogeography of roots. *Ecol. Monogr.* **2002**, *72*, 311–328. [CrossRef]
5. Pregitzer, K.S.; DeForest, J.L.; Burton, A.J.; Allen, M.F.; Ruess, R.W.; Hendrick, R.L. Fine root architecture of nine North American trees. *Ecol. Monogr.* **2002**, *72*, 293–309. [CrossRef]
6. Gao, X.; Li, H.; Zhao, X.; Ma, W.; Wu, P. Identifying a suitable revegetation technique for soil restoration on water-limited and degraded land: Considering both deep soil moisture deficit and soil organic carbon sequestration. *Geoderma* **2018**, *319*, 61–69. [CrossRef]
7. Jackson, R.B.; Canadell, J.; Ehleringer, J.R.; Mooney, H.A.; Sala, O.E.; Schulze, E.D. A global analysis of root distributions for terrestrial biomes. *Oecologia* **1996**, *108*, 389–411. [CrossRef]
8. Song, X.; Gao, X.; Wu, P.; Zhao, X.; Zhang, W.; Zou, Y.; Siddique, K.H.M. Drought responses of profile plant-available water and fine-root distributions in apple (*Malus pumila* Mill.) orchards in a loessial, semi-arid, hilly area of China. *Sci. Total Environ.* **2020**, *723*, 137739. [CrossRef]
9. Leonova, A.; Heger, A.; Vásconez Navas, L.K.; Jensen, K.; Reisdorff, C. Fine root mortality under severe drought reflects different root distribution of *Quercus robur* and *Ulmus laevis* trees in hardwood floodplain forests. *Trees* **2022**, *36*, 1105–1115. [CrossRef]
10. Zwetsloot, M.J.; Bauerle, T.L. Repetitive seasonal drought causes substantial species-specific shifts in fine-root longevity and spatio-temporal production patterns in mature temperate forest trees. *New Phytol.* **2021**, *231*, 974–986. [CrossRef]
11. Nikolova, P.S.; Bauerle, T.L.; Haberle, K.H.; Blaschke, H.; Brunner, I.; Matyssek, R. Fine-root traits reveal contrasting ecological strategies in european beech and norway spruce during extreme drought. *Front. Plant Sci.* **2020**, *11*, 1211. [CrossRef]
12. Germon, A.; Jourdan, C.; Bordron, B.; Robin, A.; Nouvellon, Y.; Chapuis-Lardy, L.; de Moraes Gonçalves, J.L.; Pradier, C.; Guerrini, I.A.; Laclau, J.-P. Consequences of clear-cutting and drought on fine root dynamics down to 17 m in coppice-managed eucalypt plantations. *For. Ecol. Manag.* **2019**, *445*, 48–59. [CrossRef]
13. Fuchs, S.; Hertel, D.; Schuldt, B.; Leuschner, C. Effects of summer drought on the fine root system of five broadleaf tree species along a precipitation gradient. *Forests* **2020**, *11*, 289. [CrossRef]
14. McDowell, N.; Pockman, W.T.; Allen, C.D.; Breshears, D.D.; Cobb, N.; Kolb, T.; Plaut, J.; Sperry, J.; West, A.; Williams, D.G.; et al. Mechanisms of plant survival and mortality during drought: Why do some plants survive while others succumb to drought? *New Phytol.* **2008**, *178*, 719–739. [CrossRef]
15. Allen, C.D.; Macalady, A.K.; Chenchouni, H.; Bachelet, D.; McDowell, N.; Vennetier, M.; Kitzberger, T.; Rigling, A.; Breshears, D.D.; Hogg, E.H.; et al. A global overview of drought and heat-induced tree mortality reveals emerging climate change risks for forests. *For. Ecol. Manag.* **2010**, *259*, 660–684. [CrossRef]
16. McDowell, N.G. Mechanisms linking drought, hydraulics, carbon metabolism, and vegetation mortality. *Plant Physiol.* **2011**, *155*, 1051–1059. [CrossRef]
17. Song, X.; Gao, X.; Zhao, X.; Wu, P.; Dyck, M. Spatial distribution of soil moisture and fine roots in rain-fed apple orchards employing a Rainwater Collection and Infiltration (RWCI) system on the Loess Plateau of China. *Agric. Water Manag.* **2017**, *184*, 170–177. [CrossRef]
18. Ma, C.; Luo, Y.; Shao, M.; Li, X.; Sun, L.; Jia, X. Environmental controls on sap flow in black locust forest in Loess Plateau, China. *Sci. Rep.* **2017**, *7*, 13160. [CrossRef]
19. Jia, X.; Zhu, Y.; Luo, Y. Soil moisture decline due to afforestation across the Loess Plateau, China. *J. Hydrol.* **2017**, *546*, 113–122. [CrossRef]
20. Böhm, W. *Methods of Studying Root Systems*; Springer: Berlin/Heidelberg, Germany, 1979.

21. Vijaya Kumar, P.; Bindi, M.; Crisci, A.; Maracchi, G. Detection of variations in precipitation at different time scales of twentieth century at three locations of Italy. *Weather Clim. Extrem.* **2013**, *2*, 7–15. [CrossRef]
22. Sperry, J.S.; Donnelly, J.R.; Tyree, M.T. A method for measuring hydraulic conductivity and embolism in xylem. *Plant Cell Environ.* **1988**, *11*, 35–40. [CrossRef]
23. Wang, Y.; Brandt, M.; Zhao, M.; Tong, X.; Xing, K.; Xue, F.; Kang, M.; Wang, L.; Jiang, Y.; Fensholt, R. Major forest increase on the Loess Plateau, China (2001–2016). *Land Degrad. Dev.* **2018**, *29*, 4080–4091. [CrossRef]
24. Ma, C.; Luo, Y.; Shao, M. Comparative modeling of the effect of thinning on canopy interception loss in a semiarid black locust (*Robinia pseudoacacia*) plantation in Northwest China. *J. Hydrol.* **2020**, *590*, 125234. [CrossRef]
25. Ma, L.H.; Liu, X.L.; Wang, Y.K.; Wu, P.T. Effects of drip irrigation on deep root distribution, rooting depth, and soil water profile of jujube in a semiarid region. *Plant Soil* **2013**, *373*, 995–1006. [CrossRef]
26. Brunner, I.; Herzog, C.; Dawes, M.A.; Arend, M.; Sperisen, C. How tree roots respond to drought. *Front. Plant Sci.* **2015**, *6*, 547. [CrossRef]
27. Ye, Z.Q.; Wang, J.M.; Wang, W.J.; Zhang, T.H.; Li, J.W. Effects of root phenotypic changes on the deep rooting of Populus euphratica seedlings under drought stresses. *PeerJ* **2019**, *7*, e6513. [CrossRef]
28. Li, H.; Luo, Y.; Sun, L.; Li, X.; Ma, C.; Wang, X.; Jiang, T.; Zhu, H. Modelling the artificial forest (*Robinia pseudoacacia* L.) root–soil water interactions in the Loess Plateau, China. *Hydrol. Earth Syst. Sci.* **2022**, *26*, 17–34. [CrossRef]
29. Ma, L.H.; Wu, P.T.; Wang, Y.K. Spatial distribution of roots in a dense jujube plantation in the semiarid hilly region of the Chinese Loess Plateau. *Plant Soil* **2011**, *354*, 57–68. [CrossRef]
30. Kleidon, A.; Heiman, M. A method of determining rooting depth from a terrestrial biosphere model and its impacts on the global water and carbon cycle. *Glob. Change Biol.* **1998**, *4*, 275–286. [CrossRef]
31. Burgess, S.S.O.; Adams, M.A.; Turner, N.C.; White, D.A.; Ong, C.K. Tree roots: Conduits for deep recharge of soil water. *Oecologia* **2001**, *126*, 158–165. [CrossRef]
32. Anderegg, W.R.; Klein, T.; Bartlett, M.; Sack, L.; Pellegrini, A.F.; Choat, B.; Jansen, S. Meta-analysis reveals that hydraulic traits explain cross-species patterns of drought-induced tree mortality across the globe. *Proc. Natl. Acad. Sci. USA* **2016**, *113*, 5024–5029. [CrossRef]
33. Song, X.; Gao, X.; Dyck, M.; Zhang, W.; Wu, P.; Yao, J.; Zhao, X. Soil water and root distribution of apple tree (*Malus pumila Mill*) stands in relation to stand age and rainwater collection and infiltration system (RWCI) in a hilly region of the Loess Plateau, China. *Catena* **2018**, *170*, 324–334. [CrossRef]

forests

Article

Root Development in *Cunninghamia lanceolata* and *Schima superba* Seedlings Expresses Contrasting Preferences to Nitrogen Forms

Haiyan Liang, Lidong Wang, Yanru Wang, Xiaoqiang Quan, Xiaoyu Li, Yaning Xiao and Xiaoli Yan *

College of Forestry, Fujian Agriculture and Forestry University, Fuzhou 350002, China
* Correspondence: ccyanxiaoli@163.com

Abstract: The inorganic nitrogen (N) that can be absorbed and utilized by plants is mainly ammonium N (NH_4^+-N) and nitrate N (NO_3^--N), which may affect seedlings' root morphology and growth through its heterogeneous distribution. Root morphology and seedling growth were investigated in a subtropical major conifer (*Cunninghamia lanceolata*) and a broadleaf tree species (*Schima superba*) under five different NH_4^+-N to NO_3^--N ratios (10:0, 0:10, 7:3, 3:7, 5:5). Results: (1) While both species developed thinner roots under the treatment with a high NO_3^--N concentration, the roots of *C. lanceolata* were longer than those of *S. superba*. In contrast, the roots of both species were thicker under the treatment with a high NH_4^+-N concentration, with those in *S. superba* being much longer than those in *C. lanceolata*. (2) The mixed NH_4^+-N and NO_3^--N treatments were more conducive to the aboveground growth and biomass accumulation of both tree species and the underground growth of *S. superba*. N sources with high NO_3^--N concentrations were more suitable for underground growth in *C. lanceolata* seedlings and aboveground growth in *S. superba* seedlings. Under the N sources with high NH_4^+-N concentrations, *C. lanceolata* tended to develop aboveground parts and *S. superba* tended to develop underground parts. (3) The roots of the two tree species adopted the expansion strategy of increasing the specific root length and reducing the root tissue density under the N sources with high NO_3^--N concentrations but the opposite with high NH_4^+-N concentrations. The root-to-shoot ratio of *C. lanceolata* increased under high NO_3^--N concentrations, while that of *S. superba* increased under high NO_3^--N concentrations. These results indicate that the responses of root morphology to different N forms are species-specific. Furthermore, according to the soil's N status, NH_4^+-N can be appropriately applied to *C. lanceolata* and NO_3^--N to *S. superba* for cultivating seedlings.

Keywords: nitrogen forms; *Cunninghamia lanceolata*; *Schima superba*; morphological characteristics of root system

Citation: Liang, H.; Wang, L.; Wang, Y.; Quan, X.; Li, X.; Xiao, Y.; Yan, X. Root Development in *Cunninghamia lanceolata* and *Schima superba* Seedlings Expresses Contrasting Preferences to Nitrogen Forms. *Forests* **2022**, *13*, 2085. https://doi.org/10.3390/f13122085

Academic Editor: Antonio Montagnoli

Received: 29 September 2022
Accepted: 5 December 2022
Published: 7 December 2022

Publisher's Note: MDPI stays neutral with regard to jurisdictional claims in published maps and institutional affiliations.

1. Introduction

Nitrogen (N) is an essential mineral element for plant growth. The inorganic nitrogen that can be absorbed and utilized by plants is mainly ammonium nitrogen (NH_4^+-N) and nitrate nitrogen (NO_3^--N). The two forms of N in soils are not uniformly distributed but instead occur in different proportions and change with time, so they are distributed heterogeneously in time and space [1,2]. Studies have shown that the ratio of NH_4^+-N and NO_3^--N in the soil ranges from 8:2 to 2:8 [3]. Heterogeneous environments with different ammonium and nitrate contents will have different effects on plant growth. In one study, mixed nitrogen culture was more conducive to the growth of the aboveground parts of *Citrus Sinensis* × *Poncirus trifoliata* seedlings than nitrogen alone, and the best effect was observed at an ammonium to nitrate ratio of 5 to 5 [4]. The growth of the aboveground and underground parts of *Pinus massoniana* seedlings cultured from tissue both reached the maximum value under the ammonium nitrogen-only treatment [5]. When plants grow in an environment with a heterogeneous distribution of different forms of N, especially for a long time, they adopt different adaptation mechanisms for the underground and aboveground parts, and the roots' regulatory pathway is particularly important.

As the main nutrient-absorbing organ in plants, the root system carries out nutrient exchange and metabolism for the underground and aboveground parts of the plant. Under natural soil nutrient adversity, tree roots can explore, acquire and utilize limited soil nutrient resources through regulating their morphology, physiology and mycorrhizal plasticity [5,6]. With the intensification of global N deposition, the plastic response of plants' root morphology and structure [root length (RL), root surface area (RSA), root average diameter (RAD), root volume (RV), specific root length (SRL), etc.] to soil N is the most direct manifestation of plants' ability to adapt to environmental changes and seek nutrients [6,7]. Under different N forms, the root morphology of plants will show different nitrogen-seeking strategies [8]. Studies have shown that mixed nitrogen can increase the surface area of the root system and increase RL and the number of root tips, thereby increasing the absorption capacity of the root system and promoting aboveground growth [9]. When the supply ratio of NO_3^--N was higher than 50%, the root quality of *Larix gmelinii* seedlings decreased significantly [10]. The biomass of shoots and lateral roots of *Cunninghamia lanceolata* seedlings was largest under NH_4^+-N treatment [11]. However, compared with the NH_4^+-N treatment, hybrid poplar plants under a NO_3^--N treatment exhibited a higher proportion of fine roots, thicker roots and bigger SRL [12]. *C. lanceolata* roots also tend to grow in NO_3^--N patches, while *Schima superba* and *P. massiniana* tend to grow more and finer roots in NH_4^+-N patches [8]. In an environment of nitrogen deficiency, plants will adjust their nitrogen-seeking strategies by increasing the SRL and root-to-shoot ratio (RSR) [13,14]. Plants' biomass allocation can reflect the nutrient utilization of plant organs [15]. According to the optimal balance theory, plants tend to allocate resources to organs that can obtain restricted resources in order to obtain more restricted resources [16,17]. The SRL, root tissue density (RTD) and RSR can reflect the carbon allocation strategy of plants from different perspectives, which also conforms to the theory of root economics [18,19]. Therefore, studying the morphological characteristics of plant roots can intuitively reveal the nutrient uptake status of plants and their growth status.

C. lanceolata is an important fast-growing afforestation tree species in southern China. In recent decades, *C. lanceolata* forests have mostly been cultivated as artificial forests. The continuous planting of forest stands and soil acidification have led to the exhaustion of forestland, declines in the yield of forest stands, and reductions in the forests' ecological services [20,21]. Some studies have found that, in view of the problem of soil fertility declining after continuous planting of coniferous species, broadleaf tree species can effectively improve soil fertility and can thus be used as soil-improving tree species for mixed planting with conifers [22]. As an evergreen broadleaf tree species, *S. superba* produces a large amount of litter that decomposes quickly and contains high amounts of nutrients, which is conducive to water storage and fertilizer conservation. Therefore, *S. superba* is mostly used for mixing with coniferous species such as *C. lanceolata* to create a mixed coniferous and broadleaf forest. The resulting economic and ecological benefits have received widespread attention.

At present, research on the effects of different N forms on plant roots has mostly been conducted in crops, herbs and fruit trees, and research on the effects on roots of woody plants such as forest tree species is relatively limited and incomplete, especially regarding the coexistence mechanism of mixed *C. lanceolata* and *S. superba* forests. We hypothesized that the N uptake strategies of the roots of *C. lanceolata* and *S. superba* are different for different N forms. In view of this, in this experiment, the potted sand culture method was used to avoid the interference of soil N and to test the effects of different N forms on the morphological plasticity of *C. lanceolata* and *S. superba* roots' responses to N uptake and their differences.

2. Materials and Methods

2.1. Plant Materials

The test materials used in this study were selected from *C. lanceolata* and *S. superba*, the two main coniferous and broadleaf tree species occurring in mixed stands in the subtropical zone. One-year-old seedlings with uniform growth that were free from disease and insect pests and had a complete root system were selected as the test seedlings. Plastic pots (26 cm diameter × 28 cm height) were used for cultivating the seedlings. The potting substrate selected was clean sand, which was repeatedly washed with distilled water until the N content in the sand was close to zero [23,24].

2.2. Experimental Setup

In April 2021, the sand culture experiments were conducted in a rainproof greenhouse with good ventilation and light transmission at Fujian Agriculture and Forestry University, China. The average temperature was 20.1 °C and the average relative humidity was about 77%. Each pot was filled with equal amount of clean sand. A matrix environment with different proportions of two N forms was constructed using different concentrations of NH_4^+-N and NO_3^--N in the nutrient solution.

According to the research results of Zhang et al. [25] on the influence of fertilization on the soil nutrient content of *C. lanceolata* and the previous research results of Meng et al. [26] and Liu et al. [27], the total N concentration was set to 2 mmol·L^{-1}. In this study, we used five different ratios of NH_4^+-N and NO_3^--N supplied as NH_4^+-N:NO_3^--N = 10:0, 0:10, 7:3, 3:7 and 5:5 (Table 1). The 10:0 and 0:10 treatments had highly heterogeneous N, 7:3 and 3:7 had medium heterogeneous N, and 5:5 had homogeneous N. NH_4^+-N was supplied as $(NH_4)_2SO_4$, while NO_3^--N was supplied as $NaNO_3$. In all treatments, while the concentration of the two forms of N were different, the content of the macronutrients (Hoagland formula) and micronutrients (Amon formula) remained the same. To prevent the conversion of ammonium to nitrate, the nitrification inhibitor dicyandiamide (7 μmol·L^{-1}) was added to the nutrient solution. The pH of the nutrient solution was adjusted to 5.8 with 2 mol·L^{-1} NaOH or HCl. Each treatment had six replicates, and the total number of pots was 60. The nutrient solutions were added to the corresponding compartments at a rate of 50 mL every 5 days. Each pot was supplied with 150 mL of distilled water every 2–3 days. We did not water the pots for 24 h before and after pouring in the nutrient solution.

Table 1. Treatments with different nitrogen forms.

Treatment (NH_4^+/NO_3^-)	Total N (mmol·L^{-1})	Different N Forms (mmol·L^{-1})	
		NH_4^+	NO_3^-
10:0	2	2.0	0
0:10	2	0	2.0
7:3	2	1.4	0.6
3:7	2	0.6	1.4
5:5	2	1.0	1.0

2.3. Plant Harvesting and Data Collection

All the seedlings were harvested after 160 days of cultivation. The initial and final plant height and ground diameter of all seedlings were measured and recorded. Plants were separated into three parts: roots, stems and leaves. The fresh stems and leaves were oven-dried at 105 °C for half an hour to deactivate the enzymes and then at 65 °C to a constant weight.

The roots in each treatment were harvested carefully, cleaned with distilled water and dried with filter paper. We imaged all the roots using an Epson Expression 12,000XL scanner (Seiko Epson Corporation, Suwa, Nagano, Japan) and analyzed the images with WinRHIZO software (Pro2017a, Regent Instruments Inc., Quebec, Canada) to obtain the roots' morphological parameters, including the total root length (TRL, cm), total root surface area (TRSA, cm^2), total root volume (TRV, cm^3) and the roots' average diameter (RAD, mm). After scanning, all the roots were oven-dried at 105 °C for half an hour to deactivate the enzymes and then at 65 °C to a constant weight. The total seedling biomass (TSB, g) of the seedlings was the sum of the root, stem and leaf biomass in each treatment.

2.4. Data Analysis

All the data were processed and analyzed in Excel 2016 (Microsoft Office, Washington, Redmond, WA, USA). The increments in seedling height (ΔH, cm) and ground diameter (ΔGD, mm) were calculated by subtracting the seedlings' initial height and ground diameter from the final height and diameter, respectively. Specific root length (SRL, $cm \cdot g^{-1}$) was calculated as total root length/total root biomass (TRB, g). Root tissue density (RTD, $g \cdot cm^{-3}$) was calculated by dividing the root biomass by the total root volume. The root-to-shoot ratio (RSR) was calculated as the ratio of root biomass to aboveground biomass.

Statistical analyses were conducted using SPSS 25.0 for Windows (SPSS Inc., Chicago, IL, USA). One-way ANOVA was conducted to evaluate the effect of nitrogen forms on the seedlings' ΔH, ΔGD, TRL, TRSA, TRV, RAD, TRB, TSB, SRL, RTD and RSR. Two-way ANOVA was performed on the effects of tree species and different N forms on the roots' morphological parameters, and plots were created using Origin 2018 (OriginLab, Northampton, USA). In this study, the ammonium and nitrate N concentrations were used as environmental factor variables, and the root and growth parameters of plants were used as biological factor variables to reflect the relationships among the parameters. The redundancy analysis (RDA) was conducted in Canoco 5.0 (Microcomputer Power, Ithaca, NY, USA).

3. Results

3.1. Significance Test of the Effects of Tree Species and Different Ratios of N Forms on the Seedlings' Growth and the Roots' Morphological Parameters

Table 2 shows that tree species and different N forms (a × b) had significant interaction effects on ΔH and ΔGD, TRL, TRSA, TRV, TRB, RTD and RSR ($p < 0.05$), while the interaction effect on RAD, TSB and SRL did not reach a significant level ($p > 0.05$). Separately, the effects of tree species (a) and different N forms (b) on all indicators reached the very significant level ($p < 0.01$), except for TRV under different ammonium to nitrate ratios, which reached the significant level ($p < 0.05$).

Table 2. Two-way ANOVA of the effects of tree species and different ratios of N forms on the seedlings' growth and the roots' morphological parameters.

Factor	*p*-Value and Significance Level		
	Tree Species (a)	NH_4^+/NO_3^- (b)	a × b
df	1	4	4
Seedling height increment (ΔH)	<0.001	<0.001	<0.001
Ground diameter increment (ΔGD)	<0.001	<0.001	<0.001
Total root length (TRL)	<0.001	<0.001	<0.01
Total root surface area (TRSA)	<0.001	<0.001	<0.01
Total root volume (TRV)	<0.001	<0.05	<0.05
Roots' average diameter (RAD)	<0.001	<0.001	0.835
Total root biomass (TRB)	<0.001	<0.01	<0.001
Total seedling biomass (TSB)	<0.001	<0.01	0.08
Specific root length (SRL)	<0.001	<0.01	0.113
Root tissue density (RTD)	<0.001	<0.01	<0.05
Root-to-shoot ratio (RSR)	<0.001	<0.001	<0.001

3.2. Effects of Different N Forms on Root Morphology in the Two Tree Species

The roots' parameters were significantly influenced by the different N forms in each treatment ($p < 0.05$) (Figure 1). Under the five different ratios of the N forms, the TRL and

TRSA of *C. lanceolata* followed the order 10:0 < 7:3 < 3:7 < 5:5 < 0:10 (Figure 1a,c). Under the NO_3^--N-only treatment, the TRL and TRSA were 16.54–70.42% and 18.58–42.68% higher, respectively, than those under the other four treatments. The TRV followed the order 10:0 < 5:5 < 7:3 < 3:7 < 0:10 (Figure 1e). It was largest under the NO_3^--N-only treatment, which was 11.48–24.64% larger than that under the other four treatments. The RAD of *C. lanceolata* followed the order 5:5 < 0:10 < 7:3 < 3:7 < 10:0 (Figure 1g), and was significantly larger under the NH_4^+-N-only treatment than under the other four treatments. In general, the roots of *C. lanceolata* preferred to grow under the treatments with higher NO_3^--N concentrations.

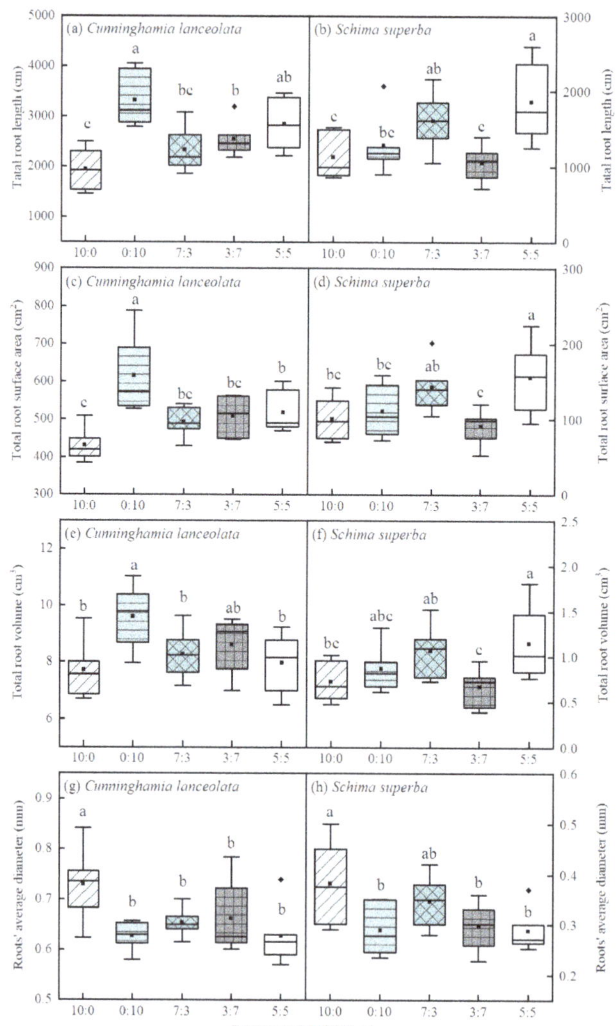

Figure 1. Effects of different NH_4^+-N and NO_3^--N ratios on the TRL (**a**,**b**), TRSA (**c**,**d**), TRV (**e**,**f**) and RAD (**g**,**h**) of the two tree species. Different lowercase letters indicate that the roots' morphological parameters were significantly different under different N forms ($p < 0.05$).

The TRL, TRSA and TRV of *S. superba* followed the order 3:7 < 10:0 < 0:10 < 7:3 < 5:5 (Figure 1b,d,f). The TRL, TRSA and TRV under the 5:5 ratio were 15.42–76.27%, 9.19–71.05%

and 7.97–72.24% larger, respectively, than those of the other four treatments. The RAD of S. *superba* followed the order 5:5 < 0:10 < 3:7 < 7:3 < 10:0 (Figure 1h). It showed the maximum value under the NH_4^+-N-only treatment, which did not differ significantly from that under the 7:3 treatment but was significantly greater than that under the other three treatments, showing a minimum value at a ratio of 5:5. The morphological parameters of the root system of *S. superba* were superior in the treatments with higher NH_4^+-N concentrations.

3.3. Effects of Different N Forms on Biomass in the Two Tree Species

The TRB of one-year-old seedlings of *C. lanceolata* ranged from 2.26 to 2.74 g·tree^{-1} after 160 days of the experimental treatment (Figure 2a), and the TSB was 8.59–9.71 g·tree^{-1} (Figure 2c). The TRB followed the order 3:7 > 0:10 > 5:5 > 7:3 >10:0, and there were no significant differences among the 3:7, 0:10 and 5:5 ratios ($p > 0.05$). Conversely, the TSB was largest under the NH_4^+-N-only treatment and followed the order 10:0 > 5:5 > 7:3 > 3:7 > 0:10. TRB accumulation was higher under the treatment with a high NO_3^--N concentration than that with a high NH_4^+-N concentration. However, the TSB in *C. lanceolata* tended to accumulate at high NH_4^+-N concentrations and homogeneous N supply.

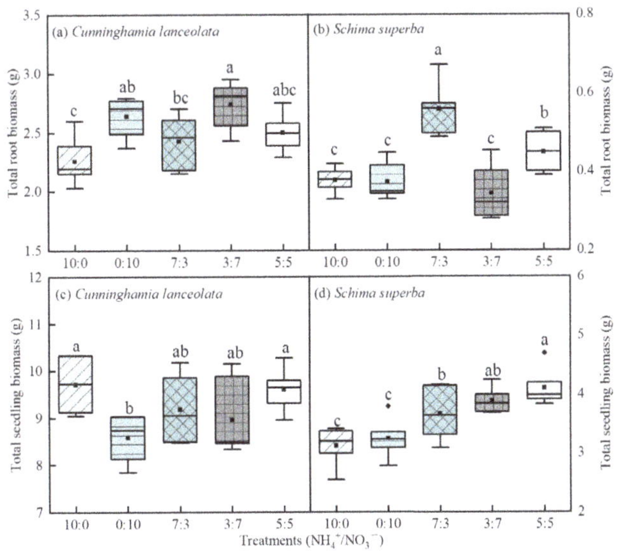

Figure 2. Effects of different NH_4^+-N and NO_3^--N ratios on the TRB (**a**,**b**) and TSB (**c**,**d**) of the two tree species. Different lowercase letters indicate that TRB and TB were significantly different under different N forms ($p < 0.05$).

The TRB of *S. superba* seedlings ranged from 0.34 to 0.56 g·tree^{-1} (Figure 2b) and the TSB was 3.12–4.10 g·tree^{-1} (Figure 2d). The TRB followed the order 7:3 > 5:5 > 10:0 > 0:10 > 3:7, while the TSB followed the order 5:5 > 3:7 > 7:3 > 0:10 > 10:0. The TRB accumulation in *S. superba* was higher under the treatment with a high NH_4^+-N concentration and homogeneous N supply than under the treatment with a high NO_3^--N concentration, while the TSB tended to accumulate under the treatments with homogeneous and medium heterogeneous N supply. The environment with highly heterogeneous N supply was the most unfavorable for the accumulation of TSB in *S. superba*.

3.4. Effects of Different N Forms on Root Abundance in the Two Tree Species

The SRL of *C. lanceolata* followed the order 0:10 > 5:5 > 7:3 > 3:7 > 10:0 (Figure 3a) and was the highest under the NO_3^--N-only treatment, which was not significantly different from that under the control treatment (5:5) ($p > 0.05$) but was significantly higher than that under the other three treatments ($p < 0.05$). The RTD of *C. lanceolata* followed the order 3:7 > 5:5 > 10:0 > 7:3 > 0:10 (Figure 3c). There were no significant differences between the 3:7 and 5:5 ratios, but both of these treatments resulted in a significantly higher RTD than the other three treatments. In contrast, the RSR followed the order 0:10 > 3:7 > 7:3 > 5:5 > 10:0 (Figure 3e), showing higher values under the treatment with a high NO_3^--N concentration.

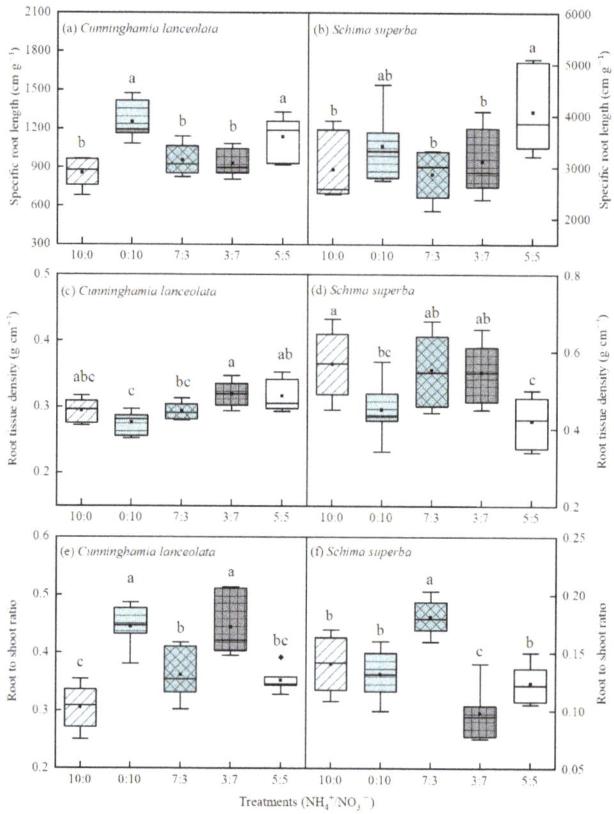

Figure 3. Effects of different NH_4^+-N and NO_3^--N ratios on SRL (**a**,**b**), RTD (**c**,**d**) and RSR (**e**,**f**) between the two tree species. Different lowercase letters indicate that the SRL, RTD and RSR were significantly different under different N forms ($p < 0.05$).

The SRL of *S. superba* followed the order 5:5 > 0:10 > 3:7 > 10:0 > 7:3 (Figure 3b), showing a higher value under the treatment with a homogeneous N supply and a high NO_3^--N level. The difference between the 5:5 and 0:10 ratios was not significant, but both of them had a significantly higher SRL than the other three treatments. The RTD of *S. superba* followed the order of 10:0 > 7:3 > 3:7 > 0:10 > 5:5 (Figure 3d), showing that the heterogeneous N supply resulted in a higher RTD than the homogeneous N supply. Moreover, the treatments with a high NH_4^+-N level resulted in a higher RTD than the treatments with a high NO_3^--N level. The RSR followed the order 7:3 > 10:0 > 0:10 > 5:5 > 3:7 (Figure 3f) and was highest under the treatments with high NH_4^+-N.

3.5. Effects of Different N Forms on Aboveground Growth in the Two Tree Species

After 160 days of cultivation, the ΔH of *C. lanceolata* and *S. superba* seedlings was 10.2–17.7 cm and 7.0–10.5 cm (Figure 4a), respectively, and the ΔGD was 3.08–4.35 mm and

2.18–3.62 mm (Figure 4b), respectively. It can be seen that the ΔH and ΔGD of *C. lanceolata* were larger than those of *S. superba*.

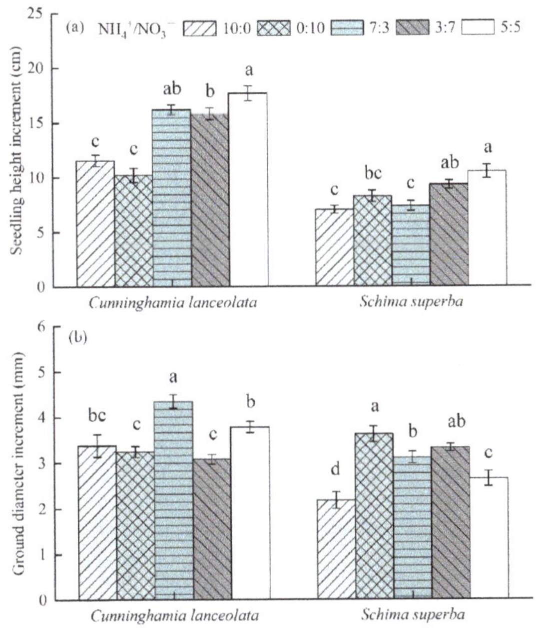

Figure 4. Effects of different NH_4^+-N and NO_3^--N ratios on ΔH (**a**) and ΔGD (**b**). Data are expressed as the mean ± standard error (SE). Different lowercase letters on the columns indicate significant differences among the treatments.

Under different N forms, the ΔH of *C. lanceolata* followed the order 5:5 > 7:3 > 3:7 > 10:0 > 0:10 and was largest under the control treatment (Figure 4a), which was 9.05–73.20% larger than that under the other treatments. The ΔH of *C. lanceolata* was greater under the homogeneous N supply than under the heterogeneous N supply, and it was smallest under the NH_4^+-N-only and NO_3^--N-only treatments. The ΔH of *S. superba* followed the order 5:5 > 3:7 > 0:10 > 7:3 > 10:0 and was largest under the control treatment (Figure 4a), which was 12.93–49.05% larger than that under the other treatments. The ΔH of *S. superba* was greater under the homogeneous N supply than the heterogeneous N supply, and it was greater under the treatment with a high NO_3^--N concentration than under the treatment with a high NH_4^+-N concentration. In the two tree species, seedling height initially increased and then decreased with the increase in NO_3^--N in the nutrients.

The ΔGD of *C. lanceolata* followed the order 7:3 > 5:5 > 10:0 > 0:10 > 3:7 and was largest at the ammonium to nitrate ratio of 7:3 (Figure 4b), which was 15.09–41.26% larger than that under the other treatments. The ΔGD of *S. superba* followed the order 0:10 > 3:7 > 7:3 > 5:5 > 10:0 (Figure 4b). It was largest at an ammonium to nitrate ratio of 0:10, which was 8.92–66.51% larger than that under the other treatments. Overall, *C. lanceolata* tended to increase the ground diameter of the seedlings under the treatment with a high NH_4^+-N concentration and a homogeneous N supply, whereas *S. superba* preferred to increase the ground diameter under the treatment with a high NO_3^--N concentration.

3.6. Redundancy Analysis of Ammonium and Nitrate N Concentrations with Root and Growth Parameters

We used N forms as the environmental factor variables, and a redundancy analysis was performed on the roots' morphological and growth parameters of *C. lanceolata* (Figure 5a) and *S. superba* (Figure 5b). A smaller angle between them (close to 0°) indicated that the positive correlation was significant, and a larger angle (close to 180°) indicated that the negative correlation was significant.

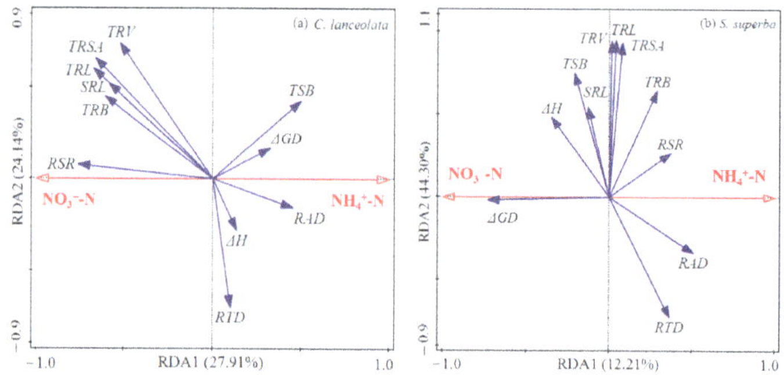

Figure 5. Redundancy analysis of the responses of roots' morphological and growth parameters to different ratios of the N forms. The proportions explained by Axis 1 and Axis 2 in (**a**) are 27.91% and 24.14% and those in (**b**) are 12.21% and 44.30%, respectively. Solid blue lines indicate the parameters. Solid red lines indicate NH_4^+-N and NO_3^--N. Abbreviations for the parameters are as follows: TRL, total root length; TRSA, total root surface area; TRV, total root volume; RAD, roots' average diameter; TRB, total root biomass; TSB, total seedling biomass; SRL, specific root length; RTD, root tissue density; RSR, root-to-shoot ratio; ΔH, height increment; ΔGD, ground diameter increment.

Figure 5a shows that the RAD, ΔGD and TSB of *C. lanceolata* had a positive correlation with NH_4^+-N and a negative correlation with NO_3^--N. However, the ΔH and RTD decreased along the Y-axis, and the correlation with NH_4^+-N was not significant. In contrast, the RSR, TRB, SRL, TRL, TRSA and TRV of *C. lanceolata* had a positive correlation with NO_3^--N and a negative correlation with NH_4^+-N. Among these, the RSR had the strongest positive correlation with NO_3^--N. The RAD and RTD had a negative correlation with SRL.

Figure 5b shows that the RSR, RAD, TRB and RTD of *S. superba* had a positive correlation with NH_4^+-N and a negative correlation with NO_3^--N. Conversely, the ΔGD, ΔH and TSB of *S. superba* had a positive correlation with NO_3^--N and a negative correlation with NH_4^+-N. However, the correlations of TRSA, TRL and TRV with NH_4^+-N and that of SRL with NO_3^--N were not significant. Among these, the ΔGD had the strongest positive correlation with NO_3^--N. The RAD and RTD had a negative correlation with SRL.

4. Discussion

4.1. Effects of Different N Forms on Root Morphology in Different Tree Species

When trees grow for a long period of time in a soil environment with a certain ratio of the N forms, they gradually form a growth pattern adapted to this soil environment, which is generally the coordinated effect of the aboveground and the underground parts. As the root is in direct contact with the soil for nutrient exchange, the change in root morphology is the intuitive embodiment of and basis for understanding the growth of forest trees. The results of this study showed that the TRL, TRSA and TRV of *C. lanceolata* were positively correlated with NO_3^--N (Figure 5a) and that the RAD was positively correlated with NH_4^+-N. The TRL, TRSA, TRV and RAD of *S. superba* tended to be greater at high NH_4^+-N concentrations, but the correlation was not significant (Figure 5b). Under a homogeneous N supply, the TRL, TRSA and TRV of *S. superba* were largest, while the RAD was smallest. *C. lanceolata* tended to grow more slender roots under the treatment with a high NO_3^--N concentration, while the roots of *S. superba* were extended under the treatment with a high NH_4^+-N concentration and an ammonium to nitrate ratio of 5:5. This is consistent with the results showing that *C. lanceolata* grows more slender roots in NO_3^--N patches and *S. superba* grows more slender roots in NH_4^+-N patches [8]. When *C. lanceolata* and *S. superba* were mixed in an environment with heterogenous nutrients, *S. superba* took advantage of the compensatory strategy of a vertical growth direction in its fine roots, and alleviated the strong competition for nutrients because of the different level of soil colonization and the niche differentiation of fine roots of the two species [28,29], which was conducive to the coordination of the relationship between the two species [30]. We infer that this opposite root growth response resulted from complementary patterns in the absorption strategies of different N forms in *C. lanceolata* and *S. superba* seedlings.

The nitrogen-seeking strategy of tree roots will change with changes in the environment and in their adaptive mechanism. The environmental factors affecting the growth of forest roots mainly include the concentration of soil nutrients (such as N), the soil's pH value, and the soil's temperature and humidity. The strategies of forest trees for obtaining nitrogen also vary according to the forest's age, which will directly or indirectly affect the nitrogen uptake of trees [31]. For example, in an environment of mild nitrogen deficiency, the TRL of plants increased, while moderate nitrogen deficiency and excessive nitrogen supply inhibited the growth of the root system [32]. In acidic environments, NH_4^+-N is the main N source for plants and is usually more readily available than NO_3^--N, especially in coniferous forests, where many conifers tend to absorb NH_4^+-N [33]. Under drought stress, the uptake of NH_4^+-N in different organs of *C. lanceolata* seedlings was significantly higher than that of NO_3^--N, showing a preference for NH_4^+ [34]. The growth and nutrient acquisition of plant roots are affected and restricted by many factors. When trees are under nutrient, water and other stress conditions, the root system will produce a series of adaptive changes in its morphological structure to obtain the limited resources and adapt to the ecological strategy of overcoming adversity through self-regulation [35]. Numerous studies have shown that in nutrient-limited environments, plants mainly use strategies such as increasing the roots' length and decreasing the roots' diameter, which expand the growth range of roots to obtain more nutrients for growth [36–38]. Nitrogen deposition causes the growth rate of the TRL and TRV of *C. lanceolata* to be faster than that of the RSA and RAD, in which the root system is elongated and dilated [39]. In another study, nitrogen deposition caused the root system of *S. superba* seedlings to show an increasing trend in the TRL, RAD, TRSA and TRV, and the root system was thick and dilated [40]. The results of this study showed that the roots of *C. lanceolata* seedlings were slender and dilated under the treatment with a high NO_3^--N concentration, while the roots of *S. superba* seedlings were sturdy and dilated under the treatment with a high NH_4^+-N concentration.

4.2. Effects of Different N Forms on Biomass and Root Abundance in Different Tree Species

The root biomass of plants is one of the manifestations of the amount of nutrients and water they absorb, and the allocation of aboveground and underground biomass also reflects the nutrient uptake and distribution in plants. The results of this study showed that in *C. lanceolata*, the accumulation of root biomass was higher under the treatment with a high NO_3^--N concentration and, the total accumulation of seedling biomass was higher under the treatment with a high NH_4^+-N concentration (Figure 5a). However, the accumulation of root biomass in *S. superba* was higher under the treatment with a high NH_4^+-N concentration. The total biomass was highest under the treatment with a homogeneous N supply, and accumulation was relatively high under the treatment with a high NO_3^--N concentration (Figure 5b). In a limited-resource environment, when plants are limited by nutrients, they allocate biomass to the roots and reduce the allocation to aboveground biomass [41]. Under low nutrient conditions, *C. lanceolata* seedlings increased the amount of nutrient absorbed, mainly by altering the root structure rather than by allocating more nonstructural carbohydrates to the roots, while nutrient alleviation driven by nitrogen addition resulted in more carbohydrate being allocated to the aboveground organs, resulting in the accumulation of structural carbohydrates in the aboveground parts [42]. This means that the addition of nitrogen leads to a reduction in fine root biomass in *C. lanceolata* seedlings [43]. The same study found that there was a significant positive correlation between the biomass of fine roots and the content of NH_4^+-N in a *S. superba* plantation [44]. The roots of *S. superba* from a heterogeneous N environment showed greater dry biomass and N absorption efficiency than those from a homogeneous N environment [45]. This is contrary to the results of this study, which may be a result of the differences in the provenance of the *S. superba* specimens, as they may have differences in their N uptake.

As important indicators of roots' morphological characteristics, the SRL and RTD reflect the ability and efficiency of plants to allocate biomass to roots to absorb nutrients and water. In general, the larger the SRL and the smaller the root diameter, the stronger the ability of the root system to absorb nutrients and water [46]. To a certain extent, RTD can be considered a reflection of root tissue's stretching force. Generally, the greater the RTD, the greater the stretching force of the root tissue [47]. Numerous studies have shown that SRL has a negative correlation with RTD and RAD [18,48]. After N application, *Castanopsis fabri* adopted a rapid absorption strategy by increasing the SRL and the growth rate of roots, while *Castanopsis carlesii* adopted a relatively conservative resource absorption strategy by increasing the tissue density of fine roots [49]. The N deposition treatment significantly increased the fine root length of *S. superba* but had no significant effect on the tissue density [50]. In this study, we found that the SRL of *C. lanceolata* and *S. superba* were negatively correlated with RTD (Figure 5a,b). Therefore, we speculated that *C. lanceolata* and *S. superba* adopt an expansion strategy of increasing their SRL and reducing their RTD in an environment with a high NO_3^--N concentration.

The RSR is one of the indicators used to measure biomass allocation in plants. Numerous studies have shown that in a normal growth environment, the aboveground and underground parts of plants have synergistic growth, and biomass accumulation is also carried out synergistically [51]. Under stresses such as drought, plants allocate more biomass to the roots and increase the RSR to enhance the absorption capacity of water and other underground resources [41,52]. A previous study found that drought reduced fine root biomass and increased the RSR [53]. This study showed that the RSR of *C. lanceolata* was positively correlated with NO_3^--N but negatively correlated with NH_4^+-N, which was consistent with the pattern shown by root biomass, and the RSR was largest under the treatments with NO_3^--N only and an ammonium to nitrate ratio of 3:7. The RSR of *S. superba* showed the opposite pattern and was consistent with the pattern shown by root biomass (Figure 5a,b), and the RSR was largest under the treatment with an ammonium to nitrate ratio of 7:3 and under the NH_4^+-N-only treatment. Therefore, the results of this study showed that the allocation of biomass to aboveground and underground parts in *C. lanceolata* and *S. superba* seedlings responds differently to different N forms, and the N uptake strategy was adjusted by increasing or decreasing the RSR. Under the treatments with high NO_3^--N concentrations, *C. lanceolata* increased the RSR to obtain more N sources, while *S. superba* increased the aboveground biomass by reducing the RSR, but the opposite was true at high NH_4^+-N concentrations. In a study of fine roots in hybrid poplar plantations, it was found that soil nutrient deficiencies caused by continuous cropping may lead to an increase in plants' carbon input to the roots [54]. Therefore, we inferred that a high NO_3^--N concentration may be a limiting nutrient for *C. lanceolata* seedlings. Correspondingly, N sources with high NH_4^+-N concentrations are more suitable for the aboveground growth of *C. lanceolata* seedlings, and N sources with high NO_3^--N concentrations are more suitable for the development of aboveground parts in *S. superba*.

4.3. Effects of Different N Forms on the Growth of Different Tree Species

As one of the limiting elements of plant growth, N occurs in different forms in natural soil, and the different forms of N occur in different ratios, which will have different effects on plant growth. It is believed that the sole application of NH_4^+-N is prone to inducing NH_4^+ toxicity, which restrains plant growth [55,56]. Moreover, many plants prefer mixed NH_4^+-N and NO_3^--N over a single N source [57]. In contrast, some studies have also found that the growth of the underground and aboveground parts of plant seedlings is positively correlated with treatments with NH_4^+-N only [5]. The results of this study showed that the ΔH of *C. lanceolata* seedlings showed higher values under ammonium to nitrate ratios of 5:5

and 7:3, and ΔGD showed higher values under ammonium to nitrate ratios of 7:3 and 5:5. The ΔH of *S. superba* was larger under ammonium to nitrate ratios of 5:5 and 3:7, and the ΔGD was higher under ammonium to nitrate ratios of 0:10 and 3:7, which both had positive correlations with NO_3^--N and TSB (Figure 5b). Overall, our study supports the idea that *C. lanceolata* and *S. superba* seedlings prefer to develop aboveground parts under mixed NH_4^+-N and NO_3^--N. Between the two, *C. lanceolata* seedlings are more inclined to absorb NH_4^+-N for aboveground growth, which is more suitable for growth under ammonium to nitrate ratios of 5:5 and 7:3. Conversely, *S. superba* seedlings are more inclined to absorb NO_3^--N, which is more suitable for aboveground growth under ammonium nitrate ratios of 5:5, 3:7 and 0:10. The ΔGD of *S. superba* seedlings was largest under the ratio of 0:10, which may be because the application of NO_3^--N alone is beneficial for the growth in the woody diameter of *S. superba*.

Because they are timber species, it is also worth considering the aboveground growth of *C. lanceolata* and *S. superba*. Because of differences in their leaves, conifer and broadleaf species respond differently to different N forms [58]. Studies suggest that conifer species tend to absorb NH_4^+-N, and broadleaf species tend to absorb NO_3^--N [58,59]. This study supports this hypothesis, which proposes that *C. lanceolata* seedlings prefer NH_4^+-N over NO_3^--N for aboveground growth, while the opposite is true for *S. superba*. It is worth noting that TRL, TRSA, TRV, TSB, ΔH and SRL all showed maximum values in *S. superba* under an ammonium to nitrate ratio of 5:5. This means that both the aboveground and underground parts grow simultaneously under a homogeneous N supply. As a result, we speculate that the roots of *S. superba* seedlings may increase their length within the root system to obtain nutrients and use nutrients for seedlings' growth in terms of height and leaves to obtain more nutrients through photosynthesis under a homogeneous N supply [60].

5. Conclusions

In conclusion, the roots of *C. lanceolata* tended to be slender and accumulate more biomass under the treatment with a high NO_3^--N concentration, while the opposite was true for *S. superba*. Under the high NO_3^--N concentration, the roots of the two tree species adopted the expansion strategy of increasing the SRL and reducing the RTD. Moreover, *C. lanceolata* allocated more biomass to the root system at the high NO_3^--N concentration, but *S. superba* allocated more biomass to the root system at the high NH_4^+-N concentration. Compared with the treatments with a single N form, the mixed NH_4^+-N and NO_3^--N treatments were more conducive to the growth and biomass accumulation of the two tree species. *C. lanceolata* tends to grow under high NH_4^+-N concentrations, while *S. superba* tends to grow under high NO_3^--N concentrations. *C. lanceolata* grows better under NH_4^+-N:NO_3^--N ratios of 5:5 and 7:3, while *S. superba* grows better under NH_4^+-N:NO_3^--N ratios of 5:5 and 3:7 ratios. Therefore, according to the soil N, we can appropriately apply more NH_4^+-N to *C. lanceolata* and more NO_3^--N to *S. superba* for cultivating seedlings.

Author Contributions: Conceptualization, H.L.; methodology and formal analysis, H.L. and L.W.; investigation, Y.W., X.Q. and X.L.; data curation, H.L. and Y.X.; writing—original draft preparation, H.L.; writing—review and editing, Y.W. and X.Q.; validation and visualization, L.W. and X.L.; supervision, X.Y.; funding acquisition, X.Y. All authors have read and agreed to the published version of the manuscript.

Funding: This research was funded by the National Natural Science Foundation of China (32171773) and the Distinguished Young Scientific Research Talent Program of Fujian Agriculture and Forestry University (KXJQ20012).

Institutional Review Board Statement: Not applicable.

Informed Consent Statement: Not applicable.

Data Availability Statement: Not applicable.

Acknowledgments: We thank the numerous students and lab staff from the College of Forestry, Fujian Agriculture and Forestry University, for their assistance in the laboratory.

Conflicts of Interest: The authors declare no conflict of interest.

References

1. Zhou, X.; Fu, Y.; Zhou, L.; Li, B.; Luo, Y. An imperative need for global change research in tropical forests. *Tree Physol.* **2013**, *33*, 903–912. [CrossRef] [PubMed]
2. Ma, Y.; Li, Y.; Nurbiye, A. Temporal and spatial distribution characteristics of soil ammonium and nitrate Nitrogen in different plant communities in the Ebinur Lake Wetland. *Henan Sci. Tech.* **2019**, *26*, 150–156.
3. Ou, J.; Liu, Y.; Zhang, J.; Cui, N.; Zhang, J.; Song, X.; Deng, C. Early responses of soil ammonium and nitrate nitrogen to forest gap harvesting of a *Pinus massoniana* plantation in the upper reaches of Yangtze River. *Chin. J. Appl. Envir. Biol.* **2015**, *21*, 147–154.
4. Sun, M.; Lu, X.; Cao, X.; Li, J.; Xiong, J.; Xie, S. Effect of different Nitrogen forms on root growth and dynamic kinetics characteristics for *Citrus sinensis* × *Poncirus trifoliata*. *Scit. Silvae Sin.* **2015**, *51*, 113–120.
5. Wang, Y.; Yao, R. Effects of different nitrogen forms and ratios on growth of tissue cultured seedlings in *Pinus massoniana*. *J. Cent. South Univ. For. Technol.* **2021**, *41*, 18–24+71.
6. Jia, L.; Chen, G.; Zhang, L.; Chen, T.; Jiang, Q.; Chen, Y.; Fan, A.; Wang, X. Plastic responses of fine root morphology and architecture traits to nitrogen addition in ectomycorrhizal and arbuscular mycorrhizal tree species in an evergreen broadleaved forest. *Chin. J. Appl. Ecol.* **2021**, *32*, 529–537.
7. Lin, S.; Shao, L.; Hui, C.; Sandhu, H.; Fan, T.; Zhang, L.; Li, F.; Ding, Y.; Shi, P. The effect of temperature on the developmental rates of seedling emergence and leaf-unfolding in two dwarf bamboo species. *Trees* **2018**, *32*, 757–763. [CrossRef]
8. Yan, X.; Hu, W.; Ma, Y.; Huo, Y.; Wang, T.; Ma, X. Nitrogen Uptake Preference of *Cunninghamia lanceolata*, *Pinus massoniana* and *Schima superba* under Heterogeneous Nitrogen Supply Environment and their Root Foraging Strategies. *Scit. Silvae Sin.* **2020**, *56*, 1–11.
9. Qiao, Y.; Miao, S.; Han, X. Effects of nitrogen form on soybean root morphological characters and H^+ release. *Soybean Sci.* **2006**, *25*, 265–269.
10. Liu, C.; Cui, X.; Guo, Y.; Zheng, H. Effects of different ratios of NH_4^+-N/NO_3^- -N on growth of *Larix gmelinii* seedlings. *J. Northeast For. Univ.* **2011**, *39*, 28–30.
11. Ye, Y.; Luo, H.; Li, M.; Liu, X.; Cao, G.; Xu, S. Effects of nitrogen forms on lateral roots development and photosynthetic characteristics in leaves of *Cunninghamia lanceoata* seedlings. *Acta Bot. Boreal. Occident. Sin.* **2018**, *38*, 2036–2044.
12. Domenicano, S.; Coll, L.; Messier, C.; Berninfer, F. Nitrogen forms affect root structure and water uptake in the hybrid poplar. *New For.* **2011**, *42*, 347–362. [CrossRef]
13. Hilbert, D. Optimization of plant root: Shoot ratios and internal nitrogen concentration. *Ann. Bot.* **1990**, *66*, 91–99. [CrossRef]
14. Gedroc, J.; McConnaughay, K.; Coleman, J. Plasticity in root/shoot partitioning: Optimal, ontogenetic, or both? *Func. Ecol.* **1996**, *10*, 44–50. [CrossRef]
15. Hermans, C.; Hammond, J.; White, P.; Verbruggen, N. How do plants respond to nutrient shortage by biomass allocation? *Trends Plant Sci.* **2006**, *11*, 610–617. [CrossRef]
16. McCarthy, M.; Enquist, B. Consistency between an allometric approach and optimal partitioning theory in global patterns of plant biomass allocation. *Func. Ecol.* **2007**, *21*, 713–720. [CrossRef]
17. Kobe, R.; Iyer, M.; Walters, M. Optimal partitioning theory revisited: Nonstructural carbohydrates dominate root mass responses to nitrogen. *Ecology* **2010**, *91*, 166–179. [CrossRef]
18. Wang, Z.; Cheng, L.; Wang, M.; Sun, J.; Zhong, Q.; Li, M.; Cheng, D. Fine root traits of woody plants in deciduous forest of the Wuyi Mountains. *Acta Ecol. Sin.* **2018**, *38*, 1–10.
19. Zhang, X.; Xing, Y.; Yan, G.; Wang, Q. Response of fine roots to precipitation change: A meta-analysis. *Chin. J. Plant Ecol.* **2018**, *42*, 164–172.
20. Tuan, D.; Shen, Y.; Kang, W.; Xiang, W.; Yan, W.; Deng, X. Characteristics of nutrient cycling in first and second rotations of Chinese fir plantations. *Acta Ecol. Sin.* **2011**, *31*, 5025–5032.
21. Suo, P.; Du, D.; Wang, Y.; Hu, Y.; Liu, X. Effects of successive rotation Chinese fir plantations on soil nitrogen content and soil enzyme activities related to nitrogen transformation. *J. For. Envir.* **2019**, *39*, 113–119.
22. Wan, X.; Huang, Z.; He, Z.; Hu, Z.; Yang, J.; Yu, Z.; Wang, M. Effects of broadleaf plantation and Chinese fir (*Cunninghamia lanceolata*) plantation on soil carbon and nitrogen pools. *Chin. J. Appl. Ecol.* **2013**, *24*, 345–350.
23. Hao, F.; Liu, X.; Fan, J. Study on key enzyme activity in nitrogen metabolism and the content of molybdenum and iron in alfalfa under different NO_3^--N/NH_4^+-N ratio. *Agric. Res. Arid. Areas* **2017**, *35*, 190–197.
24. Wu, P.; Ma, X.; Tigabu, M.; Wang, C.; Liu, A.; Odén, P. Root morphological plasticity and biomass production of two Chinese fir clones with high phosphorus efficiency under low phosphorus stress. *Can. J. For. Res.* **2011**, *41*, 228–234. [CrossRef]
25. Zhang, J.; Sheng, W.; Xiong, Y.; Wan, X. Effects of fertilization on soil nutrient content of potted Chinese Fir seedling. *Sci. Silv. Sin.* **2006**, *42*, 44–50.
26. Meng, S. Nitrogen Dynamic Uptake and Genetic Expression of Translocator of Tree Species in Fine Roots. Ph.D. Thesis, Northwest Agricultural and Forestry University, Yangling, China, 2016.

27. Liu, Z.; Lin, W.; Yang, Z.; Lin, T.; Liu, X.; Chen, Y.; Yang, Y. Effects of soil warming and nitrogen deposition on available nitrogen in a young *Cunninghamia lanceolata* stand in mid-subtropical China. *Acta Ecol. Sin.* **2017**, *37*, 44–53.
28. Ma, X. Research on Adaptability and Foraging Behavior of Different Afforestation Species to Soil Nutrient Heterogeneity. Master's Thesis, Chinese Academy of Forestry, Beijing, China, 2009.
29. Yao, J.; Chu, X.; Zhou, Z.; Tong, J.; Wang, H.; Yu, J. Effects of neighbor competition on growth, fine root morphology and distribution of *Schima superba* and *Cunninghamia lanceolata* in different nutrient environments. *Chin. J. Appl. Ecol.* **2017**, *28*, 1441–1447.
30. Zheng, C. Study on nutrient characteristics of soil in the forest and rhizosphere of *Schima Schima* and Chinese fir mixed forest. *Wuyi Sci. J.* **2006**, *22*, 123–126.
31. Xu, X.; Li, Q.; Wang, J.; Zhang, L.; Tian, S.; Zhi, L.; Li, Q.; Sun, Y. Inorganic and organic nitrogen acquisition by a fern *Dicranopteris dichotoma* in a subtropical forest in South China. *PLoS ONE* **2014**, *9*, e90075. [CrossRef]
32. Giehl, R.; Von Wirén, N. Root nutrient foraging. *Plant Physiol.* **2014**, *166*, 509–517. [CrossRef]
33. Wang, Y.; Hsu, P.; Tsay, Y. Uptake, allocation and signaling of nitrate. *Trends Plant Sci.* **2012**, *17*, 458–467. [CrossRef]
34. Li, S.; Zhou, L.; Wu, S.; Sun, M.; Ding, G.; Lin, S. Effects of different nitrogen forms on nutrient uptake and distribution of *Cunninghamia lanceolata* plantlets under drought stress. *J. Plant Nut. Fert.* **2020**, *26*, 152–162.
35. Li, L.; Zheng, S.; Xu, J.; Wu, P. Research Advance in Influence Mechanism of Tree Root Biomass Allocation. *World For. Res.* **2022**, *35*, 15–20.
36. Zhang, X.; Xing, Y.; Wang, Q.; Yan, G.; Wang, M.; Liu, G.; Wang, H.; Huang, B.; Zhang, J. Effects of long-term nitrogen addition and decreased precipitation on the fine root morphology and anatomy of the main tree species in a temperate forest. *Forest Ecol. Mana.* **2020**, *455*, 117664. [CrossRef]
37. Yan, G.; Chen, F.; Zhang, X.; Wang, J.; Han, S.; Xing, Y.; Wang, Q. Spatial and temporal effects of nitrogen addition on root morphology and growth in a boreal forest. *Geoderma* **2017**, *303*, 178–187. [CrossRef]
38. Gan, H. The Molecular and Physiological Basis Underlying Acquisition of Nitrogen and Phosphorus in Fine Roots of Poplars. Ph.D. Thesis, Northwest Agricultural and Forestry University, Yangling, China, 2016.
39. Wang, J. Root Competition and Allelopathic Effects of *Cunninghamia lanceolata* and *Phoebe chekiangensis* under the Simulated Nitrogen Deposition. Master's Thesis, Zhejiang Agriculture and Forestry University, Lin'an, China, 2021.
40. Zhang, R.; Wang, Y.; Zhou, Z.; Feng, Z. Nitrogen addition affects root growth, phosphorus and nitrogen efficiency of three provenances of *Schima superba* in barren soil. *Acta Ecol. Sin.* **2013**, *33*, 3611–3621. [CrossRef]
41. Li, B.; Zhu, Q.; Pan, Y.; Jin, A.; Gao, Y.; Pan, Y.; Wang, Z. Research progress on plant growth strategies based on biomass allocation patterns. *J. Green Sci. Tech.* **2022**, *24*, 29–32+41.
42. Wang, J.; Guan, X.; Zhang, W.; Huang, K.; Zhu, M.; Yang, Q. Responses of biomass allocation patterns to nitrogen addition of *Cunninghamia lanceolata* seedlings. *Chin. J. Plant Ecol.* **2021**, *45*, 1231–1240. [CrossRef]
43. Shi, Z.; Xiong, D.; Feng, J.; Xu, C.; Zhong, B.; Deng, F.; Chen, Y.; Chen, G.; Yang, Y. Ecophysiological effects of simulated nitrogen deposition on fine roots of Chinese fir (*Cunninghamia lanceolata*) seedlings. *Acta Ecol. Sin.* **2017**, *37*, 74–83.
44. Wang, J.; Yu, S.; Hao, Q.; Cao, Y.; Ge, Z.; Mao, L. Distribution differences of fine root biomass and morphology in subtropical secondary forest at different forest ages. *J. Northwest Agric. For. Univ.* **2021**, *49*, 38–46.
45. Yao, J.; Zhou, Z.; Chu, X.; Xu, H.; Tong, J. Effect of neighborhood competition on dry matter accumulation, nitrogen and phosphorus efficiency of three provenances of *Schima superba* in a heterogeneous nutrient environment. *Acta Ecol. Sin.* **2018**, *38*, 1780–1788.
46. Dhiman, I.; Bilheux, H.; DeCarlo, K.; Painter, S.L.; Santodonato, L.; Warren, J.M. Quantifying root water extraction after drought recovery using sub-mm in situ empirical data. *Plant Soil* **2018**, *424*, 73–89. [CrossRef]
47. Zheng, Y.; Wen, Z.; Song, G.; Ding, M. The influence of environment and phylogenic background on variation in leaf and fine root traits in the Yanhe River catchment, Shaanxi, China. *Acta Ecol. Sin.* **2014**, *34*, 2682–2692.
48. Zhang, L. Plastic Responses of Fine Root Traits to N and Pin Ectomycorrhizal and Arbuscular Mycorrhizal Tree Species in an Evergreen Broadleaved Forest. Master's Thesis, Fujian Normal University, Fuzhou, China, 2019.
49. Jia, L.; Chen, G.; Zhang, L.; Chen, T.; Jiang, Q.; Chen, Y.; Fan, A.; Wang, X. Plastic responses of fine root morphological traits of *Castanopsis fabri* and *Castanopsis carlesii* to short-term nitrogen addition. *Chin. J. Appl. Ecol.* **2019**, *30*, 4003–4011.
50. Xie, Y. Morphological and Chemometrics of *Schima superb* Seeding Fine Root in Response of Nitrogen Deposition. Master's Thesis, Fujian Normal University, Fuzhou, China, 2016.
51. Xue, Z.; Wang, S.; Ding, J.; Li, C.; Ma, B.; Rebieguri, Y. Effects of different nitrogen forms on root morphology and nitrogen uptake of walnut seedlings. *J. Shandong Agri. Univ.* **2021**, *52*, 759–763.
52. Gao, G.; Li, Z.; Ge, X.; Huang, R.; Li, A. Effects of nitrogen addition on biomass and root morphology of *Phyllostachys edulis* seedlings under drought stress. *Chin. J. Ecol.* **2022**, *41*, 858–864.
53. Zhai, D. Effects of Drought on Root Respiration and the Mechanisms. Ph.D. Thesis, East China Normal University, Shanghai, China, 2021.
54. Wang, Y.; Xu, T.; Zhu, W.; Wang, Q.; Liu, M.; Wang, H.; Li, C.; Dong, Y. Seasonal dynamics of quantitative and morphological traits of poplar fine roots and their differences between successive rotation plantations. *Chin. J. Appl. Ecol.* **2016**, *27*, 395–402.
55. Tao, S.; Hua, X.; Wang, Y.; Guo, N.; Yan, X.; Lin, J. Research advance in effects of different nitrogen forms on growth and physiology of plants. *Guizhou Agri. Sci.* **2017**, *45*, 64–68.

56. Borgognone, D.; Colla, G.; Rouphael, Y.; Cardarelli, M.; Rea, E.; Schwarz, D. Effect of nitrogen form and nutrient solution pH on growth and mineral composition of self-grafted and grafted tomatoes. *Sci. Hortic.* **2013**, *149*, 61–69. [CrossRef]
57. Sui, L.; Yi, J.; Wang, K.; Li, Y. Effects of different forms and ratios of nitrogen on physiological characteristics of *Perilla frutescens* (L.) Britt under salt stress. *Chin. J. Ecol.* **2018**, *37*, 3277–3283.
58. Uscola, M.; Oliet, J.; Villar-Salvador, P.; Dıáz-Pine´s, E.; Jacobs, D. Nitrogen form and concentration interact to affect the performance of two ecologically distinct Mediterranean forest trees. *Eur. J. For. Res.* **2014**, *133*, 235–246. [CrossRef]
59. Wang, X. Effects of Different Nitrogen Forms and Ratios on Growth Characteristics of *Phyllostachys edulis* and Evergreen Broad-leaved Forest Species Seedlings. Master's Thesis, Zhejiang A&F University, Hangzhou, China, 2018.
60. Chang, Y.; Li, B.; Zhong, Q.; Wang, G.; Shen, Q.; Xu, C.; Zhang, S. Biomass allocation strategies of three functional forest seedlings and their relation with fine root and leaf nutrient. *Chin. J. Ecol.* **2022**, *25*, 1–10.

forests

Article

The Influence of Iron Application on the Growth and Cadmium Stress Tolerance of Poplar

Mingwan Li, Changrui Liu, Dangquan Zhang, Bingwen Wang and Shen Ding *

College of Forestry, Henan Agricultural University, Zhengzhou 450002, China
* Correspondence: dingshen040033@163.com

Abstract: There is a complex cadmium (Cd) and iron (Fe) interaction in soil. To explore the influences of Fe application on the growth, Cd accumulation, and antioxidant capacity of poplar under Cd exposure, *Populus tremula* × *P. alba* '717' was treated with different concentrations of Cd (0 and 100 μM) and Fe (50 and 150 μM). In addition, the root architecture, leaf chlorophyll content, Cd accumulation, and antioxidant enzyme activity were analyzed. The results showed that the high-dose Fe (150 μM) did not change poplar biomass in zero-Cd treatment but increased the chlorophyll content, total root surface area, net photosynthetic rate, and biomass accumulation of Cd-stressed poplar. In addition, under Cd stress, high-dose Fe increased the translocation factor (TF) of Cd, decreased root and leaf malondialdehyde (MDA) content, and enhanced root and leaf SOD activity. That is, high-dose Fe could alleviate the suppression of Cd on the growth of poplar and enhance the transport of Cd to aboveground tissues and the SOD activity in roots and leaves, thus alleviating the Cd-induced oxidative stress. This study will provide reference for the remediation of Cd-contaminated soils using poplar.

Keywords: phytoremediation; root architecture; antioxidant enzyme activity; chlorophyll; photosynthesis

Citation: Li, M.; Liu, C.; Zhang, D.; Wang, B.; Ding, S. The Influence of Iron Application on the Growth and Cadmium Stress Tolerance of Poplar. *Forests* **2022**, *13*, 2023. https://doi.org/10.3390/f13122023

Academic Editor: Eustaquio Gil-Pelegrín

Received: 31 October 2022
Accepted: 28 November 2022
Published: 29 November 2022

1. Introduction

Heavy metal accumulation caused by natural processes or anthropogenic activities severely affects the life and biodiversity on earth [1]. Especially, cadmium (Cd) pollution has attracted much attention from people of many countries [2–5]. Cd in soil can be absorbed by plant roots and enters the human body through the food chain, endangering human immune, nervous, and reproductive systems [6]. To reduce the Cd accumulation in soil and the threat of Cd pollution to human health, many physical, chemical, and biological remediation technologies have been developed [7,8]. Among them, phytoremediation is an effective, low-cost, and environment-friendly way [9] to remove heavy metals from soils by plants' uptake. For example, Hu et al. [10] and Balestri et al. [11] reported that *Pteris vittata* and *Sedum alfredii* could adsorb a large amount of Cd in the soil. However, most known Cd-enriched plants are herbs. Due to the low biomass, the amount of Cd accumulation is very small compared with trees [2]. Therefore, some scholars began to study the remediation of Cd-contaminated soils using trees [12–14]. Marmiroli et al. reported that the total Cd contents extracted from *Populus* and *Thlaspi* were about 250 and 125 g ha^{-1} per year in similarly polluted soil, respectively [15]. Due to the fast growth and strong adaptability, the performance of poplar in Cd-contaminated soil remediation has been explored [16]. He et al. [17] found that due to *Populus canescens* had a large biomass: the total amount of Cd absorbed and transported to the aboveground part was considerable. The root and leaf Cd content of *P. canescens* exceeded 1000 and 100 μg g^{-1} DW, respectively. Therefore, poplar has great potential in phytoremediation.

Several studies have shown that Cd can enter plant cells through transport vectors of Fe^{2+}, Zn^{2+}, and Ca^{2+}, disrupting the uptake of minerals by plants [18–20]. The strong binding ability of Cd to proteins could replace Fe in proteins and change protein function and chloroplast structure, thus inhibiting photosynthesis and reducing biomass accumulation [21,22]. Cd can also disturb the metabolism and increase the production of reactive oxygen species (ROS), which ultimately causes oxidative stress [23]. Many studies have shown that plants can adapt to environmental stresses through changing root architecture to obtain adequate nutrients [24]. In addition, the activities of antioxidant enzymes in plant cells, such as ascorbate peroxidase (APX), ascorbate oxidase (AAO), superoxide dismutase (SOD), and catalase (CAT), could also be induced to maintain intracellular redox balance [12].

Iron (Fe) is necessary for plant growth. The redox of Fe^{2+} and Fe^{3+} participates in plant electron transport chain, photosynthesis, respiration, and chloroplast synthesis [25,26]. A previous study found that Cd can disrupt the absorption and transportation of Fe by plants, causing the iron deficiency chlorosis of plants [27]. Fe deficiency further aggravates Cd-induced oxidative stress and inhibits plant growth [28,29]. Although Fe could restore the photosynthesis of plants, it also strengthens the absorption of Cd and the allocation to the aboveground part [30,31]. However, a previous study also proved that sufficient Fe could reduce the absorption of Cd by plants [32]. Fe is a key component of the iron superoxide dismutase (Fe-SOD) subunit, which is vital for ROS scavenging [33], and most Fe in the soil is in insoluble form that cannot be directly utilized by plants. Thus, exploring the effect of high-dose Fe on poplar growth and Cd accumulation is essential for poplar phytoremediation.

The root system directly contacts with soil Cd and is the channel for Cd to enter the plant's aboveground part, so it is more sensitive to soil Cd [34]. Therefore, in this study, the effects of Fe treatments (50 and 150 μM) on the growth parameters, root architecture, Cd accumulation, and antioxidant enzyme activity of poplar under Cd exposure (100 μM) were explored, to clarify the mechanism of high-dose Fe alleviating Cd stress on poplar growth. This study hypothesized that: (1) high-dose Fe might affect the absorption and accumulation of Cd by poplar by changing poplar root architecture and (2) high-dose Fe might affect antioxidant enzyme activity to improve poplar tolerance to Cd stress. This research will provide reference for the phytoremediation of Cd-contaminated soil using poplar.

2. Materials and Methods

2.1. Plant Materials and Growth Conditions

Tissue culture plants of *Populus tremula* × *P. alba* '717' were used in this experiment. In March 2021, tissue culture plants were cultured under growth room condition (25/18 °C in the day and night; light duration: 14 h; light intensity: 150 μmol m^{-2} s^{-1}; relative humidity: 50%–60%) for one month and then transplanted in 6 L pots filled with sands. Hoagland nutrient solution (1/4 concentration, 50 mL) was irrigated into each pot every day.

In May, when the height of seedlings reached 30–35 cm, 24 seedlings with similar size were transplanted in aerated Hoagland nutrient solution without Fe and equally divided into four groups, followed by Cd (0 and 100 μM $CdCl_2$) and Fe (50 and 150 μM $FeCl_3$) treatments. The Cd and Fe levels were determined according to previous studies [14,35,36], and 50 and 150 μM Fe doses were considered normal- and high-dose Fe, respectively. There were four treatments in total, including (1) Cd0Fe50: 0 μM $CdCl_2$ + 50 μM $FeCl_3$; (2) Cd0Fe150: 0 μM $CdCl_2$ + 150 μM $FeCl_3$; (3) Cd100Fe50: 100 μM $CdCl_2$ + 50 μM $FeCl_3$; and (4) Cd100Fe150: 100 μM $CdCl_2$ + 150 μM $FeCl_3$. The nutrient solution was renewed every three days, and Cd and Fe were added again after the nutrient solution was replaced. After 5 weeks, all poplar plants were harvested (Figure 1).

Figure 1. The leaves of poplar under different Cd and Fe combined treatments. Cd0Fe50: 0 μM $CdCl_2$ + 50 μM $FeCl_3$; Cd0Fe150: 0 μM $CdCl_2$ + 150 μM $FeCl_3$; Cd100Fe50: 100 μM $CdCl_2$ + 50 μM $FeCl_3$; Cd100Fe150: 100 μM $CdCl_2$ + 150 μM $FeCl_3$. The same below.

2.2. Determination of Photosynthetic Parameters and Sampling

Three leaves of each seedling (leaf plastochron index: 7–9) were sampled, and the method of Luo et al. [37] was employed to determine the net photosynthetic rate (Pn), transpiration rate (E), and stomatal conductance (g_s) using a LI-6400 portable photosynthetic apparatus (LI-COR Inc., Lincoln, NE, USA) in the morning (9:00–11:00).

Then, the roots of each seedling were soaked in 50 mM $CaCl_2$ for 5 min, washed twice, and dried with absorbent paper, to remove root surface metal ions. A portion of the seeding roots (ca. 2 g) was used for scanning (see Section 2.3). After that, the seedlings were separated into roots, woods, barks, and leaves; weighed; and recorded. After wrapping in tin foil and freezing in liquid nitrogen, the tissues were ground at low temperature using a ball mill (Tissuelyser-64L, Jingxin, Shanghai, China) and stored at −80 °C. About 200 mg was then dried to constant weight and weighed, followed by the calculation of the dry-wet ratio.

2.3. Determination of Root Morphology

The roots sampled for scanning above mentioned were spread out in water to avoid root overlap. Then the root was scanned using the WinRHIZO root analysis system (Regent Instruments, Québec, QC, Canada) to determine the total root length (TRL), total root surface area (TRSA), total root volume (TRV), and total root tips (TRT) [24].

2.4. Assays of Chlorophyll Content

Leaf chlorophyll content (chlorophyll a (Chl a), chlorophyll b (Chl b), and carotenoid (Car)) were determined [14]. Forty milligrams of leaves was soaked in pure acetone and (400 μL) for 5 min. Then, 9.6 mL of 80% acetone solution was added. The mixture was shaken in dark conditions overnight. Fifteen minutes after the sample turned white, the absorbance at 663, 646, and 470 nm were measured using a spectrophotometer (Unico, Beijing, China).

2.5. Determination of the Concentrations of Cd and Fe

One hundred milligrams of root and leaf samples was digested using, respectively, solutions composed of perchloric acid (2 mL) and concentrated nitric acid (8 mL). Then, the ICP-MS inductively coupled plasma mass spectrometry (Agilent7800, Santa Clara, CA, USA) was used to determine the Cd and Fe concentrations [13].

2.6. Determination of MDA Content

One hundred milligrams of root and leaf samples was extracted in 10% TCA for 30 min. The supernatant was mixed with a solution composed of 0.6% TBA and 10% TCA after centrifugation. After that, the mixture was subjected to a 15 min boiling water bath, followed by cooling in ice and centrifugation. Finally, the absorbance of the supernatant was determined at 450, 532, and 600 nm [38].

2.7. Assays of Antioxidant Enzyme Activity

One hundred milligrams of root and leaf samples was extracted using an extract solution composed of polyvinylpolypyrrolidone (200 mg), potassium phosphate (0.1 M, pH 7.8), and Triton X-100 ($v/v = 0.5\%$). After centrifugation, the soluble protein content was determined using the supernatant, with bovine serum as the standard. The antioxidant enzyme activity was determined using the extracted soluble protein [39].

Following Tamás et al. [40], the reaction system for the determination of AAO (EC 1.10.3.3) activity included enzyme solution, potassium phosphate buffer (0.1 M, pH 5.6), EDTA-Na_2 (5 mM), and sodium ascorbate (50 mM). The absorbance was determined at 265 nm using a spectrophotometer. The oxidization of 1 μmol ascorbic acid (AsA) per milligram of protein per minute was defined as one unit of AAO activity.

Following He et al. [41], the reaction system for the determination of APX (EC 1.11.1.11) activity consisted of enzyme solution, phosphate buffer (0.05 M, pH 7.0), EDTA (0.1 mM), sodium ascorbate (0.1 mM), and H_2O_2 (2.5 mM). The absorbance was determined at 290 nm using a spectrophotometer. The decomposition of 1 μmoL of AsA per milligram of protein per minute was defined as one unit of APX activity.

Following Gong at al. [42], the SOD (EC1.15.1.1) activity was determined. The reaction system included enzyme solution, potassium phosphate (0.05 M, pH 7.8), methionine (0.22 M), nitro blue tetrazolium chloride monohydrate (NBT) (1.25 mM), and riboflavin (33 μM). The absorbance was determined at 560 nm using a spectrophotometer. The 50% inhibition of NBT photoreduction was defined as one unit of SOD activity.

2.8. Analysis of BCF, BAF, TF and TI

The bioconcentration factor (BCF), bioaccumulation factor (BAF), translocation factor (TF), and tolerance index (TI) were calculated [1]:

$$BCF_R = C_{CdR}/C_{Substrate} \quad (1)$$

$$BAF_W = C_{CdW}/C_{Substrate} \quad (2)$$

$$BAF_B = C_{CdB}/C_{Substrate} \quad (3)$$

$$BAF_L = C_{CdL}/C_{Substrate} \quad (4)$$

$$TF_W = C_{CdW}/C_{CdR} \quad (5)$$

$$TF_B = C_{CdB}/C_{CdR} \quad (6)$$

$$TF_L = C_{CdL}/C_{CdR} \quad (7)$$

$$TI = DBM_{Cd\text{-}treatment}/DBM_{Control} \quad (8)$$

where C_{CdR}, C_{CdW}, C_{CdB}, C_{CdR}, C_{CdL}, and $C_{Substrate}$ are the Cd concentrations in poplar root, wood, bark, leaves, and growing substrate, respectively, and $DBM_{Cd\text{-}treatment}$ and $DBM_{Control}$ are the dry biomass (DBM) of each Cd treatment and the control, respectively.

2.9. Statistical Analysis

Data normality was first analyzed. All the data were analyzed using Statgraphics software (STN, USA). The significance of the differences was determined with two-factor analyses of variance at $p < 0.05$, and for significant differences, the means were analyzed by a posteriori comparison.

3. Results

3.1. Effect of Fe Application on Cd-Stressed Poplar Growth

The biomass of roots (0.63 g), woods (0.59 g), barks (0.50 g), and leaves (0.50 g) for Cd100Fe50 decreased by 42%, 56%, 44%, and 51%, respectively, compared with those for Cd0Fe50 (Figure 2a–d). The biomass of roots (0.82 g), woods (0.71 g), barks (0.65 g), and leaves (3.79 g) for Cd100Fe150 decreased by 19%, 44%, 30%, and 30%, respectively, compared with those for Cd0Fe150 (Figure 2a–d). There was no difference between Cd0Fe50 and Cd0Fe150 (Figure 2a–d). However, the biomass of roots, woods, barks, and leaves for Cd100Fe150 increased by 29%, 20%, 29%, and 38%, respectively, compared with those for Cd100Fe50 (Figure 2a–d). The TI for Cd100Fe50 decreased by 50% compared with that for Cd0Fe50. Similarly, The TI for Cd100Fe150 decreased by 31% compared with that for Cd0Fe150 (Figure 2e). Under Cd-free condition, there was no difference in the TI between Cd0Fe150 and Cd0Fe50. The TI for Cd100Fe150 increased by 34% compared with that for Cd100Fe50 (Figure 2e).

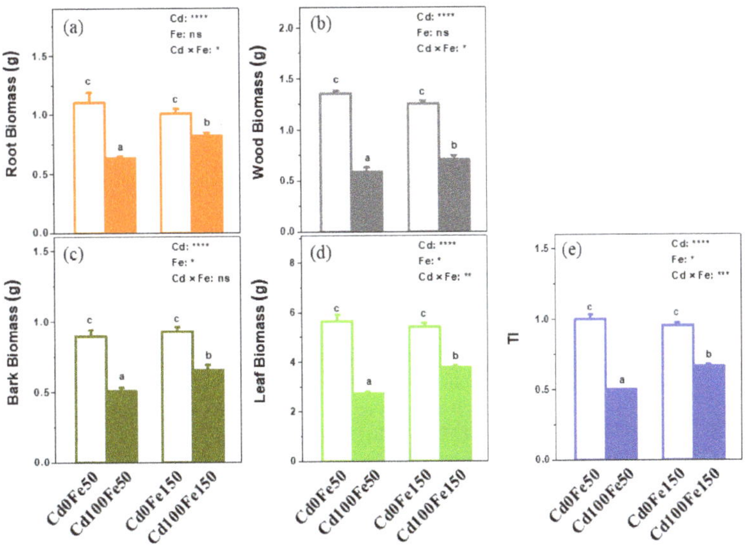

Figure 2. The biomass of root (**a**), wood (**b**), bark (**c**), leaves (**d**), and TI © of poplar under different Cd and Fe combined treatments. Data were means ± SE (n = 6). Different lowercase letters on the bars indicate significant differences in ANOVA ($p < 0.05$). p values of two-way ANOVA on Cd, Fe, and their interaction (Cd × Fe) were indicated (*, $p < 0.05$; **, $p < 0.01$; ***, $p < 0.001$; ****, $p < 0.0001$; ns, not significant). The same below.

3.2. Effect of Fe Application on Cd-Stressed Poplar Root Architecture

The TRL of poplar for Cd100Fe50 decreased by 58%, compared with that for Cd0Fe50. Similarly, the TRL of poplar for Cd100Fe150 decreased by 32% compared with that for Cd0Fe150. The TRL of poplar for Cd0Fe150 decreased by 25% compared with that for Cd0Fe50. However, there was no difference in the TRL of poplar between Cd100Fe150 and Cd100Fe50 (Figure 3a).

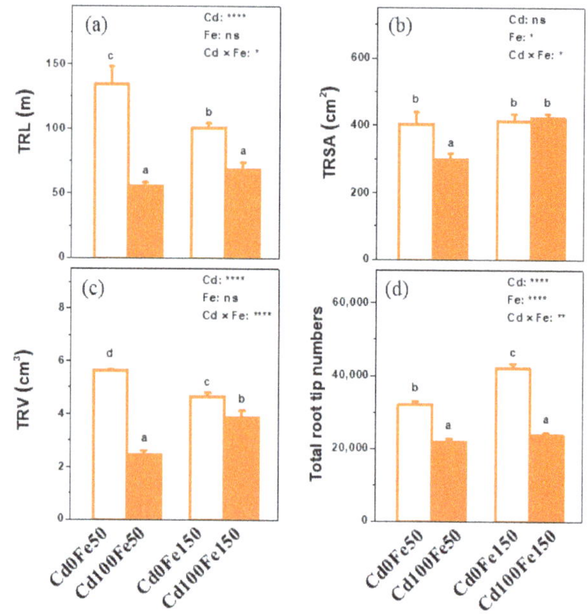

Figure 3. The TRL (**a**), TRSA (**b**), TRV (**c**) and total root tip number (**d**) of poplar under different Cd and Fe combined treatments. Data were means ± SE (n = 6). Different lowercase letters on the bars indicate significant differences in ANOVA ($p < 0.05$). p values of two-way ANOVA on Cd, Fe, and their interaction (Cd × Fe) were indicated (*, $p < 0.05$; **, $p < 0.01$; ****, $p < 0.0001$; ns, not significant).

The TRSA of poplar for Cd100Fe50 decreased by 26% compared with that for Cd0Fe50. There was no difference between Cd100Fe150 and Cd0Fe150 or between Cd0Fe50 and Cd0Fe150. However, the TRSA of poplar for Cd100Fe150 increased by 42% compared with that for Cd100Fe50 (Figure 3b).

The TRV of poplar for Cd100Fe50 decreased by 56% compared with that for Cd0Fe50. Similarly, the TRV of poplar for Cd100Fe150 decreased by 17% compared with that for Cd0Fe150. The TRV of poplar for Cd0Fe150 decreased by 17% compared with that for Cd0Fe50. However, the TRV of poplar for Cd100Fe150 increased by 56% compared with that for Cd100Fe50 (Figure 3c).

The total number of poplar root tips for Cd100Fe50 decreased by 32% compared with that for Cd0Fe50. Similarly, the total number of poplar root tips for Cd100Fe150 decreased by 43% compared with that for Cd0Fe150. However, the total number of root tips of poplar for Cd0Fe150 increased by 32% compared with that for Cd0Fe50. There was no difference between Cd100Fe50 and Cd100Fe150 (Figure 3d).

3.3. Effects of Fe on Chlorophyll Content and Photosynthetic Parameters of Cd-Stressed Poplar

The leaf Pn, E, and g_s for Cd100Fe50 decreased by 48%, 41%, and 27%, respectively, compared with those for Cd0Fe50. Similarly, the Pn, E, and g_s for Cd100Fe150 decreased by 41%, 47% and 38%, respectively, compared with those for Cd0Fe150. However, the Pn, E, and g_s of leaves for Cd0Fe150 increased by 21%, 47%, and 33%, respectively, compared with those for Cd0Fe50. In addition, the Pn, E, and g_s of leaves for Cd100Fe150 increased by 37%, 33%, and 13%, respectively, compared with those in the Cd100Fe50 (Figure 4).

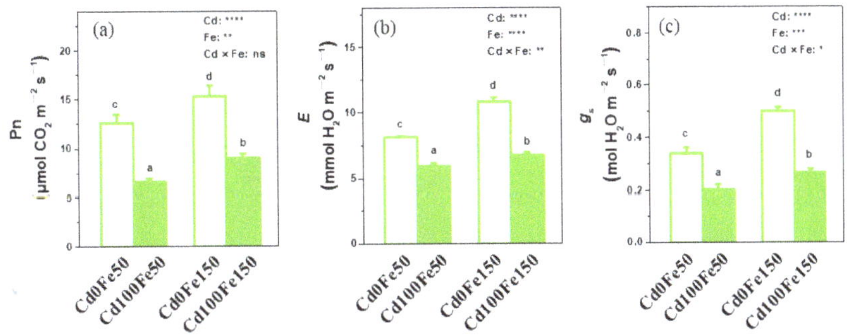

Figure 4. The Pn (**a**), E (**b**) and g_s (**c**) of poplar under different Cd and Fe combined treatments. Data were means ± SE (n = 6). Different lowercase letters on the bars indicate significant differences in ANOVA ($p < 0.05$). p values of two-way ANOVA on Cd, Fe, and their interaction (Cd × Fe) were indicated (*, $p < 0.05$; **, $p < 0.01$; ***, $p < 0.001$; ****, $p < 0.0001$; ns, not significant).

The contents of Chl a (4.61 mg g^{-1} DW), Chl b (1.01 mg g^{-1} DW), and Car (0.85 mg g^{-1} DW) in poplar leaves for Cd100Fe50 decreased by 54%, 41%, and 51%, respectively, compared with those for Cd0Fe50. Similarly, the contents of Chl a (2.82 mg g^{-1} DW), Chl b (0.77 mg g^{-1} DW), and Car (0.61 mg g^{-1} DW) in poplar leaves for Cd100Fe150 decreased by 42%, 30%, and 29%, respectively, compared with those for Cd0Fe150. However, the content of Chl b for Cd0Fe150 increased by 10%, while the contents of Chl a and Car did not change significantly compared with those for Cd0Fe50. Additionally, the contents of Chl a, Chl b, and Car in poplar leaves for Cd100Fe150 increased by 34%, 29%, and 45%, respectively, compared with those for Cd100Fe50 (Figure 5).

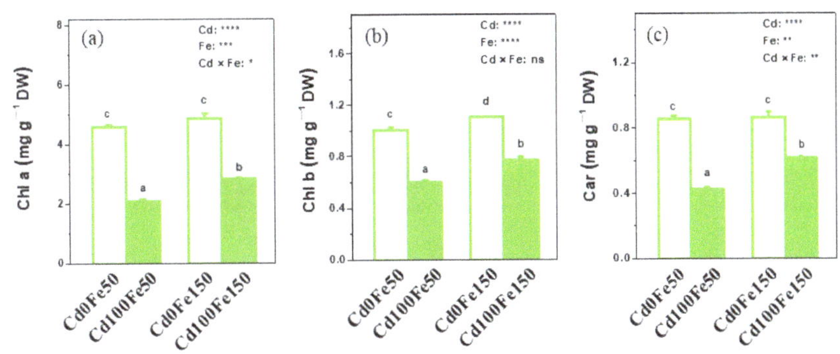

Figure 5. Chl a (**a**), Chl b (**b**), and Car (**c**) in the leaves of poplar under different Cd and Fe combined treatments. Data were means ± SE (n = 6). Different lowercase letters on the bars indicate significant differences in ANOVA ($p < 0.05$). p values of two-way ANOVA on Cd, Fe, and their interaction (Cd × Fe) were indicated (*, $p < 0.05$; **, $p < 0.01$; ***, $p < 0.001$; ****, $p < 0.0001$; ns, not significant).

3.4. Effect of Fe on the Cd Accumulation of Cd-Stressed Poplar

No Cd was found in the poplar for Cd0Fe150 and Cd0Fe50 (Figure 6a–d). The root Cd content (231.98 μg g^{-1} DW) in the Cd100Fe150 reduced by 66%, while the Cd contents in the woods (179.57 μg g^{-1} DW), barks (464.86 μg g^{-1} DW), and leaves (167.53 μg g^{-1} DW) increased by 44%, 76%, and 56%, respectively, compared with those for Cd100Fe50 (Figure 6a–d).

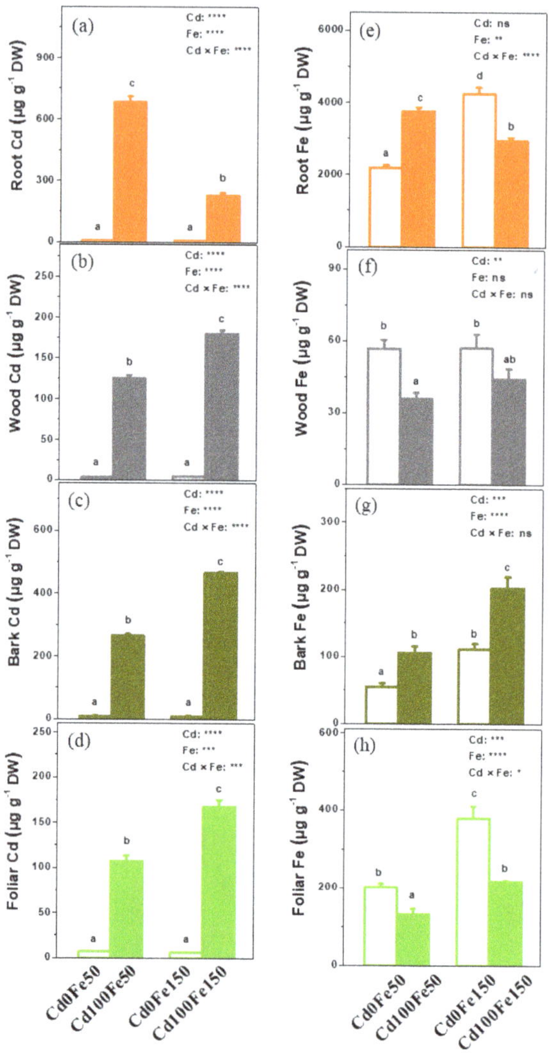

Figure 6. The Cd (**a**–**d**) and Fe (**e**–**h**) contents in the roots, woods, barks, and leaves of poplar under different Cd and Fe combined treatments. Data were means ± SE (n = 6). Different lowercase letters on the bars indicate significant differences in ANOVA ($p < 0.05$). p values of two-way ANOVA on Cd, Fe, and their interaction (Cd × Fe) were indicated (*, $p < 0.05$; **, $p < 0.01$; ***, $p < 0.001$; ****, $p < 0.0001$; ns, not significant).

The root and bark Fe content for Cd100Fe50 increased by 71% and 92%, respectively, while the wood and leaf Fe content decreased by 37% and 34%, respectively, compared with those for Cd0Fe50. The Cd content in poplar roots, woods, and leaves for Cd100Fe150 reduced by 30%, 23%, and 43%, respectively, while the content of Fe in the barks increased by 81% compared with those for Cd0Fe150. The Fe content in the roots, barks, and leaves of poplar for Cd0Fe150 increased by 92%, 103%, and 89%, respectively, compared with those for Cd0Fe50, but there was no difference in the Fe content in woods. The content of Fe in the barks and leaves for Cd100Fe150 increased by 91% and 62%, respectively, but that in the roots reduced by 22%, compared with those for Cd100Fe50 (Figure 6e–h).

The BCF_R for Cd100Fe150 (20.71) reduced by 66%, while the BAF_W, BAF_B, and BAF_L increased by 44%, 76%, and 56%, respectively, compared with those for Cd100Fe50. The TF_W (0.78), TF_B (2.02), and TF_L (0.73) for Cd100Fe150 increased by 325%, 420%, and 361%, respectively, compared with those for Cd100Fe50 (Table 1).

Table 1. BAF, BCF, and TF for Cd in the roots (R), woods (W), barks (B), and leaves (L) of poplar under different Cd and Fe combined treatments.

		BCF_R	BAF_W	BAF_B	BAF_L	TF_W	TF_B	TF_L
Cd0Fe150		61.00 ± 2.68 b	11.15 ± 0.38 a	23.64 ± 0.53 a	9.58 ± 0.54 a	0.18 ± 0.01 a	0.39 ± 0.01 a	0.16 ± 0.02 a
Cd100Fe150		20.71 ± 1.26 a	16.03 ± 0.51 b	41.51 ± 0.31 b	14.96 ± 0.70 b	0.78 ± 0.06 b	2.02 ± 0.13 b	0.73 ± 0.07 b
p-values	Fe	***	**	****	**	***	***	**

Notes: Data were means ± SE (n = 6). Different lowercase letters following the values in the same column indicate significant differences in ANOVA ($p < 0.05$). p values of one-way ANOVA on Fe were indicated (**, $p < 0.01$; ***, $p < 0.001$; ****, $p < 0.0001$; ns, not significant).

3.5. Effect of Fe application on MDA Content and Antioxidant Enzyme Activities of Cd-Stressed Poplar

The MDA content in poplar roots for Cd100Fe50 increased by 101% compared with that for Cd0Fe50 (Figure 7a). The MDA content in poplar roots for Cd100Fe150 increased by 60% compared with that for Cd0Fe150 (Figure 7a). There was no difference in the root MDA content between Cd0Fe50 and Cd0Fe150 (Figure 7a). However, the MDA content in poplar roots for Cd100Fe150 reduced by 24% compared with that for Cd100Fe50 (Figure 7a).

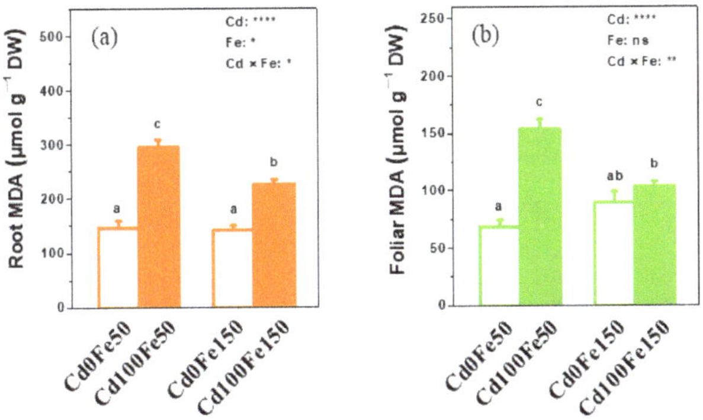

Figure 7. The contents of MDA in the roots (**a**) and leaves (**b**) of poplar under different Cd and Fe combined treatments. Data were means ± SE (n = 6). Different lowercase letters on the bars indicate significant differences in ANOVA ($p < 0.05$). p values of two-way ANOVA on Cd, Fe, and their interaction (Cd × Fe) were indicated (*, $p < 0.05$; **, $p < 0.01$; ****, $p < 0.0001$; ns, not significant).

The MDA content in poplar leaves for Cd100Fe50 increased by 124% compared with that for Cd0Fe50 (Figure 7b). There was no difference between Cd0Fe150 and Cd100Fe150 and between Cd0Fe50 and Cd0Fe150 (Figure 7b). The MDA content in poplar leaves for Cd100Fe150 reduced by 30% compared with that for Cd100Fe50 (Figure 7b).

The AAO and SOD activities in poplar roots for Cd100Fe50 reduced by 24% and 33%, respectively, compared with those for Cd0Fe50, while there was no difference in root APX activity (Figure 8a–c). The AAO, APX, and SOD activities in poplar roots for Cd100Fe150 reduced by 36%, 21%, and 23%, respectively, compared with those for Cd0Fe150 (Figure 8a–c). The APX and SOD activities in poplar roots for Cd0Fe150 increased by 9% and 11%, respectively, compared with those for Cd0Fe50, while there was no difference in root AAO activity (Figure 8a–c). The AAO and APX activities in poplar roots for Cd100Fe150 reduced by 20% and 19%, respectively, while the SOD activity in poplar roots increased by 27% compared with those for Cd100Fe50 (Figure 8a–c).

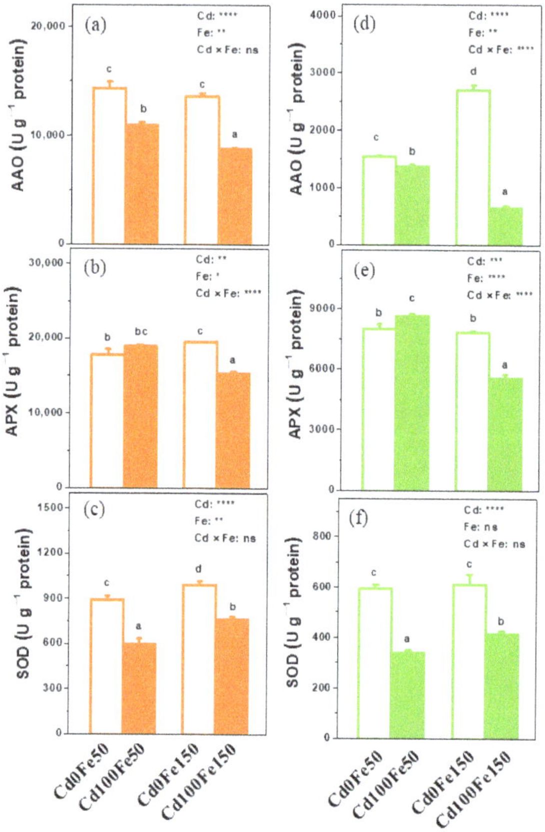

Figure 8. The activities of AAO (**a**,**d**), APX (**b**,**e**), and SOD (**c**,**f**) in the roots and leaves of poplar under different Cd and Fe combined treatments. Data were means ± SE (n = 6). Different lowercase letters on the bars indicate significant differences in ANOVA ($p < 0.05$). p values of two-way ANOVA on Cd, Fe, and their interaction (Cd × Fe) were indicated (*, $p < 0.05$; **, $p < 0.01$; ***, $p < 0.001$; ****, $p < 0.0001$; ns, not significant).

The AAO and SOD activities in poplar leaves for Cd100Fe50 reduced by 11% and 43%, respectively, while the APX activity increased by 8% compared with those for Cd0Fe50 (Figure 8d–f). The AAO, APX, and SOD activities in poplar leaves for Cd100Fe150 reduced by 76%, 29%, and 32%, respectively, compared with those for Cd0Fe150 (Figure 8d–f). The AAO activity in poplar leaves for Cd0Fe150 increased by 75% compared with that for Cd0Fe50, but there was no difference in the APX and SOD activities in leaves (Figure 8d–f). The AAO and APX activities in poplar leaves for Cd100Fe150 reduced by 53% and 35%, respectively, while the SOD activity increased by 22% compared with those for Cd100Fe50 (Figure 8d–f).

4. Discussion

4.1. Cd and Fe Interaction Affects the Growth, Photosynthesis, and Pigment Accumulation of Poplar

Cd stress can inhibit plant growth and photosynthesis, and even cause plant death [2,13,36]. This study found that Cd significantly inhibited poplar growth and reduced the biomass of roots, woods, barks, and leaves. Fe is necessary for plant growth. Sperotto et al. [43] found that proper high-dose Fe was beneficial to plant growth, but excessive Fe was toxic to plants. This study found that 150 μM Fe did not reduce the biomass of poplar roots, woods, barks, and leaves. This indicates that 150 μM is the suitable Fe level for poplar. Separately, TI can indicate a plant's resistance to Cd toxicity [1]. In this study, under Cd stress, 150 μM Fe increased the biomass accumulation, and TI of poplar compared with 50 μM Fe. These results indicate that proper high-dose Fe is helpful to enhance poplar's tolerance to Cd stress, and 150 μM is not excessive for poplar growth.

The analysis of root architecture showed that Cd significantly reduced the TRL, TRSA, TRV, and root tip number of poplar. This may be related to the toxicity of Cd [44]. In the absence of Cd, high-dose Fe decreased the TRV and TRL but increased the TRT. This indicates that high-dose Fe (150 μM) could make poplar roots shorter, thinner, and more lateral. In this study, under Cd stress, high-dose Fe increased the TRV of the poplar root system. This indicates that high-dose Fe could effectively alleviate the inhibition of Cd on poplar root growth. Meanwhile, the root system is the key part for plants to adsorb nutrients from soil, and large TRSA helps plants to adsorb nutrients [24,45]. In this study, under Cd stress, the high-dose Fe increased the TRSA. This may be due to the higher root biomass under Cd stress [45]. The increased TRSA can not only enhance the absorption of nutrients, but also enhance the absorption and accumulation of Cd by poplar.

Previous research has shown that plants with higher Pn have higher biomass [46,47]. Cd stress can lead to the reduction of biomass by inhibiting plant photosynthesis [27]. In this study, Cd significantly inhibited the Pn of poplar regardless of normal or high concentration of Fe, which was in line with the biomass results. Studies have shown that Cd could greatly inhibit the synthesis of chlorophyll, reduce the chlorophyll content, and cause the iron deficiency chlorosis of plants [38,44]. This study obtained similar results. Qureshi et al. [48] reported that Fe exists in heme- and iron-sulfur proteins as a cofactor that plays an important role in photosynthesis in chloroplasts. In addition, Fe participates in the transport of photosynthetic electrons in the photosynthetic electron transport chain [49]. In this study, under Cd stress, high-dose Fe increased the contents of Chl a, Chl b, and Car and enhanced the Pn and biomass. This indicates that high-dose Fe (150 μM) could effectively alleviate the iron deficiency chlorosis induced by Cd, enhance poplar tolerance to Cd, and restore the normal growth of poplar.

4.2. Fe application Enhances Cd Enrichment in Aboveground Parts of Poplar

Cd can be absorbed by plant roots, transported and accumulated in the aboveground tissues [8,40,50,51]. Studies have shown that the interaction between Cd and Fe in soil is complex. Because Cd^{2+} and Fe^{2+} have similar ionic radii, Cd^{2+} can enter plant cells through IRTs, ZIPs, NRAMPs, and other Fe^{2+} transport carriers [2,32,52], and Fe deficiency can increase the Cd accumulation in *Arabidopsis* [46]. In this study, under Cd stress, high-dose Fe increased the transpiration rate of poplar. Thus, more Cd could be transferred to the aboveground part with sap flow. The plant's phytoextraction capacity is defined as TF [1]. In this study, the TF_W, TF_B, and TF_L of Cd under high-dose Fe were significantly higher than those under normal-dose Fe. This indicates that high-dose Fe (150 μM) could effectively enhance the transport of Cd to the aboveground part of poplar. Separately, the Cd absorption capacity of plants from substrate/soil is defined as BCF [1]. In this study, most BAFs were lower than BCFs. It indicates that there is a limitation in the Cd transport from roots to aboveground parts, which results in more Cd accumulating in the root instead of the aboveground part [53]. However, it was found that after high-dose Fe, the BAF increased. This indicates that high-dose Fe may break some restrictions and enhance the transport and accumulation of Cd to the aboveground part of poplar. Although Cd-enriched herb plants have higher Cd contents, poplar trees can enrich a larger amount of Cd due to their larger biomass, which has important value in phytoremediation [54–56]. This study result indicates that a proper-high amount of Fe can be applied in the poplar phytoremediation of Cd-contaminated soils, so as to enhance the remediation efficiency.

As Cd^{2+} can enter plant cells through the transport protein of Fe^{2+}, it will disturb the absorption of Fe^{2+} by plants [36,57,58]. This study obtained similar results: Although high-dose Fe increased the Fe content in roots, barks, and leaves under zero Cd condition, it only increased the Fe content in barks and leaves under Cd exposure. Studies have shown that under Cd stress, increasing the Fe content of aboveground parts, especially increasing leaf Fe content, is very important for maintaining chlorophyll synthesis and improving photosynthetic rate [44,59,60]. Sarvari et al. [34] held that an appropriate Cd: Fe ratio was conducive to reducing the damage of Cd to plants. This study results showed that the poplar growing under the condition of Cd/Fe ratio = 2:3 was better than the poplar growing under the condition of Cd/Fe ratio = 2:1.

4.3. Antioxidant Defense of Poplar under Cd and Fe Interaction

Cd can induce the accumulation of a large amount of ROS after entering cells, resulting in membrane lipid peroxidation and MDA accumulation [15,61]. In this study, Cd induced the accumulation of large amounts of MDA under both normal- and high-dose Fe. In addition, in the absence of Cd, high-dose Fe did not change poplar root and leaf MDA content. This indicates that 150 μM Fe is not excessive for poplar growth and could not cause oxidative damage. Interestingly, this study found that under Cd stress, the root and leaf MDA content in the high-dose Fe group was significantly lower than that of poplar in the normal-dose Fe group. This indicates that under Cd stress, high-dose Fe could alleviate the Cd-induced oxidative damage to a certain extent and promote the growth of poplar.

To cope with the Cd-induced oxidative stress, plant antioxidant enzymes are activated to scavenge ROS, but the enzymes' responses are different under different degrees of Cd stress [12]. In this study, Cd significantly suppressed root and leaf AAO and SOD activities under normal- and high-dose Fe. However, under normal-dose Fe, the Cd-stressed poplar root and leaf APX activity increased. This indicates that APX may play a more important role in mitigating Cd-induced oxidative stress under normal-dose Fe compared with other antioxidant enzymes. In the absence of Cd, high-dose Fe increased root APX and SOD activities and leaf AAO and APX activities, thus enhancing poplar's antioxidant capacity. However, under Cd stress, high-dose Fe decreased root and leaf AAO and APX activities but increased the activity of SOD in roots and leaves, which helped to scavenge ROS. A previous study reported that under abiotic stress, plants could maintain the ROS balance by a ROS-scavenging system that mainly includes three types of SODs (Cu/Zn-SOD, Mn-SOD, and Fe-SOD) [62]. Among them, Fe-SOD is an important component in chloroplasts, peroxisomes, and mitochondria [32]. Therefore, in this study, under Cd stress, high-dose Fe may have increased the activity of Fe-SOD to enhance the ROS scavenging ability.

5. Conclusions

In this study, Cd (100 μM) could suppress the photosynthesis and reduce the total root surface area of *P. tremula* × *P. alba* '717', thus suppressing the absorption of nutrients, CO_2 assimilation, and the growth of poplar. Under Cd stress, high-dose Fe (150 μM) increased the net photosynthetic rate, total root surface area, and nutrient absorption, thus alleviating the toxicity of Cd. That is, the Fe dose of 150 μM is not excessive for poplar growth but promotes biomass accumulation. In addition, the high-dose Fe increased the BAF and TF, thus breaking some restrictions and enhancing the transport of Cd to the aboveground part of poplar. In addition, under Cd stress, high-dose Fe enhanced the antioxidant capacity of poplar by increasing the activity of Fe-SOD. Therefore, it is necessary to further explore the effects of high-dose Fe (150 μM) on the activity of Fe-SOD under Cd stress. This study has a guiding significance for the use of poplar in the high-efficiency remediation of Cd-contaminated soil.

Author Contributions: S.D. and M.L. designed the experiments, analyzed the data, drafted the manuscript and other authors revised the draft. C.L., D.Z. and B.W. carried out the experiments. S.D. finalized the manuscript. All authors have read and agreed to the published version of the manuscript.

Funding: This study was jointly supported by the National Natural Science Foundation of China (Grant No. 32001347) and the Henan Provincial Science and Technology Research Project (Grant No. 212102110190).

Data Availability Statement: All data generated or analyzed during this study are included in this published article.

Conflicts of Interest: The authors declare no conflict of interest.

References

1. Akram, M.A.; Wahid, A.; Abrar, M.; Manan, A.; Naeem, S.; Zahid, M.A.; Gilani, M.M.; Paudyal, R.; Gong, H.Y.; Ran, J.Z.; et al. Comparative study of six maize (*Zea mays* L.) cultivars concerning cadmium uptake, partitioning and tolerance. *Appl. Ecol. Environ. Res.* **2021**, *19*, 2305–2331. [CrossRef]
2. Luo, Z.B.; He, J.; Polle, A.; Rennenberg, H. Heavy metal accumulation and signal transduction in herbaceous and woody plants: Paving the way for enhancing phytoremediation efficiency. *Biotechnol. Adv.* **2016**, *34*, 1131–1148. [CrossRef] [PubMed]
3. Barrutia, O.; Artetxe, U.; Hernandez, A.; Olano, J.M.; Garcia-Plazaola, J.I.; Garbisu, C.; Becerril, J.M. Native plant communities in an abandoned Pb-Zn mining area of northern Spain: Implications for phytoremediation and germplasm preservation. *Int. J. Phytoremediat.* **2011**, *13*, 256–270. [CrossRef]
4. Nicholson, F.A.; Smith, S.R.; Alloway, B.J.; Carlton-Smith, C.; Chambers, B.J. An inventory of heavy metals inputs to agricultural soils in England and Wales. *Sci. Total Environ.* **2003**, *311*, 205–219. [CrossRef] [PubMed]
5. Herawati, N.; Suzuki, S.; Hayashi, K.; Rivai, I.F.; Koyama, H. Cadmium, copper, and zinc levels in rice and soil of Japan, Indonesia, and China by soil type. *Bull. Environ. Contam. Toxicol.* **2000**, *64*, 33–39. [CrossRef] [PubMed]
6. Jarup, L.; Akesson, A. Current status of cadmium as an environmental health problem. *Toxicol. Appl. Pharmacol.* **2009**, *238*, 201–208. [CrossRef] [PubMed]
7. Douchiche, O.; Chaibi, W.; Morvan, C. Cadmium tolerance and accumulation characteristics of mature flax, cv. Hermes: Contribution of the basal stem compared to the root. *J. Hazard. Mater.* **2012**, *235–236*, 101–107. [CrossRef]
8. He, J.; Li, H.; Ma, C.; Zhang, Y.; Polle, A.; Rennenberg, H.; Cheng, X.; Luo, Z.B. Overexpression of bacterial gamma-glutamylcysteine synthetase mediates changes in cadmium influx, allocation and detoxification in poplar. *New Phytol.* **2015**, *205*, 240–254. [CrossRef]
9. Saxena, G.; Purchase, D.; Mulla, S.I.; Saratale, G.D.; Bharagava, R.N. Phytoremediation of heavy metal-contaminated sites: Eco-environmental concerns, field studies, sustainability issues, and future prospects. *Rev. Environ. Contam. Toxicol.* **2020**, *249*, 71–131.
10. Hu, Y.; Tian, S.; Foyer, C.H.; Hou, D.; Wang, H.; Zhou, W.; Liu, T.; Ge, J.; Lu, L.; Lin, X. Efficient phloem transport significantly remobilizes cadmium from old to young organs in a hyperaccumulator *Sedum alfredii*. *J. Hazard. Mater.* **2019**, *365*, 421–429. [CrossRef]
11. Balestri, M.; Ceccarini, A.; Forino, L.M.; Zelko, I.; Martinka, M.; Lux, A.; Ruffini Castiglione, M. Cadmium uptake, localization and stress-induced morphogenic response in the fern *Pteris vittata*. *Planta* **2014**, *239*, 1055–1064. [CrossRef] [PubMed]
12. Qiu, W.; Xu, T.; Li, X.; Zhang, Y.; Ren, R.; Heng, Q.; Chen, W.; Zhang, S.; Wang, M.; Kou, L.; et al. The influence of phosphorus on leaf function, cadmium accumulation and stress tolerance of poplar leaves under cadmium exposure. *Environ. Exp. Bot.* **2022**, *204*, 105087. [CrossRef]
13. Shi, W.; Liu, W.; Ma, C.; Zhang, Y.; Ding, S.; Yu, W.; Deng, S.; Zhou, J.; Li, H.; Luo, Z.B. Dissecting microRNAs-mRNAs regulatory networks underlying sulfur assimilation and cadmium accumulation in poplar leaves. *Plant Cell Physiol.* **2020**, *61*, 1614–1630. [CrossRef] [PubMed]
14. Ding, S.; Ma, C.; Shi, W.; Liu, W.; Lu, Y.; Liu, Q.; Luo, Z.B. Exogenous glutathione enhances cadmium accumulation and alleviates its toxicity in *Populus canescens*. *Tree Physiol.* **2017**, *37*, 1697–1712. [CrossRef]
15. Marmiroli, M.; Pietrini, F.; Maestri, E.; Zacchini, M.; Marmiroli, N.; Massacci, A. Growth, physiological and molecular traits in *Salicaceae* trees investigated for phytoremediation of heavy metals and organics. *Tree Physiol.* **2011**, *31*, 1319–1334. [CrossRef]
16. Luo, J.; Liang, Z.; Wu, M.; Mei, L. Genome-wide identification of BOR genes in poplar and their roles in response to various environmental stimuli. *Environ. Exp. Bot.* **2019**, *164*, 101–113. [CrossRef]
17. He, J.; Ma, C.; Ma, Y.; Li, H.; Kang, J.; Liu, T.; Polle, A.; Peng, C.; Luo, Z.B. Cadmium tolerance in six poplar species. *Environ. Sci. Pollut. R.* **2013**, *20*, 163–174. [CrossRef]
18. Migocka, M.; Kosieradzka, A.; Papierniak, A.; Maciaszczykdziubinska, E.; Posyniak, E.; Garbiec, A.; Filleur, S. Two metal-tolerance proteins, MTP1 and MTP4, are involved in Zn homeostasis and Cd sequestration in cucumber cells. *J. Exp. Bot.* **2015**, *66*, 581–596. [CrossRef]
19. Tian, S.; Liang, S.; Qiao, K.; Wang, F.; Zhang, Y.; Chai, T. Co-expression of multiple heavy metal transporters changes the translocation, accumulation, and potential oxidative stress of Cd and Zn in rice (*Oryza sativa*). *J. Hazard. Mater.* **2019**, *380*, 120853. [CrossRef]
20. She, W.; Cui, G.; Li, X.; Su, X.; Jie, Y.; Yang, R. Characterization of cadmium concentration and translocation among ramie cultivars as affected by zinc and iron deficiency. *Acta Physiol. Plant.* **2018**, *40*, 104. [CrossRef]
21. Guo, Z.; Lv, J.; Zhang, H.; Hu, C.; Qin, Y.; Dong, H.; Zhang, T.; Dong, X.; Du, N.; Piao, F. Red and blue light function antagonistically to regulate cadmium tolerance by modulating the photosynthesis, antioxidant defense system and Cd uptake in cucumber (*Cucumis sativus* L.). *J. Hazard. Mater.* **2022**, *429*, 128412. [CrossRef] [PubMed]

22. Tian, J.; Song, Y.; Du, Q.; Yang, X.; Dong, C.; Chen, J.; Xie, J.; Li, B.; Zhang, D. Population genomic analysis of gibberellin-responsive long non-coding RNAs in *Populus*. *J. Exp. Bot.* **2016**, *67*, 2467–2482. [CrossRef] [PubMed]
23. Zhou, M.; Zhi, Y.; Dai, Y.; Lv, J.; Li, Y.; Wu, Z. The detoxification mechanisms of low-accumulating and non-low-accumulating medicinal plants under Cd and Pb stress. *RSC Adv.* **2020**, *10*, 43882–43893. [CrossRef]
24. Chen, Y.; Nguyen, T.H.N.; Qin, J.; Jiao, Y.; Li, Z.; Ding, S.; Lu, Y.; Liu, Q.; Luo, Z.B. Phosphorus assimilation of *Chinese fir* from two provenances during acclimation to changing phosphorus availability. *Environ. Exp. Bot.* **2018**, *153*, 21–34. [CrossRef]
25. Basa, B.; Lattanzio, G.; Solti, Á.; Tóth, B.; Abadía, J.; Fodor, F.; Sárvári, É. Changes induced by cadmium stress and iron deficiency in the composition and organization of thylakoid complexes in sugar beet (*Beta vulgaris* L.). *Environ. Exp. Bot.* **2014**, *101*, 1–11. [CrossRef]
26. Hu, X.; Page, M.T.; Sumida, A.; Tanaka, A.; Terry, M.J.; Tanaka, R. The iron-sulfur cluster biosynthesis protein SUFB is required for chlorophyll synthesis, but not phytochrome signaling. *Plant J.* **2017**, *89*, 1184–1194. [CrossRef]
27. Solti, Á.; Sárvári, É.; Tóth, B.; Mészáros, I.; Fodor, F. Incorporation of iron into chloroplasts triggers the restoration of cadmium induced inhibition of photosynthesis. *J. Plant Physiol.* **2016**, *202*, 97–106. [CrossRef]
28. Muneer, S.; Hakeem, K.R.; Mohamed, R.; Lee, J.H. Cadmium toxicity induced alterations in the root proteome of green gram in contrasting response towards iron supplement. *Int. J. Mol. Sci.* **2014**, *15*, 6343–6355. [CrossRef]
29. Solti, A.; Sarvari, E.; Szollosi, E.; Toth, B.; Meszaros, I.; Fodor, F.; Szigeti, Z. Stress hardening under long-term cadmium treatment is correlated with the activation of antioxidative defence and iron acquisition of chloroplasts in *Populus*. *Z. Nat. C* **2016**, *71*, 323–334. [CrossRef]
30. Nakanishi, H.; Ogawa, I.; Ishimaru, Y.; Mori, S.; Nishizawa, N.K. Iron deficiency enhances cadmium uptake and translocation mediated by the Fe^{2+} transporters OsIRT1 and OsIRT2 in rice. *Soil Sci. Plant Nutr.* **2006**, *52*, 464–469. [CrossRef]
31. Su, Y.; Liu, J.; Lu, Z.; Wang, X.; Zhang, Z.; Shi, G. Effects of iron deficiency on subcellular distribution and chemical forms of cadmium in peanut roots in relation to its translocation. *Environ. Exp. Bot.* **2014**, *97*, 40–48. [CrossRef]
32. He, X.L.; Fan, S.K.; Zhu, J.; Guan, M.Y.; Liu, X.X.; Zhang, Y.S.; Jin, C.W. Iron supply prevents Cd uptake in *Arabidopsis* by inhibiting IRT1 expression and favoring competition between Fe and Cd uptake. *Plant Soil.* **2017**, *416*, 453–462. [CrossRef]
33. Molina-Rueda, J.J.; Tsai, C.J.; Kirby, E.G. The *Populus* superoxide dismutase gene family and its responses to drought stress in transgenic poplar overexpressing a pine cytosolic glutamine synthetase (GS1a). *PLoS ONE* **2013**, *8*, e56421. [CrossRef] [PubMed]
34. Zhang, D.; Zhou, H.; Shao, L.; Wang, H.; Zhang, Y.; Zhu, T.; Ma, L.; Ding, Q.; Ma, L. Root characteristics critical for cadmium tolerance and reduced accumulation in wheat (*Triticum aestivum* L.). *J. Environ. Manag.* **2022**, *305*, 114365. [CrossRef]
35. Astolfi, S.; Ortolani, M.R.; Catarcione, G.; Paolacci, A.R.; Cesco, S.; Pinton, R.; Ciaffi, M. Cadmium exposure affects iron acquisition in barley (*Hordeum vulgare*) seedlings. *Physiol. Plantarum.* **2014**, *152*, 646–659. [CrossRef]
36. Sarvari, E.; Solti, A.; Basa, B.; Meszaros, I.; Levai, L.; Fodor, F. Impact of moderate Fe excess under Cd stress on the photosynthetic performance of poplar (*Populus jacquemontiana* var. glauca cv. Kopeczkii). *Plant Physiol. Biochem.* **2011**, *49*, 499–505. [CrossRef]
37. Luo, J.; Zhou, J.J. Growth performance, photosynthesis, and root characteristics are associated with nitrogen use efficiency in six poplar species. *Environ. Exp. Bot.* **2019**, *164*, 40–51. [CrossRef]
38. Lu, Y.; Ma, J.; Teng, Y.; He, J.; Christie, P.; Zhu, L.; Ren, W.; Zhang, M.; Deng, S. Effect of silicon on growth, physiology, and cadmium translocation of tobacco (*Nicotiana tabacum* L.) in cadmium-contaminated soil. *Pedosphere* **2018**, *28*, 680–689. [CrossRef]
39. Luo, J.; Zhou, J.J.; Masclaux-Daubresse, C.; Wang, N.; Wang, H.; Zheng, B. Morphological and physiological responses to contrasting nitrogen regimes in *Populus cathayana* is linked to resources allocation and carbon/nitrogen partition. *Environ. Exp. Bot.* **2019**, *162*, 247–255. [CrossRef]
40. Tamás, L.; Bočová, B.; Huttová, J.; Mistrík, I.; Ollé, M. Cadmium-Induced Inhibition of Apoplastic Ascorbate Oxidase in Barley Roots. *Plant Growth Regul.* **2006**, *48*, 41–49. [CrossRef]
41. He, J.; Qin, J.; Long, L.; Ma, Y.; Li, H.; Li, K.; Jiang, X.; Liu, T.; Polle, A.; Liang, Z.; et al. Net cadmium flux and accumulation reveal tissue-specific oxidative stress and detoxification in *Populus canescens*. *Physiol. Plantarum.* **2011**, *143*, 50–63. [CrossRef] [PubMed]
42. Gong, B.; Miao, L.; Kong, W.; Bai, J.G.; Wang, X.; Wei, M.; Shi, Q. Nitric oxide, as a downstream signal, plays vital role in auxin induced cucumber tolerance to sodic alkaline stress. *Plant Physiol. Biochem.* **2014**, *83*, 258–266. [CrossRef] [PubMed]
43. Sperotto, R.A.; Ricachenevsky, F.K.; Stein, R.J.; Waldow, V.; Fett, J.P. Iron stress in plants: Dealing with deprivation and overload. *Plant Stress.* **2010**, *4*, 57–69.
44. Rizwan, M.; Ali, S.; Adrees, M.; Rizvi, H.; Zia-Ur-Rehman, M.; Hannan, F.; Qayyum, M.F.; Hafeez, F.; Ok, Y.S. Cadmium stress in rice: Toxic effects, tolerance mechanisms, and management: A critical review. *Environ. Sci. Pollut. R.* **2016**, *23*, 17859–17879. [CrossRef]
45. Kawa, D.; Julkowska, M.; Sommerfeld, H.M.; Horst, A.T.; Haring, M.A.; Testerink, C. Phosphate-dependent root system architecture responses to salt stress. *Plant Physiol.* **2016**, *172*, 690–706. [CrossRef]
46. Pâques, L.E.; Lejeune, V.; Veisse, D. Do biomass partitioning and growth efficiency contribute to growth heterosis in inter-specific hybrid larch *Larix eurolepis*? *Forestry* **2022**, *95*, 466–476. [CrossRef]
47. Zhang, Y.; Kaiser, E.; Li, T.; Marcelis, L.F.M. NaCl affects photosynthetic and stomatal dynamics by osmotic effects and reduces photosynthetic capacity by ionic effects in tomato. *J. Exp. Bot.* **2022**, *73*, 3637–3650. [CrossRef]
48. Qureshi, M.I.; D'Amici, G.M.; Fagioni, M.; Rinalducci, S.; Zolla, L. Iron stabilizes thylakoid protein–pigment complexes in Indian mustard during Cd-phytoremediation as revealed by BN-SDS-PAGE and ESI-MS/MS. *J. Plant Physiol.* **2010**, *167*, 761–770. [CrossRef]

49. Thomine, S.; Vert, G. Iron transport in plants: Better be safe than sorry. *Curr. Opin. Plant Biol.* **2013**, *16*, 322–327. [CrossRef]
50. Luo, J.S.; Huang, J.; Zeng, D.L.; Peng, J.S.; Zhang, G.B.; Ma, H.L.; Guan, Y.; Yi, H.Y.; Fu, Y.L.; Han, B.; et al. A defensin-like protein drives cadmium efflux and allocation in rice. *Nat. Commun.* **2018**, *9*, 645–654. [CrossRef]
51. Shi, W.G.; Liu, W.; Yu, W.; Zhang, Y.; Ding, S.; Li, H.; Mrak, T.; Kraigher, H.; Luo, Z.B. Abscisic acid enhances lead translocation from the roots to the leaves and alleviates its toxicity in *Populus canescens*. *J. Hazard. Mater.* **2019**, *362*, 275–285. [CrossRef] [PubMed]
52. Chen, S.; Han, X.; Fang, J.; Lu, Z.; Qiu, W.; Liu, M.; Sang, J.; Jiang, J.; Zhuo, R. *Sedum alfredii* SaNramp6 metal transporter contributes to cadmium accumulation in transgenic *Arabidopsis thaliana*. *Sci. Rep.* **2017**, *7*, 13318. [CrossRef]
53. Ling, T.; Gao, Q.; Du, H.; Zhao, Q.; Ren, J. Growing, physiological responsesand cd uptake of corn (*Zea mays* L.) under different Cd supply. *Chem. Speciat. Bioavailab.* **2017**, *29*, 216–221. [CrossRef]
54. Ali, N.; Hadi, F. Phytoremediation of cadmium improved with the high production of endogenous phenolics and free proline contents in *Parthenium hysterophorus* plant treated exogenously with plant growth regulator and chelating agent. *Environ. Sci. Pollut. R.* **2015**, *22*, 13305–13318. [CrossRef] [PubMed]
55. Baldantoni, D.; Cicatelli, A.; Bellino, A.; Castiglione, S. Different behaviours in phytoremediation capacity of two heavy metal tolerant poplar clones in relation to iron and other trace elements. *J. Environ. Manag.* **2014**, *146*, 94–99. [CrossRef]
56. Capuana, M. Heavy metals and woody plants—Biotechnologies for phytoremediation. *iForest* **2011**, *4*, 7–15. [CrossRef]
57. Giehl, R.F.H.; Lima, J.E.; Nicolaus, V.W. Localized iron supply triggers lateral root elongation in *Arabidopsis* by altering the AUX1-mediated auxin distribution. *Plant Cell* **2012**, *24*, 33–49. [CrossRef]
58. Rui, H.; Chen, C.; Zhang, X.; Shen, Z.; Zhang, F. Cd-induced oxidative stress and lignification in the roots of two *Vicia sativa* L. varieties with different Cd tolerances. *J. Hazard. Mater.* **2016**, *301*, 304–313. [CrossRef]
59. Brier, N.D.; Gomand, S.V.; Donner, E.; Paterson, D.; Smolders, E.; Delcour, J.A.; Lombi, E. Element distribution and iron speciation in mature wheat grains (*Triticum aestivum* L.) using synchrotron X-ray fluorescence microscopy mapping and XANES imaging. *Plant Cell Environ.* **2016**, *39*, 1835–1847. [CrossRef]
60. Fodor, F.; Gáspár, L.; Morales, F.; Gogorcena, Y.; Lucena, J.J.; Cseh, E.; Kröpfl, K.; Abadía, J.; Sárvári, É. Effects of two iron sources on iron and cadmium allocation in poplar (*Populus alba*) plants exposed to cadmium. *Tree Physiol.* **2005**, *25*, 1173–1180. [CrossRef]
61. Wang, B.; Zhang, D.; Wang, W.; Song, Y.; Lu, M.; Ding, S. Foliar Application of Selenium Reduces Cadmium Accumulation in Walnut Seedlings. *Forests* **2022**, *13*, 1493. [CrossRef]
62. Wang, W.; Zhang, X.; Deng, F.; Yuan, R.; Shen, F. Genome-wide characterization and expression analyses of superoxide dismutase (SOD) genes in *Gossypium hirsutum*. *BMC Genom.* **2017**, *18*, 376. [CrossRef] [PubMed]

forests

MDPI

Article

Complete Genome Expression Analysis of the Auxin Response Factor Gene Family in Sandalwood and Their Potential Roles in Drought Stress

Xiaojing Liu [1,2,†], Yunshan Liu [1,2,†], Shengkun Wang [1], Fangcuo Qin [1], Dongli Wang [1], Yu Chen [2], Lipan Hu [2], Sen Meng [1] and Junkun Lu [1,*]

1 State Key Laboratory of Tree Genetics and Breeding, Research Institute of Tropical Forestry, Chinese Academy of Forestry, Guangzhou 510520, China
2 College of Biology and Food Engineering, Chongqing Three Gorges University, Wanzhou, Chongqing 404100, China
* Correspondence: junkunlu@caf.ac.cn
† These authors contributed equally to this work.

Abstract: Auxin response factors (ARFs) are essential transcription factors in plants that play an irreplaceable role in controlling the expression of auxin response genes and participating in plant growth and stress. The *ARF* gene family has been found in *Arabidopsis thaliana*, apple (*Malus domestica*), poplar (*Populus trichocarpa*) and other plants with known whole genomes. However, *S. album* (*Santalum album* L.), has not been studied. In this study, we analyzed and screened the whole genome of *S. album* and obtained 18 *S. album ARFs* (*SaARFs*), which were distributed on eight chromosomes. Through the prediction of conserved domains, we found that 13 of the 18 SaARFs had three intact conserved domains, named DBD, MR, Phox and Bem1 (PB1), while the extra five SaARFs (SaARF3, SaARF10, SaARF12, SaARF15, SaARF17) had only two conserved domains, and the C-terminal PB1 domain was missing. By establishing a phylogenetic tree, 62 *ARF* genes in *S. album*, poplar and *Arabidopsis* were divided into four subgroups, named I, II, III and IV. According to the results of collinearity analysis, we found that ten of the eighteen *ARF* genes were involved in five segmental duplication events and these genes had short distance intervals and high homology in the *SaARF* gene family. Finally, tissue-specific and drought-treatment expression of *SaARF* genes was observed by quantitative real-time polymerase chain reaction (qRT-PCR), and six genes were significantly overexpressed in haustorium. Meanwhile we found *SaARF5*, *SaARF10*, and *SaARF16* were significantly overexpressed under drought stress. These results provide a basis for further analysis of the related functions of the *S. album ARF* gene and its relationship with haustorium formation.

Keywords: *Santalum album* L.; auxin response factors; auxin; qRT-PCR; drought

Citation: Liu, X.; Liu, Y.; Wang, S.; Qin, F.; Wang, D.; Chen, Y.; Hu, L.; Meng, S.; Lu, J. Complete Genome Expression Analysis of the Auxin Response Factor Gene Family in Sandalwood and Their Potential Roles in Drought Stress. *Forests* **2022**, *13*, 1934. https://doi.org/10.3390/f13111934

Academic Editors: Jie Luo and Wentao Hu

Received: 23 October 2022
Accepted: 14 November 2022
Published: 16 November 2022

Publisher's Note: MDPI stays neutral with regard to jurisdictional claims in published maps and institutional affiliations.

1. Introduction

Santalum album L., a semiparasitic species of the genus Sandalwood in the family Sandalwood, is distributed mainly in India, Indonesia, and Australia [1,2]. In contrast to holoparasitic plants, hemiparasitic plants not only have the ability to perform photosynthesis, but also can grow autonomously, in some cases where no host plants exist [3]. Sandalwood is a rare tree species; its oil and hardwood have significant commercial value, used in perfumes, cosmetics, medicine, and aromatherapy, and recently in the prevention of skin cancer, but this has also led to a sharp decline in wild sandalwood populations [4–6]. In nature, there are more than 300 plants that can serve as sandalwood hosts and parasitic angiosperms depend on host root-derived chemical signals to control various stages of development [7,8]. The haustorium is a unique organ of parasitic plants; these parasite-specific structures penetrate host roots and connect the host and parasite xylem vessels [9]. Futhermore, their main function is to absorb water and nutrients from the host plant, especially during early phases of development [3]. However, we also found that sandalwood sometimes grow haustoria even in the absence of a host plant.

Auxin, also known as indole-3-acetic acid, is widely used as the earliest hormone found to promote plant growth and can promote the growth of plant stems, shoots and roots [10–13]. At present, several genes related to auxin transduction have been found, including *Aux/IAA* (auxin/indole-3-acetic acid), *GH3* (gretchen hagen 3), *SAUR* (small auxin-up RNA), and *ARF* (auxin response factor) [12]. *ARF*, a transcription factor that regulates the expression of auxin response genes under the influence of auxin concentration, can combine with AuxREs, the auxin response element in the promoter of target genes, to promote or inhibit the expression of auxin response genes [14]. ARF can also interact with Aux/IAA proteins, which also have auxin signal transduction functions. Auxin sensing begins with auxin binding to the TIR1 (TRANSPORT INHIBITOR RESPONSE1)/AFB (AUXIN SIGNALING F-box) receptor and leads to the subsequent degradation of Aux/IAA proteins that inhibit auxin signaling through physical interaction with ARF proteins by the ubiquitin family [15–17].

The ARF protein has three stable conserved domains, namely, the DBD, MR and Phox and Bem1 (PB1) domains [18]. The function of the DBD located at the N-terminus is to bind to AuxREs (TGTCTC cis-responsive elements), the promoters of auxin-responsive genes, to control the expression of target genes [19]. The middle MR domain is divided into an AD activation domain rich in glutamine (AtARF5, AtARF6, AtARF7, AtARF8, AtARF19) and an RD inhibition domain rich in serine and threonine (AtARF1-4, AtARF9-18, AtARF20-23) [20]. The ARF-MRs, which belong to the RD inhibitory domain, are divided into class B and C. Some MRs in class B with inhibitory domains have been found to also contain an amphiphilic inhibitory motif related to the ethylene response element binding factor the EAR motif [21], which can interact with the N-terminal part of the Aux/IAA protein. The MR domain is mainly responsible for controlling the function of ARFs. The PB1 domain located at the C-terminal is homologous to the C-terminal domain of the Aux/IAA protein, which can combine to form homologs or heterodimers [22], and its main function is to mediate protein-protein interactions, for example, between an ARF protein and Aux/IAA protein or between an ARF protein and another ARF protein. Auxin affects the activity of the ARF protein by controlling the Aux/IAA protein and then controls the expression of auxin response genes. At low auxin concentrations, the C-terminal domain of the Aux/IAA protein and the C-terminal domain of ARF protein were inhibited, and the N-terminal EAR domain of the Aux/IAA protein recruited the corepressor TOPLESS (TPL) [23], which affected the activity of the ARF protein and reduced the expression of auxin responsive genes. At high auxin concentrations, the Aux/IAA protein and a small amount of auxin are promoted to bind to the $SCF^{TIR1/AFB}$ corepressor complex, which is then decomposed by the ubiquitin family, thereby releasing more ARF protein to bind to auxin responsive genes and promote their expression [24].

As a large gene family, the *ARF* gene has been successively studied in *Arabidopsis*, maize, tomato, poplar and other plants. *Arabidopsis* contains 23 *ARF* genes [25], while rice (*Oryza sativa*), apple (*Malus*), poplar (*Populus trichocarpa*) and tomato (*Lycopersicon esculentum*) contain 25, 31, 39 and 21 *ARF* genes [26–29], respectively. However, the biological function of the *SaARF* gene family in response to auxin in sandalwood remains unclear. The related functions of *ARF* genes are generally analyzed in model plants such as *Arabidopsis thaliana* and rice [30]. For example, *AtARF2* is related to the synthesis of flavonoids and anthocyanins [31], and *AtARF7* and *AtARF19* are specifically distributed in roots and related to lateral root elongation [32]. *SlARF9* is one of the auxin-related genes differentially expressed in tomato fruit set and early fruit development [33]. The Aux/IAA9-ARF5 module regulates wood formation by coordinating the expression of HD-ZIP III transcription factors in poplar [34]. The miR167-*GmARF8* module plays a key role in the auxin-mediated nodule and lateral root formation in soybean [35]. The protein encoded by *ARF8* affects hypocotyl elongation and root behavior in *Arabidopsis* [36]. These results indicate that *ARF* is closely related to growth and development, stress response and other physiological processes in plants.

In previous studies, *BpARF1* has been shown to be associated with drought stress. Under drought stress conditions, *BpARF1* RNA interference (RNAi)-inhibited plants presented reduced reactive oxygen species (ROS) accumulation, and enhanced peroxide (POD) and superoxide dismutase (SOD) activities. On the contrary, the overexpression of *BpARF1* showed a completely opposite phenomenon [37]. However, there are few studies on the function of abiotic stress on the ARF protein in sandalwood. Therefor, in this study, we used quantitative qRT-PCR to analyze the expression profiles of *SaARF* genes under drought. At the same time, we found that the content of auxin is extremely high in the early stage of haustorium growth and plays an irreplaceable role in the formation of haustorium [2,9]. The haustorium development in *S. album* was promoted by IAA and inhibited by the auxin biosynthesis inhibitor L-Kyn (L-kynureninean, an auxin biosynthesis inhibitor) and the polar auxin transport inhibitor NPA, indicating the haustorium development in *S. album* was enhanced by auxin synthesized in the root and/or transported from shoots [9]. To systematically research the related functions and physical and chemical properties of the *S. album ARF* gene family, we conducted an analysis of the *S. album ARF* gene family through the whole genome of *S. album* and screened 18 *SaARF* genes in this study. Detailed studies on the physicochemical properties, evolutionary relationships, gene structure, chromosomal location, collinear analysis, cis-reactive elements and tissue-specific expression of *S. album ARF*s will help to better understand the auxin response mechanism in sandalwood and provide a reference for the biological functions of *ARF* genes in response to drought stress in sandalwood.

2. Materials and Methods

2.1. Plant Materials and Treatments

S. album plants were grown at 25 °C in a greenhouse with a 16/8 h light-dark cycle. Four-month-old *S. album* plants were taken, and four different tissues of the plants, including mature leaves, steam, roots and haustoria collected. Twelve plants were exposed water-limited treatment, ranging from 3 to 9 d of drought. Three biological replicates were performed. Finally, they were stored at −80 °C until RNA extraction.

2.2. Complete Genome Identification and Sequence Analysis of the ARF Gene Family in S. album

PFAM (http://pfam.xfam.org/, accessed on 12 July 2022) was used to download the protein hidden Markov model Auxin_resp.hmm (protein family: PF06507). The hmmsearch program of HMMER3.0 software was used to compare and screen the whole genome of *S. album* proteins identified from the Research Institute of Tropical Forestry, Chinese Academy of Forestry, and protein hits with an e-value of $<10^{-5}$ and sequence score of "best 1 domain" >100 were collected [38]. We used CD Search NCBI (https://www.ncbi.nlm.nih.gov/Structure/cdd/wrpsb.cgi, accessed on 18 July 2022) to determine whether the *ARF* gene structure type and quantity of fields were screened and performed a secondary screening of the *SaARF* gene families. Protparam (https://web.expasy.org/protparam/, accessed on 12 August 2022) prediction has been used to screen protein sequences related to physiological and biochemical indicators [39], including the quantity of amino acids (aa), isoelectric point (PI), molecular weight (MW), total average hydrophobicity, instability and aliphatic index, and Wolf PSORT organelles (https://www.genscript.com/wolf-psort.html, accessed on 22 August 2022), which were used to predict the *SaARF* gene positioning. We used MEME (https://meme-suite.org/meme/tools/meme, accessed on 16 July 2022) to predict *SaARF* gene family members of the conservative base sequence and set the length between 6–50 aa and the base sequence number to 15 [40].

2.3. Establishment of Phylogenetic Tree

The MUSCLE method of MEGA-X was used to compare the multiple sequences of 62 *ARF* genes in *S. album*, poplar and *Arabidopsis thaliana*. The phylogenetic tree was constructed using the Substitutions Type: Amino acid. Model: Poisson model; Rates among Sites: Uniform Rates; The Pattern among Lineages: Same (Homogeneous). After the evolutionary tree was built by iTOL, we undertook the beautification program (https://itol.embl.de/itol.cgi, accessed on 10 September 2022) [41].

2.4. Chromosome Mapping and Gene Structure Analysis

The chromosome length information of *S. album* and the location information of the *SaARF* gene were obtained from the whole genome annotation file of *S. album* determined by the Institute of Tropical Forestry, Chinese Academy of Forestry, and the chromosome location map was drawn with Map Draw software [42]. GSDS2.0 (http://gsds.gao-lab.org/index.php, accessed on 16 August 2022) was used to map the genetic structure of the *SaARF* gene family, including the number of CDS, introns, and UTR and their relative positions on the genes.

2.5. Collinearity Analysis within and between Species

In this paper, in order to more intuitively observe and analyze the evolution and genetic relationship of the *ARF* gene in different species, we not only analyzed the tandem replication and fragment replication events of the *ARF* genes in *S. album* but also compared the collinearity analysis of *ARF* among three species of *S. album*, poplar and *Arabidopsis*. The chromosome positions of *S. album* were extracted from the whole genome annotation file of *S. album*, which was provided by the Institute of Tropical Forestry, Chinese Academy of Forestry. *Arabidopsis thaliana* and poplar whole-genome files and note documents are derived from NCBI (https://www.ncbi.nlm.nih.gov/, accessed on 18 July 2022). The protein sequences of the three species were integrated, all proteins were searched by a basic local alignment search toolP (BLASTP) local search, and then tandem repeats and chromosome fragment repeats were obtained according to the results of MCScanX software [43]. Duplicate records containing the *ARF* gene were extracted by custom scripts, and finally, Dual Synteny Plotter and Advanced in TBtools (https://github.com/CJ-Chen/TBtools, accessed on 22 August 2022) were used. The Circos program was used to conduct a collinearity analysis among the three species and within the *S. album* species [44].

2.6. Analysis of Cis-Reactive Elements in the Promoter

The sandalwood gene structure annotation file and whole genome file were analyzed by TBtools, and 2000 bp upstream CDS of 18 *SaARF* genes were extracted. Submitting the sequence to the PlantCARE website (http://bioinformatics.psb.ugent.be/webtools/plantcare/html/, accessed on 22 October 2022), the promoter of cis reaction components was used to forecast and analyze and screen the required cis-reaction components [45]. The Simple BioSequence Viewer program of TBtools software was used to derive the distribution map of cis-acting elements in the promoter.

2.7. Expression Profiles of SaARF Genes in Different Plant Tissues and Drought-Treatment

An Omega kit was used to extract the total RNA from *S. album* haustoria, leaves, roots, stems and the leaves, which had been under drought conditions for 0 d, 3 d, 9 d. The quality and quantity of DNA-free total RNA was assessed using a NanoDrop ND-1000 spectrophotometer (Nanodrop Technologies, Wilmington, NC, USA). RNA samples with an A260/A280 ratio between 1.8 and 2.2 and an A260/A230 ratio greater than 2.0 were used for the subsequent analysis. According to the Takara RR036A PCR kit (Takara, San Jose, CA, USA), approximately 1–2 μL of RNA, RNase-free delH_2O, and 5X PrimeScript Rt Master were centrifuged and reverse-transcribed into cDNA in a T100^{TM} Thermal Cycler (Bio-Rad, Hercules, CA, USA). The machine program was set at 37 °C for 15 min, 85 °C for 5 s, and stored at 4 °C. The successfully synthesized cDNA was diluted with RNase-free delH_2O at a ratio of 1:10 and stored at −20 °C until later use.

Finally, using cDNA as a template, real-time PCR was performed according to SYBR qPCR Master Mix (Universal) (TOLOBIO, Shanghai, China) instructions. The 18 upstream primers and downstream primers of the *SaARF* gene were designed using Primer 3.0 (https://primer3.ut.ee/, accessed on 15 August 2022) (Table S3). The product size ranged from 100 bp to 150 bp, and the designed reaction system was as follows: 2×Q3 SYBR qPCR Master Mix 10 μL, upstream and downstream primers 0.5 μL each, cDNA template 1 μL, RNase-free delH_2O 8 μL. qRT-PCR was performed on a real-time PCR system based on the SYBR Green II method. Reaction procedure: Preincubation 95 °C for 900 s; Amplification 40 cycles of 95 °C for 10 s, 60 °C for 10 s and 72 °C for 20 s; Melting 95 °C for 10 s, 65 °C for 60 s and 97 °C for 1 s. Each experiment was performed with 3 biological replicates and 3 technical replicates, and the *S. album* housekeeping gene Actin was used as a reference. Relative gene expression was calculated using the $2^{-\Delta\Delta Ct}$ method [46]. Finally, SPSS Statistics 27 software was used for gene significance analysis.

3. Results

3.1. Identification and Sequence Analysis of the SaARF Gene Family

The known hidden Markov model of the *ARF* gene (Auxin_resp.hmm) was compared with the whole genome protein sequence of *S. album* determined by the Guangzhou Institute of Tropical Forestry by hmmsearch, and protein hits with an E-value of <10^{-5} and sequence score of >100 were collected [38]. CD Search was used to predict the conserved domains of the selected *SaARF* candidate genes, and a total of 18 *SaARF* genes were finally obtained according to the comparison of conserved domains. According to its evolutionary relationship with the *ARF* gene in *Arabidopsis thaliana*, we named them SaARF1-SaARF18.

3.2. Gene Information, Conserved Domains and Conserved Motifs of the SaARF Gene Family

The CDS length of *SaARFs* is 1771 bp-3322 bp, the encoded protein generally contains 590–1111 amino acids, and the molecular weight ranges from 65.5 kDa to 122.9 kDa. The aliphatic index of SaARF was between 63.77 and 78.33, and the total average hydrophobicity of all *SaARF* genes was negative, indicating that all *SaARF* genes were encoded hydrophilic proteins. The protein instability index of SaARF was between 40.67 and 70.11, indicating that all *SaARF* genes encoded unstable proteins. Wolf PSORT was used to predict the subcellular localization of SaARF, and all SaARFs were found to be located in the nucleus (Table 1). MEME (https://meme-suite.org/meme/tools/meme, accessed on 18 July 2022) was used to analyze the protein sequence of the *SaARF* gene family, the number of motifs was determined to be 15, and the conserved motifs were analyzed (Figure 1).

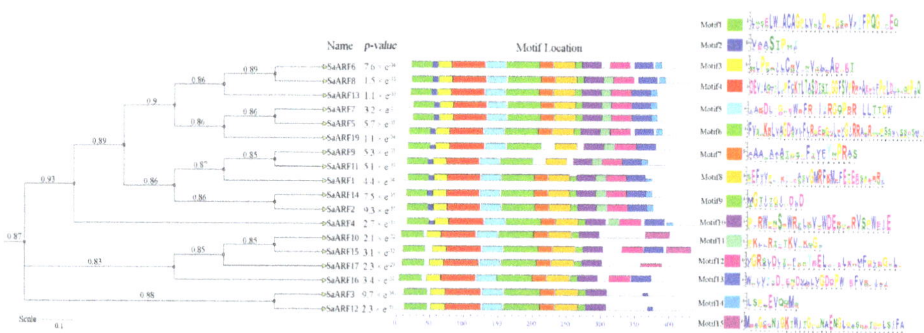

Figure 1. Conserved motif analysis and phylogenetic tree of gene families. Analysis diagram of conserved motifs of the *ARF* gene family in sandalwood. Phylogenetic tree and distribution map of conserved motifs of the *SaARF* gene family. The conserved motifs of the *SaARF* gene family were analyzed by MEME, and the length of the conserved motifs was set to 6–50 aa. The number was 15, and they were distinguished by different colors. Multiple sequence alignment of *S. album* ARF proteins was performed using MUCLE, and adjacency (NJ) trees were constructed using MEGA X to combine conserved motif analysis with evolutionary trees of the *SaARF* gene family.

Table 1. Characteristics of putative genes encoding auxin response factors in *Santalum album* L.

Gene Name	Name	CDS Length (bp) [1]	Amino Acids (aa) [2]	Molecular Weight (kDa) [3]	PI [4]	Conserved Domains	In Silico Prediction Wolf PSORT	Instability Index	Aliphatic Index	Grand Average of Hydropathicity
Sal9G26000.1	SaARF1	1993	668	74.5	6.02	DBD, MR, PB1	nucl	60.11	70.27	−0.488
Sal4G08680.1	SaARF2	2506	839	93.3	6.3	DBD, MR, PB1	nucl	55.42	63.77	−0.661
Sal6G09460.1	SaARF3	1871	626	68.9	6.45	DBD, MR	nucl	59.42	75.5	−0.34
Sal7G11880.1	SaARF4	2403	804	88.9	5.75	DBD, MR, PB1	nucl	55.05	75.12	−0.419
Sal8G13630.1	SaARF5	2413	808	89.1	5.38	DBD, MR, PB1	nucl	56.65	73.82	−0.375
Sal8G00600.1	SaARF6	2707	906	100.4	6.29	DBD, MR, PB1	nucl	70.11	70.92	−0.493
Sal1G06860.1	SaARF7	2539	850	93.9	5.58	DBD, MR, PB1	nucl	54	74.44	−0.4
Sal6G12670.1	SaARF8	2557	856	94.9	5.74	DBD, MR, PB1	nucl	66.19	72.91	−0.396
Sal4G03070.1	SaARF9	1873	628	71.1	6.33	DBD, MR, PB1	nucl	50.12	77.07	−0.488
Sal7G11650.1	SaARF10	2123	708	77.70	5.91	DBD, MR	nucl	40.67	78.33	−0.241
Sal9G20630.1	SaARF11	2067	693	76.8	5.85	DBD, MR, PB1	nucl	51.42	74.85	−0.423
Sal5G03150.1	SaARF12	2069	693	76.1	5.78	DBD, MR	nucl	52.15	69.96	−0.397
Sal5G11390.1	SaARF13	2650	887	97.90	6.13	DBD, MR, PB1	nucl	67.49	75.56	−0.398
Sal9G19410.1	SaARF14	2521	844	93.6	6.27	DBD, MR, PB1	nucl	56.52	66.5	−0.611
Sal9G26810.1	SaARF15	2111	704	77.2	7.26	DBD, MR	nucl	49.51	73.39	−0.348
Sal2G02230.1	SaARF16	1983	661	73.4	6.35	DBD, MR, PB1	nucl	55.49	72.45	−0.349
Sal6G20930.1	SaARF17	1771	590	65.5	6.59	DBD, MR	nucl	47.07	73.02	−0.447
Sal6G06080.1	SaARF18	3322	1111	122.9	6.02	DBD, MR, PB1	nucl	66.18	72.49	−0.576

[1] CDS Length; [2] Length of the amino acid sequence; [3] Molecular weight of the amino acid sequence; [4] Isoelectric point of the SaARF.

Multisequence comparison of the *SaARF* gene family was performed by DNAMAN software. CD Search in the National Center for Biotechnology Information (NCBI) search domain (https://www.ncbi.nlm.nih.gov/Structure/cdd/wrpsb.cgi, accessed on 18 July 2022) has been the conservative structure prediction with its conservative structure domain. After CD Search was used to predict the conserved domains of 18 SaARFs, SaARF3, SaARF10, SaARF12, SaARF15 and SaARF17 were found to contain only two conserved domains, while the C-terminal PB1 domain was missing. The remaining 13 SaARFs all contained three conserved domains. They are DBD, MR and PB1, respectively (Figure 2A,B). Gene structure analysis, including CDS, UTR, and intron, was performed by GSDS2.0 (http://gsds.gao-lab.org/index.php, accessed on 16 August 2022) (Figure 3).

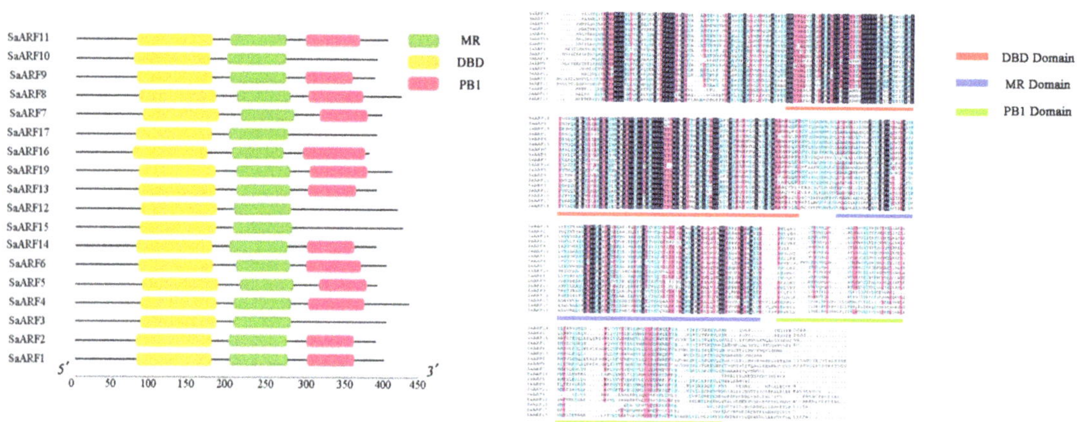

(A) (B)

Figure 2. Multiple sequence comparison and conserved domain analysis of the *SaARF* gene family. (**A**) The conserved domains of the *S. album* ARFs protein were analyzed by TBtools, where the DBD domain is shown in yellow, the MR domain is shown in green, and the PB1 domain is shown in red. (**B**) DNAMAN was used to compare multiple sequences of *S. album* ARF protein and annotate the conserved domains. The DBD domain is shown in brick red, the MR domain is shown in blue purple, and the PB1 domain is shown in green.

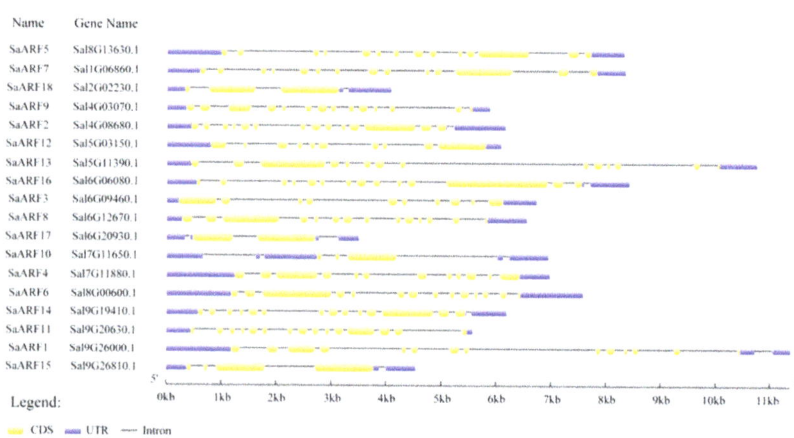

Figure 3. Genetic structure of the *ARF* gene in *S. album*. The value includes CDS, UTR, and intron. The yellow value is the CDS coding area, and the blue value is the UTR coding area.

3.3. Phylogenetic Analysis of ARF Genes in S. album, Arabidopsis thaliana and Populus trichocarpa

To better analyze the function of *S. album ARF* and its evolutionary relationship, we used the neighbor-joining (NJ), maximum likelihood and maximum parsimony method (Figures S1 and S2) to establish an evolutionary tree model of ARF proteins among *S. album*, *Arabidopsis thaliana* and *Populus trichocarpa* to observe the evolutionary relationship of *ARF*. According to the phylogenetic trees of the three species, 62 *ARF* genes were divided into four subgroups, namely, I, II, III and IV. Among them, group I included 20 *ARF* genes, including *SaARF1, 2, 9, 11, 14, 15*; there were eight *ARF* genes in group II, including *SaARF3, 4, 12*; and there were 20 *ARF* genes in group III, including *SaARF5, 6, 7, 8, 13, 18*. Group IV included 14 *ARF* genes, including *SaARF10, 16, and 17* (Figure 4).

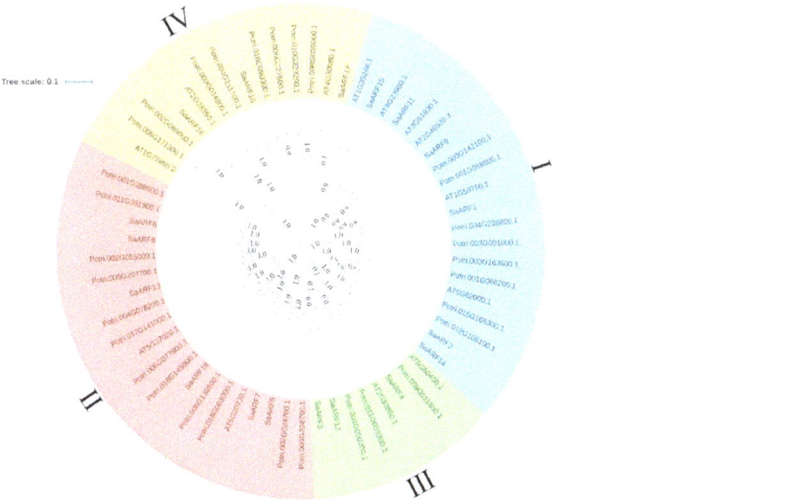

Figure 4. Phylogenetic relationships of *ARF* gene family members in *Arabidopsis thaliana*, *Populus trichocarpa* and sandalwood. The MEGA X program in the NJ connection method was used to construct a phylogenetic tree, which was used to perform beautification (https://itol.embl.de/itol.cgi, accessed on 16 August 2022).

3.4. Chromosome Mapping and Collinearity Analysis

The results of chromosome mapping showed that these genes were distributed on eight chromosomes of *S. album*, namely, chr1, 2, 4, 5, 6, 7, 8 and 9. Four *SaARF* genes were located on chr6 and chr9. Chr4, 5, 7 and 8 each contained two *SaARF* genes, chr1 and chr2 each contained an *ARF* gene, but chr3 and chr10 did not contain *ARF* genes (Figure 5). Tandem replication and fragment replication are important methods of gene expansion, and collinearity analysis can more clearly observe the results of duplication between genes. The occurrence of different members of the same gene family in the same or adjacent intergenic regions can be defined as tandem events. Through the construction of collinear maps between *S. album* and poplar and *Arabidopsis*, we found that 9 and 15 *SaARF* genes had syntenic relationships with *ARF*s of *Arabidopsis* and poplar, respectively, suggesting that some *SaARF* genes may originate from tandem or fragment replication (Table S1) (Figure 6A,B). Through the collinearity analysis among species, we found that ten of the eighteen *ARF* genes were involved in five segmental duplication events (*SaARF3/SaARF12, SaARF2/SaARF14, SaARF6/SaARF8, SaARF16/SaARF4, SaARF15/SaARF10*) in *S. album* (Table S2) (Figure 7).

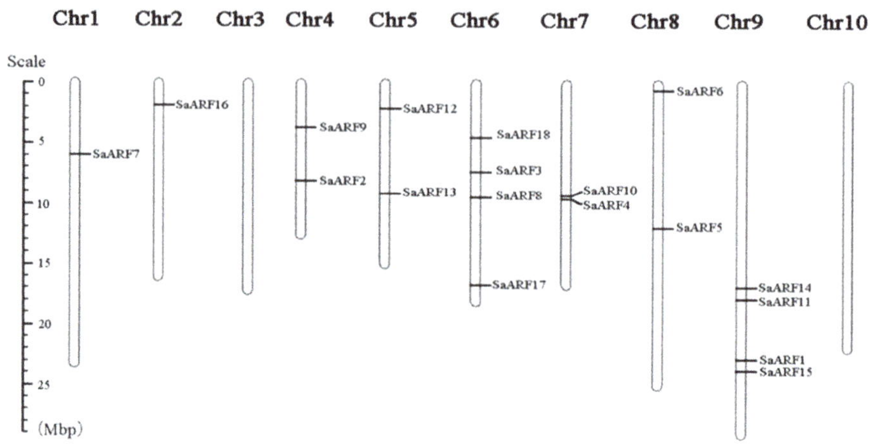

Figure 5. The chromosome location map of the *S. album ARF* gene was drawn using the MapDraw program, the gene name was marked on the right side of the chromosome, and the chromosome length unit was Mbp.

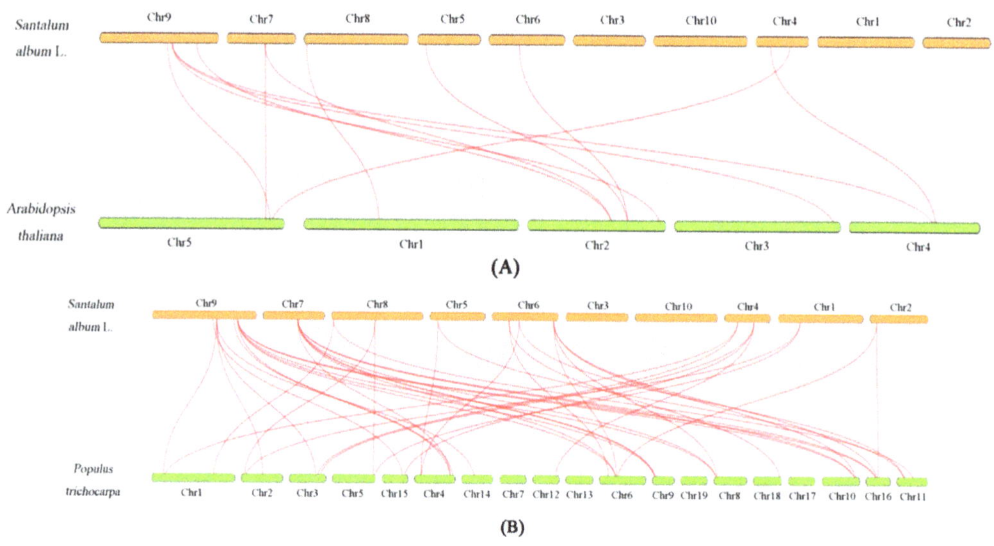

Figure 6. Collinearity contrast between *S. album* and two representative plants, poplar and *Arabidopsis*. (**A**) Collinearity analysis of *S. album* and *Arabidopsis thaliana*. (**B**) Collinearity analysis diagram of *S. album* and *Populus trichocarpa*. The gray lines in the background represent collinear blocks within the *S. album* and *Arabidopsis* and poplar genomes, while the red lines highlight collinear *ARF* gene pairs.

3.5. Analysis of Cis-Reactive Elements in the Promoter

Cis-regulatory elements in promoter sequences are essential for the temporal, spatial, and cell-specific control of gene expression, and there is evidence that genes with similar expression patterns contain the same regulatory elements in their promoters. Here, we used the 2000 bp upstream region of the 5′-UTR upstream sequence of all *S. album ARF* genes in the PlantCARE database to screen 18 cis-acting elements associated with stress, growth and development, and phytohormone responses (Figures 8 and 9). Among them, ARE cis-response elements affecting anaerobic induction were found in the promoters of 14 *SaARF* genes, and there were auxin response elements TGA-element and AuxRR in the promoters and CAT-box in the meristem formation response elements [43]. These results indicate that the *ARF* gene has a very close relationship with plant growth and development and stress.

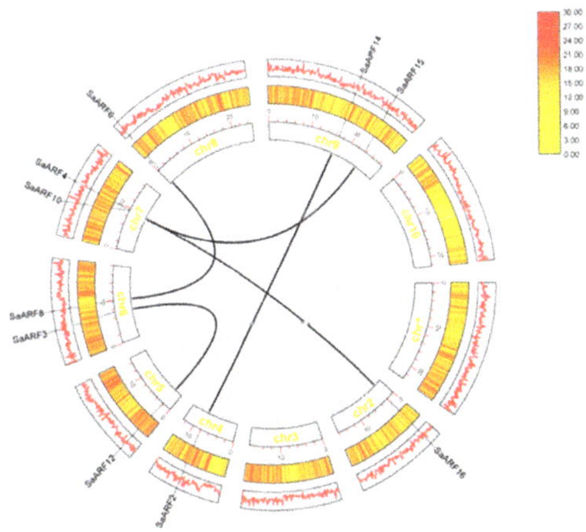

Figure 7. Interchromosomal relationship of the *ARF* gene in *S. album*. The gray lines indicate synteny blocks in the *S. album* genome, and the black lines indicate synteny blocks where the *ARF* repeat gene pairs are located. Chromosome names are shown in the middle of each chromosome, and the unit of chromosome length is Mbp. The golden yellow circle is the heatmap of gene density, and the outermost circle is the line map of gene density.

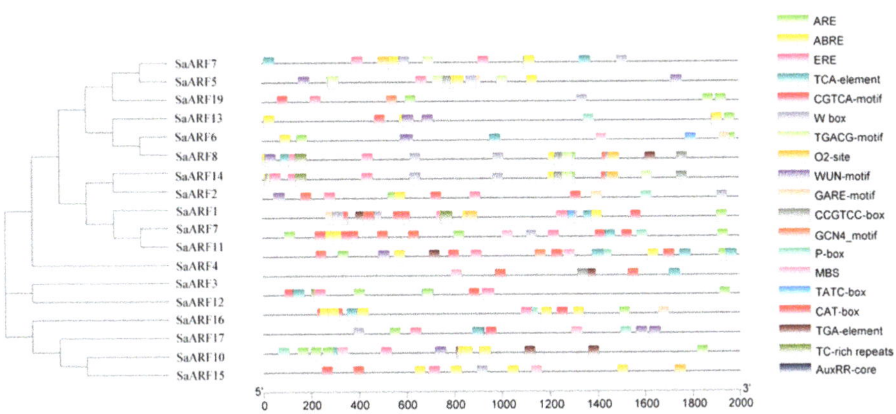

Figure 8. Distribution of cis-reactive elements in promoters of the *SaARF* gene family. There are 18 cis-reaction elements, including ARE, ABRE and ERE. ARE: anaerobic induction; ABRE: abscisic acid responsiveness; ERE: Estrogen Response Element; TCA-element: salicylic acid responsiveness; CGTCA-motif: MeJA-responsiveness; W-box: Trauma and pathogen reactivity; TGACG-motif: MeJA-responsiveness; O2-site: zein metabolism regulation; WUN-motif: wound-responsive element; GARE-motif: gibberellin-responsive element; CCGTCC-box: Meristem specific activation; GCN4-motif: endosperm expression; P-box: gibberellin-responsive element; MBS: MYB binding site involved in drought-inducibility; TATC-box: gibberellin-responsiveness; CAT-box: meristem expression; TGA-element: auxin-responsive element; TC-rich repeats: defense and stress responsiveness; AuxRR-core: auxin responsiveness (Table S4).

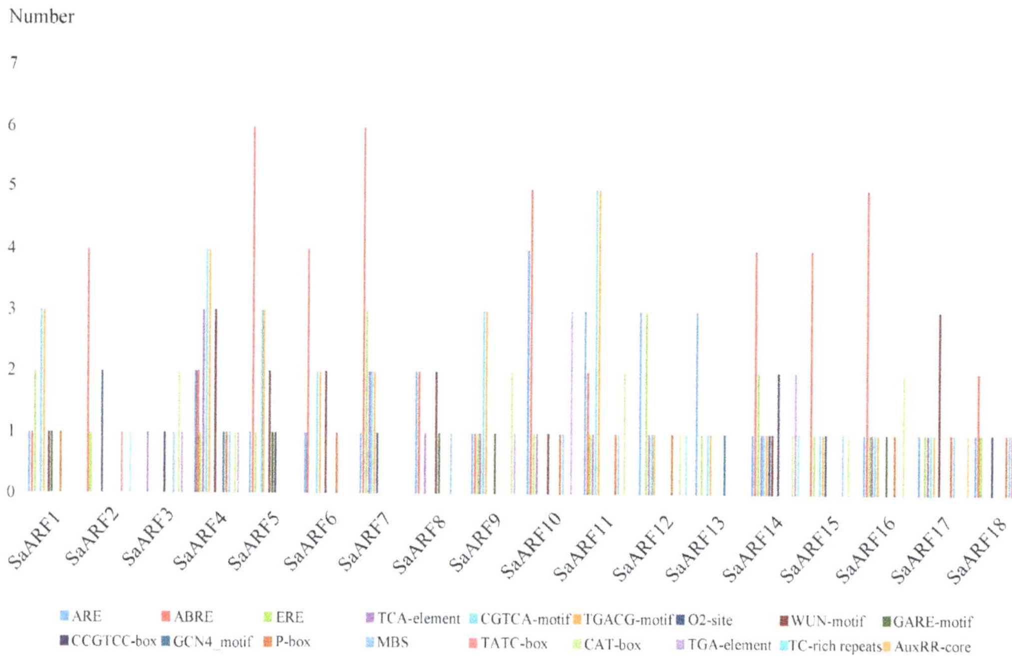

Figure 9. The number of cis-reactive elements of *ARF*s.

3.6. *Tissue-Specific Expression of ARF Gene in S. album*

To study the main function of *SaARF* in sandalwood plants, we used qRT-PCR to detect the tissue-specific expression of 18 sandalwood *ARF* genes and selected four main tissues, root, stem, leaf and haustoria, for analysis (Figure 10). After statistical analysis, we found that most *SaARF* genes were significantly expressed in the stem, and the expression level was high, mainly 10 genes, including *SaARF1*, *SaARF3*, *SaARF6*, *SaARF11*, *SaARF12*, *SaARF13*, *SaARF14*, *SaARF15*, *SaARF16* and *SaARF18*. The expression levels of the *SaARF1*, *SaARF5*, *SaARF7*, *SaARF8*, *SaARF15* and *SaARF16* genes were the highest in the haustorium, which was speculated to be related to the physiological process of haustorium formation or growth. However, *SaARF10* and *SaARF17* genes were not found to be expressed in the haustorium. The relative expression levels of *SaARF2*, *SaARF3* and *SaARF12* were the highest in leaves, while the expression levels of *SaARF9* were relatively high in roots and stems. We noticed that the expression levels of *SaARF5* and *SaARF7*, which were closely related in the evolutionary tree, were the highest in the haustorium, with significant differences. In *Arabidopsis*, *AtARF2* and *AtARF9* have been reported to control leaf senescence and promote lateral root elongation and leaf extension [47], respectively, and *SaARF2* and *SaARF9* genes in sandalwood were also highly expressed in leaves and roots, respectively. Therefore, it is reasonable to speculate that the *SaARF2* and *SaARF9* genes in sandalwood may have the same function, and *SaARF3* and *SaARF12*, both belonging to subgroup III, are also highly expressed in leaves. In addition, we noted that *AtARF7* has been shown to promote lateral root elongation [32], but after data statistics, we found that the expression of *SaARF7* in the haustorium was the highest, and it was reasonably predicted that *SaARF7* might be related to the growth or formation of the haustorium.

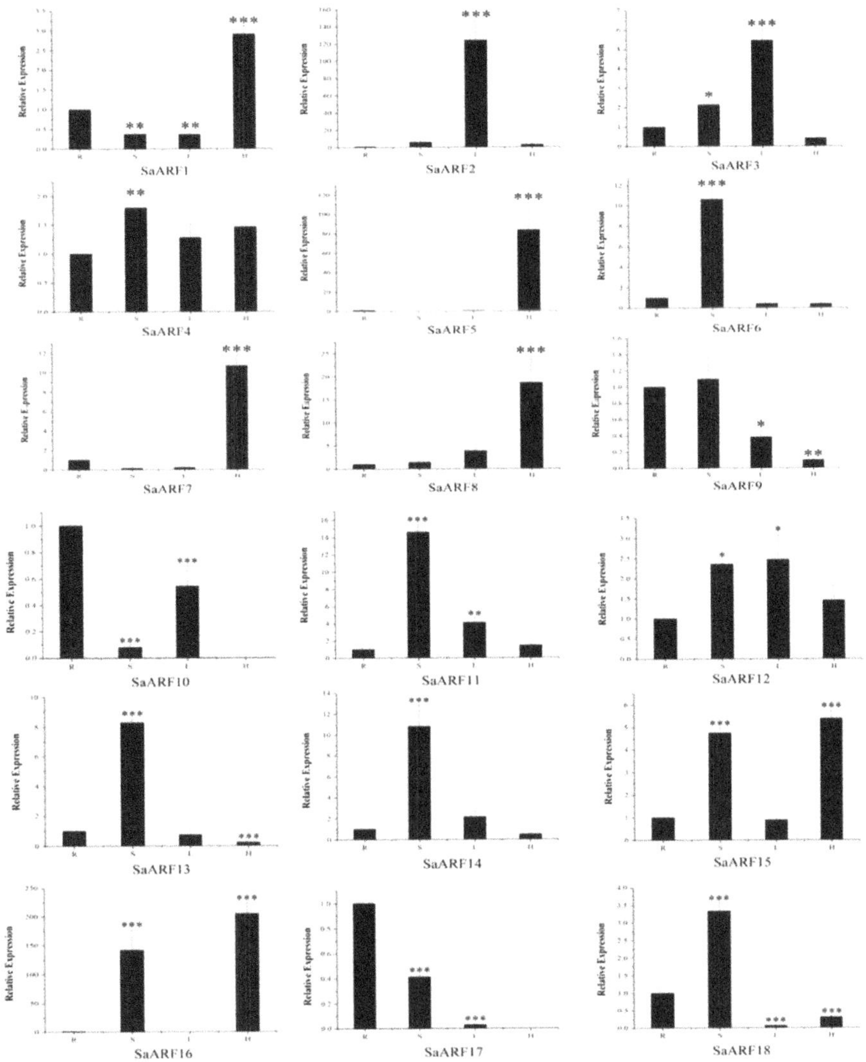

Figure 10. Expression analysis of 18 *ARF* genes in four representative tissues by qRT-PCR. Four various tissues include root, stem, leaf, and haustoria. The relative expression was calculated based on the $2^{-\Delta\Delta Ct}$ method. Error bars indicate ± standard deviation (SD) of three biological replicates. R: Root; S: Stem; L: Leave; H: Haustoria. Asterisks denote significant differences: * $p < 0.05$; ** $p < 0.01$; *** $p < 0.001$.

3.7. Expression of SaARF Genes under Drought Stress

In a previous study, *SlARF4* can affect the tomato's resistance to water shortage and *slarf4* mutants enhance plant resistance to water stress and water rehydration ability, suggesting that *SlARF4* may be an important gene in response to drought [48]. However, reports of this gene being expressed in response to drought in sandalwood are limited. Therefore, to determine whether these genes were expressed in response to drought, the expression of these 18 *SaARF* genes was also investigated during water stress. In the result, we find the expression of *SaARF2*, *SaARF5*, *SaARF10*, *SaARF16*, *SaARF17* increased significantly after 9 days (Figure 11). This suggests that it may be related to drought stress in sandalwood. And the expression of less than half of the gene, including *SaARF10*, *SaARF11*, was upregulated and then downregulated. In the meantime, we find eleven *SaARF*s' expression level continues to decline for 3 to 9 days.

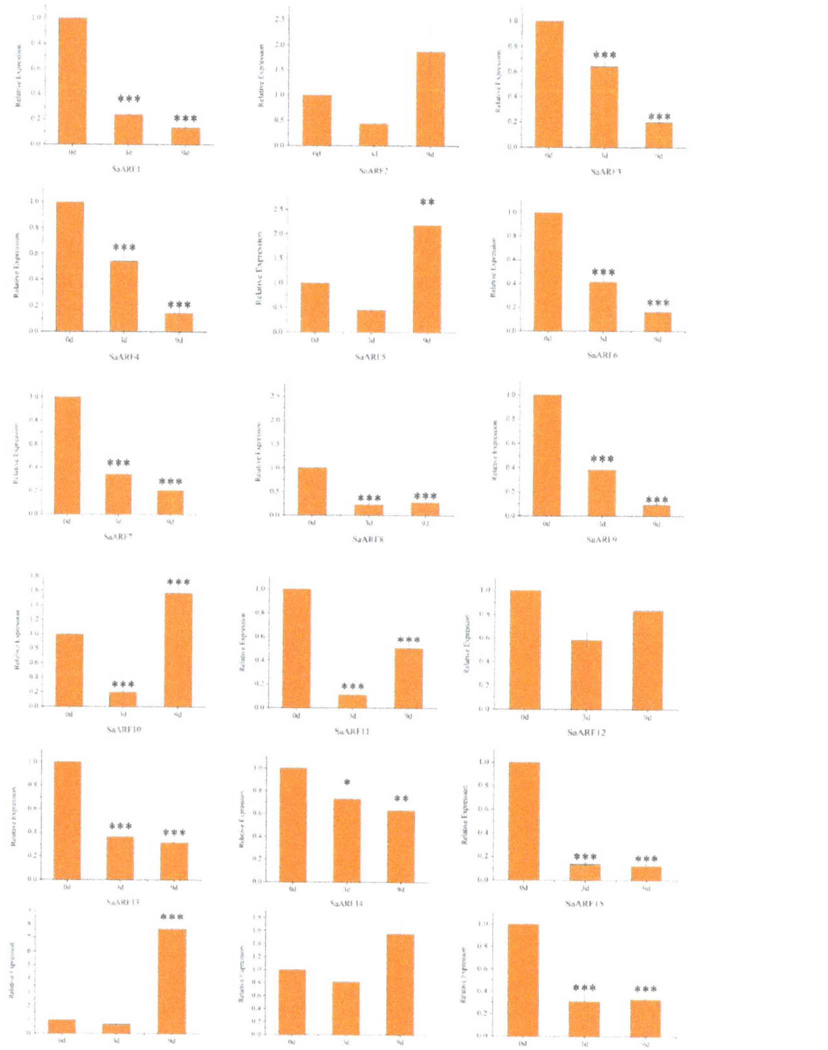

Figure 11. Expression analysis of Sa*ARF* genes in response to drought stress by qRT-PCR. Drought for 0 d, 3 d, 9 d in a greenhouse environment. The data were presented as the mean ± SD of three separate measurements. Asterisks denote significant differences: * $p < 0.05$; ** $p < 0.01$; *** $p < 0.001$.

4. Discussion

The *Arabidopsis* genome has been reported to contain 23 *ARF* genes, while the *ARF* gene families of rice, apple, poplar, tomato and eucalyptus include 25, 31, 39, 21 and 17 *ARF* genes, respectively [19,26–29]. In this study, in order to comprehend the relevant information and location of the *ARF* gene in sandalwood, we identified and characterized 18 *SaARF* transcription factors, and 18 *SaARF*s were unevenly distributed on nine chromosomes, among which chromosomes six and nine had the largest gene distribution, with four genes on each chromosome. By using Protparam to predict and analyze the physiological and biochemical indices of the sandalwood ARF protein, we found that the subcellular localization of the SaARF protein was predicted to be located in the nucleus, and *SaARF* all encoded unstable hydrophilic proteins, indicating that it played a role in different subcellular environments.

To better understand the evolution of the *ARF* gene family of *S. album*, we characterized the conserved domains, phylogenetic relationships and collinearity of SaARF. According to the conserved domain analysis, we found that 13 of the 18 identified sandalwood ARF had three conserved domains, DBD, MR, Phox and Bem1 (PB1) [33]. The MR domain is mainly responsible for controlling the function of ARFs. The PB1 domain located at the C-terminus is homologous to the C-terminal domain of the Aux/IAA protein and can bind to form homologous or heterodimers [49]. However, SaARF3, SaARF10, SaARF12, SaARF15, and SaARF17 contain only two conserved domains and lack a C-terminal PB1 domain. All 18 SaARFs have a complete DBD domain (B3), which is consistent with previous studies. Without the B3 domain, the ARF protein has been shown to be unable to bind auxin cis responsive elements in the promoter of auxin-responsive genes. The phylogenetic trees of 62 *ARF* genes in *Arabidopsis thaliana*, poplar and sandalwood were established, and they were divided into four subgroups: I, II, III and IV. The distribution of genes among the subclasses indicated that the expansion of the *ARF* family occurred before the divergence of the species. Each subgroup contained *ARF* genes of three species, indicating that the *ARF* genes among them had a relatively close evolutionary relationship. In addition, through the collinearity analysis within sandalwood species, we found that five segmental duplication events involved a total of 10 *SaARF* genes, which had short distance intervals and high homology in the *SaARF* gene family.

Synteny maps between two representative species and sandalwood were constructed to better understand the phylogenetic relationships. Through the collinearity analysis between sandalwood and poplar and *Arabidopsis*, we found there were 9 and 15 *SaARF* genes that had collinearity relationships with *ARF*s of *Arabidopsis* and poplar, respectively, so the evolutionary relationship between sandalwood and poplar was higher than that between *Arabidopsis* and sandalwood.

Cis-regulatory elements in promoter sequences are essential for temporal, spatial, and cell-specific control of gene expression, and there is evidence that genes with similar expression patterns contain the same regulatory elements in their promoters [50]. By analyzing the promoter cis-response elements within 2000 bp of the CDS upstream region of 18 *SaARF* genes, we found that the promoter contained anaerobically induced ARE cis-response elements, there were auxin response elements, TGA-element and AuxRR in the promoter, and the meristem formed response element CAT-box. The gibberellin response element P-box and abscisic acid response element ABRE are related to growth, development and stress [51], which indicates that the *ARF* gene is related to the realization of these functions. Among them, *SaARF3*, *SaARF4*, *SaARF10*, and *SaARF17* has both drought stress response elements and auxin response elements (Figures 9 and S5).

To investigate the main function of the *SaARF* genes and the roles of drought stress on it, we analyzed its expression in roots, stems, leaves, haustoria and the drought treatment of leaves for 0, 3, 9 days by qRT-PCR. Tissue-specific expression analysis showed that different *SaARF* genes played diverse role in the growth and development of plants. According to Figures 4 and 10, we found that *SaARF9*, *SaARF11*, *SaARF14* and *SaARF15* are highly expressed in stems which belong to group I. *SaARF5*, *SaARF7* and *SaARF8*, which are located in group II, were significantly overexpressed in the haustorium, with significant differences. At present, *AtARF7* in *Arabidopsis* has been confirmed to be related to lateral root growth, and *Arabidopsis AtARF5* has been identified as a key factor determining leaf initiation and vascular pattern formation [52]. *SaARF3* and *SaARF12* in group III and *SaARF10* and *SaARF17* in group IV were highly expressed in leaves and roots, respectively. Besides, most of the *SaARF* genes were significantly expressed in the stem, including 10 *SaARF* genes, such as *SaARF1*, *SaARF3* and *SaARF6*. After data analysis, we found that the relative expression levels of *SaARF9*, *SaARF10* and *SaARF17* were also high in the roots, but it was notable that neither *SaARF10* nor *SaARF17* were found in the haustorium.

By comparing the expression of *SaARF* genes, in tissues were under drought stress, we found the expression of *SaARF2, 5, 10, 16, 17* were increased in 9 d, even higher than that of 0 d, especially *SaARF10, 11, 17* which contain drought stress response elements, but most of the *SaARF* decreased in 3/9d. This suggests that *SaARF2* are associated with a drought resistance function. Therefore, this study preliminarily revealed the expression response of *SaARF* in various tissues and under drought treatment. The response and related functions of these genes need to be further verified by experiments in further studies.

5. Conclusions

This is the first study on the evolutionary relationship, expression profile and putative function of the *ARF* gene in sandalwood. In the present study, we identified 18 *ARF* genes in sandalwood by domain analysis, conserved motif analysis, phylogenetic tree construction, collinearity analysis and cis-reactive element analysis and divided them into four subgroups. By analyzing the protein structure of all members of the SaARF family, we found that all *ARF* genes in sandalwood encoded unstable hydrophilic proteins, and all *SaARF* genes were distributed equally in the nucleus. Based on the qRT-PCR results, we found that most *SaARF* genes were significantly expressed in the stem (*SaARF1, SaARF3, SaARF6, SaARF11, SaARF12, SaARF13, SaARF14, SaARF15, SaARF16* and *SaARF18*). The expression levels of the *SaARF1, SaARF5, SaARF7, SaARF8, SaARF15* and *SaARF16* genes were the highest in the haustorium. Notably, *SaARF5* and *SaARF7* were specifically distributed in the haustorium, so it is reasonable to speculate that they are related to the formation or growth of the haustorium. Similarly, we found that the *SaARF10, 16,* and *17* gene is associated with drought stress.

Supplementary Materials: The following supporting information can be downloaded at: https://www.mdpi.com/article/10.3390/f13111934/s1, Figure S1: A phylogenetic tree between sandalwood, *Arabidopsis thaliana* and *Populus trichocarpa* by Maximum likelihood method; Figure S2: A phylogenetic tree between sandalwood, *Arabidopsis thaliana* and *Populus trichocarpa* by Maximum parsimony method; Table S1: Collinearity analysis among species; Table S2: Collinearity analysis within species; Table S3: qRT-PCR primer sequence of 18 *SaARF*; Table S4: Cis-reactive elements in promoters of the *SaARF* gene family; Table S5: The number of cis-reactive elements in *ARF*s.

Author Contributions: X.L. and Y.L. designed all of the experiments and contributed to analyze the experimental results. S.M., J.L. and S.W. conceived the project. Y.C., F.Q. and D.W. helped perform the experiments. X.L., Y.L., D.W., L.H. and S.M. wrote the paper. All authors have read and agreed to the published version of the manuscript.

Funding: This work was supported by grants from the Fundamental Research Funds for the Central Non-profit Research Institution of CAF [CAFYBB2020SY018 and CAFYBB2019QD001], National Natural Science Foundation of China [31722012 and 31901304], Natural Science Foundation of Guangdong Province, China [2019A1515011595] and Natural Science Foundation of Chongqing, China, cstc2018jcyjAX0778.

Data Availability Statement: Not applicable.

Acknowledgments: We acknowledge everyone who contributed to this article.

Conflicts of Interest: The authors declare no conflict of interest.

References

1. Jones, C.G.; Keeling, C.I.; Ghisalberti, E.L.; Barbour, E.L.; Plummer, J.A.; Bohlmann, J. Isolation of cDNAs and functional characterisation of two multi-product terpene synthase enzymes from sandalwood, *Santalum album* L. *Arch. Biochem. Biophys.* **2008**, *477*, 121–130. [CrossRef] [PubMed]
2. Teixeira da Silva, J.A.; Kher, M.M.; Soner, D.; Page, T.; Zhang, X.; Nataraj, M.; Ma, G. Sandalwood: Basic biology, tissue culture, and genetic transformation. *Planta* **2016**, *243*, 847–887. [CrossRef] [PubMed]

3. Těšitel, J.; Těšitelová, T.; Minasiewicz, J.; Selosse, M.A. Mixotrophy in Land Plants: Why to Stay Green? *Trends Plant Sci.* **2018**, *23*, 656–659. [CrossRef] [PubMed]
4. Mahesh, H.B.; Subba, P.; Advani, J.; Shirke, M.D.; Loganathan, R.M.; Chandana, S.L.; Shilpa, S.; Chatterjee, O.; Pinto, S.M.; Prasad, T.S.K.; et al. Multi-Omics Driven Assembly and Annotation of the Sandalwood (*Santalum album*) Genome. *Plant Physiol.* **2018**, *176*, 2772–2788. [CrossRef] [PubMed]
5. Burdock, G.A.; Carabin, I.G. Safety assessment of sandalwood oil (*Santalum album* L.). *Food Chem. Toxicol.* **2008**, *46*, 421–432. [CrossRef]
6. Kim, T.H.; Ito, H.; Hayashi, K.; Hasegawa, T.; Machiguchi, T.; Yoshida, T. New antitumor sesquiterpenoids from *Santalum album* of Indian origin. *Tetrahedron* **2006**, *62*, 6981–6989. [CrossRef]
7. Nagaveni, H.C.; Vijayalakshmi, G. Growth performance of sandal (*Santalum album* L.) with different host species. *Sandalwood Res. Newsl.* **2003**, *18*, 1–4.
8. Zhang, X.; Teixeira da Silva, J.A.; Duan, J.; Deng, R.; Xu, X.; Ma, G. Endogenous hormone levels and anatomical characters of haustoria in *Santalum album* L. seedlings before and after attachment to the host. *J. Plant Physiol.* **2012**, *169*, 859–866. [CrossRef]
9. Meng, S.; Wang, X.; Bian, Z.; Li, Z.; Yang, F.; Wang, S.; Yoder, J.I.; Lu, J. Melatonin enhances nitrogen metabolism and haustorium development in hemiparasite *Santalum album* Linn. *Environ. Exp. Bot.* **2021**, *186*, 104460. [CrossRef]
10. Benková, E.; Michniewicz, M.; Sauer, M.; Teichmann, T.; Seifertová, D.; Jürgens, G.; Friml, J. Local, efflux–dependent auxin gradients as a common module for plant organ formation. *Cell* **2003**, *115*, 591–602. [CrossRef]
11. Ljung, K. Auxin metabolism and homeostasis during plant development. *Development* **2013**, *140*, 943–950. [CrossRef] [PubMed]
12. Woodward, A.W.; Bartel, B. Auxin: Regulation, action, and interaction. *Ann. Bot.* **2005**, *95*, 707–735. [CrossRef] [PubMed]
13. Goldental–Cohen, S.; Israeli, A.; Ori, N.; Yasuor, H. Auxin Response Dynamics During Wild–Type and entire Flower Development in Tomato. *Plant Cell Physiol.* **2017**, *58*, 1661–1672. [CrossRef] [PubMed]
14. Liscum, E.; Reed, J.W. Genetics of Aux/IAA and ARF action in plant growth and development. *Plant Mol. Biol.* **2002**, *49*, 387. [CrossRef] [PubMed]
15. Ruegger, M.; Dewey, E.; Hobbie, L.; Brown, D.; Bernasconi, P.; Turner, J.; Muday, G.; Estelle, M. Reduced naphthylphthalamic acid binding in the tir3 mutant of *Arabidopsis* is associated with a reduction in polar auxin transport and diverse morphological defects. *Plant Cell* **1997**, *9*, 745–757.
16. Ulmasov, T. ARF1, a transcription factor that binds to auxin response elements. *Science* **1997**, *276*, 1865–1868. [CrossRef]
17. Gray, W.M.; Del Pozo, J.C.; Walker, L.; Hobbie, L.; Risseeuw, E.; Banks, T.; Crosby, W.L.; Yang, M.; Ma, H.; Estelle, M. Identification of an SCF ubiquitin-ligase complex required for auxin response in *Arabidopsis thaliana*. *Genes Dev.* **1999**, *13*, 1678–1691. [CrossRef]
18. Chen, J.; Wang, S.; Wu, F.; Wei, M.; Li, J.; Yang, F. Genome-Wide Identification and Functional Characterization of Auxin Response Factor (ARF) Genes in Eggplant. *Int. J. Mol. Sci.* **2022**, *23*, 6219. [CrossRef]
19. Yu, H.; Soler, M.; Mila, I.; San Clemente, H.; Savelli, B.; Dunand, C.; Paiva, J.A.; Myburg, A.A.; Bouzayen, M.; Grima-Pettenati, J.; et al. Genome-wide characterization and expression profiling of the AUXIN RESPONSE FACTOR (ARF) gene family in Eucalyptus grandis. *PLoS ONE* **2014**, *9*, e108906. [CrossRef]
20. Cancé, C.; Martin-Arevalillo, R.; Boubekeur, K.; Dumas, R. Auxin response factors are keys to the many auxin doors. *New Phytol.* **2022**, *235*, 402–419. [CrossRef]
21. Causier, B.; Ashworth, M.; Guo, W.; Davies, B. The TOPLESS interactome: A framework for gene repression in *Arabidopsis*. *Plant Physiol.* **2012**, *158*, 423–438. [CrossRef]
22. Korasick, D.A.; Chatterjee, S.; Tonelli, M.; Dashti, H.; Lee, S.G.; Westfall, C.S.; Fulton, D.B.; Andreotti, A.H.; Amarasinghe, G.K.; Strader, L.C.; et al. Defining a two-pronged structural model for PB1 (Phox/Bem1p) domain interaction in plant auxin responses. *J. Biol. Chem.* **2015**, *290*, 12868–12878. [CrossRef] [PubMed]
23. Szemenyei, H.; Hannon, M.; Long, J.A. TOPLESS Mediates Auxin-Dependent Transcriptional Repression during *Arabidopsis* Embryogenesis. *Science* **2008**, *319*, 1384–1386. [CrossRef] [PubMed]
24. Wang, R.; Estelle, M. Diversity and specificity: Auxin perception and signaling through the TIR1/AFB pathway. *Curr. Opin. Plant Biol.* **2014**, *21*, 51–58. [CrossRef] [PubMed]
25. Guilfoyle, T.J.; Hagen, G. Auxin response factors. *Curr. Opin. Plant Biol.* **2007**, *10*, 453–460. [CrossRef]
26. Wang, D.; Pei, K.; Fu, Y.; Sun, Z.; Li, S.; Liu, H.; Tang, K.; Han, B.; Tao, Y. Genome-wide analysis of the auxin response factors (ARF) gene family in rice (*Oryza sativa*). *Gene* **2007**, *394*, 13–24. [CrossRef]
27. Luo, X.C.; Sun, M.H.; Xu, R.R.; Shu, H.R.; Wang, J.W.; Zhang, S.Z. Genomewide identification and expression analysis of the ARF gene family in apple. *J. Genet.* **2014**, *93*, 785–797. [CrossRef]
28. Kalluri, U.C.; Difazio, S.P.; Brunner, A.M.; Tuskan, G.A. Genome-wide analysis of Aux/IAA and ARF gene families in Populus trichocarpa. *BMC Plant Biol.* **2007**, *7*, 59. [CrossRef]
29. Audran–Delalande, C.; Bassa, C.; Mila, I.; Regad, F.; Zouine, M.; Bouzayen, M. Genome-wide identification, functional analysis and expression profiling of the Aux/IAA gene family in tomato. *Plant Cell Physiol.* **2012**, *53*, 659–672. [CrossRef]
30. Peng, Y.; Fang, T.; Zhang, Y.; Zhang, M.; Zeng, L. Genome-Wide Identification and Expression Analysis of Auxin Response Factor (ARF) Gene Family in Longan (*Dimocarpus longan* L.). *Plants* **2020**, *9*, 221. [CrossRef]
31. Jiang, W.; Xia, Y.; Su, X.; Pang, Y. ARF2 positively regulates flavonols and proanthocyanidins biosynthesis in *Arabidopsis thaliana*. *Planta* **2022**, *256*, 44. [CrossRef] [PubMed]

32. Okushima, Y.; Fukaki, H.; Onoda, M.; Theologis, A.; Tasaka, M. ARF7 and ARF19 regulate lateral root formation via direct activation of *LBD/ASL* genes in *Arabidopsis*. *Plant Cell* **2007**, *19*, 118–130. [CrossRef] [PubMed]
33. De Jong, M.; Wolters-Arts, M.; Schimmel, B.C.; Stultiens, C.L.; De Groot, P.F.; Powers, S.J.; Tikunov, Y.M.; Bovy, A.G.; Mariani, C.; Vriezen, W.H.; et al. *Solanum lycopersicum* AUXIN RESPONSE FACTOR 9 regulates cell division activity during early tomato fruit development. *J. Exp. Bot.* **2015**, *66*, 3405–3416. [CrossRef] [PubMed]
34. Xu, C.; Shen, Y.; He, F.; Fu, X.; Yu, H.; Lu, W.; Li, Y.; Li, C.; Fan, D.; Wang, H.C.; et al. Auxin-mediated Aux/IAA-ARF-HB signaling cascade regulates secondary xylem development in *Populus*. *New Phytol.* **2019**, *222*, 752–767. [CrossRef] [PubMed]
35. Wang, Y.; Li, K.; Chen, L.; Zou, Y.; Liu, H.; Tian, Y.; Li, D.; Wang, R.; Zhao, F.; Ferguson, B.J.; et al. MicroRNA167-Directed Regulation of the Auxin Response Factors *GmARF8a* and *GmARF8b* Is Required for Soybean Nodulation and Lateral Root Development. *Plant Physiol.* **2015**, *168*, 984–999. [CrossRef]
36. Mao, Z.; He, S.; Xu, F.; Wei, X.; Jiang, L.; Liu, Y.; Wang, W.; Li, T.; Xu, P.; Du, S.; et al. Photoexcited CRY1 and phyB interact directly with ARF6 and ARF8 to regulate their DNA-binding activity and auxin-induced hypocotyl elongation in *Arabidopsis*. *New Phytol.* **2020**, *225*, 848–865. [CrossRef]
37. Li, H.; Zhang, X.; Tong, B.; Wang, Y.; Yang, C. Expression analysis of the *BpARF* genes in *Betula platyphylla* under drought stress. *Plant Physiol. Biochem.* **2020**, *148*, 273–281. [CrossRef]
38. Wang, D.; Meng, S.; Su, W.; Bao, Y.; Lu, Y.; Yin, W.; Liu, C.; Xia, X. Genome-Wide Analysis of Multiple Organellar RNA Editing Factor Family in Poplar Reveals Evolution and Roles in Drought Stress. *Int. J. Mol. Sci.* **2019**, *20*, 1425. [CrossRef]
39. Wilkins, M.R.; Gasteiger, E.; Bairoch, A.; Sanchez, J.C.; Williams, K.L.; Appel, R.D.; Hochstrasser, D.F. Protein identification and analysis tools in the ExPASy server. *Methods Mol. Biol.* **1999**, *112*, 531–552.
40. Bailey, T.L.; Boden, M.; Buske, F.A.; Frith, M.; Grant, C.E.; Clementi, L.; Ren, J.; Li, W.W.; Noble, W.S. MEME SUITE: Tools for motif discovery and searching. *Nucleic Acids Res.* **2009**, *37*, W202–W208. [CrossRef]
41. Letunic, I.; Bork, P. Interactive Tree of Life (iTOL): An online tool for phylogenetic tree display and annotation. *Bioinformatics* **2007**, *23*, 127–128. [CrossRef] [PubMed]
42. Liu, R.H.; Meng, J.L. MapDraw: A microsoft excel macro for drawing genetic linkage maps based on given genetic linkage data. *Yi Chuan* **2003**, *25*, 317–321. [PubMed]
43. Bian, Z.; Wang, D.; Liu, Y.; Xi, Y.; Wang, X.; Meng, S. Analysis of Populus glycosyl hydrolase family I members and their potential role in the ABA treatment and drought stress response. *Plant Physiol. Biochem.* **2021**, *163*, 178–188. [CrossRef] [PubMed]
44. Chen, C.; Chen, H.; Zhang, Y.; Thomas, H.R.; Frank, M.H.; He, Y.; Xia, R. TBtools: An integrative toolkit developed for interactive analyses of big biological data. *Mol. Plant* **2020**, *13*, 1194–1202. [CrossRef] [PubMed]
45. Lescot, M.; Déhais, P.; Thijs, G.; Marchal, K.; Moreau, Y.; Van de Peer, Y.; Rouzé, P.; Rombauts, S. PlantCARE, a database of plant cis-acting regulatory elements and a portal to tools for in silico analysis of promoter sequences. *Nucleic Acids Res.* **2002**, *30*, 325–327. [CrossRef]
46. Ren, Z.; Liu, R.; Gu, W.; Dong, X. The *Solanum lycopersicum* auxin response factor *SlARF2* participates in regulating lateral root formation and flower organ senescence. *Plant Sci.* **2017**, *256*, 103–111. [CrossRef]
47. Lim, P.O.; Lee, I.C.; Kim, J.; Kim, H.J.; Ryu, J.S.; Woo, H.R.; Nam, H.G. Auxin response factor 2 (ARF2) plays a major role in regulating auxin-mediated leaf longevity. *J. Exp. Bot.* **2010**, *61*, 1419–1430. [CrossRef]
48. Chen, M.; Zhu, X.; Liu, X.; Wu, C.; Yu, C.; Hu, G.; Chen, L.; Chen, R.; Bouzayen, M.; Zouine, M.; et al. Knockout of Auxin Response Factor SlARF4 Improves Tomato Resistance to Water Deficit. *Int. J. Mol. Sci.* **2021**, *25*, 3347. [CrossRef]
49. Hagen G, Guilfoyle T: Auxin-responsive gene expression: Genes, promoters and regulatory factors. *Plant Mol. Biol.* **2002**, *49*, 373–385. [CrossRef]
50. Wu, P.; Shou, H.; Xu, G.; Lian, X. Improvement of phosphorus efficiency in rice on the basis of understanding phosphate signaling and homeostasis. *Curr. Opin. Plant Biol.* **2013**, *16*, 205–212. [CrossRef]
51. Jiang, C.; Song, X.; He, H.; Chu, L.; Zhou, H.; Zhao, Y.; Xu, Y.; Zeng, W.; Lin, X.; Lu, M.-Z. Genome-wide identification of plasma membrane aquaporin gene family in Populus and functional identification of PIP1;1 involved in osmotic stress. *Environ. Exp. Bot.* **2020**, *179*, 104200. [CrossRef]
52. Karannagoda, N.; Spokevicius, A.; Hussey, S.; Cassan-Wang, H.; Grima-Pettenati, J.; Bossinger, G. Eucalyptus grandis AUX/INDOLE-3-ACETIC ACID 13 (EgrIAA13) is a novel transcriptional regulator of xylogenesis. *Plant Mol. Biol.* **2022**, *109*, 51–65. [CrossRef] [PubMed]

forests

MDPI

Article

Foliar Application of Selenium Reduces Cadmium Accumulation in Walnut Seedlings

Bingwen Wang [†], Dangquan Zhang [†], Wenfeng Wang, Yukun Song, Mengfei Lu and Shen Ding [*]

College of Forestry, Henan Agricultural University, Zhengzhou 450002, China
[*] Correspondence: dingshen040033@163.com
[†] These authors contributed equally to this work.

Abstract: Cadmium (Cd) and selenium (Se) could jointly affect plant growth. To investigate the affect of Se on the Cd accumulation in *Juglans regia* and the physiological mechanism by which Se regulates Cd-induced oxidative stress, in this study, the effects of different foliar application doses of Se (0 (Se0), 20 (Se20), and 200 (Se200) μM) on *J. regia* (variety Xinfeng) seedlings under Cd stress (5 mM) were determined. The results show that exogenous application of Se (Se20 and Se200) increased walnut biomass compared with Se0 under Cd stress. Under Cd stress, exogenous application of 20 μM Se increased the catalase (CAT), peroxidase (POD), and ascorbate oxidase (AAO) activities in walnut roots and the CAT and AAO activities in walnut leaves, and exogenous application of 200 μM Se increased the CAT, POD, and AAO activities in walnut roots. Furthermore, under Cd stress, exogenous application of 20 and 200 μM Se both decreased the contents of superoxide ($O_2^{\bullet -}$), hydrogen peroxide (H_2O_2), and malondialdehyde (MDA) in walnut roots and the content of MDA in walnut leaves. Moreover, application of 20 and 200 μM Se both reduced the accumulation of Cd in the root, wood, bark, and leaves of walnuts, and application of 200 μM Se enhanced Se concentration in the root, wood, bark, and leaves. Overall, exogenous application of Se, especially 200 μM Se, could reduce Cd accumulation and enhance CAT, POD, and AAO activities in Cd-stressed walnut roots, thus alleviating Cd stress. This study provides technical guidance for reducing the effects of Cd stress on walnut growth.

Keywords: *Juglans regia*; biomass; soil contamination; oxidative stress; antioxidant enzymes

Citation: Wang, B.; Zhang, D.; Wang, W.; Song, Y.; Lu, M.; Ding, S. Foliar Application of Selenium Reduces Cadmium Accumulation in Walnut Seedlings. *Forests* **2022**, *13*, 1493. https://doi.org/10.3390/f13091493

Academic Editor: Giovanbattista De Dato

Received: 18 August 2022
Accepted: 13 September 2022
Published: 15 September 2022

Publisher's Note: MDPI stays neutral with regard to jurisdictional claims in published maps and institutional affiliations.

1. Introduction

About 30,000 tons of cadmium (Cd) are released into the environment every year worldwide due to human activities, such as mining, the use of pesticides and fertilizers, and the combustion of fossil fuels [1]. Cd in soil can be absorbed by plant roots and transported to the aerial parts through the xylem, which inevitably causes ion imbalance, lowered photosynthetic capacity, and oxidative stress in plants [2]. Therefore, Cd stress always reduces agricultural productivity, and threatens human health via food chains [1,3–6]. Previous studies estimated that more than 80% of the Cd accumulated in human bodies is from vegetables and crops grown in Cd-contaminated soils [7]. A recent survey report on soil contamination in China shows that the area of Cd-contaminated soil ranks first among the heavy metal-contaminated soils, accounting for 7% [8]. At present, Cd is detected in rice [9], peanuts [10], apples [6], pears, grape, peach-shaped plums, and oranges [3].

Walnut (*Juglans regia* L.) is one of the four major nuts in the world. Due to the high nutrition content and economic value, walnut is widely cultivated in Asia, Europe, America, South Africa, Australia, and New Zealand [11,12]. However, Peng found that soil Cd contamination greatly affected walnut growth and yield [13]. Previous studies show that the highest Cd concentration in walnut kernels reaches 0.71 mg kg^{-1}, which exceeds the safety standard [14–17].

Selenium (Se) is widely distributed worldwide. However, there are significant differences in the concentration and chemical forms of Se in natural soils in different regions [18,19]. Se is a non-essential element for plants, but an appropriate amount of Se could promote plant growth. Several studies found that Se can increase chloroplast size, enhance chloroplast ultrastructure [20] and antioxidant capacity [21], as well as regulate the uptake and translocation of essential elements [22], improving crop yield and quality [23]. Moreover, some studies also show that Se could reduce Cd accumulation and reactive oxygen species (ROS) contents by regulating carbon and nitrogen metabolism in potatoes [7,24,25]. However, Yu et al. found that foliar application of Se enhanced the accumulation of Cd in tobacco [26]. Yu et al. reported that Se application decreased Cd concentration in *Brassica chinensis* shoots treated with 10 μM Cd, but increased the Cd concentration in shoots treated with 50 μM Cd [27]. Therefore, the application of different concentrations of Se have different effects on Cd accumulation in plants [27]. This may be caused by the differences in plant species, Se concentration, Se application method, and Cd dosage. However, the effect of different application rates of Se on Cd accumulation in walnut is unknown at present.

Besides, Se is an essential mineral element for both humans and animals [28]. Studies show that Se deficiency is associated with Keshan disease (KD) and white muscle disease [29]. According to the report of the World Health Organization (WHO), China is one of the 40 Se-deficient countries in the world [30]. More than half of China's land is deficient in Se, and over 105 million people's health is threatened by Se deficiency [19]. Therefore, improving the Se concentration in foods is urgent for improving the health of Chinese people.

Previous studies show that exogenous application of appropriate amount of Se may reduce Cd accumulation and increase Se concentration in rice under Cd stress [31]. Therefore, in this study, the effects of different doses of Se (0, 20, and 200 μM) on ROS contents, antioxidant enzyme activities, and Cd accumulation in different organs of *J. regia* (variety Xinfeng) were analyzed under Cd stress (5 mM), aiming to clarify the physiological mechanism by which Se regulates Cd-induced oxidative stress. We hypothesized that: (1) Se application might reduce Cd accumulation in walnut; and (2) Se application might cause physiological responses of walnut that reduce oxidative stress caused by Cd. The study will provide guidance for reducing Cd concentration and increasing Se concentration in walnuts through agronomic measures to provide high-quality Se-enriched walnuts.

2. Materials and Methods

2.1. Plant Material and Growth Conditions

The experiments were conducted at Henan Agricultural University, Zhengzhou, China (34°47′8″ N, 113°39′50″ E). Seedlings of *Juglans regia* (variety Xinfeng) were cultivated in a greenhouse (day/night temperature: 35/25 °C; relative humidity: 50–60%; natural light). After germination, seedlings were planted in plastic pots filled with sands, and the quarter-strength Hoagland solution (100 mL) was added to each pot every three days. After 12 weeks, 36 seedlings with a height of about 25 cm were selected, and divided equally into six groups.

Subsequently, walnut plants in the six groups were exposed to 0 (-Cd) or 5 (+Cd) mM $CdCl_2$ together with 0 (Se0), 20 (Se20), or 200 (Se200) μM Na_2SeO_3. Briefly, this experiment contained six treatments including Se0Cd0 (0 μM Na_2SeO_3 + 0 mM $CdCl_2$), Se0Cd5 (0 μM Na_2SeO_3 + 5 mM $CdCl_2$), Se20Cd0 (20 μM Na_2SeO_3 + 0 mM $CdCl_2$), Se20Cd5 (20 μM Na_2SeO_3 + 5 mM $CdCl_2$), Se200Cd0 (200 μM Na_2SeO_3 + 0 mM $CdCl_2$), and Se200Cd5 (200 μM Na_2SeO_3 + 5 mM $CdCl_2$) (Table 1). Cd and Se doses were determined according to our pre-experiment and previous studies [27,31,32]. Our pre-experiment results show that walnuts cultured in sand could tolerate higher concentrations of Cd than walnuts cultured in nutrient solution. Furthermore, in previous studies, the Se concentrations for treating herbaceous plants were 3–20 μM [27,31]. For woody plants, such as walnut, the Se concentration should be higher. One hundred milliliters of quarter-strength Hoagland solution were added every three days. For Se0Cd5, Se20Cd5, and Se200Cd5 treatments,

Cd was supplied every three days with the nutrient solution. Exogenous Se was sprayed on walnut leaves after being dissolved in distilled water every three days. Seedlings in Se0Cd0 and Se0Cd5 treatments were sprayed with distilled water.

Table 1. Experimental design.

Treatment	Se (μM)	Cd (mM)
Se0Cd0	0	0
Se0Cd5	0	5
Se20Cd0	20	0
Se20Cd5	20	5
Se200Cd0	200	0
Se200Cd5	200	5

After a 60-day cultivation, the root system of each plant was washed carefully using 50 mM $CaCl_2$ for 3 min to remove Cd^{2+}. The shoots were rinsed with distilled water three times. Then, each plant was separated into root, wood, bark, and leaves. The fresh weight of each sample was recorded before wrapping with tinfoil and freezing in liquid nitrogen. The frozen samples were ground into fine powder in liquid nitrogen and stored at −80 °C for further analysis. About 150 mg powder of each organ of each plant was dried at 70 °C to determine the fresh-to-dry mass ratio, which was used to calculate the biomass of each organ.

2.2. Determination of Cd and Se

One hundred and fifty milligrams of powder of each organ were digested in a mixture containing 8 mL of concentrated HNO_3 and 2 mL of concentrated $HClO_4$ at 170 °C [33]. The Cd and Se concentrations in the extract were determined by inductively coupled plasma mass spectrometry (ICP-MS, Agilent7800, Santa Clara, CA, USA).

2.3. Analysis of $O_2^{\bullet-}$, H_2O_2, and MDA

The superoxide ($O_2^{\bullet-}$) content in the roots and leaves of the Xinfeng walnut were determined spectrophotometrically at 530 nm as described by Lei [34]. Briefly, 100 mg of powder was extracted in potassium phosphate buffer (50 mM, pH 7.8) and centrifuged (10,000× *g*, 4 °C, 10 min). Then, the supernatant was mixed with the potassium phosphate buffer (0.9 mL, 50 mM, pH 7.8) and hydroxylamine hydrochloride (0.1 mL, 10 mM). The mixture was then incubated at 25 °C for 20 min, before adding p-aminobenzene sulfonic acid (1 mL, 17 mM) and α-naphthylamine (1 mL, 7 mM). The absorbance of the mixture was determined at 530 nm after an incubation at 25 °C for 20 min.

The hydrogen peroxide (H_2O_2) contents in the roots and leaves were determined according to the method of He [35]. Briefly, 50 mg of powder was extracted in 5% trichloroacetic acid (TCA) for 20 min and centrifuged (10,000× *g*, 4 °C, 10 min). Then, the supernatant was mixed with 20% $TiCl_4$ and 25% aqueous ammonia. After that, the precipitate was collected and dissolved in H_2SO_4 (1 M). The absorbance of the solution was determined spectrophotometrically at 410 nm.

The malondialdehyde (MDA) contents in the roots and leaves were determined according to the method of Lei [34]. Briefly, 100 mg of powder was extracted in 10% TCA for 30 min and centrifuged (10,000× *g*, 4 °C, 10 min). Then, the supernatant was mixed with 0.6% thiobarbituric acid (TBA) and 10% TCA. After reacting in boiling water for 15 min, the mixture was rapidly cooled in ice water and centrifuged (10,000× *g*, 4 °C, 10 min). The absorbance of the solution was determined spectrophotometrically at 450, 532, and 600 nm.

2.4. Assays of Antioxidant Enzyme Activities

The soluble proteins in the roots and leaves were extracted based on the method of Luo [36]. Briefly, frozen powder was homogenized in a cold extraction buffer containing 100 mM potassium phosphate (pH 7.8), 200 mg polyvinylpolypyrrolidone, and 0.5% (*v*/*v*) Triton X-100. Then, the mixture was incubated for 15 min in an ice bath and centrifuged

(15,000× g, 4°C, 30 min). The supernatant was eluted through Sephadex G-25 columns (PD-10 column, Pharmacia, Freiburg, Germany). The soluble proteins in the eluent were determined according to the Bradford method, using bovine serum albumin (Interchim, Montluçon, France) as the standard. The soluble protein extracts were used for the assays of enzyme activities.

The activity of catalase (CAT, EC 1.11.1.6) was measured according to the method of He [37], and the absorbance of the reaction system was determined spectrophotometrically at 240 nm. The reaction system contained 50 mM potassium phosphate buffer (pH 7.0), 40 mM H_2O_2, and the protein extract. One unit of CAT was defined as the amount of the enzyme that is needed to decompose 1 mmol of H_2O_2 per min at 25 °C.

The activity of peroxidase (POD, EC 1.11.1.7) was determined according to the method of Chen [38], and the absorbance of the reaction system was determined spectrophotometrically at 436 nm. The reaction system contained 50 mM potassium phosphate buffer (pH 6.5), 40 mM guaiacol, 10 mM H_2O_2, and the protein extract. One unit of the enzyme was defined as the amount of POD that is needed to oxidize 1 mmol of guaiacol min^{-1} mg^{-1} protein.

The activity of ascorbate peroxidase (APX, EC 1.11.1.11) was measured according to the method of He [37], and the absorbance of the reaction system was determined spectrophotometrically at 290 nm. The reaction system contained 50 mM potassium phosphate buffer (pH 7.0), 0.1 mM EDTA, 0.1 mM sodium ascorbate, 2.5 mM H_2O_2, and the protein extract. One unit of APX was defined as the amount of the enzyme that is needed to degrade 1 μmol of ascorbate min^{-1} mg^{-1} protein.

The activity of ascorbate oxidase (AAO, EC 1.10.3.3) was determined according to the method of Tamás [39], and the absorbance of the reaction system was determined spectrophotometrically at 265 nm. The reaction system contained 100 mM potassium phosphate buffer (pH 5.6), 5 mM EDTA-Na_2, 50 mM sodium ascorbate, and the protein extract. One unit of AAO was defined as the amount of the enzyme that is needed to oxidize 1 μmol of ascorbate min^{-1} mg^{-1} protein.

2.5. Statistical Analysis

Data were analyzed using Statgraphics (STN, St Louis, MO, USA). Two-way analysis of variances (ANOVA) was performed to determine the significant differences. If it was significant, a posteriori comparison of means was performed. The normality of the data was tested before statistical analysis. Differences between means were considered significant when the p-value was less than 0.05 according to the ANOVA F-test.

3. Results

3.1. Effect of Se and Cd on the Growth of Xinfeng Walnut

Under Cd-free condition, the application of 20 μM Se increased the biomass of wood, bark, and leaves (Table 2). The biomass of the wood, bark, and leaves in the Se20Cd0 treatment was 4.02, 2.76, and 12.65 g, respectively, which was 20%, 17%, and 18% higher than those in the Se0Cd0 treatment, respectively, and there was no difference in root biomass (Table 2). However, under Cd-free condition, the application of 200 μM Se inhibited the root and leaf growth of Xinfeng walnut (Table 2). The biomass of roots and leaves in the Se200Cd0 treatment was 9.58 and 8.96 g, respectively, which was 13% and 17% lower than those in the Se0Cd0 treatment, respectively, and there was no difference in the biomass of wood and bark (Table 2).

Table 2. The biomass of the roots, wood, bark, and leaves of *J. regia* (variety Xinfeng) in different treatments.

Se (μM)	Cd (mM)	Root (g)	Wood (g)	Bark (g)	Leaves (g)
0	0	10.99 ± 0.24 d	3.34 ± 0.12 b	2.35 ± 0.13 b	10.79 ± 0.62 c
	5	6.37 ± 0.30 a	2.99 ± 0.24 ab	1.96 ± 0.07 a	7.16 ± 0.12 a
20	0	11.25 ± 0.80 d	4.02 ± 0.15 c	2.76 ± 0.14 c	12.65 ± 0.42 d
	5	8.26 ± 0.30 b	2.79 ± 0.24 a	2.01 ± 0.07 a	7.86 ± 0.94 ab
200	0	9.58 ± 0.39 c	3.18 ± 0.13 ab	2.17 ± 0.09 ab	8.96 ± 0.27 b
	5	9.63 ± 0.42 c	3.11 ± 0.13 ab	2.14 ± 0.10 ab	8.86 ± 0.36 b
p-values	Cd	****	***	****	****
	Se	*	ns	*	*
	Cd × Se	****	**	**	***

Notes: Data were means ± SE (n = 6). Different lowercase letters in the same column indicate significant difference between treatments. *p*-values of the ANOVAs for Cd, Se, and their interaction (Cd × Se) are also indicated. *, $p < 0.05$; **, $p < 0.01$; ***, $p < 0.001$; ****, $p < 0.0001$; and ns, insignificant.

The biomass of the roots, wood, bark, and leaves was lower in the Se0Cd5 and Se20Cd5 treatments than in the Se0Cd0 and Se20Cd0 treatments, respectively (Table 2), and there was no difference between Se200Cd5 treatment and Se200Cd0 treatment (Table 2). Under Cd stress, the application of Se (Se20 and Se200) increased the biomass of the roots and leaves of the Xinfeng walnut (Table 2). The biomass of the root and leaves in the Se20Cd5 treatment was 8.26 and 7.86 g, respectively, which were 30% and 10% higher than those in the Se0Cd5 treatment, respectively (Table 2). The biomasses of the roots and leaves in the Se200Cd5 treatment were 9.63 and 8.86 g, respectively, which were 51% and 24% higher than those in the Se0Cd5 treatment, respectively (Table 2).

3.2. The Se and Cd Accumulations in Walnuts

Cd was not detected in the walnuts without Cd exposure (Figure 1a–d). The Cd concentrations in root and wood in the Se20Cd5 treatment were 1782.80 and 12.58 μg g^{-1} dry weight, respectively, which were 12% and 14% lower than those in the Se200Cd5 treatment, respectively (Figure 1a,b). Furthermore, there was no difference in the Cd concentration in the bark and leaves between Se20Cd5 treatment and Se0Cd5 treatment (Figure 1c,d). The Cd concentrations in the root, wood, bark, and leaves in the Se200Cd5 treatment were 1526.33, 10.28, 347.46, and 34.42 μg g^{-1} dry weight, respectively, which were 25%, 26%, 21%, and 23% lower than those in the Se0Cd5 treatment, respectively (Figure 1a,d).

Under Cd exposure or Cd-free conditions, the application of 20 μM Se increased Se concentration in bark and leaves, but did not alter Se concentration in root and wood compared with Se0 (Figure 1e–h). Under Cd-exposure or Cd-free conditions, the application of 200 μM Se increased the Se concentration in root, wood, bark, and leaves, compared with Se0 (Figure 1e–h).

3.3. Effects of Se Application on $O_2^{\bullet-}$, H_2O_2, and MDA Contents in Xinfeng Walnut

The $O_2^{\bullet-}$ contents in the roots of the Xinfeng walnut in the Se0Cd5, Se20Cd5, and Se200Cd5 treatments were 6.48, 5.72, and 5.28 μmol g^{-1} dry weight, respectively, which were 13%, 10%, and 21% higher than that in the Se0Cd0, Se20Cd0, and Se200Cd0 treatments, respectively, and the root $O_2^{\bullet-}$ content in the Se20Cd5 and Se200Cd5 treatments was 5.72 and 5.28 μmol g^{-1} dry weight, respectively, was were 12% and 8% lower than that in the Se0Cd5 treatment (Figure 2a). The leaf $O_2^{\bullet-}$ content in the Se0Cd5 treatment was 27% higher than that in the Se0Cd0 treatment, and the leaf $O_2^{\bullet-}$ content in the Se20Cd5 and Se200Cd5 treatments was 16% and 26% lower than that in the Se0Cd5 treatment (Figure 2b).

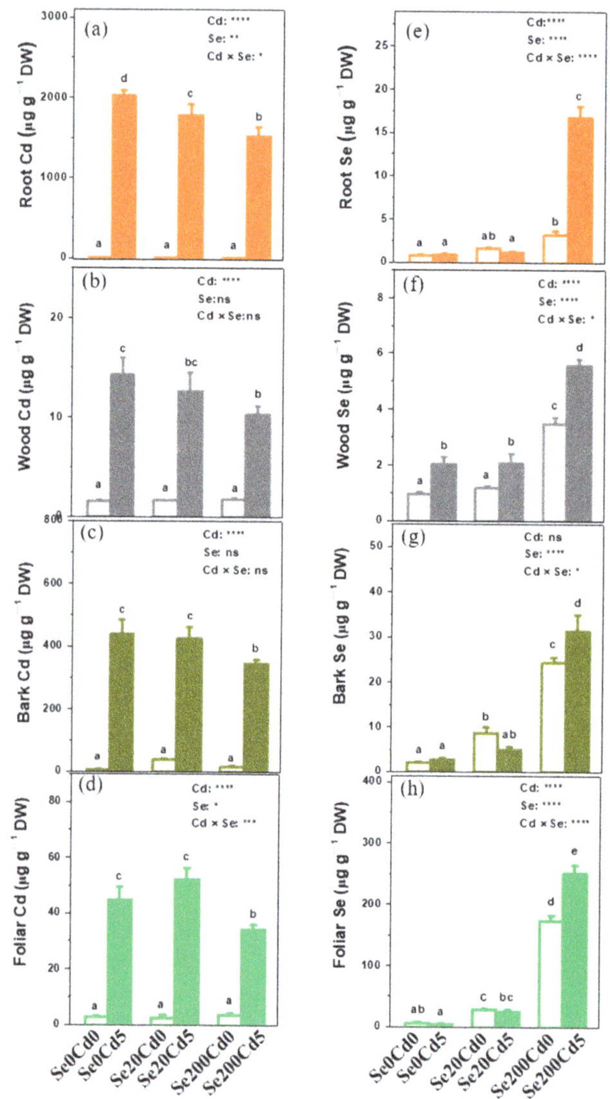

Figure 1. The Cd (**a–d**) and Se (**e–h**) concentrations in the roots, wood, bark, and leaves of *J. regia* (variety Xinfeng) exposed to either 0 (Cd0) or 5 mM (Cd5) Cd together with one of three Se levels (0 (Se0), 20 (Se20), or 200 (Se200) μM Se). Data were means ± SE (n = 6). Different lowercase letters on the bars indicate significant difference at $p < 0.05$ (F-test). *p*-values of the ANOVAs for Cd, Se, and their interaction (Cd × Se) are also indicated. *, $p < 0.05$; **, $p < 0.01$; ***, $p < 0.001$; ****, $p < 0.0001$; and ns, insignificant.

The root H_2O_2 content in the Se0Cd5, Se20Cd5, and Se200Cd5 treatments was 473.22, 365.60, and 345.35 μmol g^{-1} dry weight, respectively, which was 38%, 35%, and 27% higher than that in the Se0Cd0, Se20Cd0, and Se200Cd0 treatments, respectively (Figure 2c). The root H_2O_2 content in the Se20Cd5 and Se200Cd5 groups was 365.60 and 345.35 μmol g^{-1} dry weight, respectively, which was 23% and 27% lower than that in the Se0Cd5 treatment, respectively (Figure 2c). The leaf H_2O_2 content in the Se0Cd5 treatment was 164.63 μmol g^{-1} dry weight, which was 35% higher than that in the Se0Cd0 treatment

(Figure 2d). The leaf H_2O_2 content in the Se20Cd5 and Se200Cd5 treatments had no difference with that in the Se0Cd5 treatment (Figure 2d).

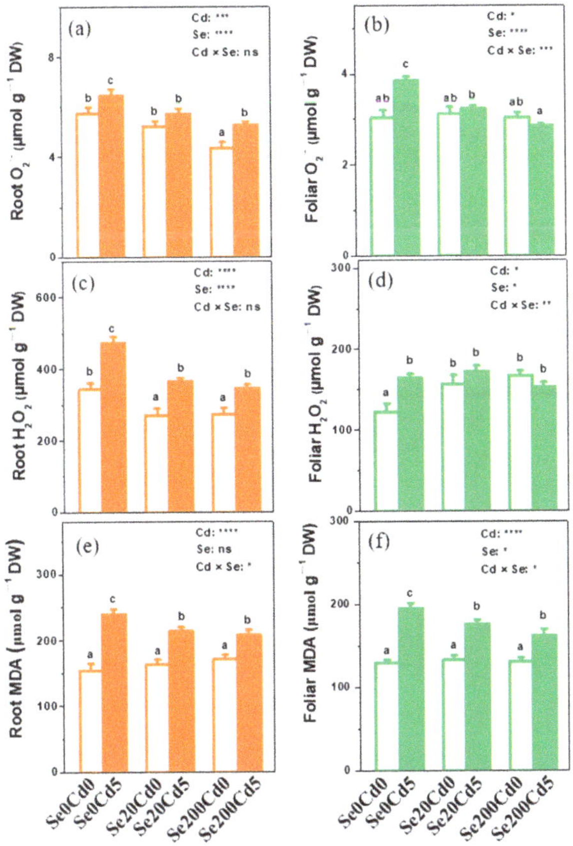

Figure 2. The contents of $O_2^{\bullet-}$ (**a**,**b**), H_2O_2 (**c**,**d**), and MDA (**e**,**f**) in the roots and leaves of *J. regia* (variety Xinfeng) exposed to either 0 (Cd0) or 5 mM (Cd5) Cd together with one of three Se levels (0 (Se0), 20 (Se20) or 200 (Se200) μM Se). Data were means ± SE (n = 6). Different lowercase letters on the bars indicate significant difference at $p < 0.05$ (F-test). *p*-values of the ANOVAs for Cd, Se and their interaction (Cd × Se) are also indicated. *, $p < 0.05$; **, $p < 0.01$; ***, $p < 0.001$; ****, $p < 0.0001$; and ns, insignificant.

The root MDA content in the Se0Cd5, Se20Cd5, and Se200Cd5 treatments was 240.21, 213.60, and 208.01 μmol g^{-1} dry weight, respectively, which was 55%, 31%, and 21% higher than those in the Se0Cd0, Se20Cd0, and Se200Cd0 treatments, respectively (Figure 2e). The root MDA content in the Se20Cd5 and Se200Cd5 treatments was 213.60 and 208.01 μmol g^{-1} dry weight, respectively, which was 11% and 13% lower than that in the Se0Cd5 treatment, respectively (Figure 2e). The leaf MDA content in the Se0Cd5, Se20Cd5, and Se200Cd5 treatments was 194.62, 175.99, and 162.23 μmol g^{-1} dry weight, respectively, which was 50%, 32%, and 24% higher than those in the Se0Cd0, Se20Cd0, and Se200Cd0 treatments, respectively (Figure 2f). The leaf MDA content in the Se20Cd5 and Se200Cd5 treatments was 175.99 and 162.23 μmol g^{-1} dry weight, respectively, which was 10% and 17% lower than that in the Se0Cd5 treatment, respectively (Figure 2f). However, there were no differences in root and leaf MDA contents between Se20Cd5 treatment and Se200Cd5 treatment (Figure 2e,f).

3.4. Effects of Se Application on the Antioxidant Enzyme Activities in Xinfeng Walnut

The root CAT activity in the Se0Cd5 treatment was lower than that in the Se0Cd0 treatment (Figure 3a). The root CAT activity in the Se20Cd5 and Se200Cd5 treatments was 302.26 and 220.86 U g^{-1} protein, respectively, which was 127% and 66% higher than that in the Se0Cd5 treatment (Figure 3a). The leaf CAT activity in the Se0Cd5 treatment was 46.47 U g^{-1} protein, which was 33% lower than that in the Se0Cd0 treatment (Figure 3b). The leaf CAT activity in the Se20Cd5 and Se200Cd5 treatments was 78.96 and 72.00 U g^{-1} protein, respectively, which was 77% and 55% higher than that in the Se0Cd5 treatment, respectively (Figure 3b).

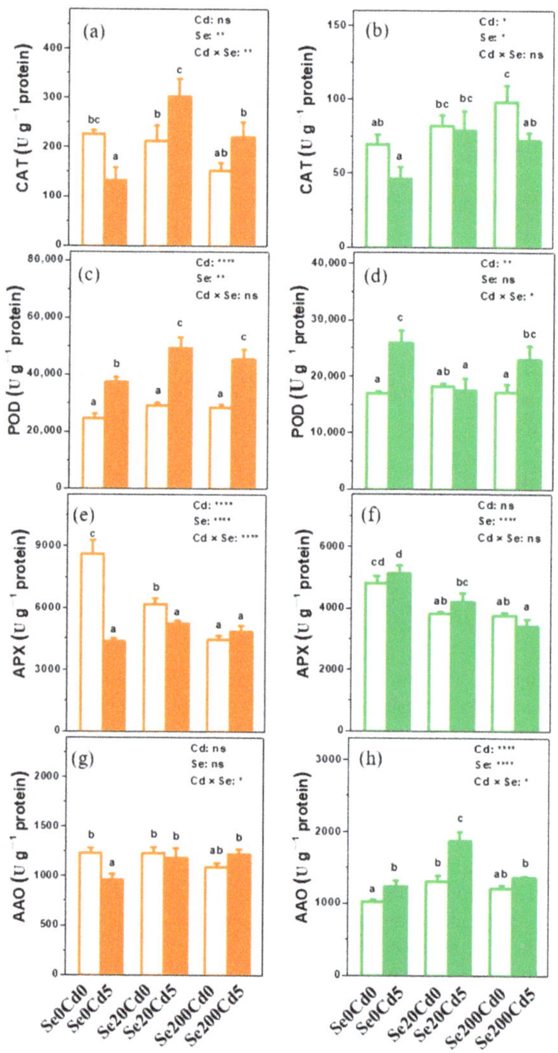

Figure 3. The activities of CAT (**a**,**b**), POD (**c**,**d**), APX (**e**,**f**), and AAO (**g**,**h**) in the roots and leaves of Xinfeng walnut exposed to either 0 (Cd0) or 5 mM (Cd5) Cd together with one of three Se levels (0 (Se0), 20 (Se20), or 200 (Se200) μM Se). Data were means ± SE (n = 6). Different lowercase letters on the bars indicate significant difference at $p < 0.05$ (F-test). p-values of the ANOVAs for Cd, Se and their interaction (Cd × Se) are also indicated. *, $p < 0.05$; **, $p < 0.01$; ****, $p < 0.0001$; and ns, insignificant.

The root POD activity in the Se0Cd5 treatment was 37,324.80 U g^{-1} protein, which was 51% higher than that in the Se0Cd0 treatment, and the root POD activity in the Se20Cd5 and Se200Cd5 treatments was 49,309.73 and 45,284.17 U g^{-1} protein, respectively, which was 21% and 32% higher than that in the Se0Cd5 treatment, respectively (Figure 3c). The leaf POD activity in the Se0Cd5 treatment was 25,872.33 U g^{-1} protein, which was 52% higher than that in the Se0Cd0 treatment (Figure 3c), and the leaf POD activity in the Se20Cd5 treatment was 17,489.86 U g^{-1} protein, which was 32% lower than that in the Se0Cd5 treatment (Figure 3d). There was no difference in the leaf POD activity between Se200Cd0 and Se0Cd0 treatments and between Se20Cd0 and Se0Cd0 treatments (Figure 3d).

The root APX activity in the Se0Cd5 treatment was 4359.23 U g^{-1} protein, which was 50% lower than that in the Se0Cd0 treatment (Figure 3e), and there was no difference in the root APX activity between Se20Cd5 and Se200Cd5 treatments (Figure 3e). There was also no difference in the leaf APX activity between Se0Cd5 treatment and Se0Cd0 treatment (Figure 3f). The leaf APX activities in the Se20Cd5 and Se200Cd5 treatments were 4225.99 and 3406.56 U g^{-1} protein, respectively, which were 18% and 34% lower than that in the Se0Cd5 treatment, respectively (Figure 3f).

The root AAO activity in the Se0Cd5 treatment was 968.49 U g^{-1} protein, which was 22% lower than that in the Se0Cd0 treatment. The leaf AAO activity in the Se0Cd5 treatment was 1243.63 U g^{-1} protein, which was 22% higher than that in the Se0Cd0 treatment (Figure 3g,h). The root AAO activities in the Se20Cd5 and Se200Cd5 treatments were 1182.46 and 1224.03 U g^{-1} protein, which were 22% and 26% higher than that in the Se0Cd5 treatment, respectively (Figure 3g). The leaf AAO activity in the Se20Cd5 treatment was 1870.53 U g^{-1} protein, which was 50% higher than that in the Se0Cd5 treatment, but there was no difference between Se200Cd5 treatment and Se0Cd5 treatment (Figure 3h).

4. Discussion

4.1. Se Application Promoted Xinfeng Walnut Growth and Reduced Cd Accumulation

Se is beneficial to plant growth. An appropriate dosage of Se could promote plant growth and increase crop yield [7,23]. However, excessive Se in soil is toxic to plants and inhibits plant growth [18]. In this study, the biomass of the wood, bark, and leaves of the Xinfeng walnut increased in the Se20Cd0 treatment, but reduced in the Se200Cd0 treatment. This indicates that 20 μM Se may be an appropriate dosage for Xinfeng walnut growth, while 200 μM Se is excessive. Previous studies show that excessive Cd could lead to stunted plant growth, chlorosis, and even cell death [1,40]. In this study, Cd addition decreased the biomass of Xinfeng walnut. This indicates that 5 mM of Cd could inhibit walnut growth. However, after the application of Se, the Cd-induced inhibition on roots and leaves was alleviated (Table 2). This indicates that Se application (Se20 and Se200) is beneficial for the growth of walnuts with Cd exposure.

In previous studies, plants with leaf Cd concentration higher than 100 μg g^{-1} dry weight were defined as hyperaccumulating plants [1,41]. In some herbaceous Cd hyperaccumulators, Cd concentration was as high as 3000 μg g^{-1} dry weight [42]. Several studies found that some woody plants, such as *Populus* and *Salix* species, can accumulate considerable levels of Cd in their aerial parts [43,44]. The Cd concentration reached 116 μg g^{-1} dry biomass in the leaves of *Salix caprea* [45]. In some poplar species, Cd concentrations in root, bark, and leaves reached 1000, 300, and 100 μg g^{-1} dry weight, respectively [35,37]. In this study, Cd concentrations in the bark and root reached 2022 and 439 μg g^{-1} dry biomass, respectively, which were higher than those in poplar bark and root, respectively, in previous studies [37,46]. This may be related to differences in plant species, Cd concentration, and treatment duration.

Cd in soil could be absorbed by plant roots and translocated to the aerial organs via xylem [47]. Previous studies show that exogenous application of Se could reduce the accumulation of Cd in cucumber and rice [24,25]. In this study, application of 20 μM Se reduced the Cd concentration in root and wood, and application of 200 μM Se reduced Cd concentration in root, wood, bark, and leaves. Similarly, in *Helianthus annuus*, the application of 5 and 10 μM Se decreased the Cd accumulation in roots and leaves, while the application of 20 μM Se did not alter the Cd concentration in roots and leaves [48]. This indicates that the application of Se could reduce Cd accumulation in walnuts, but the impact of Se on Cd accumulation varies with the change in Se concentration.

Cd may hinder nutrient uptake by plants, resulting in an imbalance of elements [49]. This study found that Cd addition enhanced the accumulation of Se in walnut wood treated with 20 μM Se, and the Se concentration in roots, wood, bark, and leaves of walnuts treated with 200 μM Se (Figure 1). Previous studies show that the application of inorganic Se fertilizers could effectively increase Se concentration in crops [18,50]. In this study, application of 20 μM Se increased Se concentration in walnut bark and leaves, and application of 200 μM Se increased Se concentration in walnut root, wood, bark, and leaves compared with non-Se treatment. This suggests that the Se sprayed on walnut leaves may preferentially accumulate in leaves and bark before being transported to wood and root.

4.2. Se Application Enhanced Antioxidant Enzyme Activities and Alleviated Cd-Induced Oxidative Stress

Cd is known to stimulate the formation of free radicals, which disrupts the plant defense system, resulting in the excessive accumulation of $O_2^{\bullet-}$ and H_2O_2 [51]. The content of MDA is used to assess the oxidative stress of chemical pollutants, including heavy metals [51,52]. In this study, under Cd exposure condition, the application of exogenous Se reduced the contents of ROS and MDA in walnut roots and leaves. This indicates that the oxidative stress induced by Cd could be relieved by Se. To reduce oxidative damage, plants evolved an antioxidant defense system consisting of a verity of enzymes, such as POD, CAT, APX, and AAO [51,53]. POD plays an important role in scavenging ROS produced under oxidative stress [54], and CAT, APX, and AAO could convert H_2O_2 to H_2O and O_2 [51,55]. For example, Ding found that a reduced activity of CAT could result in the accumulation of ROS in poplars [33]. In this study, the activities of CAT, APX, and AAO in walnut roots and CAT in leaves decreased after Cd exposure. The decreased activities of antioxidant enzymes could lead to the accumulation of ROS and oxidative stress. Furthermore, in this study, the application of Se increased the activities of CAT, POD, and AAO in root and CAT in the leaves of walnuts under Cd exposure. This suggests that Se could enhance the activities of antioxidant enzymes in walnut roots and leaves, thereby enhancing the ROS scavenging ability and reducing the Cd-induced oxidative stress.

In this study, exogenous application of 200 μM Se decreased the Cd accumulation in walnuts, which alleviated the growth inhibition and ROS accumulation induced by Cd. Furthermore, under Cd stress, the application of 20 μM Se increased the activities of CAT, POD, and AAO in walnut root and the activities of CAT and AAO in leaves, and the application of 200 μM Se enhanced the activities of CAT, POD, and AAO in walnut roots and CAT activity in leaves. Thus, Se application could alleviate the Cd-induced oxidative stress by increasing the ROS scavenging ability.

5. Conclusions

Exogenous application of Se could reduce the Cd concentration in the root, wood, bark, and leaves of Cd-stressed Xinfeng walnut, and the performance at the Se dose of 200 μM was better than that at the Se dose of 20μM. Furthermore, foliar application of 200 μM Se could remarkably increase Se concentration in walnuts compared with foliar application of 20 μM Se. Spraying Se on Cd-stressed walnuts could effectively reduce the growth inhibition induced by Cd. In addition, foliar-sprayed Se can be transported into the edible part of walnuts. The determination of appropriate Se dose will provide a practical guidance for the planting of Se-enriched walnuts. It should be noted that for walnuts grown in non-Cd-contaminated soils, the application rate of 200 μM Se may be excessive, and it is necessary to further explore the appropriate dose of Se.

Author Contributions: S.D. and B.W. designed the experiments, analyzed the data, drafted the manuscript and other authors revised the draft. B.W., D.Z., W.W., Y.S. and M.L. carried out the experiments. S.D. finalized the manuscript. All authors have read and agreed to the published version of the manuscript.

Funding: This study was jointly supported by the National Natural Science Foundation of China (Grant No. 32001347) and the Henan Provincial Science and Technology Research Project (Grant No. 212102110190).

Data Availability Statement: All data generated or analyzed during this study are included in this published article.

Conflicts of Interest: The authors declare no conflict of interest.

References

1. Luo, Z.B.; He, J.; Polle, A.; Rennenberg, H. Heavy metal accumulation and signal transduction in herbaceous and woody plants: Paving the way for enhancing phytoremediation efficiency. *Biotechnol. Adv.* **2016**, *34*, 1131–1148. [CrossRef] [PubMed]
2. Zhou, J.; Wan, H.; Qin, S.; He, J.; Lyu, D.; Li, H. Net cadmium flux and gene expression in relation to differences in cadmium accumulation and translocation in four apple rootstocks. *Environ. Exp. Bot.* **2016**, *130*, 95–105. [CrossRef]
3. Fang, B.; Zhu, X.Q. High content of five heavy metals in four fruits: Evidence from a case study of Pujiang County, Zhejiang Province, China. *Food Control* **2014**, *39*, 62–67. [CrossRef]
4. Liang, Y.; Zeng, T.; Tian, J.; Yan, J.; Lan, Z.; Chen, J.; Xin, X.; Lei, B.; Cai, Z. Long-term environmental cadmium exposure induced serum metabolic changes related to renal and liver dysfunctions in a female cohort from Southwest China. *Sci. Total Environ.* **2021**, *798*, 149379. [CrossRef] [PubMed]
5. Rizwan, M.; Ali, S.; Adrees, M.; Rizvi, H.; Zia-Ur-Rehman, M.; Hannan, F.; Qayyum, M.F.; Hafeez, F.; Ok, Y.S. Cadmium stress in rice: Toxic effects, tolerance mechanisms, and management: A critical review. *Environ. Sci. Pollut. Res.* **2016**, *23*, 17859–17879. [CrossRef] [PubMed]
6. Wang, Q.; Liu, J.; Cheng, S. Heavy metals in apple orchard soils and fruits and their health risks in Liaodong Peninsula, Northeast China. *Environ. Monit. Assess.* **2015**, *187*, 4178. [CrossRef] [PubMed]
7. Shahid, M.A.; Balal, R.M.; Khan, N.; Zotarelli, L.; Garcia-Sanchez, F. Selenium impedes cadmium and arsenic toxicity in potato by modulating carbohydrate and nitrogen metabolism. *Ecotoxicol. Environ. Saf.* **2019**, *180*, 588–599. [CrossRef] [PubMed]
8. Wan, Y.; Yu, Y.; Wang, Q.; Qiao, Y.; Li, H. Cadmium uptake dynamics and translocation in rice seedling: Influence of different forms of selenium. *Ecotoxicol. Environ. Saf.* **2016**, *133*, 127–134. [CrossRef]
9. Tian, S.; Liang, S.; Qiao, K.; Wang, F.; Zhang, Y.; Chai, T. Co-expression of multiple heavy metal transporters changes the translocation, accumulation, and potential oxidative stress of Cd and Zn in rice (*Oryza sativa*). *J. Hazard. Mater.* **2019**, *380*, 120853. [CrossRef]
10. Xu, L.L.; Fan, Z.Y.; Dong, Y.J.; Kong, J.; Bai, X.Y. Effects of exogenous salicylic acid and nitric oxide on physiological characteristics of two peanut cultivars under cadmium stress. *Biol. Plant.* **2015**, *59*, 171–182. [CrossRef]
11. Aradhya, M.K.; Potter, D.; Simon, C.J. Cladistic Biogeography of Juglans (*Juglandaceae*) Based on Chloroplast DNA Intergenic Spacer Sequences. In *Darwins Harvest. New Approaches Origins*; Columbia University Press: New York, NY, USA, 2004; pp. 143–170.
12. Martínez-García, P.; Crepeau, M.W.; Puiu, D.; Gonzalez-Ibeas, D.; Whalen, J.; Stevens, K.A.; Paul, R.; Butterfield, T.S.; Britton, M.T.; Reagan, R.L. The walnut (*Juglans regia*) genome sequence reveals diversity in genes coding for the biosynthesis of non-structural polyphenols. *Plant J.* **2016**, *87*, 507–532. [CrossRef]
13. Peng, Y.; Yao, J. Investigation and evaluation on heavy metal content and pesticide residues of three edible forest products in central Sichuan province. *Agric. Biotechnol.* **2019**, *8*, 108–112.
14. Arpadjan, S.A.; Momchilova, S.A.; Elenkova, D.B.; Blagoeva, E.C. Essential and toxic microelement profile of walnut (*Juglans regia* L.) cultivars grown in industrially contaminated area—Evaluation for human nutrition and health. *J. Food Nutr. Res.* **2013**, *52*, 121–127.
15. Muller, A.; Muller, C.C.; Lyra, F.; Mello, P.A.; Mesko, M.F.; Muller, E.I.; Flores, E. Determination of Toxic Elements in Nuts by Inductively Coupled Plasma Mass Spectrometry after Microwave-Induced Combustion. *Food Anal. Methods* **2013**, *6*, 258–264. [CrossRef]
16. Kafaoglu, B.; Fisher, A.; Hill, S.; Kara, D. Determination and evaluation of element bioaccessibility in some nuts and seeds by in-vitro gastro-intestinal method. *J. Food Compos. Anal.* **2016**, *45*, 58–65. [CrossRef]
17. Moreda-Pineiro, J.; Herbello-Hermelo, P.; Dominguez-Gonzalez, R.; Bermejo-Barrera, P.; Moreda-Pineiro, A. Bioavailability assessment of essential and toxic metals in edible nuts and seeds. *Food Chem.* **2016**, *205*, 146–154. [CrossRef]
18. White, P.J. Selenium accumulation by plants. *Ann. Bot.* **2016**, *117*, 217–235. [CrossRef]
19. Dinh, Q.T.; Cui, Z.; Huang, J.; Tran, T.A.T.; Wang, D.; Yang, W.; Zhou, F.; Wang, M.; Yu, D.; Liang, D. Selenium distribution in the Chinese environment and its relationship with human health: A review. *Environ. Int.* **2018**, *112*, 294–309. [CrossRef]
20. Filek, M.; Gzyl-Malcher, B.; Zembala, M.; Bednarska, E.; Laggner, P.; Kriechbaum, M. Effect of selenium on characteristics of rape chloroplasts modified by cadmium. *J. Plant Physiol.* **2010**, *167*, 28–33. [CrossRef]

21. Handa, N.; Kohli, S.K.; Sharma, A.; Thukral, A.K.; Bhardwaj, R.; Alyemeni, M.N.; Wijaya, L.; Ahmad, P.A. Selenium ameliorates chromium toxicity through modifications in pigment system, antioxidative capacity, osmotic system, and metal chelators in *Brassica juncea* seedlings. *S. Afr. J. Bot.* **2018**, *119*, 1–10. [CrossRef]
22. Zhao, Y.; Hu, C.; Wang, X.; Qing, X.; Wang, P.; Zhang, Y.; Zhang, X.; Zhao, X. Selenium alleviated chromium stress in Chinese cabbage (*Brassica campestris* L. ssp. Pekinensis) by regulating root morphology and metal element uptake. *Ecotoxicol. Environ. Saf.* **2019**, *173*, 314–321. [CrossRef]
23. Nawaz, F.; Ashraf, M.Y.; Ahmad, R.; Waraich, E.A.; Shabbir, R.N.; Bukhari, M.A. Supplemental selenium improves wheat grain yield and quality through alterations in biochemical processes under normal and water deficit conditions. *Food Chem.* **2015**, *175*, 350–357. [CrossRef]
24. Gao, M.; Zhou, J.; Liu, H.; Zhang, W.; Hu, Y.; Liang, J.; Zhou, J. Foliar spraying with silicon and selenium reduces cadmium uptake and mitigates cadmium toxicity in rice. *Sci. Total Environ.* **2018**, *631–632*, 1100–1108. [CrossRef]
25. Sun, H.; Wang, X.; Wang, Y.; Wei, Y.; Wang, G. Alleviation of cadmium toxicity in cucumber (*Cucumis sativus*) seedlings by the application of selenium. *Span. J. Agric. Res.* **2016**, *133*, 114–126. [CrossRef]
26. Yu, Y.; Wan, Y.; Wang, Q.; Li, H. Effect of humic acid-based amendments with foliar application of Zn and Se on Cd accumulation in tobacco. *Ecotoxicol. Environ. Saf.* **2017**, *138*, 286–291. [CrossRef]
27. Yu, Y.; Yuan, S.; Zhuang, J.; Wan, Y.; Wang, Q.; Zhang, J.; Li, H. Effect of selenium on the uptake kinetics and accumulation of and oxidative stress induced by cadmium in *Brassica chinensis*. *Ecotoxicol. Environ. Saf.* **2018**, *162*, 571–580. [CrossRef]
28. Harvey, M.A.; Erskine, P.D.; Harris, H.H.; Brown, G.K.; Pilon-Smits, E.; Casey, L.W.; Echevarria, G.; Ent, A. Distribution and chemical form of selenium in *Neptunia amplexicaulis* from Central Queensland, Australia. *Metallomics* **2020**, *12*, 514–527. [CrossRef]
29. Shi, Z.; Pan, P.; Feng, Y.; Kan, Z.; Li, Z.; Wei, F. Environmental water chemistry and possible correlation with Kaschin-Beck Disease (KBD) in northwestern Sichuan, China. *Environ. Int.* **2017**, *99*, 282–292. [CrossRef] [PubMed]
30. Wang, J.; Li, H.R.; Yang, L.S.; Li, Y.H.; Wei, B.G.; Yu, J.P.; Feng, F.J. Distribution and translocation of selenium from soil to highland barley in the Tibetan Plateau Kashin-Beck disease area. *Environ. Geochem. Health* **2017**, *39*, 221–229. [CrossRef]
31. Hussain, B.; Lin, Q.; Hamid, Y.; Sanaullah, M.; Di, L.; Hashmi, M.; Khan, M.B.; He, Z.; Yang, X. Foliage application of selenium and silicon nanoparticles alleviates Cd and Pb toxicity in rice (*Oryza sativa* L.). *Sci. Total Environ.* **2020**, *712*, 136497. [CrossRef]
32. Esringu, A.; Turan, M.; Cangnul, A. Remediation of Pb and Cd polluted soils with fulvic acid. *Forests* **2021**, *12*, 1608. [CrossRef]
33. Ding, S.; Ma, C.; Shi, W.; Liu, W.; Lu, Y.; Liu, Q.; Luo, Z.B. Exogenous glutathione enhances cadmium accumulation and alleviates its toxicity in *Populus canescens*. *Tree Physiol.* **2017**, *37*, 1697–1712. [CrossRef] [PubMed]
34. Lei, Y.; Korpelainen, H.; Li, C. Physiological and biochemical responses to high Mn concentrations in two contrasting *Populus cathayana* populations. *Chemosphere* **2007**, *68*, 686–694. [CrossRef]
35. He, J.; Ma, C.; Ma, Y.; Li, H.; Kang, J.; Liu, T.; Polle, A.; Peng, C.; Luo, Z.B. Cadmium tolerance in six poplar species. *Environ. Sci. Pollut. Res.* **2013**, *20*, 163–174. [CrossRef]
36. Luo, J.; Zhou, J.-J.; Masclaux-Daubresse, C.; Wang, N.; Wang, H.; Zheng, B. Morphological and physiological responses to contrasting nitrogen regimes in *Populus cathayana* is linked to resources allocation and carbon/nitrogen partition. *Environ. Exp. Bot.* **2019**, *162*, 247–255. [CrossRef]
37. He, J.; Qin, J.; Long, L.; Ma, Y.; Li, H.; Li, K.; Jiang, X.; Liu, T.; Polle, A.; Liang, Z.; et al. Net cadmium flux and accumulation reveal tissue-specific oxidative stress and detoxification in *Populus canescens*. *Physiol. Plant* **2011**, *143*, 50–63. [CrossRef]
38. Chen, L.; Han, Y.; Jiang, H.; Korpelainen, H.; Li, C. Nitrogen nutrient status induces sexual differences in responses to cadmium in *Populus yunnanensis*. *J. Exp. Bot.* **2011**, *62*, 5037–5050. [CrossRef]
39. Tamás, L.; Bočová, B.; Huttová, J.; Mistrík, I.; Ollé, M. Cadmium-induced inhibition of apoplastic ascorbate oxidase in barley roots. *Plant. Growth Regul.* **2006**, *48*, 41–49. [CrossRef]
40. Gallego, S.M.; Pena, L.B.; Barcia, R.A.; Azpilicueta, C.E.; Iannone, M.F.; Rosales, E.P.; Zawoznik, M.S.; Groppa, M.D.; Benavides, M.P. Unravelling cadmium toxicity and tolerance in plants: Insight into regulatory mechanisms. *Environ. Exp. Bot.* **2012**, *83*, 33–46. [CrossRef]
41. Reeves, R.D.; Baker, A.J.M.; Jaffre, T.; Erskine, P.D.; Echevarria, G.; van der Ent, A. A global database for plants that hyperaccumulate metal and metalloid trace elements. *New Phytol.* **2017**, *218*, 407–411. [CrossRef]
42. Zhou, W.; Qiu, B. Effects of cadmium hyperaccumulation on physiological characteristics of *Sedum alfredii* Hance (Crassulaceae). *Plant Sci.* **2005**, *169*, 737–745. [CrossRef]
43. Konlechner, C.; Tuerktas, M.; Langer, I.; Vaculík, M.; Wenzel, W.W.; Puschenreiter, M.; Hauser, M.T. Expression of zinc and cadmium responsive genes in leaves of willow (*Salix caprea* L.) genotypes with different accumulation characteristics. *Environ. Pollut.* **2013**, *178*, 121–127. [CrossRef] [PubMed]
44. Robinson, B.H.; Mills, T.M.; Petit, D.; Fung, L.E.; Green, S.R.; Clothier, B.E. Natural and induced cadmium-accumulation in poplar and willow: Implications for phytoremediation. *Plant Soil* **2000**, *227*, 301–306. [CrossRef]
45. Unterbrunner, R.; Puschenreiter, M.; Sommer, P.; Wieshammer, G.; Tlustos, P.; Zupan, M.; Wenzel, W.W. Heavy metal accumulation in trees growing on contaminated sites in Central Europe. *Environ. Pollut.* **2007**, *148*, 107–114. [CrossRef]
46. Zhang, Z.C.; Chen, B.X.; Qiu, B.S. Phytochelatin synthesis plays a similar role in shoots of the cadmium hyperaccumulator Sedum alfredii as in non-resistant plants. *Plant Cell Environ.* **2010**, *33*, 1248–1255. [CrossRef]
47. He, X.L.; Fan, S.K.; Zhu, J.; Guan, M.Y.; Liu, X.X.; Zhang, Y.S.; Jin, C.W. Iron supply prevents Cd uptake in Arabidopsis by inhibiting IRT1 expression and favoring competition between Fe and Cd uptake. *Plant Soil* **2017**, *41*, 446–453. [CrossRef]

48. Saidi, I.; Chtourou, Y.; Djebali, W. Selenium alleviates cadmium toxicity by preventing oxidative stress in sunflower (*Helianthus annuus*) seedlings. *J. Plant Physiol.* **2014**, *171*, 85–91. [CrossRef]
49. Fodor, F.; Gáspár, L.; Morales, F.; Gogorcena, Y.; Lucena, J.J.; Cseh, E.; Kröpfl, K.; Abadía, J.; Sárvári, E. Effects of two iron sources on iron and cadmium allocation in poplar (*Populus alba*) plants exposed to cadmium. *Tree Physiol.* **2005**, *25*, 1173–1180. [CrossRef]
50. Alfthan, G.; Eurola, M.; Ekholm, P.; Venalainen, E.; Root, T.; Korkalainen, K.; Hartikainen, H.; Salminen, P.; Hietaniemi, V.; Aspila, P.; et al. Effects of nationwide addition of selenium to fertilizers on foods, and animal and human health in Finland: From deficiency to optimal selenium status of the population. *J. Trace. Elem. Med. Biol.* **2015**, *31*, 142–147. [CrossRef]
51. Li, S.; Chen, J.; Islam, E.; Wang, Y.; Wu, J.; Ye, Z.; Yan, W.; Peng, D.; Liu, D. Cadmium-induced oxidative stress, response of antioxidants and detection of intracellular cadmium in organs of moso bamboo (*Phyllostachys pubescens*) seedlings. *Chemosphere* **2016**, *153*, 107–114. [CrossRef]
52. Del, R.D.; Stewart, A.J.; Pellegrini, N. A review of recent studies on malondialdehyde as toxic molecule and biological marker of oxidative stress. *Nutr. Metab. Cardiovas.* **2005**, *15*, 316–328. [CrossRef]
53. Kanazawa, S.; Sano, S.; Koshiba, T.; Ushimaru, T. Changes in antioxidative enzymes in cucumber cotyledons during natural senescence: Comparison with those during dark-induced senescence. *Physiol. Plant.* **2000**, *109*, 211–216. [CrossRef]
54. Sun, Y.; Zhou, Q.; Lin, W.; Liu, W. Cadmium tolerance and accumulation characteristics of *Bidens pilosa* L. as a potential Cd-hyperaccumulator. *J. Hazard. Mater.* **2009**, *161*, 808–814. [CrossRef]
55. Pignocchi, C.; Fletcher, J.M.; Wilkinson, J.E.; Barnes, J.D.; Foyer, C.H. The function of ascorbate oxidase in tobacco. *Plant Physiol.* **2003**, *132*, 1631–1641. [CrossRef] [PubMed]

forests

MDPI

Article

Analysis of the NAC Gene Family in *Salix* and the Identification of *SpsNAC005* Gene Contributing to Salt and Drought Tolerance

Haifeng Yang *, Lijiao Fan, Xingwang Yu, Xinqian Zhang, Pu Hao, Dongshan Wei and Guosheng Zhang

College of Forestry, Inner Mongolia Agricultural University, Hohhot 010018, China;
fanlijiao@emails.imau.edu.cn (L.F.); yuxingwang_0104@163.com (X.Y.); zxq17614885517@163.com (X.Z.);
h1692146316@163.com (P.H.); wdsh0812@imau.edu.cn (D.W.); zgs1960@imau.edu.cn (G.Z.)
* Correspondence: haifeng@imau.edu.cn; Tel.: 86+134-8851-8872

Abstract: The NAC gene family is of great value for plant stress resistance and development. In this study, five NAC genes with a typical NAM domain were isolated from *Salix psammophila*, which is a stress-resistant willow endemic to western China. Two hundred sixty-two NAC genes from *Salix psammophila*, *Salix purpurea*, and *Arabidopsis* were used to construct the phylogenetic tree to examine the phylogenetic relationship. Five NAC genes in *Salix psammophila* were the focus of bioinformatics analysis and conserved structural domain analysis. The *SpsNAC005* gene was overexpressed in *Populus hopeiensis*, and the transgenic lines were subjected to salt and simulated drought stress to analyze their phenotype changes and tolerance to stress. The results showed that transgenic poplar height and leaf area increased by 29.73% and 76.36%, respectively, compared with those of wild-type plants. Under stress treatment, the height growth rates and ground diameter growth rates of the transgenic lines were significantly higher than those of the wild-type, whereas their fresh weight and dry weight were decreased compared to those of the wild-type. The SOD activities, POD activities, and Pro contents of the transgenic plants were significantly increased, and the accumulation of MDA was significantly lower than that in the wild-type, and the transgenic lines showed clear tolerance to salt and drought. The expressions of the *SOS1*, *MPK6*, *HKT1*, and *P5CS1* genes were downregulated in the transgenic lines. The expression of the *PRODH1* gene was downregulated in the transgenic lines. These results indicate that overexpression of the *SpsNAC005* gene in transgenic plants can promote plant growth and development and improve tolerance to salt and drought.

Keywords: SpsNAC005 gene; *Populus × hopeiensis Hu & Chow*; Salix; stress tolerance

Citation: Yang, H.; Fan, L.; Yu, X.; Zhang, X.; Hao, P.; Wei, D.; Zhang, G. Analysis of the NAC Gene Family in *Salix* and the Identification of *SpsNAC005* Gene Contributing to Salt and Drought Tolerance. *Forests* **2022**, *13*, 971. https://doi.org/10.3390/f13070971

Academic Editors: Jie Luo and Wentao Hu

Received: 27 May 2022
Accepted: 20 June 2022
Published: 21 June 2022

1. Introduction

NAC (No apical meristem (*NAM*), *Arabidopsis* transcription activation factor (*ATAF*), Cup-shaped cotyledon (*CUC*)) transcription factors have an important regulatory role in the processes of plant growth, development and defensive responses. The first genes identified in the NAC family were *ATAF1* and *ATAF2* in 1993 [1]. *NAM* was cloned from petunia in 1996 [2], followed by the discovery of the *CUC2* genes in *Arabidopsis* [3]. The NAC structural domain is a sequence structure unique to the NAC transcription factor family; it consists of approximately 160 amino acids and is a highly conserved structural domain located at the N-terminal end with the function of binding to DNA or proteins [4]. The NAC structural domains are generally composed of five structural subdomains, A, B, C, D, and E; three substructural domains, A, C, and D, are highly conserved, whereas the B and E substructural domains are less conserved [5].The C-terminus of NAC protein also has a highly differentiated transcriptional regulatory region with transcriptional activation or repression activity, which determines its functional differences, enabling NAC transcription factors to regulate plant growth and development, the formation of meristems and border organs, lateral root development, plant senescence, biotic and abiotic stress responses [6], cell wall formation, and apoptosis [7]. The NAC transcription factor family in several species, including *Arabidopsis thaliana*, *Oryza sativa*, and *Populus trichocarpa*, has been systematically analyzed. Functional differences among NAC gene family members were assessed using comprehensive expression profiling [8,9].

Responding to abiotic stress is one of the important functions of NAC transcription factors. Ectopic expression of the *HaNAC1* gene in *Arabidopsis* resulted in a significant increase in proline content and enhanced drought resistance [10]. The transcriptional regulator *ANAC032* positively regulates leaf senescence by regulating the production of reactive oxygen species and can promote the production of H_2O_2 under auxin and salt stress [11]. In addition, *ANAC032* negatively regulates the expression of anthocyanin biosynthesis genes in *Arabidopsis* in response to abiotic stresses such as high light, salinity, and oxidative stress [12]. The *CaNAC23* gene cloned from *Capsicum annuum* was significantly induced to express under drought and salt stress [13]. The expression of the *MfNAC37* gene in *Medicago falcata* under salt stress was increased in three tissue parts of roots, stems, and leaves [14]. Overexpression of the *MdNAC1* gene in apple (*Malus domestica*) enhances the activity of plant photosynthesis and ROS scavenging enzymes, making apple plants tolerant to drought stress [15]. In rice, overexpression of the *SNAC1* gene increased seed setting by 21–34% under drought conditions and also improved salt tolerance in rice [16].

Salix psammophila C. Wang et Chang Y. Yang is an important wind break and sand fixation shrub species in western China, and it has a wide range of adaptation and strong resistance to adversity [17]. *S. psammophila* plays a vital role in vegetation restoration and afforestation, especially in sandy areas of northwest China. *S. psammophila* is not only an excellent tree species for wind prevention and sand fixation in western China, but also a pioneer tree species in arid and saline areas of northwest China, which is of great significance to maintaining the stability of the ecosystem and improving the regional ecological environment [18]. Consistently, *S. psammophila* interlaced through underground roots to form a strong sand-stabilizing barrier and built a protective forest belt in a short time, which improved the overall level of land desertification control [19]. *Populus hopeiensis* is one of the common tree species in northern China [20]. It is tolerant of soil conditions and has good adaptability. It is a tree species for soil and water conservation or timber forests. Our research group established the genetic transformation system of *P. hopeiensis* in the early stage. On the one hand, it is intended to be used as a model woody plant to study the gene function of woody plants. On the other hand, it is also intended to obtain transgenic lines of *P. hopeiensis* with stronger adaptability for future application.

2. Materials and Methods

2.1. *Salix NAC Gene Family Analysis and Functional Prediction of SpsNAC005 Gene*

2.1.1. Bioinformatics Analysis of Genes in Salix

The tolerance of *S. psammophila* to harsh environments directly proved that there were excellent resistance genes in *S. psammophila*. Therefore, five potential stress resistance genes, *SpsNAC005*, *SpsNAC034*, *SpsNAC041*, *SpsNAC042*, and *SpsNAC120*, were screened separately from *S. psammophila*. We obtained the sequence information on the *SpsNAC005*, *SpsNAC034*, *SpsNAC041*, *SpsNAC042*, and *SpsNAC120* genes from *Salix psammophila*. Prosite ExPASy server (http://web.expasy.org/protparam/, accessed on 20 March 2022) was used to analyze the amino acids, molecular weight (kD), theoretical pI, instability index, and aliphatic index. Hydrophobicity was analyzed by https://web.expasy.org/protscale/, accessed on 20 March 2022) and used to predict physicochemical characteristics of the SpsNAC proteins. iLoc-LncRNA (http://lin-group.cn/server/iLoc-LncRNA/predictor.php, accessed on 22 March 2022) was used to predict subcellular localization. SignalP-5.0 (https://services.healthtech.dtu.dk/service.php?SignalP-5.0, accessed on 22 March 2022) was used to predict signal peptides. The open reading frames (ORF) of *SpsNACs* were determined by ORF Finder (https://www.ncbi.nlm.nih.gov/orffinder/, accessed on 24 March 2022). Clustal Omega (https://www.ebi.ac.uk/Tools/msa/clustalo/, accessed on 24 March 2022) was used for amino acid multiple sequence alignment. MEME (http:

//meme-suite.org/, accessed on 24 March 2022) was used for base-sequence analysis of *SpsNACs* by (maximum number of motifs = 7).

2.1.2. Evolutionary Tree Construction of Genes in Salix and Arabidopsis

The complete genome annotation information and sequences of *Salix* and *Arabidopsis* were obtained from the phytozome database (https://phytozome-next.jgi.doe.gov/, accessed on 26 March 2022). The Pfam database (http://pfam.xfam.org, accessed on 26 March 2022) Hidden Markov Model (HMM) was used to obtain the NAC specific structural domain and HMMER (https://www.ebi.ac.uk/Tools/hmmer/, accessed on 26 March 2022) was used to identify the NAC genes of *Salix purpurea*. The genes containing NAC specific structural domains were further identified using the protein conserved structural domain prediction tool SMART (http://smart.embl-heidelberg.de/, accessed on 26 March 2022), and redundant *Salix purpurea* NAC gene family members were manually removed. Multiple sequence alignment of NAC proteins was carried out using MAFFT online website (version 7 https://mafft.cbrc.jp/alignment/software/, accessed on 20 March 2022). The phylogenetic tree was constructed with MEGA X (MEGA X 10.2.5) using the neighbor-joining (NJ) method with 1000 bootstrap test replicates.

2.1.3. Subfamilies Analysis and Functional Prediction of *SpsNAC005*

The protein structural domains of *SpsNAC005* sequences were analyzed using Pfam online software, and homologous genes of *Salix purpurea*, *Populus trichocarpa*, *Populus deltoides*, *Citrus sinensis*, *Arabidopsis*, *Oryza sativa*, *Zea mays*, and *Glycine max* were obtained from the phytozome database. These were compared with *SpsNAC005* cloned fragments (Figure 1d) using MAFFT online software for sequence alignment using MEGA X to construct a phylogenetic tree for family analysis of *SpsNAC005*. The homologous structural domains of *SpsNAC005*, *Salix purpurea* (Sapur.005G052900.1), and *Populus trichocarpa* (Potri.005G069500.1) homologous genes were compared by MEME with the structural domain parameter of 7. The structural domain differences between them were analyzed and compared to further predict the function of *SpsNAC005*.

2.2. Experimental Materials and Vector Construction

Salix psammophila used in this experiment was sampled from the Germplasm Resources gene bank of *S. psammophila* in Ordos Dalad, the Inner Mongolia Autonomous Region of China (E 110°38′59.1″, N 40°14′15.5″). The branches of 1-year-old *S.psammophila* clone (11–30) were collected as experimental materials for cloning of *SpsNAC005* gene. The tissue culture seedlings of *P. hopeiensis* were preserved by the forest genetics and breeding department of Forestry College, Inner Mongolia Agricultural University. The *SpsNAC005* gene was cloned in overexpression vector pMDC32 driven by 2 × 35 s promoter and transformed into *P. hopeiensis* [21]. *Agrobacterium tumefaciens* strain GV3101 carrying CaMV 35S::*SpsNAC005* plasmid was stored in a refrigerator at −80 °C.

2.3. *Agrobacterium*-Mediated Transformation of *P. hopeiensis* Leaf Discs and PCR Assay

Agrobacterium tumefaciens containing the 35S::SpsNAC005 recombinant plasmid was activated and transformed into wild-type *P. hopeiensis* via leaf disc transformation [22].

The total DNA of resistant *P. hopeiensis* was extracted using the CTAB method [23], and the target gene fragments were amplified using PCR with the primers SpsNAC005-F1 (5′-ATGAAGGGAAATGGATCGGC-3′) and SpsNAC005-R1 (5′-CACCATACAGTGCCATTTCTGG-3′). The PCR amplification procedure was 94 °C for 5 min; 35 cycles of 94 °C for 30 s, 62 °C for 30 s, and 72 °C for 1 min; and then 72 °C for 7 min followed by 1% agarose gel detection after amplification. The specific primers were designed by Primer Version 5.00 (PREMIER Biosoft International, Palo Alto, CA, USA) and Primer 3 input (v. 0.4.0) (Whitehead Institute for Biomedical Research, Cambridge, MA, USA, https://bioinfo.ut.ee/primer3-0.4.0/ accessed on 26 March 2022).

2.4. Tissue Expression Specificity of SpsNAC005 in Salix psammophila and Relative Expression Level of SpsNAC005 in Overexpression Lines

RNA (TIABGEN RNA prep pure, Tiangen Biotech Co., Ltd., Beijing, China) was extracted from six different tissues of *S. psammophila* (leaves, roots, shoots, soft stems, semi-lignified stems, and mature stems) and reverse transcribed into cDNA (SuperScript IV First-Strand Synthesis System, Tiangen Biotech Co., Ltd., Beijing, China). The relative expression of the *SpsNAC005* gene in different tissues of *S. psammophila* was detected by qRT-PCR. The primers were UBQ-F (5′-AAGCCCAAGAAGATCAAGCA-3′) and UBQ -R (5′-ACCACCAGCCTTCTGGTAAA-3′) [24]. The primers for SpsNAC005 gene expression analysis were SpsNAC005-F1 (5′-ATGAAGGGAAATGGATCGGC-3′) and SpsNAC005-R1 (5′-CACCATACAGTGCCATTTCTGG-3′). They were identical to the primer sequence mentioned in Section 2.3. The qRT-PCR was conducted with three replicates on a Roche Light Cycler 480 II (Roche, Shanghai, China), and the relative gene expression levels were calculated using the $2^{-\Delta\Delta Ct}$ method [25].

RNA was extracted from the leaves of different overexpression lines and reverse transcribed to cDNA, and qRT-PCR was performed to detect expression of the *SpsNAC005* gene in the positive lines. The primers for the quantitative detection of the *SpsNAC005* gene were SpsNAC005-F1 and SpsNAC005-R1, and the primers for the internal reference gene of *P. hopeiensis* were PtrActin-F (5′-AAACTGTAATGGTCCTCCCTCCG-3′) and PtrActin-R (5′-AGCATCATCACAATCACTCTCCGA-3′) [24]. The PtrActin gene was used as an internal control to normalize the data, and the relative gene expression levels were calculated using the $2^{-\Delta\Delta Ct}$ method.

2.5. Determination of Salt and Drought Stress and Related Indices in Overexpression Lines

2.5.1. Transplanting before Stress and Determination of Related Growth Indices

Transgenic seedlings with the same growth vigor and wild-type tissue culture seedlings of *P. hopeiensis* were selected for seedling refining, and a mixture of grass charcoal soil and vermiculite (2:1) was used as the substrate. The temperature of the culture chamber was 25 °C, and the light/dark cycle was 16 h/8 h. After 2 weeks, the seedlings were transferred to the greenhouse.

The plant height and ground diameter were measured by ruler and Vernier calipers on the 30th day of seedling refining, and the leaf area was measured by CAD [26] (four and five internode leaf measurements). On the 1st day before stress treatment and the 5th day after stress treatment, the plant height and ground diameter of *P. hopeiensis* were measured. Each treatment was repeated in three pots, and each pot was measured three times. The relative growth rates of plant height and ground diameter of *P. hopeiensis* during the stress period were calculated. On the 5th day of stress treatment, the fresh weight of the plant was measured with a balance, and then the plant was dried to a constant weight in a 50 °C oven.

The relative growth rates of plant height (%) = $(H_1 - H_0)/H_0 \times 100\%$

H_1: Plant height on day 5 of stress (cm)

H_0: Plant height one day before stress (cm)

The relative growth rates of ground diameter (%) = $(D_1 - D_0)/D_0 \times 100\%$

D_1: Ground diameter on day 5 of stress (cm)

D_0: Ground diameter one day before stress (cm)

2.5.2. Salt and Drought Treatments and Determination of SOD, POD, MDA, and Pro

The wild-type *P. hopeiensis* and three lines (OE-4, OE-3, and OE-1) of transgenic *P. hopeiensis* were each treated with one of three solutions (water, a 0.9% NaCl solution, or 10% PEG) at 30 d and 35 d, with three biological replicates of each treatment [27]. Stress was stopped when the treated plants showed signs of leaf yellowing and wilting. On day 40, mixed samples of functional leaves (3–6 leaves) with normal growth were collected, snap-frozen in liquid nitrogen, and stored at −80 °C for the subsequent determination of superoxide dismutase (SOD), peroxidase (POD), malondialdehyde (MDA), and proline

(Pro); three replicate measures were made. These physiological parameters were measured using Solarbio biochemical kits (Beijing Solarbio Science & Technology Co., Ltd. Beijing, China, item numbers in order: BC0170, BC0090, BC0020, BC0290).

2.5.3. Expression Analysis of the SOS1, MPK6, HKT1, P5CS1, and PRODH1 Genes under Drought and Salt Stress

The expression of the *SOS1*, *MPK6*, *HKT1*, *P5CS1*, and *PRODH1* genes under different treatments was detected via qRT-PCR using cDNA obtained by the method described in Section 2.4. The primers for gene quantification were SOS1-F1 (5′-GGCTGTTGTTGCTCTGTTGA-3′), SOS1-R1 (5′-TTATGGCACCCGAGGTAAAG-3′), MPK6-F1 (5′-CTGCAAACGTCCTGCATAGA-3′), MPK6-R1 (5′-AACACACAACCCACTGACCA-3′), HKT1-F1 (5′-TGGTTCAGTGCCTGTTGTTC-3′), and HKT1-R1 (5′-CATCCTTGCACGAGCTATCA-3′), P5CS1-F1 (5′-GTGTTGGCACCCTCTTTCAT-3′), P5CS1-R1 (5′-CATCAGCTACGTCCAGCAAA-3′), PRODH1-F1 (5′-ATGGCACGATTCAAGCCTAC-3′), PRODH1-R1 (5′-TTCAAGCATGAACGAAGCAC-3′), and the internal reference primer of *P. hopeiensis* genes was PtrActin. The primer sequences of the *SOS1*, *MPK6*, HKT1, P5CS1, and *PRODH1* genes were designed by Primer 3 input (Whitehead Institute for Biomedical Research, MA, USA, https://bioinfo.ut.ee/primer3-0.4.0/ accessed on 26 March 2022).

2.6. Statistical Analysis

All experiments were conducted at least three times independently, and statistical significance level analysis of the data was performed using Excel 2019 (Microsoft Office, Washington, Redmond, WA, USA), F-test (two-sample ANOVA), *t*-test (two-sample equal/heteroskedasticity hypothesis), and one-way ANOVA for randomized block design [28]. The results of one-way ANOVA for randomized block design were labeled using lowercase letters. Other significance was designated as follows: *** $p < 0.001$, extremely significant; ** $0.001 \leq p < 0.01$, very significant; * $0.01 \leq p < 0.05$, significant. The tissue expression of *SpsNAC005* in *S. psammophila* and the relative gene expression of *SpsNAC005* positive lines were analyzed by one-way ANOVA for randomized block design. Data on *P. hopeiensis* height and leaf area, plant height growth rate, ground diameter growth rate, fresh weight, dry weight, SOD activities, POD activities, MDA contents, and Pro contents, and relative expression level of *SOS1*, *MPK6*, *HKT1*, *P5CS1*, and *PRODH1* genes were tested for equality of variance with F-test before *t*-test.

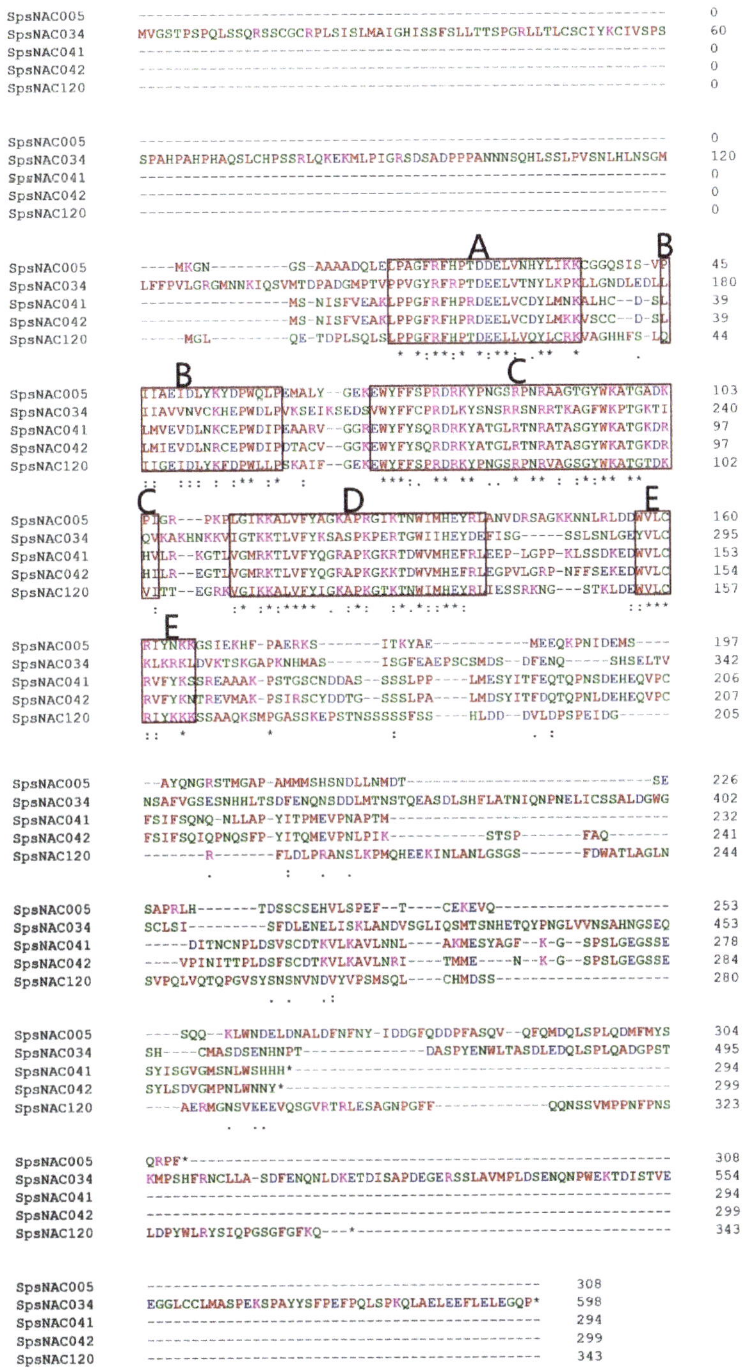

(**a**)

Figure 1. *Cont.*

(**b**)

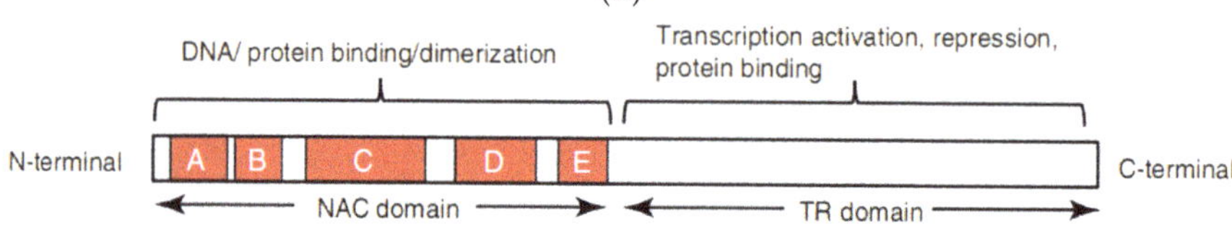

(**c**)

```
1     ATGAAGGGAAATGGATCGGCGGCAGCAGCAGATCAGCTGGAGTTACCAGCTGGGTTTAGA
1      M  K  G  N  G  S  A  A  A  A  D  Q  L  E  L  P  A  G  F  R
61    TTTCATCCAACAGATGACGAGTTGGTGAATCATTACTTAATCAAGAAATGTGGAGGCCAA
21     F  H  P  T  D  D  E  L  V  N  H  Y  L  I  K  K  C  G  G  Q
121   TCAATTTCTGTTCCTATTATTGCTGAGATTGATCTTTACAAGTATGATCCTTGGCAGCTT
41     S  I  S  V  P  I  I  A  E  I  D  L  Y  K  Y  D  P  W  Q  L
181   CCAGAAATGGCACTGTATGGTGAAAAGGAATGGTACTTCTTTTCCCCGAGGGATAGAAAA
61     P  E  M  A  L  Y  G  E  K  E  W  Y  F  F  S  P  R  D  R  K
241   TATCCAAACGGTTCAAGGCCGAACCGGGCGGCCGGAACCGGGTATTGGAAAGCAACCGGA
81     Y  P  N  G  S  R  P  N  R  A  A  G  T  G  Y  W  K  A  T  G
301   GCGGATAAACCCATTGGTCGGCCCAAACCACTTGGCATAAAAAAAGCACTTGTATTTTAT
101    A  D  K  P  I  G  R  P  K  P  L  G  I  K  K  A  L  V  F  Y
361   GCTGGCAAAGCTCCCAGAGGAATCAAAACCAATTGGATCATGCATGAATACCGCCTTGCT
121    A  G  K  A  P  R  G  I  K  T  N  W  I  M  H  E  Y  R  L  A
421   AATGTTGATAGGTCTGCTGGCAAGAAAAATAACCTAAGGCTTGATGATTGGGTATTGTGT
141    N  V  D  R  S  A  G  K  K  N  N  L  R  L  D  D  W  V  L  C
481   AGAATATACAACAAGAAAGGAAGCATAGAGAAGCATTTTCCTGCAGAAAGAAAATCAATA
161    R  I  Y  N  K  K  G  S  I  E  K  H  F  P  A  E  R  K  S  I
541   ACCAAGTATGCTGAAATGGAAGAGCAAAAGCCCAACATAGATGAAATGTCTGCATACCAG
181    T  K  Y  A  E  M  E  E  Q  K  P  N  I  D  E  M  S  A  Y  Q
601   AATGGAAGGTCAACAATGGGAGCTCCAGCAATGATGATGAGTCATAGCAATGACTTGTTG
201    N  G  R  S  T  M  G  A  P  A  M  M  M  S  H  S  N  D  L  L
661   AATATGGACACGTCTGAGTCAGCACCGAGGTTGCATACAGACTCAAGCTGTTCAGAGCAC
221    N  M  D  T  S  E  S  A  P  R  L  H  T  D  S  S  C  S  E  H
721   GTGCTGTCACCTGAGTTCACTTGCGAGAAGGAGGTTCAGAGCCAGCAGAAATTATGGAAT
241    V  L  S  P  E  F  T  C  E  K  E  V  Q  S  Q  Q  K  L  W  N
781   GATGAGCTGGATAATGCACTCGATTTCAACTTCAATTACATTGACGATGGCTTCCAAGAT
261    D  E  L  D  N  A  L  D  F  N  F  N  Y  I  D  D  G  F  Q  D
841   GACCCTTTCGCCTCGCAAGTTCAGTTCCAAATGGACCAGCTCTCACCTCTGCAGGACATG
281    D  P  F  A  S  Q  V  Q  F  Q  M  D  Q  L  S  P  L  Q  D  M
901   TTCATGTATTCACAGAGGCCCTTTTGA
301    F  M  Y  S  Q  R  P  F  *
```

(**d**)

Figure 1. The five NAC genes was highly conserved in *Salix psammophila*. (**a**) Amino acid sequence alignment of *SpsNAC005*, *SpsNAC034*, *SpsNAC041*, *SpsNAC042*, and *SpsNAC120*. (**b**) Motif analysis of five *S.psammophila* NAC proteins. (**c**) Schematic diagram of NAC transcription factor structure [29]. (**d**) *SpsNAC005* gene sequence and its translated amino acid sequence.

3. Results

3.1. Identification of NAC Genes Obtained by Cloning in Salix psammophila

The five NAC family genes *SpsNAC005, SpsNAC034, SpsNAC041, SpsNAC042, and SpsNAC120* were isolated from *S. psammophila* and named after the classification of NAC genes in *Populus trichocarpa* because of the close genetic relationship between poplar and willow [9].

These five NACs encode proteins ranging from 294 amino acids (33.09 kD) to 598 amino acids (65.82 kD), and the predicted isoelectric points varied widely from 5.32 to 8.56. The predicted aliphatic index ranged from 60.91 to 71.92. The results of the predicted subcellular localizations revealed that *SpsNAC005, SpsNAC034*, and *SpsNAC120* all were localized in the nucleolus, nucleus, and nucleoplasm. *SpsNAC041* and *SpsNAC042* were localized in cytoplasm and cytosol. All five NAC sequences are unstable proteins, and all are hydrophilic proteins without signal peptides (Table 1).

Table 1. Basic information on *SpsNAC005, SpsNAC034, SpsNAC041, SpsNAC042, SpsNAC120* proteins in *Salix psammophila*.

Gene-ID	Amino Acids	Molecular Weight (kD)	Theoretical pI	Instability Index	Aliphatic Index	Hydrophobicity	Subcellular Localization
SpsNAC005	308	35.24	6.02	43.68	60.91	Hydrophilic	Nucleolus, Nucleus, Nucleoplasm
SpsNAC034	598	65.82	5.32	68.83	71.92	Hydrophilic	Nucleolus, Nucleus, Nucleoplasm
SpsNAC041	294	33.09	6.46	46.12	65.34	Hydrophilic	Cytoplasm, Cytosol
SpsNAC042	299	34.19	6.45	51.39	68.09	Hydrophilic	Cytoplasm, Cytosol
SpsNAC120	343	38.30	8.56	40.74	67.53	Hydrophilic	Nucleolus, Nucleus, Nucleoplasm

Multiple sequence alignment of the five *S. psammophila* NAC genes revealed that the NAC proteins all contained the complete NAM structural domain (A–E) (Figure 1a). To further analyze the diversity of NAC proteins in *S. psammophila*, the conserved motif analysis of five SpsNAC proteins was mapped using the MEME tool (Figure 1b). The NAC transcription factor N-terminal has a NAC structural domain consisting of approximately 150 amino acid residues, and the NAC domain can be divided into five sub-structural domains [29] (Figure 1c). The highly conserved positively charged subdomains C and D bind to DNA, while subdomain A may be involved in functional dimer formation, whereas the divergent subdomains B and E may be responsible for the functional diversity of NAC genes [5]. The results showed that motif 3 (subdomain A), motif 4 (subdomain B), motif 1 (subdomain C), motif 2 (subdomain D), and motif 5 (subdomain E) were the typical structural domains of NAC proteins with highly conserved *NAM* structural domains. In addition, the motif 7 is the characteristic domain in *SpsNAC041* and *SpsNAC042*.

3.2. Classification of Salix NAC Gene Family Members and Phylogenetic Tree Analysis

To examine the phylogenetic relationship between NAC proteins in *Arabidopsis* and *Salix*, a phylogenetic tree was constructed by sequence alignment of full-length NAC proteins (Figure 2). The phylogenetic tree divided NAC family proteins into 21 subfamilies. Among them, Sapur.014G031200.1 was distinguished from other NAC members and formed an individual clade. For simplicity, subfamily names are based on the tree topology alphabet letter series (NAC-a to NAC-r).

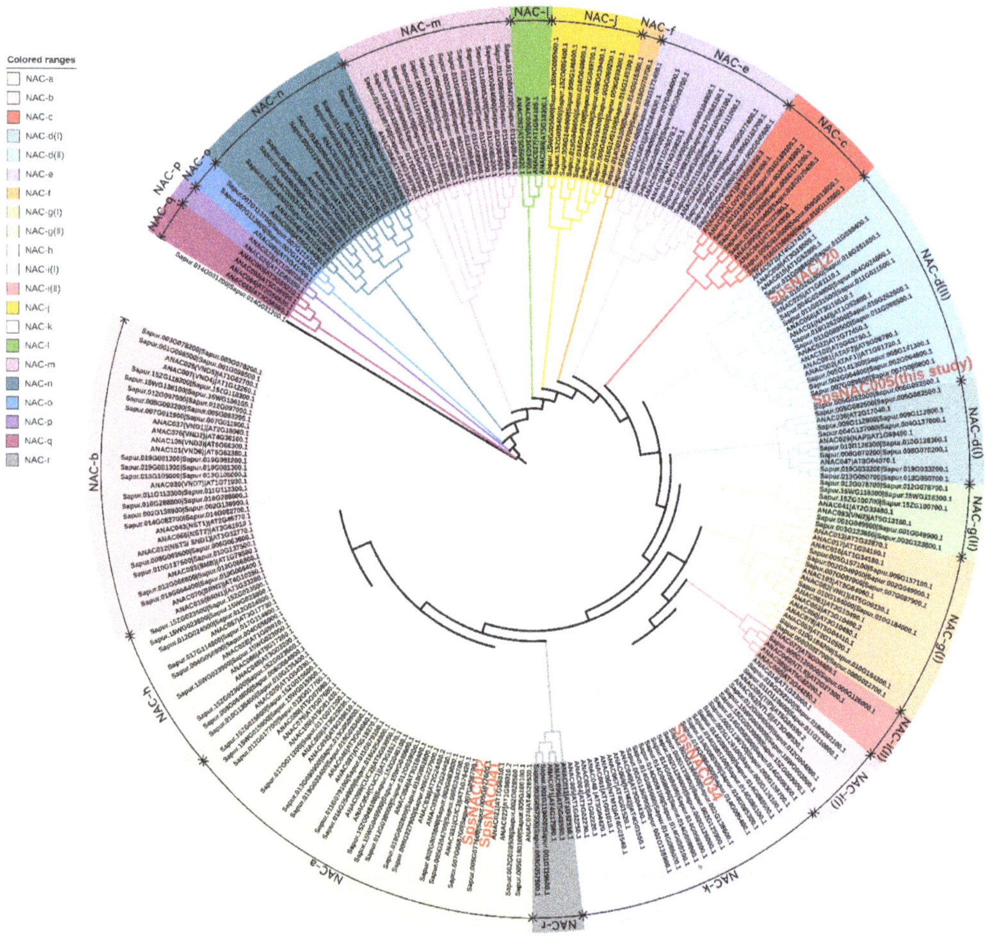

Figure 2. Phylogenetic analysis of NAC proteins from *Salix* and *Arabidopsis thaliana*.

NAC members are distributed in most subfamilies, but NAC genes with the same function showed a tendency to belong to one subfamily [9,30]. The NAC-a subfamily contains 21 NAC genes of *Salix*. Related studies have shown that this subfamily is mainly involved in the regulation of bud organ boundary division. *Arabidopsis CUC1/NAC054* and *CUC2/NAC098* genes play an indispensable role in the development of bud tip meristem [3,31]. The NAC proteins associated with secondary wall formation during fiber and vascular vessel development are divided into three independent subfamilies: NAC-b, NAC-n and NAC-e. The three subfamilies contain 18, 9 and 10 *Salix NAC* genes, respectively. All *NSTs* (*NST1*, *NST2*, *NST3/SND1*) and *VNDs* (*VND1–VND7*) were classified as NAC-b subfamily, while *SND2* and *SND3* were classified as NAC-n [32,33]. Another well-characterized member, *XND1* in secondary wall formation process, was assigned to the NAC-e subfamily. At5g64530 (*ANAC104*) encodes NAC structural domain 1 (*XND1*) of xylem, which negatively regulates the synthesis of secondary cell wall and programmed cell death, so as to antagonize differentiation of xylem [34]. NAC-d(I) and NAC-d(II) subfamilies play a role in the establishment of shoot apical meristem (SAM). The two subfamilies contain a total of 17 NAC genes in *Salix*. *ATAF* in NAC-d(II) subfamilies may be related to abiotic stress, and the *ANAC019*, *ANAC055*, and *ANAC072* genes, are induced by abiotic stresses (ABA, drought, and salinity) and enhance tolerance to drought when ectopically overexpressed [35]. *NAP* is involved in regulation of leaf senescence in NAC-d(I) [36]. Membrane-associated NAC proteins that mediate cytokinin signaling or endoplasmic reticulum stress responses during cell division cluster into the NAC-i(I), NAC-i(II), and NAC-k subfamilies. A total of six NAC genes were found from NAC-i(I) and NAC-i(II) subfamilies in *Salix*. The NAC members named NTLs in these subfamilies were bound by membrane to ensure that plants respond to developmental changes and environmental stimuli rapidly. Related studies have shown that the plasma membrane-associated transcription factor *ANAC062* (*NTL6*) is another important unfolded protein response (UPR) mediator in *Arabidopsis* plants. *ANAC062* transfers endoplasmic reticulum (ER) stress signals from plasma membrane to nucleus and plays an important role in regulating the expression of UPR downstream genes [37]. The NAC-k subfamily has 15 NAC genes from *Salix* and *NTM* from *Arabidopsis thaliana*. Studies have shown that ANAC069 (*NTM2*) responds to abiotic stress physiology and regulation. At the same time, during Arabidopsis seed germination, *NTM2* integrates auxin signal into salt stress signaling, which can regulate seed germination under high salinity [38]. Additionally, the NAC-f and NAC-j subfamilies only contain NAC proteins from *Salix*, indicating that they may be lost in *Arabidopsis*, but from the other side they can be acquired de novo in *Salix*. These gene differences may also be due to the perennial woody habit of *Salix*. Phylogenetic analysis also revealed NAC-1, NAC-p, and NAC-q subfamilies, which only contain *Arabidopsis* NAC sequences, indicating that NAC members in these subgroups were different in herbs and woody plants.

3.3. Functional Prediction of SpsNAC005

To clarify the evolutionary relationship and possible function of *SpsNAC005* gene in woody plants, the homologous genes from *S. psammophila*, *Salix purpurea*, *Populus trichocarpa*, *Populus deltoides*, *Citrus sinensis*, *Arabidopsis*, *Oryza sativa*, *Zea mays*, and *Glycine max* were compared to construct an evolutionary tree (Figure 3A). The results showed that *S. psammophila*, *S. purpurea*, *P. trichocarpa*, *P. deltoides*, and *C. sinensis* were in the same evolutionary branch, *G.max* was in one branch, and *Arabidopsis* was in one branch, indicating that the homologous gene of *SpsNAC005* was differentiated in woody and herbaceous plants. The comparison results for homologous structural domains indicated that the NAC protein sequences of *SpsNAC005*, Sapur.005G052900.1, Potri.005G069500.1 (*PNAC005*), and Podel.05G075500.1 have the same structural domain and the gene functions may be extremely similar (Figure 3B). Sapur.005G052900.1 is a member of the NAC-d(II) subfamily in the family evolutionary tree (Figure 2). Previous studies have shown that expression of the *Arabidopsis ATAF1* gene, which is also in the NAC-d(II) subfamily, was significantly induced by high salinity and abscisic acid (ABA) [39]; furthermore, *ATAF1* transgenic rice had significantly increased salt tolerance and was more insensitive to ABA, both of which could verify that *ATAF1* plays an important role in response to salt stress. Through the functional analysis of related genes, it was predicted that *SpsNAC005* may enhance drought and salt stress tolerance in *Salix*.

3.4. Tissue-Specific Expression of the SpsNAC005 Gene in Salix psammophila

The expression of the *SpsNAC005* gene was analyzed using qRT-PCR in the tissues of leaves, roots, shoots, soft stems, semi-lignified stems, and mature stems of *S. psammophila*. The results showed that the relative expression of the *SpsNAC005* gene was highest in soft stems, followed by leaves, shoots, semi-lignified stems, and mature stems, with the lowest expression in roots (Figure 4). Data analysis showed there were significant differences in *SpsNAC005* relative expression ($p < 0.05$) in leaves, roots, shoots, and soft stems, but it was not significant between semi-lignified stems, shoots, and mature stems.

3.5. PCR Identification and Expression Analysis of Transgenic *P. hopeiensis*

The PCR identification of *P. hopeiensis* transgenic lines showed that seven lines were positive, and the size of the identified target bands was 1049 bp (Figure 5A), which was consistent with the expected band size.

qRT-PCR expression analysis of *SpsNAC005* gene was performed on the seven transgenic lines of *P. hopeiensis*. The relative expression levels of OE-3 and OE-4 were significantly higher than those of all other overexpression lines. This results indicated that all seven lines obtained in this study expressed the target gene (Figure 5B); the three lines with the high expression, OE-4, OE-3, and OE-1, were selected for subsequent stress treatment and related data measurements.

3.6. Phenotypic Analysis of Transgenic *P. hopeiensis* Lines

3.6.1. Changes in Plant Height in SpsNAC005 Overexpression Lines

Statistical comparisons of height were performed for the OE-4, OE-3, and OE-1 lines. The results showed that the average plant heights of OE-4, OE-3, and OE-1 lines were 26.1 ± 0.81 cm, 23.6 ± 0.97 cm, and 22.3 ± 0.61 cm, respectively, and the average plant height of the wild-type was 18.5 ± 0.65 cm (Figure 6A). Based on these results, the plant heights of three overexpression lines were significantly higher than that of the wild-type (Figure 6B), indicating that *SpsNAC005* gene can enhance the height of the transgenic lines.

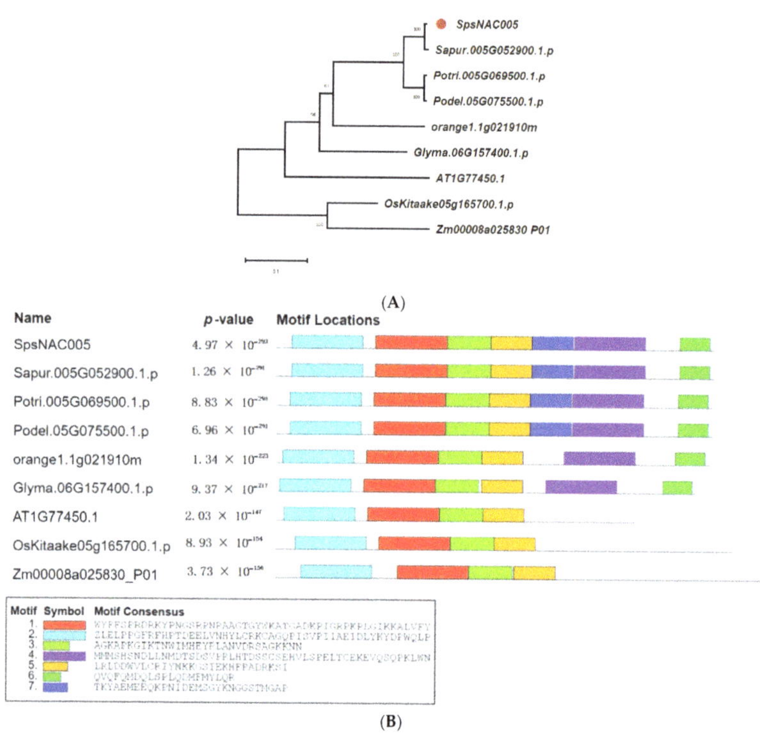

Figure 3. Evolution and protein structure of *SpsNAC005* homologous gene. (**A**) Phylogenetic analysis of *SpsNAC005* in monocotyledonous and dicotyledonous plants. The phylogenetic trees were constructed with MEGA X software using the neighbor-joining (NJ) method with 1000 bootstrap test replicates. Bar = 0.1 substitutions per site. (**B**) The homology structure domain alignment of amino acid sequence of NAC-like gene family.

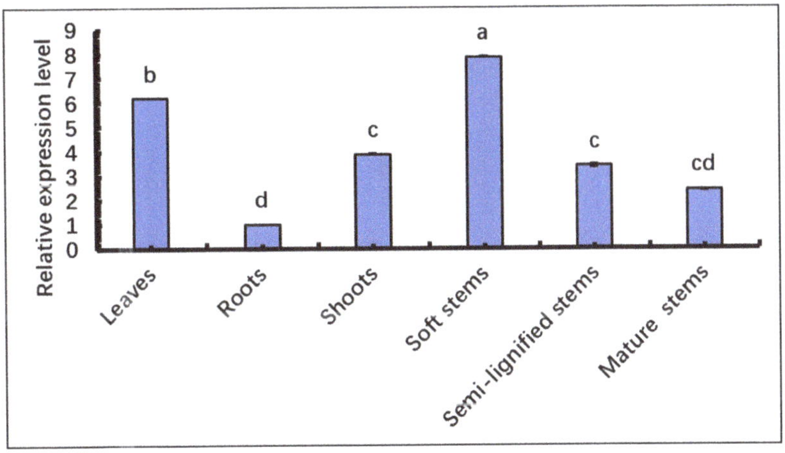

Figure 4. Tissue expression patterns of *SpsNAC005* in *S. psammophila*. qRT-PCR was used to analyze the relative expression level of *SpsNAC005* in different tissues. Error bars represent ± SD from three biological repeats. These samples used the One-way ANOVA for randomized block design, as long as there was one same marked letter, the difference was not significant, and the difference was significant if there is a different marked letter. Generally, lowercase letters indicate the significant level α = 0.05.

(**A**)

(**B**)

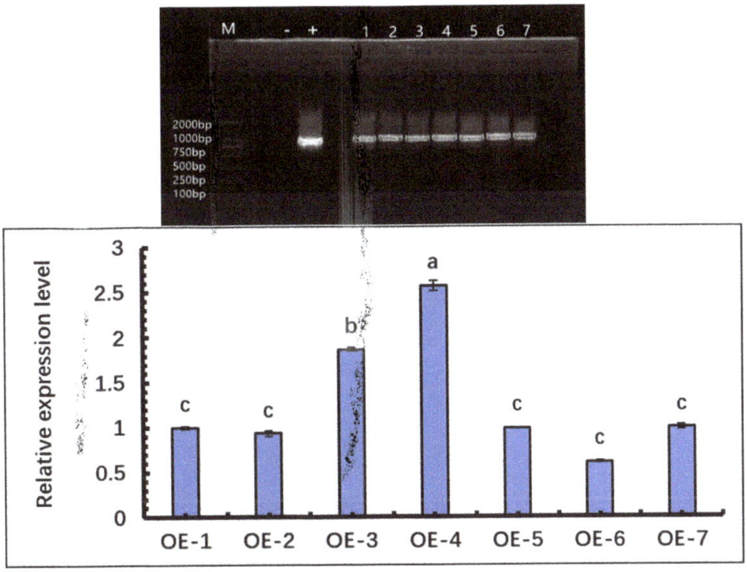

Figure 5. Identification of transgenic lines of *P. hopeiensis*. (**A**) Agarose gel assay of PCR amplified target genes from *P. hopeiensis* overexpression lines. (**B**) Determination of relative gene expression in *SpsNAC005* positive lines. These samples used the One-way ANOVA for randomized block design, as long as there was one same marked letter, the difference was not significant, and the difference was significant if there is a different marked letter. Generally, lowercase letters indicate the significant level α = 0.05.

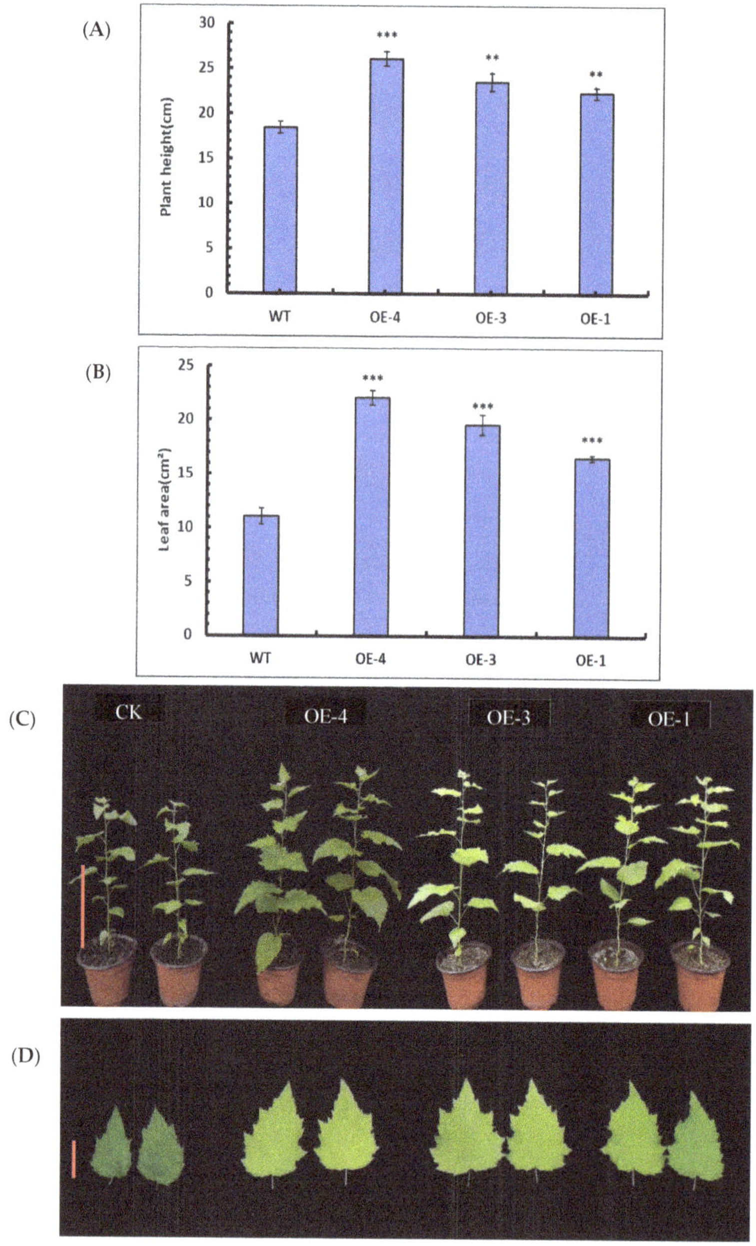

Figure 6. Phenotypic analysis of transgenic *P. hopeiensis*. (**A**) Plant height of WT and *SpsNAC005* overexpression lines of *P. hopeiensis*. (**B**) Leaf areas of wild-type and *SpsNAC005* overexpression lines of *P. hopeiensis*. (**C**) The 30-day-old wild-type and overexpression lines of *P. hopeiensis* without stress. (**D**) Leaf areas of 30-day-old wild-type and overexpression lines of *P. hopeiensis* without stress. Whole plant scale = 10 cm, leaf scale = 3 cm. (Means ± S.D.) These averages are the average of three individual measurements. The significance of differences was determined based on *t*-test and F-test. *** $p < 0.001$, extremely significant; ** $0.001 \leq p < 0.01$, very significant.

3.6.2. Leaf Area Changes in SpsNAC005 Overexpression Lines

The leaf areas of OE-4, OE-3, and OE-1 were measured. The average leaf areas of OE-4, OE-3, and OE-1 were 22.1 ± 0.68 cm^2, 19.6 ± 0.96 cm^2, and 16.5 ± 0.27 cm^2, respectively, whereas the average leaf area of the wild-type was 11.0 ± 0.75 cm^2 (Figure 6C). The results showed that the average leaf area was significantly increased in overexpression lines (Figure 6D). Consequently, the *SpsNAC005* gene may play a role by positively regulating the leaf growth of *P. hopeiensis*.

3.7. Morphological Changes in Transgenic P. hopeiensis under Stress Treatment

The overexpression lines, OE-4, OE-3, and OE-1 were treated with 0.9% NaCl and 10% PEG to evaluate the tolerance of OE lines for salt and drought stress. The relative growth rates of height and ground diameter in wild-type and overexpression lines were calculated under salt and simulated drought separately. The results showed that the height and ground diameter growth rates of wild-type and overexpression lines were decreased under stress. However, the height and ground diameter growth in overexpression lines were significantly faster than those in the wild-type under stresses (Figure 7A,B). After 5 d of salt stress, the basal leaves of the WT turned yellow, while most of the overexpression lines remained green and fresh (Figure 7C). After 5 d of drought stress, most of the leaves of WT died due to water loss (Figure 7D), whereas most of the transgenic poplar leaves were still fresh and alive. These results showed that under stress, the relative growth rate of height and ground diameter of *SpsNAC005* overexpression lines was significantly higher than that of wild-type, and overexpression lines had resistance to stress.

(A)

(B)

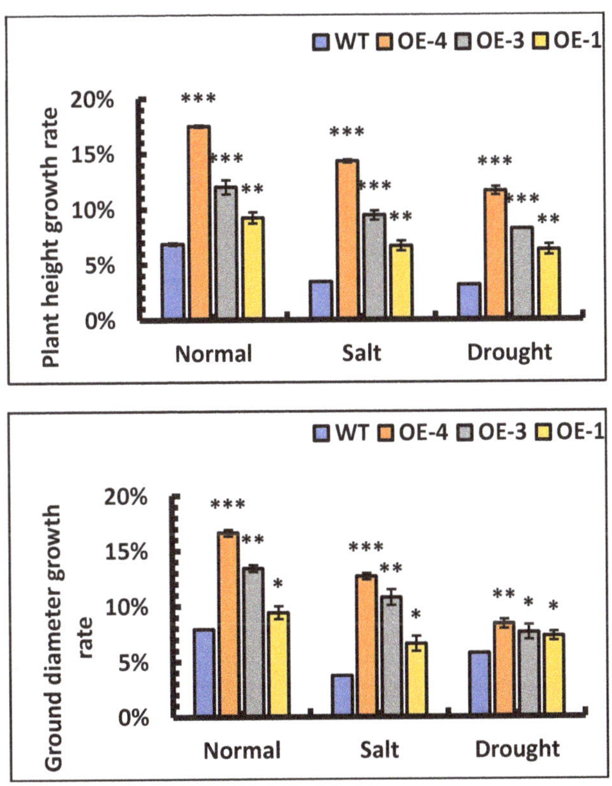

Figure 7. *Cont.*

(C)

(D)

Figure 7. Morphological changes in transgenic *P. hopeiensis* under stress. (**A**) The relative growth rate of height of *P. hopeiensis* under stress. (**B**) The relative growth rate of the ground diameter of *P. hopeiensis* under stress. (**C**) Morphological changes under salt stress. (**D**) Morphological changes under simulated drought stress. Phenotypes of natural leaf growth of overexpressed and wild-type lines after 30 days of seedling refinement. Whole plant scale = 10 cm. (Means ± S.D.) These averages are the average of three individual measurements. The significance of differences was determined using the *t*-test and F-test. *** $p < 0.001$, extremely significant; ** $0.001 \leq p < 0.01$, very significant; * $0.01 \leq p < 0.05$, significant.

3.8. The Fresh Weight and Dry Weight of *P. hopeiensis* under Stress Treatment

The fresh weight and dry weight of each line under stress treatment were measured. The results of fresh weight measurements showed that OE-4, OE-3, OE-1 and wild-type under salt stress decreased by 7.11%, 7.46%, 4.23% and 13.27%, respectively, compared with those of the control lines. Under simulated drought stress, the fresh weight of OE-4, OE-3, OE-1 and wild-type decreased by 13.26%, 3.39%, 3.38% and 33.20%, respectively, compared with their control lines. Among them, the fresh weight of wild-type decreased significantly under stress (Figure 8A). The results of dry weight measurements showed that OE-4, OE-3, OE-1 and wild-type under salt stress decreased by 12.22%, 5.00%, 2.98% and 27.93%, respectively, compared with those of their control lines. Under simulated drought stress, the dry weight of the OE-4, OE-3, OE-1 and wild-type decreased by 15.16%, 3.33%, 7.72% and 42.30%, respectively, compared with their control lines. Under stress treatment, the dry weight of wild-type was significantly reduced compared to that of the control group (Figure 8B). The above results showed that *SpsNAC005* gene stress could reduce the water loss of plants and enhance their resistance to stress.

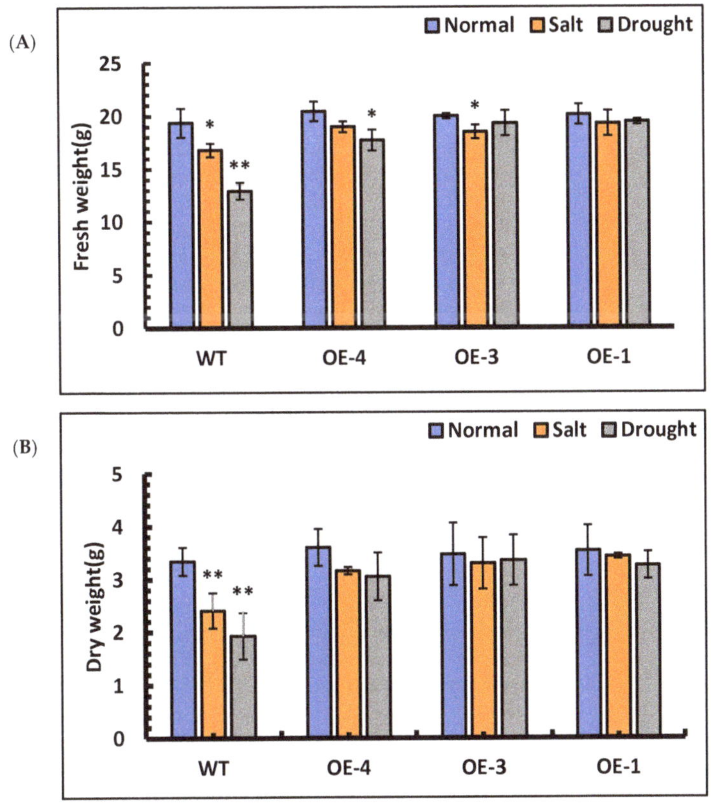

Figure 8. The fresh weight and dry weight of *P. hopeiensis* under salt and drought stress. (**A**) The fresh weight of *P. hopeiensis* under stress treatment. (**B**) The dry weight of *P. hopeiensis* under stress treatment. Values are the mean ± standard deviation of three independent experiments. The significance of differences was determined using *t*-test and F-test. ** $0.001 \leq p < 0.01$, very significant; * $0.01 \leq p < 0.05$, significant.

3.9. Response of Antioxidant System and Osmotic Regulation of Transgenic P. hopeiensis to Stress

To further characterize the resistance of overexpressed lines, the SOD activities, POD activities, MDA contents, and Pro contents in OE-4, OE-3, OE-1, and wild-type plants were quantified after stress treatment. The mean differences in the SOD activities of OE-4, OE-3, OE-1, and wild-type plants compared with those of the control lines after salt stress were 356.66%, 345.09%, 271.96%, and 213.31%, respectively, and the POD activities increased by 310.25%, 273.61%, 177.53%, and 126.64%, respectively, compared with those of the control lines. Under drought stress, the differences in SOD activities of OE-4, OE-3, OE-1, and the wild-type were 481.73%, 455.42%, 290.38%, and 233.70%, respectively, and the POD activities increased by 149.59% and 147. 59%, 93.67%, and 71.03%, respectively, compared with those of the control lines. The results showed that the SOD and POD activities increased significantly in each line when subjected to salt and drought stress (Figure 9A,B). Among the lines, the SOD and POD activities were significantly higher in overexpression lines than wild-type, indicating that the overexpression of the *SpsNAC005* gene enhanced the ability of plants to scavenge oxygen radicals and break down hydrogen peroxide, thus improving the stress resistance of the transgenic line.

The changes in the MDA contents of the transgenic lines after stress were quantified. The results showed that the MDA contents of OE-4, OE-3, OE-1, and wild-type increased by 56.74%, 78.59%, 101.20%, and 134.85%, respectively, under salt stress compared with those of the control lines. Under drought stress, the MDA contents of OE-4, OE-3, OE-1, and wild-type plants increased by 128.91%, 192.92%, 216.11%, and 290.89%, respectively, compared with control lines. Based on these results, the MDA content was significantly higher in all lines under salt and drought stresses, indicating that free radicals acted on lipids in the plant, resulting in peroxidation reactions, and plant cells were damaged to some extent. However, under the same stress, the MDA content in the transgenic plants was significantly lower than that in the wild-type (Figure 9C), indicating that the free radicals in the transgenic plants produced lipid peroxidation to a lesser extent, indirectly indicating that there was significantly less cell damage in the transgenic plants than in the wild-type plants, and the overexpression of the *SpsNAC005* gene in plants resulted in greater resistance to stress.

The proline content in overexpression lines and wild-type under stress was quantified. The results showed that under salt stress, the Pro content in OE-4, OE-3, OE-1, and wild-type increased by 249.21%, 176.47%, 115.03% and 71.84%, respectively. Under simulated drought stress, the Pro content in OE-4, OE-3, OE-1, and wild-type increased by 332.73%, 147.45%, 101.56% and 84.54%, respectively. The results showed that the Pro content in overexpression lines increased significantly under stress, indicating that overexpression of the *SpsNAC005* gene protects the membrane system of cells in plants to some extent, slows down the degradation of intracellular proteins, and enhances the adaptability to stress.

3.10. Up-Regulated Expression of SOS1, MPK6, HKT1, P5CS1, and PRODH1 Genes in Transgenic P. hopeiensis under Stress

To analyze the possible regulatory pathways of *SpsNAC005* genes, changes in the expressions of three genes, *SOS1*, *MPK6*, and *HKT1*, were examined. Under salt and drought stress, the expression levels of *SOS1*, *MPK6*, and *HKT1* were significantly higher in the overexpression lines than in wild-type. Under salt stress, the relative expression of *SOS1* genes in the OE-4, OE-3, OE-1, and wild-type plants was 9.51, 8.16, 6.89, and 5.14, respectively. Under drought stress, the relative expression of *SOS1* genes in the OE-4, OE-3, OE-1, and wild-type plants was 17.60, 15.22, 13.17, and 5.81, respectively. These results indicate that the expression of the *SOS1* gene under stress was increased compared with that under the water treatment, but the expression pattern of the overexpression lines was significantly higher than that of wild-type plants under the same treatment (Figure 10A). Therefore, *SpsNAC005* promoted the expression of the *SOS1* gene, maintained the dynamic balance of Na+/H+ in the transgenic line under stress, and improved the stress resistance of transgenic *P. hopeiensis*.

(A)

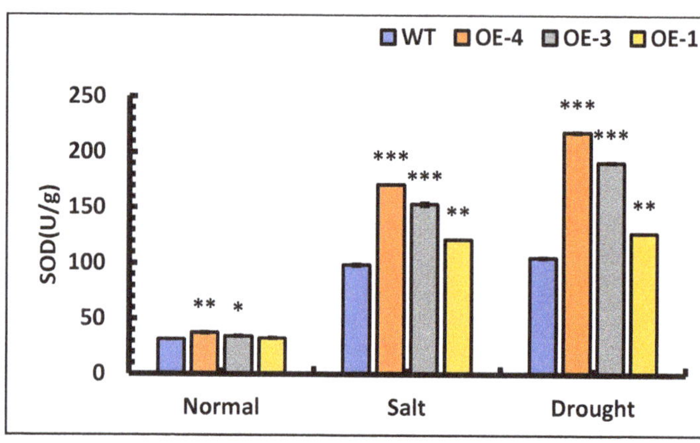

Figure 9. *Cont.*

(B)
(C)
(D)

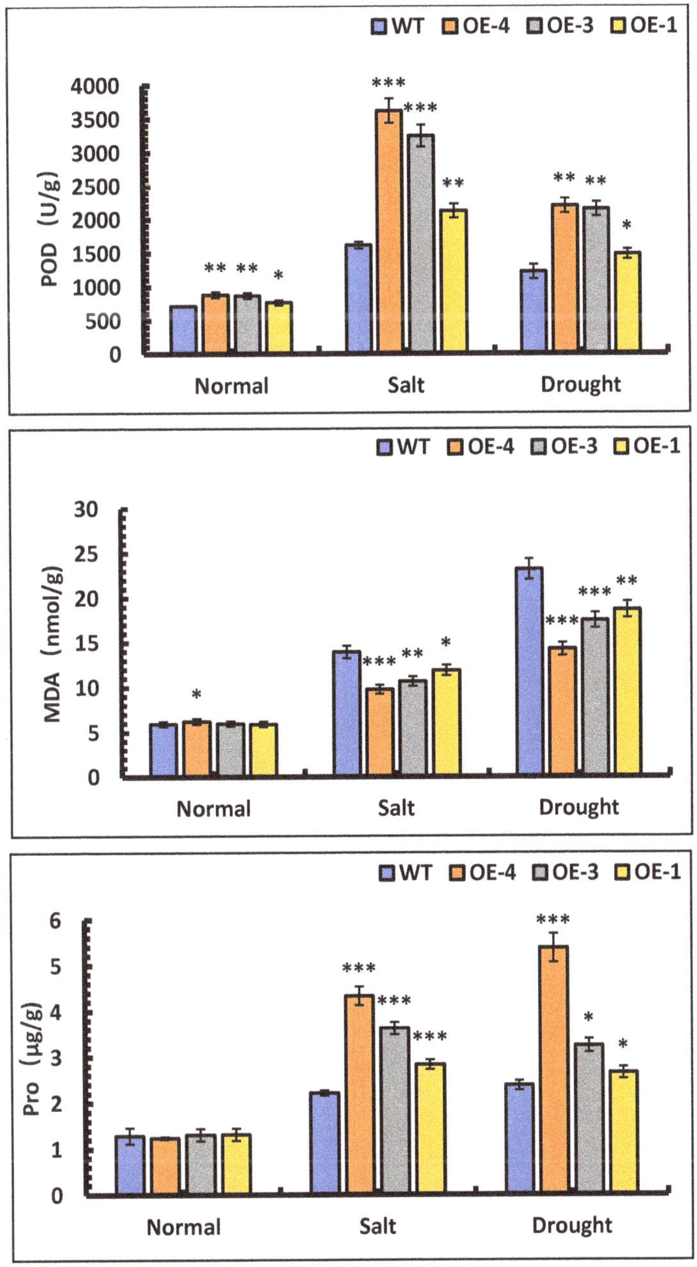

Figure 9. Changes in physiological indicators of *P. hopeiensis* transgenic lines under salt and drought stress. (**A**) SOD activity of *P. hopeiensis* under stress treatment. (**B**) POD activity of *P. hopeiensis* under stress treatment. (**C**) MDA content in *P. hopeiensis* under stress treatment. (**D**) Pro content in *P. hopeiensis* under stress treatment. Values are the mean ± standard deviation of three independent experiments. The significance of differences was determined using *t*-test and F-test. *** $p < 0.001$, extremely significant; ** $0.001 \leq p < 0.01$, very significant; * $0.01 \leq p < 0.05$, significant.

(A)

(B)

(C)

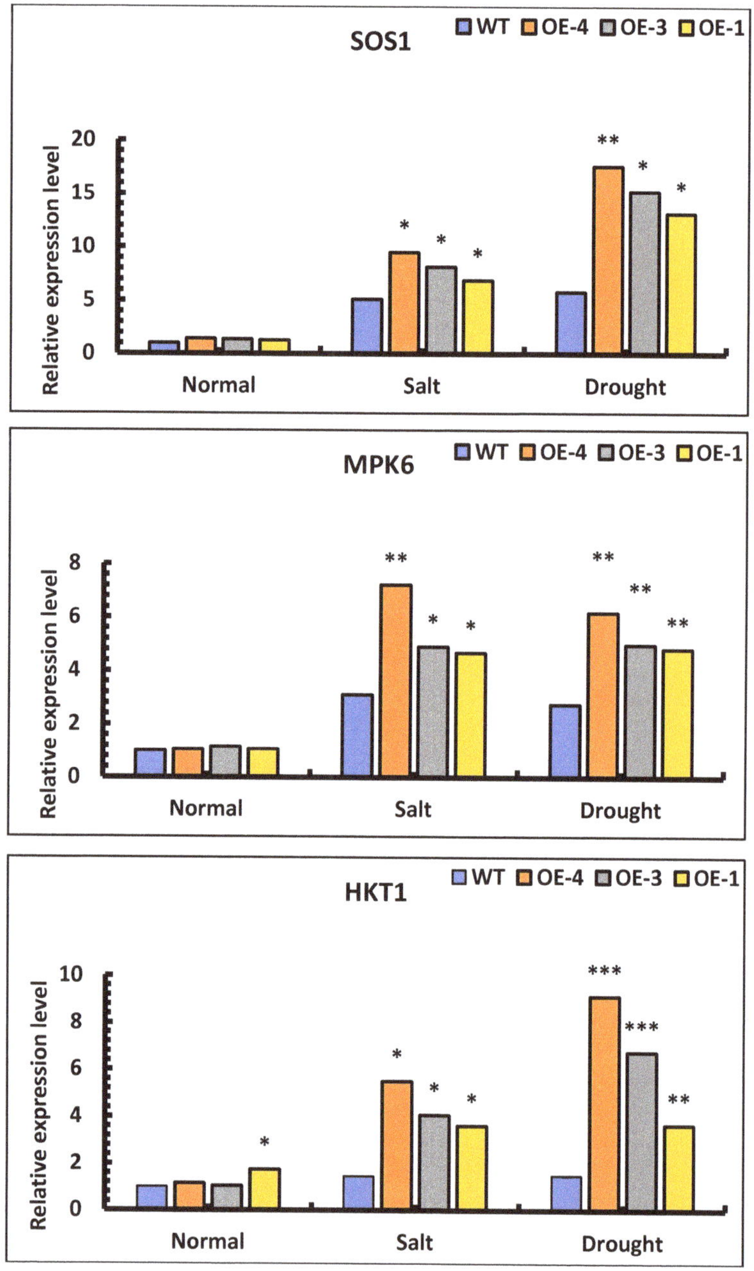

Figure 10. *Cont.*

(D)

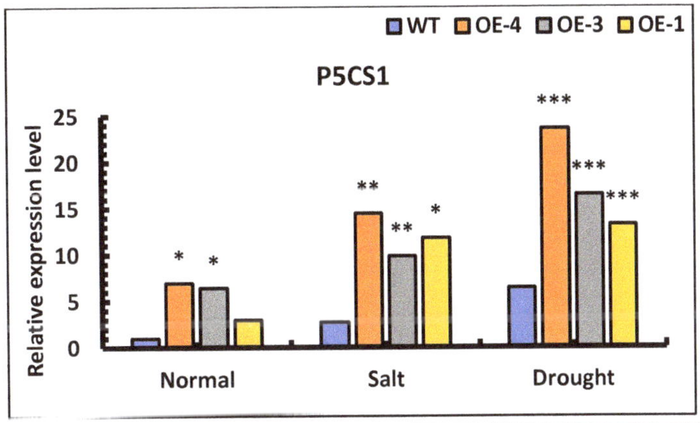

(E)

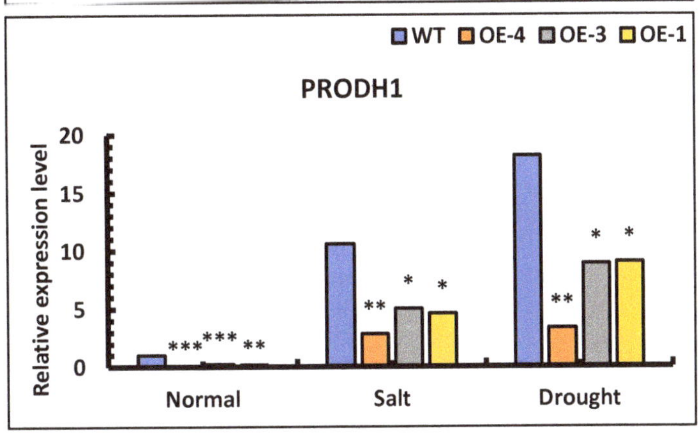

Figure 10. Expression analysis of *SOS1*, *MPK6*, and *HKT1* genes in transgenic lines of *P. hopeiensis* under salt and drought stress: (**A**) *SOS1* gene expression in wild-type and overexpression lines of *P. hopeiensis* under salt and drought stress; (**B**) *MPK6* gene expression in wild-type and overexpression lines of *P. hopeiensis* under salt and drought stress; (**C**) *HKT1* gene expression in wild-type and overexpression lines of *P. hopeiensis* under salt and drought stress; (**D**) *P5CS1* gene expression in wild-type and overexpression lines of *P. hopeiensis* under salt and drought stress; (**E**) *PRODH1* gene expression in wild-type and overexpression lines of *P. hopeiensis* under salt and drought stress. Values are the mean ± standard deviation of three independent experiments. The significance of differences was determined using *t*-test and F-test. *** $p < 0.001$, extremely significant; ** $0.001 \leq p < 0.01$, very significant; * $0.01 \leq p < 0.05$, significant.

An analysis of *MPK6* expression in overexpression lines showed that the expression of the transgenic lines was 3–6 times higher than that of wild-type under salt stress. Under drought stress, the expression of the transgenic lines was 4–5 times higher than that of wild-type. Based on the results of qRT-PCR, the expression of the *MPK6* gene under stress was increased compared to level with water treatment, and the expression of the transgenic lines was significantly higher than that of wild-type plants under the same treatment (Figure 10B). Therefore, the *MPK6* gene was activated under salt and drought stress, which eventually affected the redox status of the plant cells. The expression of *SpsNAC005* increased the relative expression of the *MPK6* gene, making transgenic *P. hopeiensis* significantly more resistant to stress than wild-type.

The analysis of *HKT1* expression in transgenic lines showed that the relative expression in OE-4, OE-3, OE-1, and wild-type plants increased 4.81-fold, 3.90-fold, 2.08-fold, and

1.45-fold, respectively, under salt stress compared with water treatments. Under drought stress, the relative expression in OE-4, OE-3, OE-1, and wild-type plants increased 7.97-fold, 6.47-fold, 2.11-fold, and 1.48-fold, respectively, compared with water treatments. The results indicate that the expression of the *MPK6* gene in all transgenic lines was significantly higher than that in the wild-type plants under the same stress treatment (Figure 10C). Therefore, *SpsNAC005* upregulated the expression of the *HKT1* gene, which enabled the plants to maintain the sodium–potassium ratio during osmoregulation and improved the tolerance of transgenic *P. hopeiensis* to stress.

The expression analysis of *P5CS1* in overexpression lines showed that the relative expression level in transgenic lines under salt stress was 3–5 times higher than that in wild type. Under drought stress, the relative expression level in transgenic lines was 2–4 times higher than in wild type. The results showed that expression of the *P5CS1* gene in overexpression lines was up-regulated under stress and that its relative expression in transgenic lines was significantly higher than in wild-type under the same treatment (Figure 10D). The expression analysis of *PRODH1* in overexpression lines showed that the relative expression level in transgenic lines under salt stress was 2–3 times lower than in wild type. Under drought stress, the expression level in transgenic lines was 2–4 times lower than in wild type. The results showed that the relative expression of the *PRODH1* gene was down-regulated under stress and that its relative expression in transgenic lines was significantly lower than in wild-type under the same treatment (Figure 10E). Therefore, the *P5CS1* and *PRODH1* genes can be induced by salt and drought stress, promote the synthesis of proline in plants, and protect plants from injury under stress.

4. Discussion

4.1. *SpsNAC005 Gene Promotes the Growth and Development of Plants*

The *SpsNAC005* gene promotes the growth and development of woody plants. The *SpsNAC005* gene is an orthologous gene with *ANAC032* (AT1G77450.1), Potri.005G069500.1, and Sapur.005G052900.1 [40]. *ANAC032* overexpression of *Arabidopsis* causes plants to accumulate more biomass, and their leaf area was significantly larger than that in the control and chimeric repressor lines [41]. The NAC transcription factor *HaNAC1* was isolated from *Haloxylon ammodendron*, and ectopic expression in *Arabidopsis* promoted the growth of *Arabidopsis* [10]. Among 74 NAC transcription factors obtained from the whole *Solanum lycopersicum* genome, *SNAC4-9* genes were involved in the growth and development of *S. lycopersicum* [42]. In this study, the *SpsNAC005* gene was overexpressed in *P. hopeiensis*, and a significant increase in plant height and ground diameter was found in transgenic lines. At the same time, the results showed that the leaf area of plants increased, which was consistent with the phenotype of *ANAC032* overexpression plants in *Arabidopsis*. NAC genes can positively regulate plant growth and development. So, we think the *SpsNAC005* gene had a role in promoting plant growth and development in woody plants.

4.2. *SpsNAC005 Gene Enhances Stress Tolerance and Antioxidant Capacity of Transgenic P. hopeiensis*

The *SpsNAC005* gene can enhance plant tolerance under stress. As demonstrated in other NAC-related gene studies, numerous *NAC* transcription factors are involved in the expression of drought, salinity, low temperature, and high temperature responsive genes, which play an important role in plant resistance to stress [43]. Through microarray analysis, *ANAC032* was isolated from *Arabidopsis thaliana* under multiple biological stresses (salt, drought, and cold stress). *ANAC032* was identified as a valuable candidate gene that could be used to cultivate feed crops resistant to multiple biological stresses through genetic modification of a single gene [41]. The overexpression of *OSNAC10* [44] and *OSNAC045* [45] genes in *Oryza sativa* improved resistance. The *Tartary buckwheat* NAC transcription factor *FtNAC17* responds to abiotic stresses such as drought, low temperature, and salt [46]. Transient genetic transformation of the *LoNAC18* gene in *Larix olgensis* indicated that the gene was involved in regulating the response of *L. olgensis* to PEG simulated drought stress [47]. *HaNAC20* was cloned in *Haloxylon ammodendron*, and an in-depth analysis verified that this gene can respond to abiotic stress [48]. The *PsnNAC030* gene was cloned, and it was verified through relevant experiments using *Populus simonii* × *P. nigra* as experimental material that the gene's expression could be effectively induced by salt, drought, and high and low temperature [49]. In *Populus simonii* × *P. nigra* overexpressing *ERF76*, Potri.005G069500.1 was significantly up-regulated under NaCl, KCl, $CdCl_2$, and PEG stress, indicating that Potri.005G069500.1 had a synergistic effect on improving the tolerance of transgenic plants [50]. In our study, we found that the height, leaf area, and growth rate of the transgenic lines were significantly higher than in wild-type during stress. Under salt and simulated drought treatment, the overexpression lines remained alive, and the antioxidant capacity of transgenic poplar was significantly higher than that of wild-type. In conclusion, the overexpression of *SpsNAC005* enhanced the tolerance of *P. hopeiensis* to the stress treatments.

The *SpsNAC005* gene enhanced the antioxidant capacity of *P. hopeiensis*. In related studies, the cloned sweet potato *IbNAC72* gene showed significantly higher SOD activity, significantly lower MDA content, and significantly higher drought resistance in transgenic plants under prolonged drought conditions [51]. In a study on drought tolerance genes in *Zea mays*, *ZmNAC33* participated in the regulation of ABA signaling pathway genes by regulating the expression of drought tolerance genes in plants, improving the activity of antioxidant enzymes and the content of osmotic protection substances in plants, and enhancing the drought tolerance of plants [52]. In *Tamarix chinensis*, overexpression of *ThNAC12* lines enhanced antioxidant enzyme activities and improved salt tolerance in transgenic plants under salt stress [53]. In this study, the SOD and POD activities of the overexpression lines were significantly higher than those of wild-type, and the MDA contents were significantly lower under salt and simulated drought stress. These results indicate that overexpression of the *SpsNAC005* gene in *P. hopeiensis* enhanced the antioxidant capacity of *P. hopeiensis*. The *SpsNAC005* gene high expression line had higher SOD and POD and lower MDA; these results further indicate that the antioxidant capacity of the *SpsNAC005* gene high expression line was stronger, and that the expression of the *SpsNAC005* gene was positively correlated with the antioxidant capacity of poplar. The abovementioned studies fully demonstrated that *SpsNAC005* could enhance the antioxidant enzyme activity in transgenic *P. hopeiensis* and mitigate the extent of damage to plant lipid membranes from osmotic stress.

4.3. *SpsNAC005 Gene Regulates Tolerance to Stress through* Na^+ *Transporters Pathway*

The *SpsNAC005* gene enhanced the expression of the *SOS1*, *MPK6*, and *HKT1* genes. *SOS1*, which encodes a plasma membrane Na^+/H^+ reverse transporter protein, is a major member of the ion homeostasis regulatory pathway during plant salt stress signal transduction [54]. Moreover, overexpression of the *SOS1* gene in *Arabidopsis* and *Chrysanthemum* resulted in enhanced salt tolerance [55,56]. In our study, we found that the expression level of the *SOS1* gene was significantly higher in overexpression lines of *P. hopeiensis* than in wild-type under stress. The overexpression of the *SpsNAC005* gene in this study likely regulated the *SOS1* gene directly or indirectly to attenuate the toxic effects of osmotic stress on transgenic lines by regulating the ion concentration around the cell membrane and improved the resistance of transgenic *P. hopeiensis* to stress.

MPK is a large family of serine/threonine protein kinases and is one of the biological signal transduction systems [57]. When plants are subjected to abiotic stress, plant cells can activate defense genes that affect the redox status, ultimately leading to the activation of *MPK* with a series of metabolic changes [58]. *Arabidopsis* activates *MPK6* under salt stress [59]. In the present study, the expression of the *MPK6* gene was significantly increased under both salt and drought stresses. The *MPK6* gene was positively correlated with the expression of the *SpsNAC005* gene under stress conditions, indicating that the *SpsNAC005* gene may positively regulate the *MPK6* gene, thus activating the defense mechanism of the plant and making transgenic *P. hopeiensis* more tolerant of osmotic stress.

HKT is a protein transporter with a high affinity for K^+ that is also capable of transporting Na^+, which maintains a normal sodium–potassium ratio in plants when they are subjected to osmoregulatory stress and ion partitioning [60]. The salt tolerance of the *HKT1* gene has been verified in *Triticum aestivum* [61], *Hordeum vulgare* [62], *Oryza sativa* [63], and *Vitis vinifera* [64], which reduced the accumulation of Na^+ in plants and reduced the osmotic imbalance to the greatest extent. When subjected to stress, the expression of the *HKT1* gene was significantly higher in overexpression lines than in wild-type. The present study showed that under stress, *SpsNAC005* overexpression in *P. hopeiensis* may alleviate the accumulation of Na^+ in plants by regulating the expression of the *HKT1* gene in the medium, attenuating the toxic effect of Na^+ on plants and enhancing the tolerance of transgenic *P. hopeiensis*.

4.4. SpsNAC005 Gene Responds to Stress by Regulating Proline Synthesis

Under different stress conditions caused by factors such as drought, salinization, heavy metals and ultraviolet radiation, proline accumulation in plants can be induced, playing a protective role in plants. In higher plants, proline synthesis has two pathways, with glutamic acid and ornithine as precursors, respectively. Among them, the glutamate pathway is mainly responsible for proline accumulation under stress conditions [65]. Pyrroline-5-carboxylic acid synthase (P5CS) catalyzes the first step of the glutamate synthesis pathway and is the key enzyme in proline synthesis pathway [66]. PRODH is the rate-limiting enzyme in the proline decomposition pathway [67]. Early studies have improved drought and salt tolerance in Arabidopsis by antisense inhibition of the *PRODH* gene [68]. In *Arabidopsis*, the *P5CS1* gene is expressed in most organs and is induced by high salt, drought and ABA stress, but not in rapidly dividing cells [69]. The stress tolerance of transgenic *AtP5CS1* plants was analyzed, showing that under the stress of 150 mmol/L NaCl and 15% PEG6000, the proline content was significantly increased, and *AtP5CS1* transgenic plants had a better phenotype than the control; that is, the expression of *AtP5CS1* gene in *Brassica oleracea* significantly improved the salt tolerance and drought tolerance of transgenic plants. In our study, under salt and simulated drought stress, the *P5CS1* gene of *SpsNAC005* transgenic lines was significantly up-regulated, the *PRODH1* gene was significantly down-regulated, and the proline content was significantly increased. These results indicate that overexpression of the *SpsNAC005* gene may directly or indirectly regulate *P5CS1* and *PRODH1* genes under pressure stress and improve the resistance of transgenic *Populus hopeiensis* to stress by catalyzing the glutamate synthesis pathway and limiting proline decomposition.

5. Conclusions

We obtained seven transgenic lines using an Agrobacterium-mediated transformation technique, and their expression levels were determined. The three lines with the highest relative expression were selected for phenotypic observation and subsequent stress treatment, and it was found that the height and leaf area of the transgenic plants were significantly increased compared with the wild-type plants, and the growth rates of their plant height and ground diameter were significantly higher than those of the wild-type. The fresh weight and dry weight of wild-type were significantly lower than their control. Under salt and drought stress, the SOD activities, POD activities and Pro contents in the transgenic plants were significantly increased, the accumulation of MDA was significantly lower than that in the wild type, and the transgenic strains clearly showed tolerance to salt and drought. Meanwhile, it was found that the expressions of *SOS1*, *MPK6*, and *HKT1* genes were also up-regulated by quantitative measurements. The expression of the *PRODH1* gene was down-regulated based on quantitative measurements. In conclusion, overexpression of the *SpsNAC005* gene in transgenic plants can promote plant growth and development and improve tolerance to salt and drought.

Author Contributions: H.Y., L.F. wrote sections of the manuscript. L.F. performed the experiments and carried out the statistical analysis. X.Y. completed the vector construction. X.Z. and P.H. assisted in the sample collection and sample grinding. D.W. and G.Z. provided guidance and advice on stress methods in earlier stages of this study. All authors contributed to manuscript revision and approved the submitted version. All authors have read and agreed to the published version of the manuscript.

Funding: This work was supported by the National Key Program on Transgenic Research (2018ZX08020002-005-005), the Key Technology Research Project of Inner Mongolia Autonomous Region (2021GG0075) and the National Natural Science Foundation of China (31660216).

Institutional Review Board Statement: Not applicable.

Informed Consent Statement: Not applicable.

Data Availability Statement: Not applicable.

Acknowledgments: We acknowledge everyone who contributed to this article.

Conflicts of Interest: The authors declare no conflict of interest.

Abbreviations

qRT-PCR	Quantitative real-time PCR
cDNA	Complementary DNA

References

1. Christianson, J.A.; Dennis, E.S.; Llewellyn, D.J.; Wilson, I.W. ATAF NAC transcription factors: Regulators of plant stress signaling. *Plant. Signal. Behav.* **2010**, *5*, 428–432. [CrossRef] [PubMed]
2. Souer, E.; van Houwelingen, A.; Kloos, D.; Mol, J.; Koes, R. The No Apical Meristem Gene of Petunia Is Required for Pattern Formation in Embryos and Flowers and Is Expressed at Meristem and Primordia Boundaries. *Cell* **1996**, *85*, 159–170. [CrossRef]
3. Aida, M.; Ishida, T.; Fukaki, H.; Fujisawa, H.; Tasaka, M. Genes involved in organ separation in Arabidopsis: An analysis of the cup-shaped cotyledon mutant. *Plant. Cell* **1997**, *9*, 841–857. [CrossRef]
4. Ernst, H.A.; Olsen, A.N.; Skriver, K.; Larsen, S.; Leggio, L.L. Structure of the conserved domain of ANAC, a member of the NAC family of transcription factors. *EMBO Rep.* **2004**, *5*, 297–303. [CrossRef] [PubMed]
5. Chen, Q.; Wang, Q.; Xiong, L.; Lou, Z. A structural view of the conserved domain of rice stress-responsive NAC1. *Protein Cell* **2011**, *2*, 55–63. [CrossRef] [PubMed]
6. Olsen, A.N.; Ernst, H.A.; Leggio, L.L.; Skriver, K. NAC transcription factors: Structurally distinct, functionally diverse. *Trends Plant. Sci.* **2005**, *10*, 79–87. [CrossRef]
7. Li, W.; Han, L.; Qian, Y.Q.; Sun, Z.Y. Characteristics and Functions of NAC Transcription Factors in Plants. *Chin. J. Appl. Environ. Biol.* **2011**, *17*, 596–606. Available online: https://en.cnki.com.cn/Article_en/CJFDTotal-YYHS201104030.htm (accessed on 13 December 2021).
8. Ooka, H.; Satoh, K.; Doi, K.; Nagata, T.; Otomo, Y.; Murakami, K.; Matsubara, K.; Osato, N.; Kawai, J.; Carninci, P.; et al. Comprehensive Analysis of NAC Family Genes in Oryza sativa and Arabidopsis thaliana. *DNA Res.* **2003**, *10*, 239–247. [CrossRef]
9. Hu, R.; Qi, G.; Kong, Y.; Kong, D.; Gao, Q.; Zhou, G. Comprehensive Analysis of NAC Domain Transcription Factor Gene Family in Populus trichocarpa. *BMC Plant. Biol.* **2010**, *10*, 145. [CrossRef]
10. Gong, L.; Zhang, H.; Liu, X.; Gan, X.; Nie, F.; Yang, W.; Zhang, L.; Chen, Y.; Song, Y.; Zhang, H. Ectopic expression of HaNAC1, an ATAF transcription factor from Haloxylon ammodendron, improves growth and drought tolerance in transgenic Arabidopsis. *Plant. Physiol. Biochem.* **2020**, *151*, 535–544. [CrossRef]
11. Mahmood, K.; El-Kereamy, A.; Kim, S.-H.; Nambara, E.; Rothstein, S.J. ANAC032 Positively Regulates Age-Dependent and Stress-Induced Senescence in Arabidopsis thaliana. *Plant. Cell Physiol.* **2016**, *57*, 2029–2046. [CrossRef] [PubMed]
12. Mahmood, K.; Xu, Z.; El-Kereamy, A.; Casaretto, J.A.; Rothstein, S.J. The Arabidopsis Transcription Factor ANAC032 Represses Anthocyanin Biosynthesis in Response to High Sucrose and Oxidative and Abiotic Stresses. *Front. Plant Sci.* **2016**, *7*, 1548. [CrossRef] [PubMed]
13. Fan, L.; Fan, C.H. Cloning and Expression Analysis of CaNAC23 Gene in *Capsicum annuum* L. *Mol. Plant Breed.* **2020**, *18*, 3854–3861.
14. Zhang, L.Q.; Jia, X.H.; Zhao, J.W.; Liu, Y.X.; Hasiagula. Cloning and Expression Analysis of MfNAC37 from Medicago falcata under Salt Stress. *Chin. J. Grassl.* **2019**, *41*, 10–14+166. [CrossRef]
15. Jia, D.; Jiang, Q.; van Nocker, S.; Gong, X.; Ma, F. An apple (*Malus domestica*) NAC transcription factor enhances drought tolerance in transgenic apple plants. *Plant Physiol. Biochem.* **2019**, *139*, 504–512. [CrossRef]
16. Hu, H.; Dai, M.; Yao, J.; Xiao, B.; Li, X.; Zhang, Q.; Xiong, L. Overexpressing a NAM, ATAF, and CUC (NAC) transcription factor enhances drought resistance and salt tolerance in rice. *Proc. Natl. Acad. Sci. USA* **2006**, *103*, 12987–12992. [CrossRef]
17. Jia, H.; Zhang, J.; Li, J.; Sun, P.; Zhang, Y.; Xin, X.; Lu, M.; Hu, J. Genome-wide transcriptomic analysis of a desert willow, Salix psammophila, reveals the function of hub genes SpMDP1 and SpWRKY33 in drought tolerance. *BMC Plant Biol.* **2019**, *19*, 1–15. [CrossRef]

18. Li, W.X.; Liu, Z.X.; Yan, W.; Zhang, X.J.; Miu, S.G.; Huo, J.F.; Wang, Y.S. Study on Good Family Breeding of Salix Psammophila. *J. Desert Res.* **2008**, *28*, 679–684.
19. Gu, M.H.; Xie, Z.H. Soil moisture characteristics of eight types of shelter forest in Kubuqi desert. *Pratacult. Sci.* **2017**, *34*, 2437–2444.
20. Luo, W.X.; Liu, G.Q.; Li, J.J. Cultivation Techniques for Main Tree Species in Northwest China. *China For. Press.* **2007**, *8*, 682–687.
21. Yu, X.Y. Cloning of NAC in *Salix psammophila*, Bioinformatics Analysis and Construction of Expression. Master's Thesis, Inner Mongolia agricultural University, Hohhot, China, 2020.
22. Horsch, R.B.; Fry, J.E.; Hoffmann, N.L.; Wallroth, M.; Eichholtz, D.; Rogers, S.G.; Fraley, R.T. A Simple and General Method for Transferring Genes into Plants. *Science* **1985**, *227*, 1229–1231. [CrossRef] [PubMed]
23. Chen, K.-S.; Li, F.; Xu, C.-J.; Zhang, S.-L.; Fu, C.-X. An efficient macro-method of genomic DNA isolation from Actinidia chinensis leaves. *Hereditas* **2004**, *26*, 529–531. [PubMed]
24. Yang, H.F.; Bo, G.F. Cloning and expression analysis of *SpsNAC042* gene in Salix psammophila. *J. Northwest For. Univ.* **2021**, *36*, 11–17+27.
25. Livak, K.J.; Schmittgen, T.D. Analysis of Relative Gene Expression Data Using Real-Time Quantitative PCR and the $2^{-\Delta\Delta CT}$ Method. *Methods* **2001**, *25*, 402–408. [CrossRef] [PubMed]
26. Tian, Q.; Cao, Z.Z. Digital camera and Auto CAD based method for measuring the leaf area of landscape plants. *Grassl. Turf.* **2008**, *3*, 25–28.
27. Cai, X.; Jiang, Z.; Tang, L.; Zhang, S.; Li, X.; Wang, H.; Liu, C.; Chi, J.; Zhang, X.; Zhang, J. Genome-wide characterization of carotenoid oxygenase gene family in three cotton species and functional identification of GaNCED3 in drought and salt stress. *J. Appl. Genet.* **2021**, *62*, 527–543. [CrossRef]
28. Ning, H.L. *Field Experiment and Statistical Methods*; Science Press: Beijing, China, 2012. Available online: http://ss.zhizhen.com/detail_38502727e7500f26a81075d227d4798ebf0734766bdfb21d1921b0a3ea25510134114c969f2eae5c34597dee38cfeccc2f9303951adbee66eb7ea3d7539ba9795c5481485155be4acdc25c707daf213c? (accessed on 4 March 2022).
29. Puranik, S.; Sahu, P.P.; Srivastava, P.S.; Prasad, M. NAC proteins: Regulation and role in stress tolerance. *Trends Plant Sci.* **2012**, *17*, 369–381. [CrossRef]
30. Shen, H.; Yin, Y.; Chen, F.; Xu, Y.; Dixon, R.A. A Bioinformatic Analysis of NAC Genes for Plant Cell Wall Development in Relation to Lignocellulosic Bioenergy Production. *Bioenerg. Res.* **2009**, *2*, 217–232. [CrossRef]
31. Hibara, K.; Takada, S.; Tasaka, M. *CUC1* gene activates the expression of SAM-related genes to induce adventitious shoot formation. *Plant J.* **2003**, *36*, 687–696. [CrossRef]
32. Zhong, R.; Richardson, E.A.; Ye, Z.-H. Two NAC domain transcription factors, SND1 and NST1, function redundantly in regulation of secondary wall synthesis in fibers of Arabidopsis. *Planta* **2007**, *225*, 1603–1611. [CrossRef]
33. Mitsuda, N.; Iwase, A.; Yamamoto, H.; Yoshida, M.; Seki, M.; Shinozaki, K.; Ohme-Takagi, M. NAC Transcription Factors, NST1 and NST3, Are Key Regulators of the Formation of Secondary Walls in Woody Tissues of Arabidopsis. *Plant Cell* **2007**, *19*, 270–280. [CrossRef] [PubMed]
34. Zhao, C.; Avci, U.; Grant, E.H.; Haigler, C.H.; Beers, E.P. XND1, a member of the NAC domain family in Arabidopsis thaliana, negatively regulates lignocellulose synthesis and programmed cell death in xylem. *Plant J.* **2008**, *53*, 425–436. [CrossRef] [PubMed]
35. Tran, L.-S.P.; Nakashima, K.; Sakuma, Y.; Simpson, S.D.; Fujita, Y.; Maruyama, K.; Fujita, M.; Seki, M.; Shinozaki, K.; Yamaguchi-Shinozaki, K. Isolation and Functional Analysis of Arabidopsis Stress-Inducible NAC Transcription Factors That Bind to a Drought-Responsive cis-Element in the early responsive to dehydration stress 1 Promoter[W]. *Plant Cell* **2004**, *16*, 2481–2498. [CrossRef] [PubMed]
36. Guo, Y.; Gan, S. AtNAP, a NAC family transcription factor, has an important role in leaf senescence. *Plant J.* **2006**, *46*, 601–612. [CrossRef]
37. Yang, Z.-T.; Lu, S.-J.; Wang, M.-J.; Bi, D.-L.; Sun, L.; Zhou, S.-F.; Song, Z.-T.; Liu, J.-X. A plasma membrane-tethered transcription factor, NAC062/ANAC062/NTL6, mediates the unfolded protein response in Arabidopsis. *Plant J.* **2014**, *79*, 1033–1043. [CrossRef]
38. Jung, J.-H.; Park, C.-M. Auxin modulation of salt stress signaling in Arabidopsis seed germination. *Plant Signal. Behav.* **2011**, *6*, 1198–1200. [CrossRef]
39. Liu, Y.; Sun, J.; Wu, Y. Arabidopsis ATAF1 enhances the tolerance to salt stress and ABA in transgenic rice. *J. Plant Res.* **2016**, *129*, 955–962. [CrossRef]
40. Yang, H.F.; Yu, X.W.; Jin, N.; He, Y.J.; Li, A.Y.; Zhang, X.; Wang, Y.Z.; Gao, Y.Y.; Wang, L. Cloning and bioinformatics analysis of *SpsNAC005* gene in Salix psammophila. *J. Inn. Mong. Agric Univ.* **2021**, *42*, 65–71.
41. Sun, L.; Zhang, P.; Wang, R.; Wan, J.; Ju, Q.; Rothstein, S.J.; Xu, J. The SNAC-A Transcription Factor ANAC032 Reprograms Metabolism in Arabidopsis. *Plant Cell Physiol.* **2019**, *60*, 999–1010. [CrossRef]
42. Wang, S. Cloning and Expression Analysis of Solanum Lycopersicum NAC Transcription Factors. Master's Thesis, Tianjin University, Tianjin, China, 2014.
43. Shao, H.; Wang, H.; Tang, X. NAC transcription factors in plant multiple abiotic stress responses: Progress and prospects. *Front. Plant Sci.* **2015**, *6*, 902. [CrossRef]
44. Jeong, J.S.; Kim, Y.S.; Baek, K.H.; Jung, H.; Ha, S.-H.; Do Choi, Y.; Kim, M.; Reuzeau, C.; Kim, J.-K. Root-Specific Expression of OsNAC10 Improves Drought Tolerance and Grain Yield in Rice under Field Drought Conditions. *Plant Physiol.* **2010**, *153*, 185–197. [CrossRef] [PubMed]

45. Yu, S.; Huang, A.; Li, J.; Gao, L.; Feng, Y.; Pemberton, E.; Chen, C. OsNAC45 plays complex roles by mediating POD activity and the expression of development-related genes under various abiotic stresses in rice root. *Plant Growth Regul.* **2018**, *84*, 519–531. [CrossRef]
46. Rong, Y.P.; Tang, B.; Li, P.; Zhang, J.Q.; Chen, Q.F.; Zhu, L.W.; Deng, J.; Huang, J. Identification and Expression of NAC Transcription Factor FtNAC17 in Tartary Buckwheat. *Biotechnol. Bull.* **2021**, *37*, 174. [CrossRef]
47. Zhang, L.; Xiong, H.H.; Cao, Q.; Zhao, J.L.; Zhang, H.G. Drought Resistance of Larch NAC Gene by Transient Genetic Transformation. *Bull. Bot. Res.* **2020**, *40*, 394–400.
48. Zhou, L.D.; Yao, Z.P. Cloning and characterization of the *HaNAC20* gene of P. sylvestris. *Acta Agric. Boreali-Occident. Sin.* **2021**, *30*, 1556–1564.
49. Wang, X.Y.; Zhang, X.M. Function analysis of poplar transcription factor *PsnNAC030* gene. *J. Northeast. For. Univ.* **2021**, *49*, 1–8.
50. Yao, W.; Zhou, B.; Zhang, X.; Zhao, K.; Cheng, Z.; Jiang, T. Transcriptome analysis of transcription factor genes under multiple abiotic stresses in *Populus simonii* × *P.nigra*. *Gene* **2019**, *707*, 189–197. [CrossRef]
51. Zhang, H.; Yang, N.K. Cloning and functional analysis of drought-related gene *IbNAC72* in sweet potato. *Acta Agron. Sin.* **2020**, *46*, 1649–1658.
52. Liu, W.P. Mining of Drought-Resistant Related Genes in Maize and Functional Identification of Drought-Resistant Genes. Ph.D. Thesis, Northeastern Agricultural University, Harbin, China, 2020.
53. Wang, R.; Zhang, Y.; Wang, C.; Wang, Y.-C.; Wang, L.-Q. ThNAC12 from Tamarix hispida directly regulates ThPIP2;5 to enhance salt tolerance by modulating reactive oxygen species. *Plant Physiol. Biochem.* **2021**, *163*, 27–35. [CrossRef]
54. Foster, K.J.; Miklavcic, S.J. A Comprehensive Biophysical Model of Ion and Water Transport in Plant Roots. II. Clarifying the Roles of SOS1 in the Salt-Stress Response in Arabidopsis. *Front. Plant Sci.* **2019**, *10*, 1121. [CrossRef]
55. Shi, H.; Lee, B.; Wu, S.-J.; Zhu, J.-K. Overexpression of a plasma membrane Na^+/H^+ antiporter gene improves salt tolerance in Arabidopsis thaliana. *Nat. Biotechnol.* **2003**, *21*, 81–85. [CrossRef] [PubMed]
56. Gao, J.; Sun, J.; Cao, P.; Ren, L.; Liu, C.; Chen, S.; Chen, F.; Jiang, J. Variation in tissue Na^+ content and the activity of SOS_1 genes among two species and two related genera of Chrysanthemum. *BMC Plant Biol.* **2016**, *16*, 98. [CrossRef] [PubMed]
57. Mishra, N.S.; Tuteja, R.; Tuteja, N. Signaling through MAP kinase networks in plants. *Arch. Biochem. Biophys.* **2006**, *452*, 55–68. [CrossRef] [PubMed]
58. Zhang, S.; Liu, Y. Activation of Salicylic Acid–Induced Protein Kinase, a Mitogen-Activated Protein Kinase, Induces Multiple Defense Responses in Tobacco. *Plant Cell* **2001**, *13*, 1877–1889. [CrossRef]
59. Yu, L.; Nie, J.; Cao, C.; Jin, Y.; Yan, M.; Wang, F.; Liu, J.; Xiao, Y.; Liang, Y.; Zhang, W. Phosphatidic acid mediates salt stress response by regulation of MPK6 in Arabidopsis thaliana. *New Phytol.* **2010**, *188*, 762–773. [CrossRef]
60. Rubio, F.; Gassmann, W.; Schroeder, J.I. Sodium-Driven Potassium Uptake by the Plant Potassium Transporter HKT1 and Mutations Conferring Salt Tolerance. *Science* **1995**, *270*, 1660–1663. [CrossRef]
61. Waters, S.; Gilliham, M.; Hrmova, M. Plant High-Affinity Potassium (HKT) Transporters Involved in Salinity Tolerance: Structural Insights to Probe Differences in Ion Selectivity. *Int. J. Mol. Sci.* **2013**, *14*, 7660–7680. [CrossRef]
62. Van Bezouw, R.F.H.M.; Janssen, E.M.; Ashrafuzzaman, M.; Ghahramanzadeh, R.; Kilian, B.; Graner, A.; Visser, R.G.F.; van der Linden, C.G. Shoot sodium exclusion in salt stressed barley (*Hordeum vulgare* L.) is determined by allele specific increased expression of HKT1;5. *J. Plant Physiol.* **2019**, *241*, 153029. [CrossRef]
63. Suzuki, K.; Yamaji, N.; Costa, A.; Okuma, E.; Kobayashi, N.I.; Kashiwagi, T.; Katsuhara, M.; Wang, C.; Tanoi, K.; Murata, Y.; et al. OsHKT1;4-mediated Na^+ transport in stems contributes to Na^+ exclusion from leaf blades of rice at the reproductive growth stage upon salt stress. *BMC Plant Biol.* **2016**, *16*, 22. [CrossRef]
64. Liu, C.; Gao, Z. Functional Identification of Grape Potassium Ion Transporter VviHKT1;7 Under Salt Stress. *Sci. Agric. Sinina* **2021**, *54*, 1952–1963.
65. Delauney, A.; Hu, C.; Kishor, P.; Verma, D. Cloning of ornithine delta-aminotransferase cDNA from Vigna aconitifolia by trans-complementation in Escherichia coli and regulation of proline biosynthesis. *J. Biol. Chem.* **1993**, *268*, 18673–18678. [CrossRef]
66. Kishor, P.B.K.; Sangam, S.; Amrutha, R.N.; Laxmi, P.S.; Naidu, K.R.; Rao, K.R.S.S.; Rao, S.; Reddy, K.J.; Theriappan, P.; Sreenivasulu, N. Regulation of proline biosynthesis, degradation, uptake and transport in higher plants: Its implications in plant growth and abiotic stress tolerance. *Curr. Sci.* **2005**, *88*, 424–438.
67. Funck, D.; Eckard, S.; Müller, G. Non-redundant functions of two proline dehydrogenase isoforms in Arabidopsis. *BMC Plant Biol.* **2010**, *10*, 70. [CrossRef] [PubMed]
68. Nanjo, T.; Kobayashi, M.; Yoshiba, Y.; Kakubari, Y.; Yamaguchi-Shinozaki, K.; Shinozaki, K. Antisense suppression of proline degradation improves tolerance to freezing and salinity in Arabidopsis thaliana. *FEBS Lett.* **1999**, *461*, 205–210. [CrossRef]
69. Yoshiba, Y.; Kiyosue, T.; Katagiri, T.; Ueda, H.; Mizoguchi, T.; Yamaguchi-Shinozaki, K.; Wada, K.; Harada, Y.; Shinozaki, K. Correlation between the induction of a gene for Δ^1-pyrroline-5-carboxylate synthetase and the accumulation of proline in *Arabidopsis thaliana* under osmotic stress. *Plant J.* **1995**, *7*, 751–760. [CrossRef]

forests

MDPI

Article

Chemical Compositions of Walnut (*Juglans* Spp.) Oil: Combined Effects of Genetic and Climatic Factors

Hanbo Yang [1], Xu Xiao [1,2], Jingjing Li [1], Fang Wang [1], Jiaxuan Mi [1], Yujie Shi [1], Fang He [1], Lianghua Chen [1], Fan Zhang [3] and Xueqin Wan [1,*]

1 Forestry Ecological Engineering in the Upper Reaches of the Yangtze River Key Laboratory of Sichuan Province & National Forestry and Grassland Administration Key Laboratory of Forest Resources Conservation and Ecological Safety on the Upper Reaches of the Yangtze River & Rainy Area of West China Plantation Ecosystem Permanent Scientific Research Base, Institute of Ecology & Forestry, Sichuan Agricultural University, Chengdu 611130, China; yanghanbo6@sicau.edu.cn (H.Y.); 2020304074@stu.sicau.edu.cn (X.X.); 2020304092@stu.sicau.edu.cn (J.L.); 2020204014@stu.sicau.edu.cn (F.W.); 2020104002@stu.sicau.edu.cn (J.M.); 2019104006@stu.sicau.edu.cn (Y.S.); 14686@sicau.edu.cn (F.H.); chenlh@sicau.edu.cn (L.C.)

2 Sichuan Surveying and Planning Institute of Forestry and Grassland, Chengdu 311400, China

3 College of Landscape Architecture, Sichuan Agricultural University, Chengdu 611130, China; 13305@sicau.edu.cn

* Correspondence: wanxueqin@sicau.edu.cn; Tel.: +86-18281366168

Abstract: Walnut oil is a high-value oil product. Investigation of the variation and the main climatic factors affecting the oil's chemical composition is vital for breeding and oil quality improvement. Therefore, the fatty acid, micronutrients, and secondary metabolites compositions and contents in walnut oil were determined in three species: *Juglans regia* L. (common walnut), *J. sigillata* Dode (iron walnut), and their hybrids (*Juglans sigillata* Dode × *J. regia* L.), which were cultivated aat different sites. The major fatty acids were linoleic (51.39–63.12%), oleic (18.40–33.56%), and linolenic acid (6.52–11.69%). High variation in the contents of fatty acids, micronutrients, and secondary metabolites was found between both species and sites. Interestingly, myristic, margaric, and margaroleic acid were only detected in the hybrids' walnut oil, yet α-tocopherol was only detected in common and iron walnut oil. Climatic factors significantly affected the composition and content of fatty acid, whereas δ-tocopherol was mostly dependent on the genetic factors. The average relative humidity explained the most variation in the fatty acids, micronutrients, and secondary metabolites, which showed a significant positive and negative effect on the monounsaturated fatty acids and polyunsaturated fatty acids, respectively. These findings contribute to the provision of better guidance in matching sites with walnut trees, and improvement of the nutritional value of walnut oil.

Keywords: walnut oil; fatty acid; micronutrients; secondary metabolites; variation; genetic effect; climatic factors

Citation: Yang, H.; Xiao, X.; Li, J.; Wang, F.; Mi, J.; Shi, Y.; He, F.; Chen, L.; Zhang, F.; Wan, X. Chemical Compositions of Walnut (*Juglans* Spp.) Oil: Combined Effects of Genetic and Climatic Factors. *Forests* **2022**, *13*, 962. https://doi.org/10.3390/f13060962

Academic Editors: Jie Luo and Wentao Hu

Received: 12 May 2022
Accepted: 15 June 2022
Published: 20 June 2022

1. Introduction

Walnut (*Juglans* spp.) is one of the four famous tree nuts (walnut, almond, chestnut, and cashew) that have been consumed as rich nutritious food in many countries around the world [1]. Walnut oil is a high-value oil product that is used widely in food and health care [2]. Walnut oil is naturally rich in polyunsaturated fatty acids (PUFAs), mainly linoleic and linolenic acids, and is consequently poor in monounsaturated fatty acids (MUFAs), represented by oleic acid, and saturated fatty acids (SFAs) [3–5]. The fatty acid profile indicates that nutritionally, unsaturated fatty acid has a significant effect on the regulation of blood lipid, cleaning of thrombus, and immunoregulation [6]. Walnut oil is also rich in micronutrients and secondary metabolites, such as tocopherol, flavone, and polyphenols, which have been reported to exhibit numerous beneficial effects, such as antidiabetic, antioxidative, anti-inflammatory, and anti-proliferative effects in cancer [7,8]. With the increasing emphasis on health care, walnut oil has become more popular among consumers for its particular health functions [2]. The nutritional value of walnut oil is mainly related to the species and genotypes of *Juglans*, and the environment factors [2,6]. Many researchers that have described the major and minor compositions of vegetable oils, which have been produced and published in standard texts [1–3,6,9–12]. The oil composition and content are determined by genetic control of plants [13]. For instance, the concentrations and compositions of tocopherol in almond oil are under genetic control [10]. Significant variation in the linoleic, linolenic, and oleic acid contents also exists naturally due to the genotypes in walnut oil [14]. Despite genetic control, environmental factors, such as latitude, temperature, and drought, may also affect the oil composition and concentration [3]. Much of the variation recorded in the fatty acid profile was associated with the cultivated site [14]. An increase in the oleic/linoleic acid rate with increasing temperature has been widely reported in several oil crops, such as sunflower [15], soybean [16], and walnut [3]. Hot summers also induce higher concentrations of tocopherol in almond and many other plant species [17]. However, little information has been reported about the combined effects of environmental factors during the maturing stage on different species or genotypes of walnut oil, such as temperature, rainfall, altitude, and latitude, etc.

Changes in the oil composition and content are currently the goal of many oilseed crop breeding programs [3]. The genus *Juglans* includes approximately 21 species that are widely distributed around the world [2,18]. Common walnut (*J. regia* L.) and iron walnut (*J. sigillata* Dode) are the two main species that are cultivated for nut production in China [19]. The two species can be used to extract and produce walnut oil. There are significant differences in the lipid composition and minor composition contents between common walnut and iron walnut [2]. Mating between the common walnut and iron walnut is compatible, and their hybrids are also widely cultivated in China. The hybrids may have a wider range of cultivation and higher quality of oil than the parents according to heterosis. There are different interaction strength effects of the genetic or environmental factors on the compositions and concentrations of vegetable oil. For instance, the site was the main effect on γ-tocopherol and δ-tocopherol, but for α-tocopherol, the site effect was dependent on the genotype [10]. The walnut species may affect the fatty acid composition, and the temperature affects the fatty acid content in walnut oil [2,3]. Therefore, comprehensive analysis of the differences in the fatty acid composition and content of walnut oil between species/genotypes and cultivation sites is important for breeding programs and cultivation (matching sites with species or varieties). Consequently, the aim of this work was (i) to distinguish walnut oil from different species/genotypes, (ii) determine the effect of genotypes and climatic factors on walnut oil, and (iii) evaluate the effect of the main climatic factors during nut development on the fatty acid.

2. Materials and Methods

2.1. Plant Materials and Experimental Area

The walnut samples consisted of *Juglans regia* L. (Common walnut), *J. sigillata* Dode (Iron walnut), and *J. sigillata* Dode × *J. regia* L. at nine cultivation sites. The commercial cultivar was 'Yanyuanzao' (a hybrid of *J. sigillata* Dode × *Juglans regia* L., widely cultivated in China), which was cultivated at six sites (the main distribution area of 'Yanyuanzao') under Hengduan Mountains. *J. regia* was cultivated at three sites, and *J. sigillata* Dode was cultivated at two sites. Information of the accessions and cultivation sites is shown in Table 1. In 2019, 3 replicates of 50 nuts were randomly collected from around the canopy of a healthy tree after open pollination for each accession at each site. The meteorological data (minimum, average, and maximum daily temperature; accumulated precipitation; sunshine duration; average and minimum relative humidity during fruit development) for each site were obtained from meteorological stations at the cultivated site.

Table 1. Geographic site and cultivated species/genotypes of the walnut samples.

Cultivation Site	Species	Longitude/°	Latitude/°	Altitude/m
Batang, Ganzi, China (RBT)	Common walnut (*J. regia* L.)	99.013	29.778	2493
Derong, Ganzi, China (RDR)	Common walnut (*J. regia* L.)	99.376	29.047	3196
Jiulong, Ganzi, China (SJL)	Iron walnut (*J. sigillata* Dode)	101.723	28.528	2304
Jiulong, Ganzi, China (SRJL)	Hybrids (*J. sigillata* Dode × *J. regia* L.)			
Leibo, Liangshan, China (SLB)	Iron walnut (*J. sigillata* Dode)	103.434	28.258	989
Leibo, Liangshan, China (SRLB)	Hybrids (*J. sigillata* Dode × *J. regia* L.)			
Xiangcheng, Ganzi, China (RXC)	Common walnut (*J. regia* L.)	99.467	29.087	2814
Dechang, Liangshan, China (SRDC)	Hybrids (*J. sigillata* Dode × *J. regia* L.)	102.017	27.050	1596
Luding, Ganzi, China (SRLD)	Hybrids (*J. sigillata* Dode × *J. regia* L.)	102.017	29.050	1412
Mianning, Liangshan, China (SRMN)	Hybrids (*J. sigillata* Dode × *J. regia* L.)	102.683	28.033	2004
Yanyuan, Liangshan, China (SRYY)	Hybrids (*J. sigillata* Dode × *J. regia* L.)	101.050	27.033	2527

Note: Common walnut and iron walnut were cultivated by seedlings, *J. sigillata* Dode × *J. regia* L. was cultivated by grafting with a commercial cultivar of 'Yanyuanzao'.

2.2. Fatty Acid Composition

The nuts were immediately transported to the laboratory after harvest, dried in an oven at 45 °C for 3 days, and dehusked using a hammer. The walnut oil was extracted from the kernels using the Soxhlet method [20]. The fatty acid was analyzed using an 8890 gas chromatograph (GC) (Agilent, Shanghai, China) and a CP-Sil 88 FAME capillary column (0.20 μm, 100 m × 0.25 mm, Agilent, Santa Clara, CA, USA). The operating conditions were as follows: nitrogen as the carrier gas with a linear velocity of 0.7 mL/min, flame ionization detector (FID) temperature of 280 °C and inlet temperature of 270 °C, split ratio of 100:1, and injection volume of 1.0 μL. The oven was held at 100 °C for 13 min and then programmed at 10 °C/min to 180 °C and held for 6 min, then programmed at 1 °C/min to 200 °C and held for 20 min, and finally increased to 230 °C at 4 °C/min and held for 10.5 min. The samples were identified by comparing the retention times of the sample peaks with those of a mixture of FAME standards. The fatty acid contents were expressed as the relative in terms of the percentage of individual fatty acids.

2.3. Micronutrients and Secondary Metabolites Determination

In total, 4.00 g of oil was weighted and diluted with 4 mL of *n*-hexane in a 10 mL volumetric flask and the content of tocopherols determined by high-performance liquid chromatography [21]. The ultraviolet detector was performed using high-performance liquid chromatography (LC-20A, Shimadzu, Tokyo, Japan) installed with a Waters Spherisorb Silica Column (250 mm × 4.6 mm, 5 μm). The injection volume of the sample was 10 μL, the column temperature was 20 °C, and the mobile phase was methanol with a rate of 0.8 mL/min. A determining wavelength of 294 nm was used. Through comparison of the standards, α-, β-, γ-, and δ-tocopherols were identified and quantified. The tocopherol compositions were the mean values of three replicates from each sample and were expressed as mg/kg oil. The polyphenols and flavone contents of walnut oil were determined using Folin's reagent and the aluminum nitrate-sodium nitrite colorimetric method, respectively [20]. The absorbance of the solution was measured at 510 nm for flavone and 765 nm for polyphenols. The flavone and polyphenols contents were the mean values of three replicates from each sample and were expressed as mg/kg oil.

2.4. Statistical Analysis

Statistical analysis was performed using R version 4.0.5 (https://www.r-project.org/, accessed on 31 December 2021). The functions of the Shapiro test and Bartlett test were used to calculate the normal distribution and the homogeneity of variance test. Then, one-way ANOVA was used to determine the difference in the chemical composition of walnut oil between different cultivation sites. The mean separation was assessed with the LSD test at $p \leq 0.05$. Further, two-way ANOVA based on the data of two sites (Jiulong and Leibo) of two species (iron walnut and hybrids) was used to explore the effect of species, cultivation sites, and the species–sites interaction on the chemical composition and content of walnut oil (Table 1). Pearson correlation coefficients between the chemical characteristics were calculated ($\alpha = 0.05$). The correlation between sites/species was assessed by Pearson's test according to the chemical compositions of walnut oil. The correlation between the variation in the chemical composition of walnut oil and climatic factor variables was calculated by linear regression analysis. Further, to assess the relative importance of each climatic factor to explain the variation in walnut oil, full subset regression analysis and multiple regression models were implemented using ordinary least squares (OLS). All variables of the climatic factors were standardized before conducting the regression analysis.

3. Results

3.1. Fatty Acid Composition

Data on oil the fatty acid compositions and contents are reported in Figure 1. In all samples, the oil was mainly composed of five fatty acids: palmitic acid (C16:0, 4.01–5.41%), stearic acid (C18:0, 2.41–3.10%), oleic acid (C18:1, 18.40–33.56%), linoleic acid (C18:2, 51.39–63.12%), and linolenic acid (C18:3, 6.52–11.69%). Polyunsaturated acids (PUFAs) were the main group of fatty acids in the walnut oil in all three species, followed by monounsaturated fatty acids (MUFAs). The results of the one-way ANOVA showed that the content of PUFAs, MUFAs, linoleic, and linolenic acid were significantly different in the three walnut species. The contents of linoleic and linolenic acid in the hybrids (*J. sigillata* × *J. regia*) were significantly higher than common walnut (*J. regia*). Therefore, the PUFAs content of the hybrids was significantly higher than common walnut ($F = 27.62$, $p < 0.001$), which would be due to the heterosis of iron and common walnut. Oleic acid (C18:1) was the major MUFA, and the content of oleic acid in common walnut was significantly higher than that in hybrids and iron walnut ($F = 9.81$, $p < 0.001$). Notably, there was a negative correlation between the content of oleic (C18:1) and linoleic acid (C18:2) ($r = -0.979$, $p < 0.001$) (Figure 2). PUFA showed a significant negative correlation with the content of oleic acid (C18:1) and MUFAs but a significant positive correlation with the content of linoleic acid (C18:2). A significant positive correlation between the content of margaroleic acid (C17:1) and myristic (C14:0) and margaric acid (C17:0) was observed. There was a significant negative correlation between the content of eicosenoic acid (C20:1) and palmitic (C16:0) and palmitoleic acid (C16:1). Saturated fatty acids (SFAs) were the lesser group in walnut oil, ranging from 6.91% to 8.91%, with palmitic acid (C16:0) and stearic acid (C18:0) the two main SFAs present, totaling, on average, 4.31% and 2.63%, respectively (Figure 1). It should be noted that the hybrids' walnut oil contained some minor fatty acids, such as myristic acid (C14:0, 0.01–0.02%), margaric acid (C17:0, 0.02–0.03%), and margaroleic acid (C17:1, 0.02%), that were not detected in the common and iron walnut oil. We studied the effects of species, climatic factors, and the genotype–environment interaction on the fatty acid composition and content based on the design of two cultivation sites (Jiulong and Leibo) of two species (iron walnut and hybrids). The results of two-way ANOVA showed significant differences in the linolenic acid (C18:3) content between the sites and species (Table 2). A significant difference between species was shown regarding margaroleic acid (C17:1) ($p < 0.001$). The content of myristic acid (C14:0) was significantly influenced by species, cultivation sites, and the species–sites interaction.

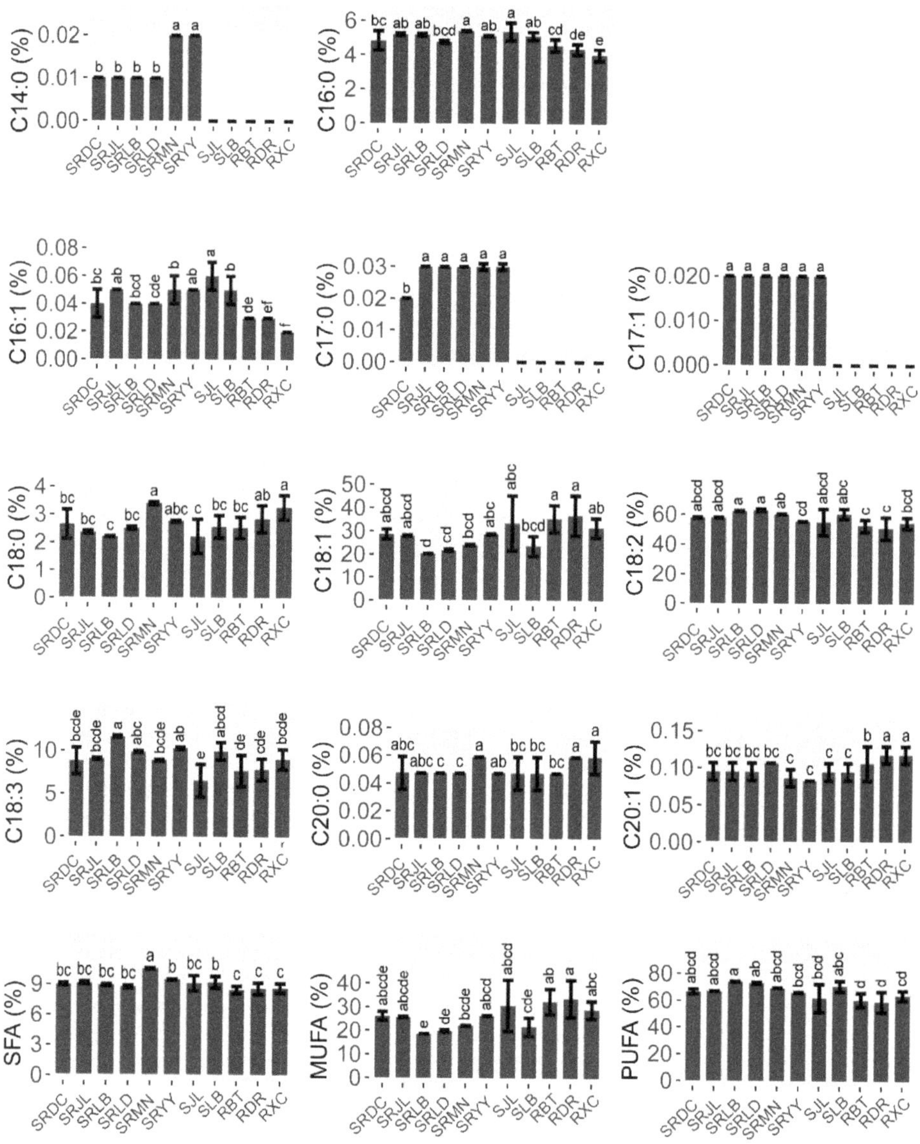

Figure 1. Fatty acid composition and contents of walnut oils. Note: Values are the mean ± standard deviation, and different letters indicate significant differences ($p < 0.05$) between different species/sites. SFAs, saturated fatty acids (C14:0 + C16:0 + C17:0 + C18:0 + C20:0); MUFAs, monounsaturated fatty acids (C:16:1 + C17:1 + C18:1 + C20:1); PUFAs, polyunsaturated fatty acid (C18:2 + C18:3), the same as below.

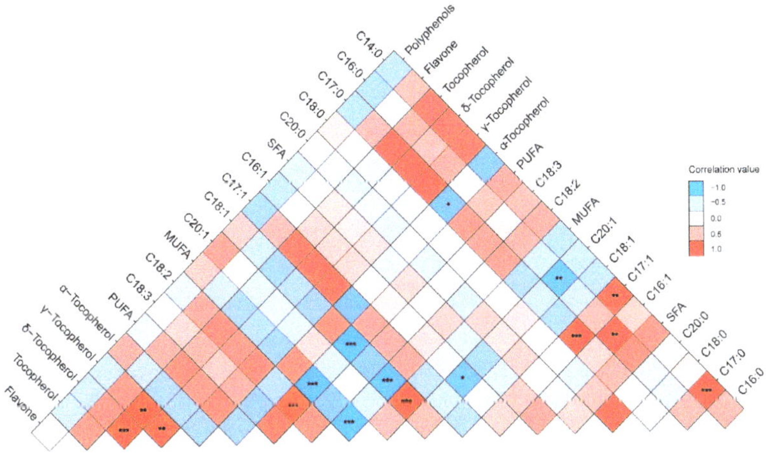

Figure 2. The correlation between the chemical characteristics of walnut oil. Note: *, **, and *** indicate significant correlations ($p < 0.05$, 0.01, and 0.001) between the 19 chemical characteristics of walnut oil, respectively.

Table 2. Two-way analysis of variance (two-way ANOVA) of fatty acid.

Source	Sum of Square	Mean Square	*F* Value	Source	Sum of Square	Mean Square	*F* Value
C14:0				C18:1			
Species	4.51×10^{-4}	4.51×10^{-4}	3464.16 ***	Species	19.50	19.50	0.474
Sites	1.40×10^{-6}	1.40×10^{-6}	10.74 *	Sites	179.00	179.00	4.348
Species × Sites	1.40×10^{-6}	1.40×10^{-6}	10.74 *	Species × Sites	1.50	1.50	0.037
Residuals	1.00×10^{-6}	1.00×10^{-7}		Residuals	330.00	41.20	
C16:0				C20:1			
Species	1.55×10^{-1}	1.55×10^{-1}	1.551	Species	1.96×10^{-4}	1.96×10^{-4}	1.69
Sites	2.64×10^{-1}	2.64×10^{-1}	2.654	Sites	3.10×10^{-6}	3.10×10^{-6}	0.027
Species × Sites	2.44×10^{-1}	2.44×10^{-1}	2.449	Species × Sites	2.59×10^{-5}	2.59×10^{-5}	0.223
Residuals	7.97×10^{-1}	9.96×10^{-2}		Residuals	9.28×10^{-4}	1.16×10^{-4}	
C18:0				MUFA			
Species	5.20×10^{-3}	5.20×10^{-6}	0.073	Species	19.30	19.30	0.467
Sites	3.85×10^{-2}	3.85×10^{-2}	0.544	Sites	180.00	180.00	4.36
Species × Sites	1.82×10^{-1}	1.82×10^{-1}	2.569	Species × Sites	1.50	1.50	0.037
Residuals	5.66×10^{-1}	7.08×10^{-2}		Residuals	330.00	41.30	
C20:0				C18:2			
Species	9.18×10^{-6}	9.18×10^{-6}	0.317	Species	1.41	1.41	0.052
Sites	7.72×10^{-5}	7.72×10^{-5}	2.662	Sites	73.70	73.70	2.694
Species × Sites	2.64×10^{-6}	2.64×10^{-6}	0.091	Species × Sites	4.90×10^{-1}	4.90×10^{-1}	0.018
Residuals	2.32×10^{-4}	2.90×10^{-5}		Residuals	219.00	27.30	
SFA				C18:3			
Species	6.32×10^{-2}	6.32×10^{-2}	2.425	Species	12.60	12.60	8.001 *
Sites	1.07×10^{-1}	1.07×10^{-1}	4.12	Sites	26.60	26.60	16.921 **
Species × Sites	4.20×10^{-3}	4.20×10^{-3}	0.161	Species × Sites	3.76×10^{-1}	3.76×10^{-1}	0.239
Residuals	2.09×10^{-1}	2.61×10^{-2}		Residuals	12.60	1.57	
C16:1				PUFA			
Species	2.98×10^{-4}	2.98×10^{-4}	3.516	Species	22.40	22.40	0.554
Sites	3.76×10^{-4}	3.76×10^{-4}	4.435	Sites	189.00	189.00	4.666
Species × Sites	3.74×10^{-5}	3.74×10^{-5}	0.442	Species × Sites	1.70	1.70	0.043
Residuals	6.77×10^{-4}	8.47×10^{-5}		Residuals	324.00	40.50	

Table 2. *Cont.*

Source	Sum of Square	Mean Square	*F* Value	Source	Sum of Square	Mean Square	*F* Value
C17:1							
Species	1.14×10^{-3}	1.14×10^{-3}	607.382 ***				
Sites	8.20×10^{-6}	8.20×10^{-6}	4.375				
Species × Sites	8.20×10^{-6}	8.20×10^{-6}	4.375				
Residuals	1.50×10^{-5}	1.90×10^{-6}					

Note: The data of two cultivation sites (Jiulong and Leibo) of two species (iron walnut and hybrids) were used in two-way ANOVA. "*", "**", and "***" indicate significant differences ($p < 0.05$, 0.01, and 0.001) between species/sites, respectively. The same as below.

3.2. Micronutrient and Secondary Metabolites Levels

The contents of tocopherol, polyphenols, and flavone in walnut oil are shown in Figure 3. Three forms (α, γ, and δ) of tocopherol were detected. The total tocopherol values ranged from 109.55 (RXC) to 428.81 mg/kg (SRLD). The results clearly indicated that the predominant tocopherol form of walnut oil of iron walnut, common walnut, and the hybrid walnut trees was γ-tocopherol (86.11–350.89 mg/kg), with a proportion of more than 70%. The γ- and δ-tocopherol contents of the hybrid walnut oil were significantly higher than that of common and iron walnut oil while α-tocopherol was only detected in the iron and common walnut oil. The results of the correlation analysis showed a significant positive correlation among the contents of δ-tocopherol, γ-tocopherol, and tocopherol (Figure 2). We also studied the effects of species, climatic factors, and genotype–environment interaction on the micronutrient and secondary metabolite contents based on the design of two cultivation sites (Jiulong and Leibo) of two species (iron walnut and hybrids). The results of the two-way ANOVA showed that only the species effect was significant, which indicated that genetic determinism may be the main factor affecting the tocopherol content in walnut oil (Table 3). The flavone content ranged from 0.60–5.78 mg/kg and showed similar trends to tocopherol (Figure 3). The content of flavone was not influenced by the genotypic variation or cultivation sites alone but was affected by the interaction of species × sites (Table 3). The content of polyphenols was significantly influenced by the cultivation sites and the interaction of species × sites and showed opposite trends to tocopherol and flavone (Figure 3, Table 3). The polyphenols content of the common walnut oil was significantly higher than that of the hybrid walnut oil ($F = 17.63$, $p < 0.001$), and 1.2 times higher than that of the iron walnut oil ($F = 1.187$, $p > 0.05$) (Figure 3).

Table 3. Two-way analysis of variance (two-way ANOVA) of the micronutrients and secondary metabolites.

Source	Sum of Square	Mean Square	*F* Value
α-tocopherol			
Species	47,926	47,926	9.432 *
Sites	1561	1561	0.307
Species × Sites	16,301	16,301	3.208
Residuals	40,648	5081	
δ-tocopherol			
Species	3029.9	3029.9	18.279 **
Sites	344.5	344.5	2.078
Species × Sites	300.8	300.8	1.815
Residuals	1326.1	165.7625	
Tocopherol			
Species	61,329	61,329	8.943 *
Sites	219	219	0.032
Species × Sites	22,853	22,853	3.332
Residuals	54,864	6858	

Table 3. *Cont.*

Source	Sum of Square	Mean Square	*F* Value
Flavone			
Species	0.0312	0.0312	0.419
Sites	0.2115	0.2115	2.842
Species × Sites	1.6337	1.6337	21.950 **
Residuals	0.5958	0.0744	
Polyphenols			
Species	182.1	182.1	2.823
Sites	745.1	745.1	11.550 **
Species × Sites	1176.7	1176.7	18.239 **
Residuals	516.1	64.5	

"*" and "**" indicate significant differences ($p < 0.05$ and 0.01) between species/sites, respectively.

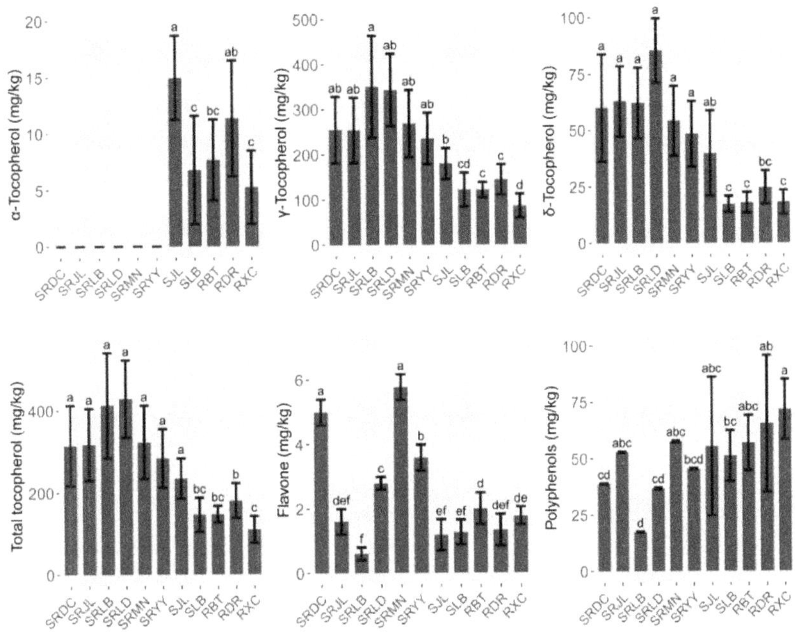

Figure 3. Micronutrient and secondary metabolites composition and contents of walnut oils. different letters indicate significant differences ($p < 0.05$) between different species/sites.

3.3. Correlation Analysis

The results of correlation between sites/species also showed that high level of correlations were found between the same species, even though cultivated in different sites (Figure 4). For instance, a highly relationship existed between the cultivated site of Leibo and Jiulong with the same species (SRLB vs. SRJL), yet a lowly relationship existed between *J. sigillata* and *J. sigillata* × *J. regia* which cultivated in the same site (SRLB vs. SLB). Similarly results also occurred in the cultivated site of Jiulong, the relationship between SRJL and SJL was lower than SRJL vs. SRMN, SRJL vs. SRDC, and SRJL vs. SRYY.

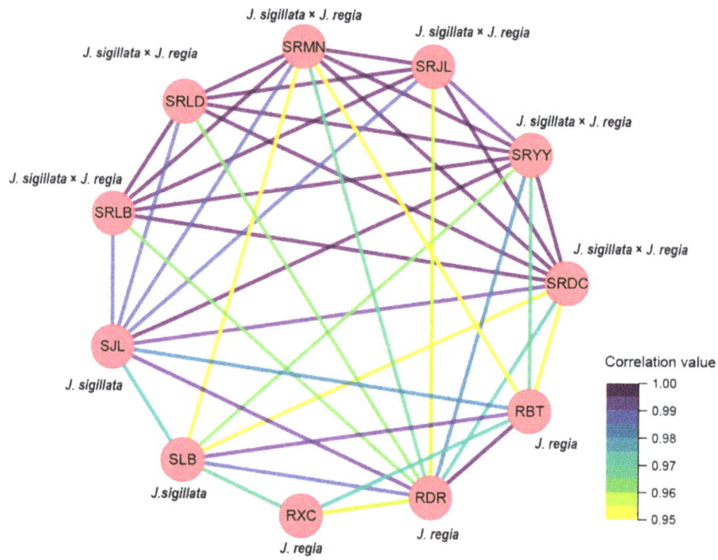

Figure 4. The related network diagram of cultivation sites/species and the main composition of walnut oil.

3.4. Climatic Factors and Fatty Acid Composition

The effects of climatic factors on the fatty acid composition throughout the fruit development period from the end of flowering to physiological maturity were analyzed. Precipitation showed a significant positive correlation with myristic acid (C14:0), palmitic acid (C16:0), margaric acid (C17:0), and SFAs, and a significant negative correlation with eicosenoic acid and polyphenols (Figure 5). Minimum relative humidity (MinRH) and average relative humidity (AvgRH) showed a significant negative correlation with oleic acid (C18:1) and MUFAs. Nevertheless, MinRH had a positive significant effect on linoleic acid (C18:3) and PUFA. There was also significant negative correlation between palmitic acid (C16:0), margaric acid (C17:0), and margaroleic acid (C17:1) and between average relative humidity (AvgRH) and oleic acid (C18:1) and MUFAs. Further, after multiple regression model comparison and averaging, the model explained 85%, 91%, and 65% of the variation in oleic acid (C18:1), linoleic acid (C18:3), and linolenic (C18:2), respectively. The average relative humidity (AvgRH) explained the most variation in both oleic (C18:1) (positive) and linolenic acid (C18:2) (negative) (Figure 6). Latitude explained the most in linoleic acid (C18:3), and other variables had a significant but small effect on linoleic acid (C18:3). The model explained 83%, 85%, and 97% of the variation in PUFAs, MUFAs, and SFAs, respectively, The AvgRH also explained the most variation in PUFAs (negative), MUFAs (positive), and SFAs (positive). Other variables with a small effect included temperature, precipitation, altitude, etc. Regarding the micronutrients and secondary metabolites, AvgRH was most important (negative) in tocopherol and flavone, and latitude explained the most variation (positive) in polyphenols. The model explained 52% and 61% of the variation in polyphenols and flavone, respectively. Other variables had a small effect on them, e.g., the estimated value of the minimum temperature (MinTem) on flavone was 11.33, which is lower than the absolute estimated value of AvgRH (193.62).

27. Amaral, J.S.; Alves, M.R.; Seabra, R.M.; Oliveira, B.P.P. Vitamin E composition of walnuts (*Juglans regia* L.): A 3-year comparative study of different cultivars. *J. Agric. Food Chem.* **2005**, *53*, 5467–5472. [CrossRef]
28. Gharibzahedi, S.M.T.; Mousavi, S.M.; Hamedi, M.; Khodaiyan, F. Determination and characterization of kernel biochemical composition and functional compounds of persian walnut oil. *J. Food Sci. Technol.* **2014**, *51*, 34–42. [CrossRef]
29. Slatnar, A.; Mikulic-Petkovsek, M.; Stampar, F.; Veberic, R.; Solar, A. Identification and quantification of phenolic compounds in kernels, oil and bagasse pellets of common walnut (*Juglans regia* L.). *Food Res. Int.* **2015**, *67*, 255–263. [CrossRef]

forests

MDPI

Article

Water Uptake and Hormone Modulation Responses to Nitrogen Supply in *Populus simonii* under PEG-Induced Drought Stress

Zhen Li [1,†], Xiaoling Wang [2,†], Yunshan Liu [3,4], Yangyan Zhou [4], Zhiliang Qian [5], Zequn Yu [6], Na Wu [1] and Zhan Bian [1,3,*]

1 Institute of Applied Biotechnology, School of Agronomy and Life Science, Shanxi Datong University, Datong 037009, China; lizhen90426@foxmail.com (Z.L.); zhenlingna90@126.com (N.W.)
2 Institute of Biological Resources, Jiangxi Academy of Sciences, Nanchang 330096, China; wangxiaoling1979@126.com
3 State Key Laboratory of Tree Genetics and Breeding, Research Institute of Tropical Forestry, Chinese Academy of Forestry, Guangzhou 510520, China; lys061949@163.com
4 Salver Academy of Botany, Rizhao 262300, China; yangyan_zhou@pku.edu.cn
5 College of Life Sciences, South China Agricultural University, Guangzhou 510642, China; qzl@stu.scau.edu.cn
6 Shanghai Gardening-Landscaping Construction Co., Ltd., Shanghai 200333, China; zequn2020@126.com
* Correspondence: bianzhanzhan@163.com
† These authors contributed equally to this work.

Abstract: In the present study, the effects of nitrogen (N) supply on water uptake, drought resistance, and hormone regulation were investigated in *Populus simonii* seedlings grown in hydroponic solution with 5% polyethylene glycol (PEG)-induced drought stress. While acclimating to drought, the *P. simonii* seedlings exhibited a reduction in growth; differential expression levels of aquaporins (AQPs); activation of auxin (IAA) and abscisic acid (ABA) signaling pathways; a decrease in the net photosynthetic rate and transpiration rate; and an increase in stable nitrogen isotope composition ($\delta^{15}N$), total soluble substances, and intrinsic water use efficiency (*WUEi*), with a shift in the homeostasis of reactive oxygen species (ROS) production and scavenging. A low N supply (0.01 mM NH_4NO_3) or sufficient N supply (1 mM NH_4NO_3) exhibited distinct morphological, physiological, and transcriptional responses during acclimation to drought, primarily due to strong responses in the transcriptional regulation of genes encoding AQPs; higher soluble phenolics, total N concentrations, and ROS scavenging; and lower transpiration rates, IAA content, ABA content, and ROS accumulation with a sufficient N supply. *P. simonii* can differentially manage water uptake and hormone modulation in response to drought stress under deficient and sufficient N conditions. These results suggested that increased N may contribute to drought tolerance by decreasing the transpiration rate and O_2^- production while increasing water uptake and antioxidant enzyme activity.

Keywords: gene; water deficit; fertilization; aquaporins; auxin; abscisic acid; poplar

Citation: Li, Z.; Wang, X.; Liu, Y.; Zhou, Y.; Qian, Z.; Yu, Z.; Wu, N.; Bian, Z. Water Uptake and Hormone Modulation Responses to Nitrogen Supply in *Populus simonii* under PEG-Induced Drought Stress. *Forests* **2022**, *13*, 907. https://doi.org/10.3390/f13060907

Academic Editor: Ilona Mészáros

Received: 14 April 2022
Accepted: 8 June 2022
Published: 10 June 2022

1. Introduction

Drought is a major limiting factor for plant growth on the Loess Plateau [1]. Forests in such areas often suffer from drought stress due to continued deficit of rain [2]. Increased forest mortality may be due to persistent drought stresses around the world [3]. In the past few decades, drought stress has caused widespread die-off of *Populus tremuloides* [4]. To ensure the sustainable development of forest plantations, it is essential to better understand the responses of woody species growing on poor soils.

P. simonii is an important poplar species on the Loess Plateau of China and is widely used for afforestation. However, the soil is often poor and water deficiency is also frequent in these regions [4]. Recently, we found that drought stress stunts nitrogen assimilation and N deficiency stimulated root development in *P. simonii*. However, how water uptake responds to N supply and drought stress remains unclear.

Drought stress often leads to physiological and molecular responses in plants [5,6]. A schematic model of responses to drought in plants has been well documented [7–9]. Poplar species (*Populus* spp.) are very sensitive to water availability and drought stress often decreases photosynthesis, alters root morphology, and inhibits enzyme activities. For instance, drought-induced stomatal closure often leads to an inhibited CO_2 assimilation rate [10,11]. Drought stress also can trigger the overproduction of reactive oxygen species (ROS) and eventually lead to oxidative damage. Against the harmful effects of ROS, plants' defence systems can rely on the oxygen scavenging systems (SOD, APX, POD, and CAT) and osmotic adjustment compounds (soluble sugars, soluble phenolics, and amino acid compounds) [12,13]. In addition, water stress triggers signal transduction pathways, including the abscisic acid (ABA) and auxin (IAA) signaling pathway [14,15]. The enzyme 9-cis-epoxycarotenoid dioxygenase (NCED) and the protein phosphatase 2C (PP2C) play key roles in ABA biosynthesis and ABA-mediated signaling pathways [8,16]. In auxin biosynthesis, flavin-containing monooxygenase (YUCC) and tryptophan aminotransferase (TAA) are essential components. Transcriptional regulation of the key genes involved in the ABA and/or IAA signal transduction chains are important for detecting changes in soil water availability [7,17]. Furthermore, plants transport water by apoplastic and/or cell-to-cell (symplastic and transmembrane) pathways, and aquaporins (AQPs) play crucial roles in transmembrane water uptake [18]. The expression levels of AQPs, including the plasma membrane intrinsic proteins (PIPs), tonoplast intrinsic proteins (TIPs), and small basic intrinsic proteins (SIPs) can be modulated to regulate the water uptake in various soil water availability [19,20]. However, the role of AQPs in water transport under drought condition and N supply is unclear.

Previous studies have found that carbon and nitrogen metabolism can be altered by drought stress. Indeed, soluble sugars, soluble phenolics, total C, and total N can be altered and play important roles in osmoregulation in response to drought stress [8,21]. In addition, N uptake is often coupled with water uptake because NH_4^+ and/or NO_3^- must be dissolved in water, and then absorbed by roots [8]. The N supply (via irrigation) affects plant growth under drought conditions by increasing soluble sugars and the activities of antioxidant enzymes [22].

P. simonii is widely distributed on the Loess Plateau, and the soil in this area suffers from drought and N deficiencies. In a previous study, it was found that an increase in N contributed to drought tolerance by increasing NH_4^+ uptake and decreasing nitrogen metabolism in *P. simonii* [2]. We investigated several morphological (root characteristics), physiological (photosynthesis, C metabolism, N metabolism, ROS accumulation, and scavenging), and molecular (transcript levels of representative genes involved in water uptake and hormone metabolism) changes in response to N supply under drought condition. We hypothesized that the negative effects of drought stress on water uptake and hormone metabolism could be eased by a high N supply.

2. Materials and Methods

2.1. Plant Materials and Treatments

P. simonii cuttings (approximately 15 cm in length and 20 mm in diameter) were rooted in fine sandy soil and vermiculite (1:1). The plants were cultivated in a greenhouse as described previously [23]. One sprout was left for each plant. After 4 weeks, the substrate were removed and similar saplings (approximately 20 cm) were cultivated under hydroponic conditions in Hoagland solution (10 μM EDTA·FeNa, 0.5 μM H_2MoO_4, 30 μM H_3BO_3, 1 μM $CuSO_4{\cdot}5H_2O$, 1 μM $ZnSO_4{\cdot}7H_2O$, 5 μM $MnSO_4{\cdot}H_2O$, 1 mM $MgSO_4{\cdot}7H_2O$, 1 mM KH_2PO_4, 1 mM Na_2SO_4, and 1 mM $CaCl_2$) containing 0.01 mM NH_4NO_3 (low N supply) or 1 mM NH_4NO_3 (sufficient N supply), and the pH was adjusted to 5.5 for each solution. Subsequently, a total of 36 plants were subjected to drought stress via the addition of 5% PEG-6000 (osmotic stress, −0.362 MPa) to the modified Hoagland's nutrient solution; 36 plants that had been grown in the nutrient solution without PEG-6000 served as the control. During the experiment, the aerated nutrient solutions were renewed every two days. The treatments were maintained under hydroponic cultivation for 4 weeks prior to harvest.

2.2. Measurement of Gas Exchange, Photosynthetic Pigments, and Root Growth

Gas exchange was determined on three mature leaves (leaf plastochron index = 8–10) with a portable photosynthesis system (Li-Cor-6400, Li-Cor, Inc., Lincoln, NE, USA). The measurements of net photosynthetic rates (A), stomatal conductance (g_s), and transpiration rates were carried out from 9:00 to 11:00. The instantaneous water use efficiency (WUE_i) was calculated as the ratio between A and g_s. The chlorophyll content was measured with a portable chlorophyll meter (SPAD 502 Meter, Minolta Corporation, Tokyo, Japan).

Parts of roots (approximately 2 g) were excised from the root system, scanned with a scanner (HP Scanjet G4050), and analyzed with a WinRHIZO root analyzer system (WinRHIZO version 2007b, Regent Instruments, Quebec, QC, Canada). Six leaves from each plant were submerged in water for 24 h to measure the leaf saturated weight (SW). The remaining roots and leaves were wrapped with tinfoil, ground into fine powder in liquid N with a mortar and pestle, and stored at −80 °C. Frozen powder (c. 100 mg) from each plant was dried at 60 °C for 72 h to determine dry biomass and the fresh mass to dry mass ratio.

2.3. RWC, MDA Content, and Rate of Superoxide Anion Radical (O_2^-) Production

The RWC was defined according to Guo et al. [23]. RWC (%) = (FW − DW)/(SW − DW) × 100%, where FW, DW, and SW are the dry weight, fresh weight, and saturated weight, respectively.

The rate of O_2^- production was quantified according to Meng et al. [10]. Fine powder (100 mg) was homogenized in 50 mM potassium phosphate buffer (pH 7.8). After centrifugation (10,500× g, 20 °C,10 min), 1 mL of 1 mM hydroxylamine hydrochloride and 0.5 mL of 50 mM phosphate buffer (pH 7.8) were added to 1 mL supernatant and the mixture was incubated at 25 °C for 60 min. Then, 1 mL of 17 mM sulfanilamide and 1 mL of 7 mM α-naphthylamine were added to the mixture. After 20 min, the O_2^- concentration was quantified spectrophotometrically at 530 nm.

The MDA content was determined according to the methods of Dhinsa and Matowe [24]. The fine powder (100 mg) was extracted in 10% (w/v) trichloroacetic acid (TCA) and centrifuged (10,000× g, 20 °C, 10 min). Then, 2 mL of TCA containing 0.6% (w/v) thiobarbituric acid (TBA) was added and mixed with 1 mL of the supernatant. After incubating at 95 °C for 30 min, the mixture was centrifugated (10,000× g, 4 °C, 10 min). The absorbance of the supernatant was measured at 532 and 600 nm.

2.4. Determination of Soluble Sugars, Soluble Phenolics, and Amino Acid Compounds

The concentration of total soluble sugars was analyzed using the anthrone method [12]. Briefly, approximately 100 mg fine powder was homogenized in 4 mL of 80% ethanol. The mixture was incubated at 80 °C for 30 min. After centrifugation (6000× g, 25 °C, 10 min), 10 mg of activated charcoal was added for 30 min at 80 °C to decolorize the supernatant. Then, 5 mL of anthrone reagent was added. The mixture was incubated in boiling water for 10 min. After the mixture was cooled to room temperature, the absorbance of the supernatant was determined spectrophotometrically at 620 nm (Shimadzu, UV-3600, Kyoto, Japan).

The soluble phenolics in the plant materials were determined as described by Pritchard [25]. Approximately 100 mg fine powder was extracted in 1.5 mL of 80% methanol and shaken (150 rpm, 25 °C, 12 h) in the dark. After centrifugation (12,000× g, 25 °C, 10 min), the supernatant was collected. The soluble phenolics were determined spectrophotometrically at 765 nm.

The soluble amino acid compounds were determined as described previously [26]. A total of 100 mg of the powder was extracted in an extraction solution (methanol/ chloroform (7:3, v/v), 20 mM HEPES, 5 mM ethylene glycol tetraacetic acid, 10 mM NaF, pH 7.0) and incubated on ice for 30 min. Then, the mixture was extracted with distilled water, freeze-dried (Alpha 2–4, Christ, Osterode, Germany), and finally dissolved in lithium citrate buffer (pH 2.2). Free amino acids in the samples were analyzed using an automated amino acid analyzer (L-8900, Hitachi High-Technologies Corporation, Kyoto, Japan).

2.5. Analysis of Total C, Total N, ^{13}C, and ^{15}N

The samples were dried in an oven at 80 °C. The isotopic ratio was analyzed using an elemental analyzer (NA 1110, CE Instruments, Rodano, Italy) coupled to a GVI IsoPrime isotope ratio mass spectrometer (IRMS).

The stable C isotope composition was calculated as:

$$\delta^{13}C = (Rsac - Rsdc)/Rsdc \times 1000\ [‰]$$

where Rsac and Rsdc are the ratios of ^{13}C to ^{12}C in the sample and the standard, respectively.

The stable N isotope composition was calculated as:

$$\delta^{15}N = (Rsan - Rsdn)/Rsdn \times 1000\ [‰]$$

where Rsan and Rsdn are the ratios of ^{15}N to ^{14}N in the sample and the standard, respectively. The reference standard was N_2 in air.

2.6. Antioxidant Enzymes Activities

Due to space limitations, the detailed methods for determining antioxidant enzyme activities (SOD, POD, CAT, and APX) are provided in Supplementary Materials and Methods.

2.7. IAA and ABA Contents

IAA and ABA were extracted, purified, and quantified according to the method described by Diego [27]. The fine powder (approximately 0.5 g) was extracted in 4.0 mL of isopropanol/hydrochloric acid and shaken for 30 min at 4 °C before 10 mL of dichloromethane was added. The mixture was shaken again for 30 min at 4 °C. After centrifugation (13,000 rpm, 5 min, 4 °C), the lower organic phase was dried under N_2, dissolved in 150 μL of methanol (0.1% methane acid), and passed through a 0.22 μm membrane. Then, the filtrate was subjected to high-performance liquid chromatography–tandem mass spectrometry (HPLC–MS/MS) using a ZORBAX SB-C18 (Agilent Technologies, Santa Clara, CA, USA) column. The MS conditions were as follows: the spray voltage was 4500 V; the pressure of the air curtain, nebulizer, and aux gas were 15, 65, and 70 psi, respectively; and the atomizing temperature was 400 °C.

To quantify ABA, endogenous ABA was extracted from 50 mg of each sample using 0.5 mL of homogenization buffer (70% methanol and 0.1% formic acid); as an internal standard, 2 ng of ABA-d6 (Olchemim, Olomouc, Czech Republic) was added to the extracts. The mixture was diluted two times with deionized water, and the ABA content was analyzed with the ultrahigh performance liquid chromatography-triple TOF 5600+ system (Sciex, Concord, ON, Canada).

2.8. Analysis of Transcript Levels

Total RNA was extracted and purified with a plant RNA extraction kit (R6827, Omega Bio-Tek, Norcross, GA, USA); trace genomic DNA was digested by DNase I (E1091, Omega Bio-Tek); and 1 μg of total RNA was used to synthesize cDNA with the PrimeScript RT reagent kit (DRR037S, Takara, Dalian, China). Quantitative PCR was performed in a 20 μL solution reaction volume using 10 μL 2× SYBR Green Premix Ex Taq II, 2 μL of cDNA, and 0.2 μM primers (Supplementary Table S1) in a real-time platform (LightCycler 96 System, Roche, Basel, Swiss). The conservative actin2/7 was used as a reference gene [28]. Key genes involved in water uptake and hormone modulation were selected for the gene transcription analysis [8,17]. Primers were designed using the Primer 3 online tool. The qRT-PCR experiments were performed in triplicate for each sample. The efficiencies of the PCR reactions ranged between 95 and 105% (Supplementary Table S1).

2.9. Data Processing and Statistical Analysis

Statistical tests were performed with the SPSS software (version 20.0, SPSS Inc., Chicago, IL, USA). The effects of N supply and drought treatments on variables were analyzed by two-way analysis of variance (ANOVA). The means were compared on the basis of least significant differences ($p = 0.05$). The gene expression heatmap was generated by the command heatmap.2() using the 'gplots' package in R, as described by Luo et al. [13].

3. Results

3.1. Growth Parameters

Drought stress led to serious growth inhibition (decreased height and biomass) under both 0.01 mM NH_4NO_3 and 1 mM NH_4NO_3 (Supplementary Table S2 and Figure S1). Drought stress also markedly reduced the total length, surface area, and volume of the roots. N availability had a different effect on plant growth under drought stress and control conditions (Supplementary Table S2 and Figure S1). Under drought stress, the plants supplied with sufficient N had a greater biomass, chlorophyll content, and root development as compared with the plants supplied with less N (Supplementary Table S2 and Figure S1). Unexpectedly, plants under 0.01 mM NH_4NO_3 grew better under control conditions (without drought stress), exhibiting greater height and increased root development when they were supplied with 0.01 mM NH_4NO_3 than when they were supplied with 1 mM NH_4NO_3.

3.2. RWC, Photosynthesis, MDA, and the Rate of O_2^- Production

Drought stress significantly decreased the RWC, net photosynthetic rates, stomatal conductance, and transpiration rate under both 0.01 mM and 1 mM NH_4NO_3 (Table 1). The *WUEi*, MDA, and O_2^- in leaves increased significantly in response to drought stress in the plants under 0.01 mM NH_4NO_3, whereas these variables were not significantly affected in the plants under 1 mM NH_4NO_3 (Table 1). Drought stress increased the root MDA content under both 0.01 mM and 1 mM NH_4NO_3 (Table 1).

Table 1. The RWC, photosynthesis, MDA, and the rate of O_2^- production of *P. simonii* as affected by drought and nitrogen supply.

N Supply (mM)	Drought Treatment	RWC (%)	*A* (μmol CO_2 m^{-2} s^{-1})	g_s (mmol m^{-2} s^{-1})	*E* (mmol H_2O m^{-2} s^{-1})	*WUEi* (μmol CO_2 mol^{-1} H_2O)	MDA in Roots (μmol g^{-1} FW)	MDA in Leaves (μmol g^{-1} FW)	O_2^- in Roots (nmol g^{-1} FW min^{-1})	O_2^- in Leaves (nmol g^{-1} FW min^{-1})
0.01	Control	88.65 ± 0.95a	11.9 ± 0.4a	221.67 ± 9.10a	4.85 ± 0.15a	53.87 ± 2.51b	12.18 ± 1.90b	11.22 ± 1.47c	0.11 ± 0.02a	0.04 ± 0.01b
	Drought	77.57 ± 1.34b	2.7 ± 0.2c	33.33 ± 4.22b	1.28 ± 0.16b	83.54 ± 6.44a	35.75 ± 3.65a	28.62 ± 2.12a	0.12 ± 0.01a	0.14 ± 0.02a
1	Control	90.41 ± 1.29a	9.8 ± 0.2b	255.00 ± 9.92a	4.97 ± 0.19a	38.48 ± 1.63bc	15.23 ± 3.09b	15.57 ± 1.07bc	0.14 ± 0.02a	0.04 ± 0.01b
	Drought	80.01 ± 1.10b	2.4 ± 0.4c	71.67 ± 7.49b	0.26 ± 0.02c	33.60 ± 4.77c	30.32 ± 2.08a	20.63 ± 1.65b	0.12 ± 0.01a	0.06 ± 0.00b
p-values	N	ns	*	***	**	***	ns	ns	ns	**
	D	***	***	***	***	**	***	***	ns	***
	N × D	ns	ns	ns	**	***	ns	**	ns	**

Data indicate mean ± SE ($n = 6$). Different letters in the same column indicate significant differences. *p*-values of the ANOVAs of drought, nitrogen supply, and their interaction are indicated. *, $p < 0.05$; **, $p < 0.01$; ***, $p < 0.001$; ns, not significant.

Under control conditions, N supply had minor effect on RWC, photosynthesis, electrolyte leakage, MDA, and ROS (Table 1). However, under PEG-induced drought stress, 0.01 mM NH_4NO_3 resulted in significant increases in transpiration rate, *WUEi*, MDA, and O_2^- in leaves as compared with 1 mM NH_4NO_3 (Table 1).

3.3. Soluble Sugars, Soluble Phenolics, and Amino Acid Compounds

Soluble sugars and phenolics were significantly increased in response to drought stress in both the roots and leaves (Figure 1). N level had no significant effect on soluble sugars in roots, whereas 0.01 mM NH_4NO_3 decreased foliar soluble sugars (Figure 1A,B). Soluble phenolics (particularly in the leaves) were increased in response to 0.01 mM NH_4NO_3, regardless of the drought treatment (Figure 1C,D).

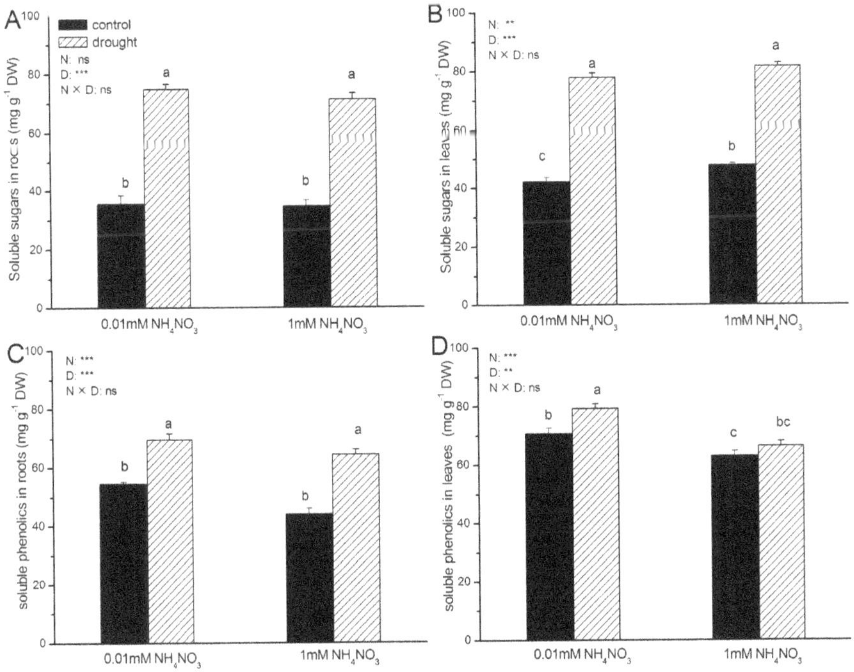

Figure 1. Soluble sugars and soluble phenolics in roots (**A**,**C**) and leaves (**B**,**D**) of *Populus simonii* as affected by nitrogen supply and drought stress. Data indicate mean ± SE (n = 6). Different letters in the same column indicate significant differences. N, N supply; D, drought treatment; N × D, interaction of N supply and drought treatment. **, $p < 0.01$; ***, $p < 0.001$; ns, not significant.

The total amino N concentrations and the composition of amino compounds changed in response to osmotic stress and N concentration (Supplementary Table S3). The 17 common amino acids were sorted into five groups based on biosynthetic origin [26]. In the 0.01 mM NH_4NO_3 treatment, the increase in the total amino acid content in response to osmotic stress in the roots was primarily due to a significant increase in the aspartate group (Figure 2A). This proportional increase in the aspartate group (from 28% to 37%) was compensated by a proportional decrease in the glutamate group (from 31% to 25%). Asparagine and glutamine were the main components of the aspartate group and the glutamate group, respectively (Figure 2A). Similarly, the total amount of amino compounds was higher in response to osmotic stress in the roots of plants grown with 1 mM NH_4NO_3, and the increase was primarily due to the changes in the aspartate, alanine, and glutamate groups. The proportion of aspartate compounds increased (from 28% to 39%), whereas the proportions of alanine compounds and glutamate compounds decreased (from 15% to 11% and from 34% to 29%, respectively) under drought conditions (Figure 2A); under both drought and control conditions, 0.01 mM NH_4NO_3 decreased the total amino acid content but had a little effect on the relative proportion of groups (Figure 2A).

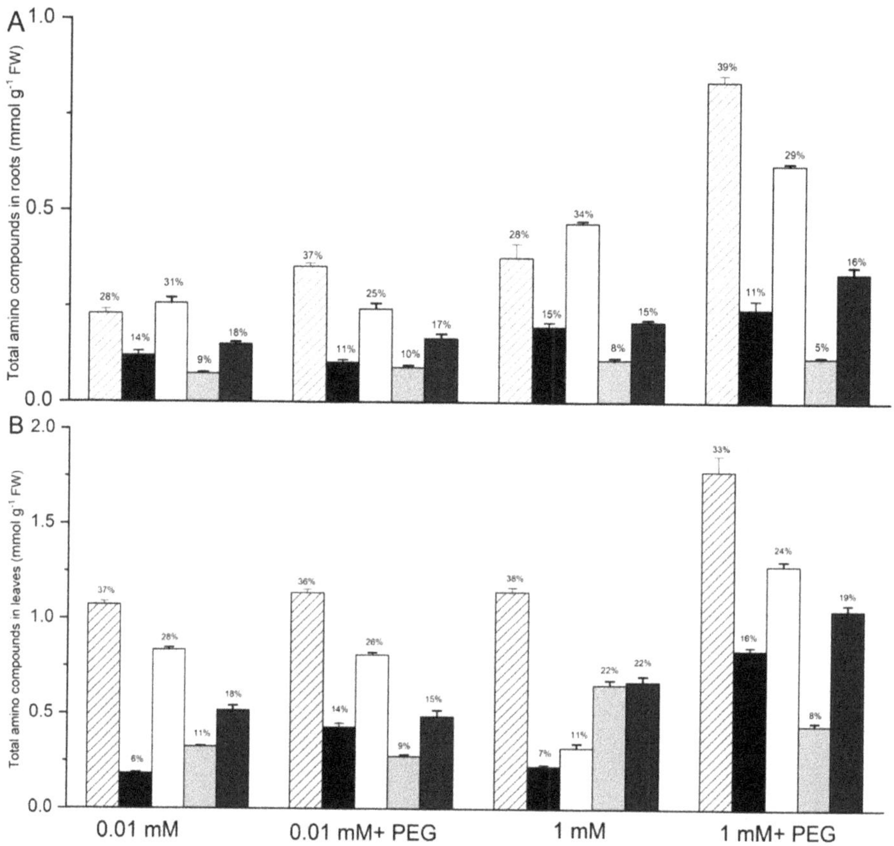

Figure 2. Amino compounds in roots (**A**) and leaves (**B**) of *Populus simonii* as affected by nitrogen supply and drought stress. Amino acids deriving from the same pathway are grouped together. The percent distributions of these biosynthetic groups are presented, as well as the mean values of sums of amino compounds of each group of three plants and SE. (Striped) aspartate, threonine, isoleucine, methionine, lysine; (black) serine, glycine, cysteine; (white) glutamate, histidine, arginine, proline; (light grey) phenylalanine, tyrosine; (dark grey) alanine, leucine, valine.

In the leaves, the serine group showed a higher absolute (1.5-fold) and proportional (from 6% to 14%) content in response to drought stress under 0.01 mM NH_4NO_3 (Figure 2B). Osmotic stress also increased the total amino acid content of most biosynthetic groups (except the tryptophan group) under 1 mM NH_4NO_3 (Figure 2B). Under 1 mM NH_4NO_3, the proportions of the serine group and the glutamate group increased (from 7% to 16% and from 11% to 24%, respectively), whereas the proportion of the tryptophan group decreased (from 22% to 8%) in response to drought stress (Figure 2B). Under the control conditions, N supply had no significant effect on the total amino acid concentration and relative content (Figure 2B). However, under the drought treatment, the total amino concentrations were higher for all biosynthetic groups under the 1 mM NH_4NO_3 treatment than under the 0.01 mM NH_4NO_3 treatment (Figure 2B).

3.4. Total C, Total N, ^{13}C, and ^{15}N

Drought stress significantly decreased the total C in roots, whereas it had no effect on foliar total C and $\delta^{13}C$ under 0.01 mM and 1 mM NH_4NO_3 (Figure 3). The total C and $\delta^{13}C$ in roots and leaves were not affected by N levels (Figure 3).

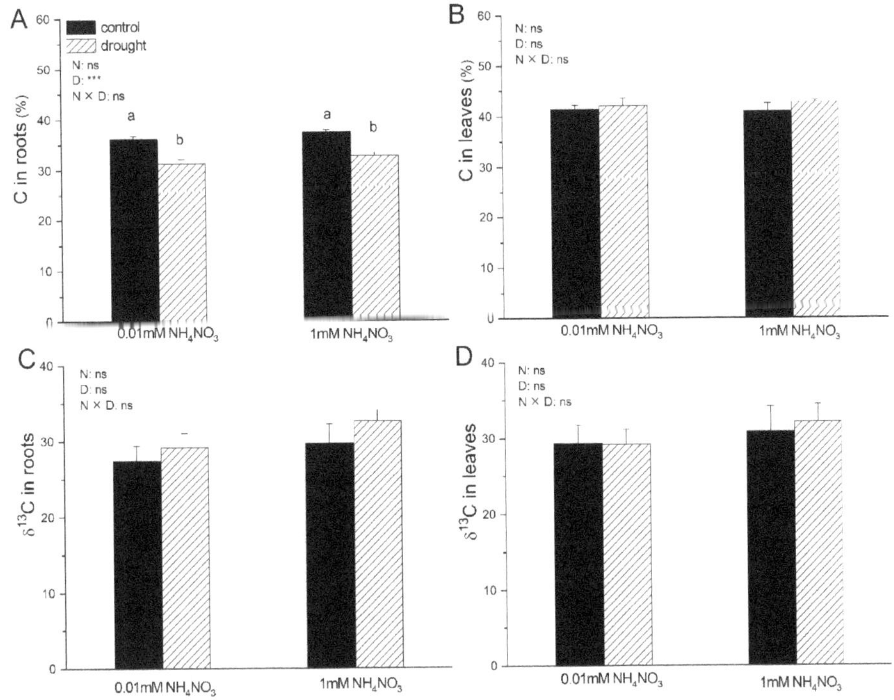

Figure 3. Total C concentration and $\delta^{13}C$ in roots (**A**,**C**) and leaves (**B**,**D**) of *Populus simonii* as affected by nitrogen supply and drought stress. Data indicate mean ± SE (n = 6). Different letters in the same column indicate significant differences. N, N supply; D, drought treatment; N × D, interaction of N supply and drought treatment. ***, $p < 0.001$; ns, not significant.

Drought stress had little effect on the total N concentration of roots, while the foliar total N concentration under 1 mM NH_4NO_3 decreased in response to the drought treatment (Figure 4A,B). There was a significant increase in $\delta^{15}N$ under the drought treatment as compared with the control in both roots and leaves (Figure 4C,D).

The N levels also affected the total N concentration of both roots and leaves (Figure 4A,B). Generally, the total N concentration of both roots and leaves was lower under 0.01 mM NH_4NO_3 as compared with that under 1 mM NH_4NO_3 under both drought stress and control conditions (Figure 4A,B). In contrast, 0.01 mM NH_4NO_3 led to higher $\delta^{15}N$ in roots (Figure 4C), 0.01 mM NH_4NO_3 also led to higher $\delta^{15}N$ in the leaves under control conditions, whereas it had no significant effect on $\delta^{15}N$ under drought stress (Figure 4D).

3.5. Antioxidant Enzymes Activities

The SOD activity in roots was not affected by drought stress, whereas drought stress significantly increased SOD activity under both 0.01 and 1 mM NH_4NO_3 treatments (Figure 5A,B). Drought stress also led to increased APX activity in both roots and leaves, except, root APX activity under 0.01 mM NH_4NO_3 remained unaltered in response to drought stress (Figure 5C,D).

Root SOD activity was not affected by N availability under either drought or control conditions (Figure 5A). N levels also had no impact on SOD activity in the control treatment; however, the 1 mM NH_4NO_3 treatment led to a higher SOD activity under drought stress as compared with the 0.01 mM NH_4NO_3 treatment (Figure 5B). Similar results were observed for root APX activity (Figure 5C). The APX activity in leaves was not affected by N levels under either drought or control conditions (Figure 5D).

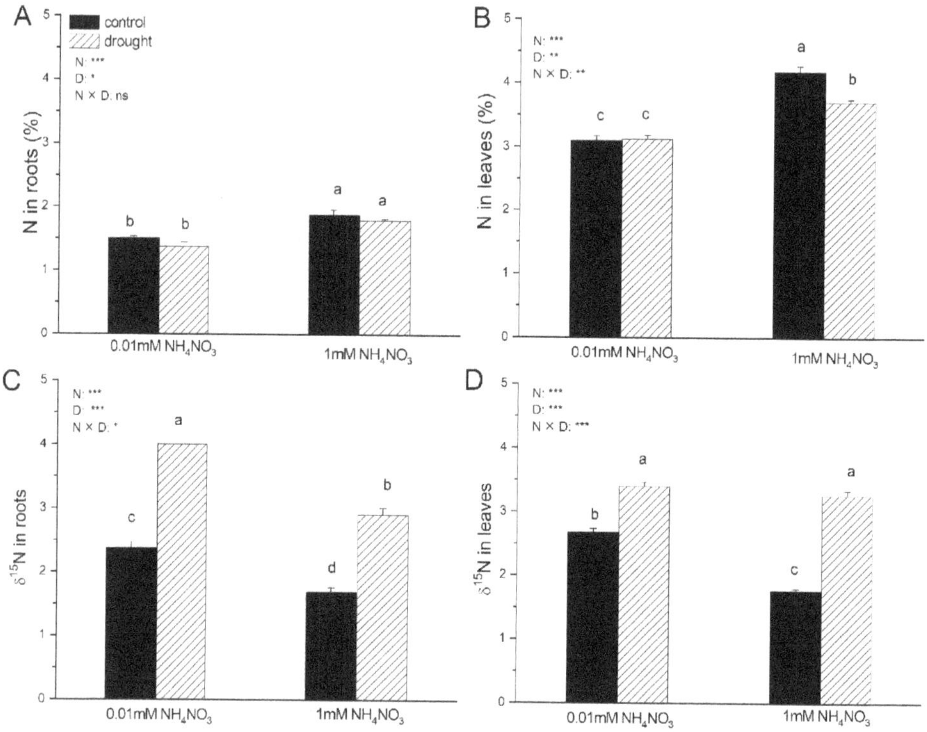

Figure 4. Total N concentration and δ^{15}N in roots (**A**,**C**) and leaves (**B**,**D**) of *Populus simonii* as affected by nitrogen supply and drought stress. Data indicate mean ± SE (n = 6). Different letters in the same column indicate significant differences. N, N supply; D, drought treatment; N × D, interaction of N supply and drought treatment. *, $p < 0.05$; **, $p < 0.01$; ***, $p < 0.001$; ns, not significant.

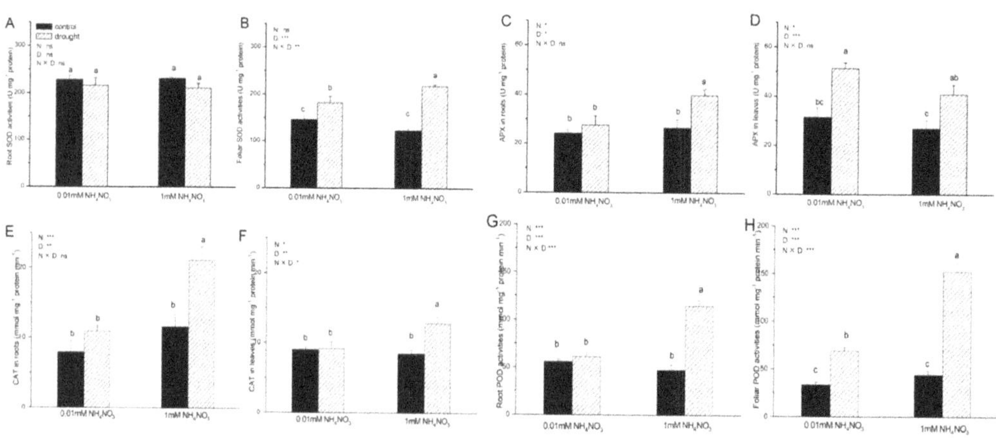

Figure 5. Activities of SOD, APX, CAT, and POD enzymes in roots (**A**,**C**,**E**,**G**) and leaves (**B**,**D**,**F**,**H**) of *Populus simonii* as affected by nitrogen supply and drought stress. Data indicate mean ± SE (n = 6). Different letters in the same column indicate significant differences. N, N supply; D, drought treatment; N × D, interaction of N supply and drought treatment. *, $p < 0.05$; **, $p < 0.01$; ***, $p < 0.001$; ns, not significant.

Drought stress increased CAT activity in roots and leaves under the 1 mM NH_4NO_3 treatment but had no effect under 0.01 mM NH_4NO_3 (Figure 5E,F). The POD activity in roots and leaves was also stimulated by drought stress, except, the root POD activity remained unchanged under 0.01 mM NH_4NO_3 (Figure 5G,H). In the PEG-treated plants, CAT and POD activities increased significantly when the N supply was increased. However, CAT and POD activities were not affected by N availability in the control plants (Figure 5E,H).

3.6. IAA and ABA Concentrations

Drought stress had no effect on root IAA concentration under 0.01 mM NH_4NO_3 (Figure 6A). However, the root IAA concentration under 1 mM NH_4NO_3 was increased under the drought stress (Figure 6A). IAA concentration was decreased by drought stress under 0.01 mM NH_4NO_3 but remained unaltered under 1 mM NH_4NO_3 (Figure 6B). ABA concentrations in both roots and leaves were higher under drought stress as compared with the control conditions (Figure 6C,D).

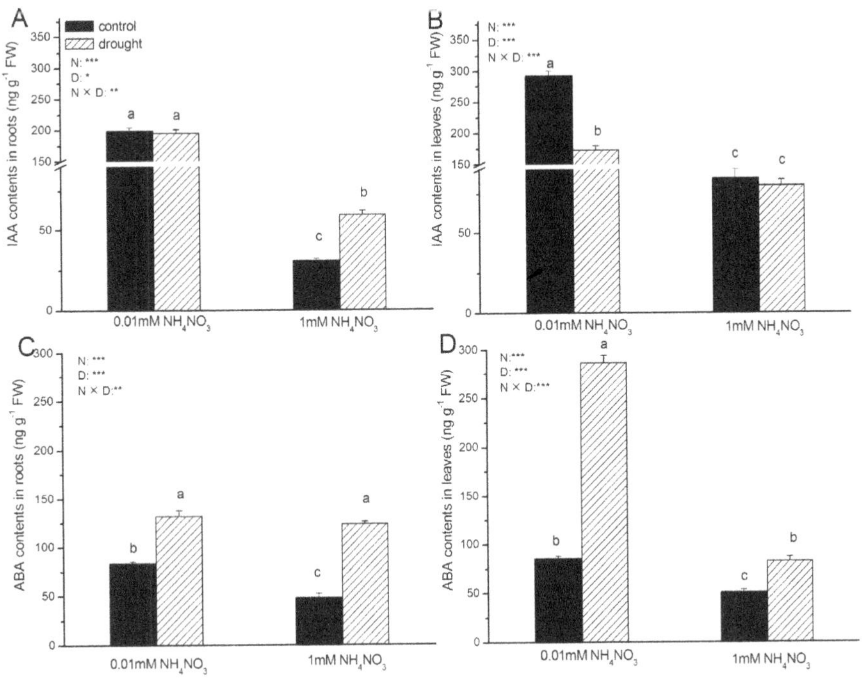

Figure 6. IAA and ABA contents in roots (**A**,**C**) and leaves (**B**,**D**) of *Populus simonii* as affected by nitrogen supply and drought stress. Data indicate mean ± SE (n = 6). Different letters in the same column indicate significant differences. N, N supply; D, drought treatment; N × D, interaction of N supply and drought treatment. *, $p < 0.05$; **, $p < 0.01$; ***, $p < 0.001$; ns, not significant.

The 0.01 mM NH_4NO_3 treatment induced a higher IAA concentration in both PEG-treated and control plants (Figure 6A,B). Additionally, ABA concentrations in both roots and leaves were higher in response to 0.01 mM NH_4NO_3 under the control conditions (Figure 6C,D). Under drought stress, root ABA concentration was not influenced by N supply; however, 0.01 mM NH_4NO_3 significantly increased foliar ABA concentration (Figure 6C,D).

3.7. PCA of Morphological and Physiological Responses

PC1 and PC2 accounted for 49 and 18% of the variation, respectively (Figure 7 and Supplementary Table S4). PC1 uncoupled the effect of drought, and PC2 clearly separated the variation due to N supply. Root ABA content and the soluble sugar content of roots and leaves were key contributors to PC1, whereas both total N concentration and the IAA content of roots and leaves were important factors influencing PC2.

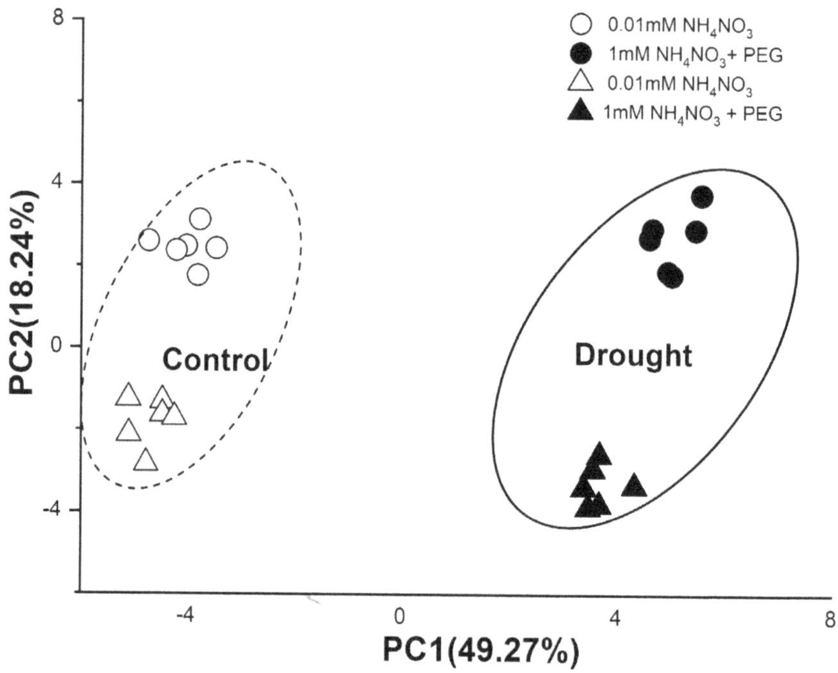

Figure 7. PCA plots of the first two principal components in roots and leaves of *Populus simonii* as affected by nitrogen supply and drought stress. PCA was conducted based on data (both values were averaged in the same treatment) of parameters presented in Supplementary Table S4.

3.8. Transcript Levels of AQPs and Representative Genes Involved in IAA and ABA Signaling Pathways

The cluster analysis separated the effects of drought stress and N availability based on the responsiveness of transcript levels (Figure 8). Under 1 mM NH_4NO_3, the mRNA levels of most genes (except *TIP1;6*) in roots were upregulated in response to drought stress (Figure 8A). The transcript abundance of *NCED3*, *PIP2;1*, *TIP2;1*, *PP2C*, and *PIP1;2* in roots was similarly increased by drought conditions under 0.01 mM NH_4NO_3 (Figure 8A). However, under the 0.01 mM NH_4NO_3 treatment, *YUCCA1*, *YUCCA3*, and *TIP1;6* were suppressed, whereas *TAA1* and *SIP1;2* remained unchanged in response to drought stress in roots (Figure 8A). Additionally, the mRNA levels of genes related to auxin biosynthesis (*TAA1*, *YUCCA1*, and *YUCCA3)* increased under 0.01 mM NH_4NO_3 as compared with under 1 mM NH_4NO_3 under control conditions, whereas this effect was not significant under drought conditions (Figure 8A).

In the leaves, the transcript abundances of almost all of the investigated genes (except *NCED3* and *TAA1*) was also increased by drought stress under the 1 mM NH_4NO_3 treatment (Figure 8B). Under the 0.01 mM NH_4NO_3 treatment, drought stress also upregulated the expression of *TIP2;1*, *PIP2;1*, *TIP1;6*, and *PP2C* but had little effect on the mRNA levels of other genes in the leaves (Figure 8B). Under control conditions, 0.01 mM NH_4NO_3 resulted in a significant decrease in the *PP2C* and *TIP1;6* transcript levels in the leaves. However, the mRNA levels of *YUCCA1* were increased by 0.01 mM NH_4NO_3 under control conditions (Figure 8B). Under drought conditions, the transcript abundances of almost all of the investigated genes (*YUCCA1*, *NCED3*, and *TAA1*) was higher under 1 mM NH_4NO_3 than under 0.01 mM NH_4NO_3 (Figure 8B).

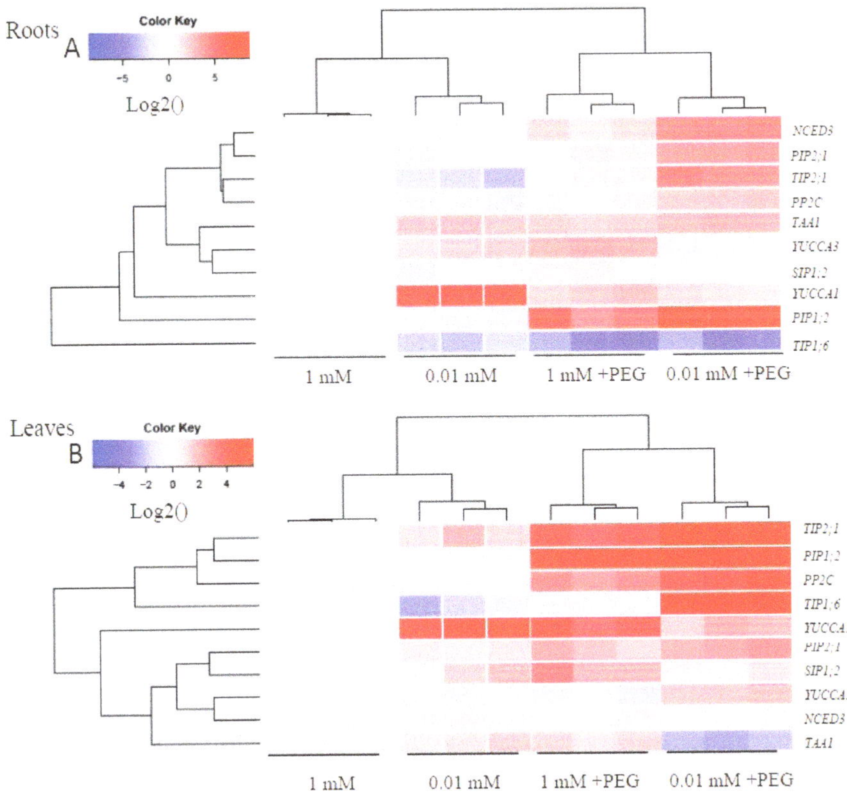

Figure 8. Fold changes of transcript levels of *AQPs* and representative genes involved in IAA and ABA signaling pathways in roots (**A**) and leaves (**B**) of *Populus simonii* as affected by nitrogen supply and drought stress. Signal intensities were calibrated according to a constitutively expressed poplar actin gene [10].

4. Discussion

Plants have developed strategies to adapted to drought stress. For instance, the inhibition of physiological activity may change the redistribution of limited resources and improve plant survival [29,30]. Consistent with many previous studies, drought stress has detrimental effects on the root morphology and photosynthesis of *P. simonii* [31–33]. Quite unexpectedly, under control conditions (no drought stress), nitrogen deficiency stimulated the root growth of *P. simonii* (Supplementary Table S2 and Figure S1). Generally, high nitrogen supply significantly increases root growth parameters in plants [2,30]. In this study, however, *P. simonii* seedlings showed better root development under 0.01 mM NH_4NO_3 under control conditions, exhibiting a greater root biomass and root surface area. These results may be explained by adaptation to the environment that *P. simonii* often grows on the Loess Plateau under conditions of nitrogen deficiency in Northwest China. *P. simonii* may have developed a mechanism (such as root development) to take up nitrogen efficiently, even from nutrient-deficient soil [2]. However, the PEG-treated plants showed improved growth when the N level was increased, suggesting that an increased N supply may help *P. simonii* survive under drought stress.

Drought stress often induces changes in cell membranes (increased permeability and decreased stability) [34]. Thus, RWC was used to indicate the degree of dehydration and membrane stability. In this study, decreased RWC was observed in the leaves of *P. simonii* under drought stress, which could reduce turgor pressure and enhance membrane permeability. The increase in MDA and O_2^- in leaves in response to drought stress reflects oxidative damage to membrane lipids and other vital substances. These results indicated that drought stress damaged sensitive biological macromolecules, impaired their functions, and harmed membrane [35]. To maintain turgor pressure in the cytoplasm, the accumulation of soluble substances plays an important role in osmoregulation. In our study, the notable increase in soluble sugars, phenolics, and amino acids suggested that more soluble substances were synthesized. Significant increases were observed in the aspartate, serine, and glutamate biosynthesis groups in response to drought stress. Aspartate and glutamine play important roles in the storage and/or transport of N from source to sink tissues [26,36], an intensive N reallocation under stress. Cysteine (the precursor of which is serine), glutamate, and glycine are used for the synthesis of glutathione (GSH) and the elimination of free radicals under stress [37]. Additionally, proline (belonging to the glutamate group) is a non-toxic osmoprotectant which is 300 times more soluble in water than other amino acids and may play an important role in cell pressure adjustment [38].

In this study, 1 mM NH_4NO_3 compensated for the negative effects of drought on osmolytes, O_2^- production, and the MDA content of leaves. Many studies have also found that nitrogen supply can decrease the MDA concentration, adjust osmotic pressure, and increase membrane stability to prevent damage caused by drought stress [39,40]. For example, N supply (100 and 200 kg/ha) prevented cell membrane damage and enhanced osmotic regulation in *Agrostis palustris* Huds. under water stress [41]. Nitrogen fertilizer could decrease MDA content of maize (*Zea mays* L.) under drought conditions [42]. These results indicated that the detrimental effects of drought stress on *P. simonii* were somehow eased by 1 mM NH_4NO_3.

After detecting reduced soil water availability, altered homeostasis of ROS production and scavenging can occur in plants during acclimation to drought [43]. In the enzymatic systems, SOD and APX can convert O_2^- to H_2O_2 [44], and POD and CAT can decompose H_2O_2 to H_2O at different cellular locations. The ROS production and the activities of oxygen scavenging systems determine whether oxidative signaling and/or damage will occur [45]. In this study, high activities of SOD, POD, CAT, and APX were observed (particularly under the 1 mM NH_4NO_3 treatment) in response to drought stress. Under drought conditions, the activities of most antioxidant enzymes were significantly increased in response to 1 mM NH_4NO_3 as compared with 0.01 mM NH_4NO_3, indicating that N supply improved the drought resistance of *P. simonii*.

Metabolites of C and N in plants can also act as osmoprotectants or signal molecules under drought stress [12,46]. Water deprivation can lead to a reduction in total C concentration as a result of decreased CO_2 assimilation in plants [47,48]. In the present study, long-term drought treatment also led to decreased concentrations of total C and N in *P. simonii*. Plants face a dilemma under drought stress: On the one hand, plants must prevent water loss via stomatal closure, and thus, the photosynthesis is limited. on the other hand, they need photosynthates to support key processes necessary for survival [49,50]. Thus, plants must increase *WUE* during acclimation to drought conditions. The $\delta^{15}N$ at natural abundance levels acts as a tracer, and significant discrimination is positive in most biological systems [8]. Plants retard their N uptake and assimilation processes and are forced to utilize the ^{15}N to meet N demands and are enriched in the tissues under 0.01 mM NH_4NO_3. The $\delta^{13}C$ and $\delta^{15}N$ values show positive linear correlations with *WUEi* in plants and $\delta^{13}C$ and $\delta^{15}N$ enrichment are closely associated with water consumption and can be considered to be indicators of *WUE* [51,52]. The $\delta^{13}C$ of roots and leaves was not affected by drought stress. However, $\delta^{15}N$ increased in the roots and leaves of *P. simonii* under drought conditions as compared with the control. These results reflect the acclimation of *P. simonii* to drought conditions via increased *WUE*.

The transcriptional regulation of *PIPs* is crucial for water transport across the plasma membrane and acclimation to drought in poplar plants [53–55]. Most of the *AQP* genes in roots and/or leaves have been reported to be upregulated in response to drought stress, suggesting that these *AQPs* can play key roles in facilitating water movement and redistribution under dynamic environmental water conditions. However, *AQPs* have been reported to be upregulated, downregulated, or unchanged in various poplar genotypes in response to drought stress [8,53,56]. In this study, the mRNA levels of *AQPs* (especially *PIP1;2* and *TIP2;1*) in *P. simonii* were significantly upregulated in response to drought stress and similar results were found in *Populus cathayana* [8]. These results indicate that *P. simonii* has developed the *AQPs* expression modulation strategy to acclimate to drought stress on the Loess Plateau and *PIP1;2* and *TIP2;1* may play decisive roles in the strategy.

Poplar plants can perceive changes in soil water availability via the IAA and ABA signaling pathways [9]. At physiological concentrations, auxin induces the opening of stomata [57,58]. To minimize water loss, root auxin concentrations are often increased and leaf auxin concentrations are often decreased during acclimation to drought conditions [59,60]. Indeed, in the present study, the root IAA concentrations under 1 mM NH_4NO_3 were increased, whereas the IAA concentrations were decreased by drought stress. In addition, 0.01 mM NH_4NO_3 induced higher IAA concentrations in the roots and leaves of *P. simonii* under both drought and control conditions, which was consistent with the root growth stimulated by N deficiency. In auxin biosynthesis, TAAs and YUCs play essential roles: The TAA of amino transferases converts tryptophan to indole-3-pyruvate (IPA), and then the IPA is converted to IAA by the YUCCA (YUC) family of flavin monooxygenases [61]. Indeed, the representative genes (*TAA1*, *YUCCA, 1* and *YUCCA3*) in *P. simonii* were induced by drought stress in the roots of *P. simonii* under 1 mM NH_4NO_3. However, the mRNA levels of these genes in leaves (except *YUCCA1*) under 0.01 mM NH_4NO_3 were significantly decreased by drought stress and were consistent with the IAA concentration. In particular, the high levels of *TAA1*, *YUCCA1*, and *YUCCA3* observed under 0.01 mM NH_4NO_3 suggest that *P. simonii* might be sensitive to N deficiency and thus, may accelerate IAA synthesis to stimulate root development, which is consistent with the observed root characteristics.

Additionally, the ABA concentrations in both roots and leaves of *P. simonii* were higher under drought stress as compared with control conditions. The ABA concentrations in poplars are often increased by drought stress and can recover to previous levels after re-watering [62–64]. The significant increase in ABA concentration in response to drought stress may be an important mechanism for transmitting a chemical signal that indicates the water status of the soil and leads to a cascade of reactions via regulation of leaf stomatal conductance [65]. Accordingly, *NCED* and *PP2C*, which are key genes involved in the ABA signaling pathway, were downregulated in response to drought treatment. A similar result was reported in *Populus euramericana* and *Populus cathayana*, which suggests that *NCED3* and *PP2C* may play decisive roles in the ABA signaling pathway during acclimation to limited water availability [65].

In summary, *P. simonii* demonstrated decreased height; differential expression of aquaporins (AQPs); activation of the IAA and ABA signaling pathways; decreased net photosynthetic rate and transpiration rate; increased $\delta^{15}N$, total soluble substances and *WUEi*; and altered homeostasis of reactive oxygen species (ROS) production and scavenging (Figure 9). These data suggest that *P. simonii* slows the process of water acquisition during acclimation to drought stress. Moreover, plants with a sufficient N supply exhibited more soluble compounds and greater antioxidant enzyme activities, lower transpiration rates (E), IAA content, ABA content and ROS production, and greater responsiveness of transcriptional regulation of 10 genes involved in water uptake and hormone modulation than plants with a low N supply under drought stress (Figure 9). These results suggest that *P. simonii* can differentially manage water uptake and hormone modulation under different N supply conditions. This is important when selecting poplar species for different soil conditions.

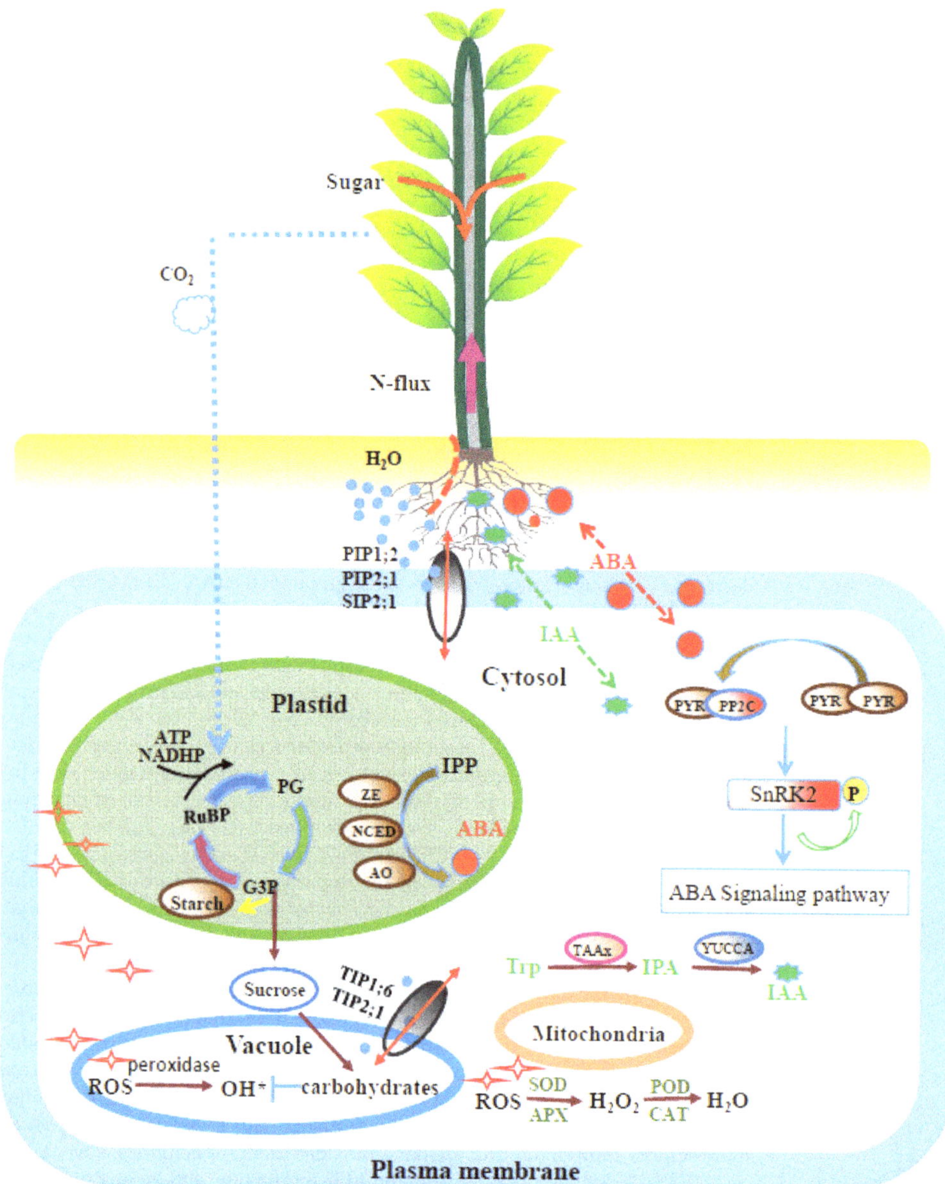

Figure 9. A schematic model of responses to nitrogen supply in *Populus simonii* under drought stress. While acclimating to drought, the *P. simonii* exhibited a reduction in growth; a decrease in the photosynthetic rate and transpiration rate; and an increase in $\delta^{15}N$, soluble substances, ROS scavenging; and water use efficiency. Strong response in the transcriptional regulation of *AQPs*; higher soluble phenolics; lower transpiration rates, IAA and ABA content, and ROS accumulation and scavenging were observed with a sufficient N supply.

Supplementary Materials: The following are available online at https://www.mdpi.com/article/10.3390/f13060907/s1, Supplementary materials and methods, Supplementary Table S1: Primers used for qRT-PCR, Supplementary Table S2: Growth parameters of *P. simonii*, Supplementary Figure S1: The typical root state of *P. simonii*, Supplementary Table S3: The concentrations of 17 free amino acids in fine roots and leaves, Supplementary Table S4: PCA of morphological and physiological responses. References [66–69] are cited in Supplementary Materials.

Author Contributions: Writing—original draft preparation, Z.L., X.W., Y.L., Z.Q. and Z.B.; writing—review and editing, Z.L., X.W. and Y.Z.; visualization, Z.L., X.W., Z.L. and N.W. Resources, Z.Y. All authors have read and agreed to the published version of the manuscript.

Funding: This research was supported by the Key Research Project of RiZhao Shandong Province, China (2021ZDYF010121), the National Natural Science Foundation of China (31901227), the Basic Research Program of Shanxi Province (20210302124247), and the Science and Technology Innovation Project of Higher Education Institutions in Shanxi Province (2021L376, 2021L373).

Institutional Review Board Statement: Not applicable.

Informed Consent Statement: Not applicable.

Data Availability Statement: All data relevant to the study are included in the article.

Acknowledgments: We thank numerous students and lab staff from the Chinese Academy of Forestry for their assistance in laboratory.

Conflicts of Interest: The authors declare no conflict of interest.

Abbreviations

AQP, aquaporin; IAA, auxin; ABA, abscisic acid; δ^{13}C, stable carbon isotope composition; δ^{15}N, stable nitrogen isotope composition; *WUEi*, intrinsic water use efficiency; ROS, reactive oxygen species; O_2^-, superoxide anion radical; SOD, superoxide dismutase; APX, ascorbate peroxidase; CAT, catalase; POD, peroxidase; MDA, malonaldehyde; NCED3, 9-cis-epoxycarotenoid dioxygenase 3; PP2C, protein phosphatase 2C; YUCC, flavin-containing monooxygenase; TAA, tryptophan aminotransferases; PCA, principal component analysis.

References

1. Wang, J.; Fu, B.; Lu, N.; Zhang, L. Seasonal variation in water uptake patterns of three plant species based on stable isotopes in the semi-arid Loess Plateau. *Sci. Total Environ.* **2017**, *609*, 27–37. [CrossRef] [PubMed]
2. Meng, S.; Zhang, C.; Su, L.; Li, Y.; Zhao, Z. Nitrogen uptake and metabolism of *Populus simonii* in response to PEG-induced drought stress. *Environ. Exp. Bot.* **2016**, *123*, 78–87. [CrossRef]
3. Keen, R.M.; Voelker, S.L.; Wang, S.S.; Bentz, B.J.; Goulden, M.L.; Dangerfield, C.R.; Reed, C.C.; Hood, S.M.; Csank, A.Z.; Dawson, T.E.; et al. Changes in tree drought sensitivity provided early warning signals to the California drought and forest mortality event. *Glob. Chang. Biol.* **2022**, *28*, 1119–1132. [CrossRef] [PubMed]
4. Anderegg, W.R.; Berry, J.A.; Smith, D.D.; Sperry, J.S.; Anderegg, L.D.; Field, C.B. The roles of hydraulic and carbon stress in a widespread climate-induced forest die-off. *Proc. Natl. Acad. Sci. USA* **2012**, *109*, 233–237. [CrossRef]
5. Gupta, A.; Rico-Medina, A.; Caño-Delgado, A.I. The physiology of plant responses to drought. *Science* **2020**, *368*, 266–269. [CrossRef]
6. Bian, Z.; Wang, D.; Liu, Y.; Xi, Y.; Wang, X.; Meng, S. Analysis of Populus glycosyl hydrolase family I members and their potential role in the ABA treatment and drought stress response. *Plant Physiol. Biochem.* **2021**, *163*, 178–188. [CrossRef]
7. Soon, F.F.; Ng, L.M.; Zhou, X.E.; West, G.M.; Kovach, A.; Tan, M.H.; Suino-Powell, K.M.; He, Y.; Xu, Y.; Chalmers, M.J.; et al. Molecular mimicry regulates ABA signaling by SnRK2 kinases and PP2C phosphatases. *Science* **2012**, *335*, 85–88. [CrossRef]
8. Cao, X.; Jia, J.; Zhang, C.; Li, H.; Liu, T.; Jiang, X.; Polle, A.; Peng, C.; Luo, Z.B. Anatomical, physiological and transcriptional responses of two contrasting poplar genotypes to drought and re-watering. *Physiol. Plant.* **2012**, *151*, 480–494. [CrossRef]
9. Lu, M.; Chen, M.; Song, J.; Wang, Y.; Pan, Y.; Wang, C. Anatomy and transcriptome analysis in leaves revealed how nitrogen (N) availability influence drought acclimation of populus. *Trees* **2019**, *33*, 1003–1014. [CrossRef]
10. Meng, S.; Cao, Y.; Li, H.; Bian, Z.; Wang, D.; Lian, C.; Yin, W.; Xia, X. *PeSHN1* regulates water-use efficiency and drought tolerance by modulating wax biosynthesis in poplar. *Tree Physiol.* **2019**, *39*, 1371–1386. [CrossRef]
11. Bian, Z.; Wang, X.; Lu, J.; Wang, D.; Zhou, Y.; Liu, Y.; Wang, S.; Yu, Z.; Xu, D.; Meng, S. The yellowhorn AGL transcription factor gene *XsAGL22* contributes to ABA biosynthesis and drought tolerance in poplar. *Tree Physiol.* **2021**. [CrossRef] [PubMed]

12. Luo, J.; Li, H.; Liu, T.; Polle, A.; Peng, C.; Luo, Z.B. Nitrogen metabolism of two contrasting poplar species during acclimation to limiting nitrogen availability. *J. Exp. Bot.* **2013**, *64*, 4207–4224. [CrossRef] [PubMed]
13. Nadarajah, K.K. ROS homeostasis in abiotic stress tolerance in plants. *Int. J. Mol. Sci.* **2020**, *21*, 5208. [CrossRef] [PubMed]
14. Farooq, M.; Wahid, A.; Kobayashi, N.S.M.A.; Fujita, D.B.S.M.A.; Basra, S.M.A. Plant drought stress: Effects, mechanisms and management. In *Sustainable Agriculture*; Springer: Dordrecht, The Netherlands, 2009; pp. 153–188.
15. Muhammad Aslam, M.; Waseem, M.; Jakada, B.H.; Okal, E.J.; Lei, Z.; Saqib, H.; Yuan, W.; Xu, W.; Zhang, Q. Mechanisms of Abscisic Acid-Mediated Drought Stress Responses in Plants. *Int. J. Mol. Sci.* **2022**, *23*, 1084. [CrossRef]
16. Ali, F.; Qanmber, G.; Li, F.; Wang, Z. Updated role of ABA in seed maturation, dormancy, and germination. *J. Adv. Res.* **2022**, *35*, 199–214. [CrossRef] [PubMed]
17. Popko, J.; Hänsch, R.; Mendel, R.R.; Polle, A.; Teichmann, T. The role of abscisic acid and auxin in the response of poplar to abiotic stress. *Plant Biol.* **2010**, *12*, 242–258. [CrossRef]
18. Pawłowicz, I.; Masajada, K. Aquaporins as a link between water relations and photosynthetic pathway in abiotic stress tolerance in plants. *Gene* **2019**, *687*, 166–172. [CrossRef]
19. Danielson, J.Å.; Johanson, U. Unexpected complexity of the aquaporin gene family in the moss Physcomitrella patens. *BMC Plant Biol.* **2008**, *8*, 45. [CrossRef]
20. Kapilan, R.; Vaziri, M.; Zwiazek, J.J. Regulation of aquaporins in plants under stress. *Biol. Res.* **2018**, *51*, 4. [CrossRef]
21. Ozturk, M.; Turkyilmaz Unal, B.; García-Caparrós, P.; Khursheed, A.; Gul, A.; Hasanuzzaman, M. Osmoregulation and its actions during the drought stress in plants. *Physiol. Plant.* **2021**, *172*, 1321–1335. [CrossRef]
22. Guo, J.; Yang, Y.; Wang, G.; Yang, L.; Sun, X. Ecophysiological responses of Abies fabri seedlings to drought stress and nitrogen supply. *Physiol. Plant.* **2010**, *139*, 335–347. [PubMed]
23. Meng, S.; Su, L.; Li, Y.; Wang, Y.; Zhang, C.; Zhao, Z. Nitrate and ammonium contribute to the distinct nitrogen metabolism of *Populus simonii* during moderate salt stress. *PLoS ONE* **2016**, *11*, e0150354. [CrossRef] [PubMed]
24. Dhindsa, R.S.; Matowe, W. Drought tolerance in two mosses: Correlated with enzymatic defence against lipid peroxidation. *J. Exp. Bot.* **1981**, *32*, 79–91. [CrossRef]
25. Pritchard, S.; Peterson, C.; Runion, G.B.; Prior, S.; Rogers, H. Atmospheric CO_2 concentration, N availability, and water status affect patterns of ergastic substance deposition in longleaf pine (*Pinus palustris* Mill.) foliage. *Trees* **1997**, *11*, 494–503. [CrossRef]
26. Ehlting, B.; Dluzniewska, P.; Dietrich, H.; Selle, A.; Teuber, M.; Hänsch, R.; Nehls, U.; Polle, A.; Schnitzler, J.P.; Rennenberg, H.; et al. Interaction of nitrogen nutrition and salinity in Grey poplar (*Populus tremula* × *alba*). *Plant Cell Environ.* **2007**, *30*, 796–811. [CrossRef]
27. De Diego, N.; Rodríguez, J.L.; Dodd, I.C.; Pérez-Alfocea, F.; Moncaleán, P.; Lacuesta, M. Immunolocalization of IAA and ABA in roots and needles of radiata pine (*Pinus radiata*) during drought and rewatering. *Tree Physiol.* **2013**, *33*, 537–549. [CrossRef]
28. Brunner, A.M.; Yakovlev, I.A.; Strauss, S.H. Validating internal controls for quantitative plant gene expression studies. *BMC Plant Biol.* **2004**, *4*, 14. [CrossRef]
29. Deng, X.; Xiao, W.; Shi, Z.; Zeng, L.; Lei, L. Combined effects of drought and shading on growth and non-structural carbohydrates in *Pinus massoniana* Lamb. seedlings. *Forests* **2019**, *11*, 18. [CrossRef]
30. Meng, S.; Ma, H.B.; Li, Z.S.; Yang, F.C.; Wang, S.K.; Lu, J.K. Impacts of nitrogen on physiological interactions of the hemiparasitic *Santalum album* and its N_2-fixing host *Dalbergia odorifera*. *Trees* **2021**, *35*, 1039–1051. [CrossRef]
31. Álvarez, S.; Navarro, A.; Bañón, S.; Sánchez-Blanco, M.J. Regulated deficit irrigation in potted *Dianthus* plants: Effects of severe and moderate water stress on growth and physiological responses. *Sci. Hortic.* **2009**, *122*, 579–585. [CrossRef]
32. Lei, Y.; Yin, C.; Li, C. Differences in some morphological, physiological, and biochemical responses to drought stress in two contrasting populations of *Populus przewalskii*. *Physiol. Plant.* **2006**, *127*, 182–191. [CrossRef]
33. Yang, Y.; Han, C.; Liu, Q.; Lin, B.; Wang, J. Effect of drought and low light on growth and enzymatic antioxidant system of *Picea asperata* seedlings. *Acta Physiol. Plant.* **2008**, *30*, 433–440. [CrossRef]
34. Blokhina, O.; Virolainen, E.; Fagerstedt, K.V. Antioxidants, oxidative damage and oxygen deprivation stress: A review. *Ann. Bot.* **2003**, *91*, 179–194. [CrossRef] [PubMed]
35. Yang, X.; Lu, M.; Wang, Y.; Wang, Y.; Liu, Z.; Chen, S. Response mechanism of plants to drought stress. *Horticulturae* **2021**, *7*, 50. [CrossRef]
36. Tegeder, M.; Masclaux-Daubresse, C. Source and sink mechanisms of nitrogen transport and use. *New Phytol.* **2018**, *217*, 35–53. [CrossRef] [PubMed]
37. Rajput, V.D.; Harish Singh, R.K.; Verma, K.K.; Sharma, L.; Quiroz-Figueroa, F.R.; Meena, M.; Gour, V.S.; Minkina, T.; Sushkova, S.; Mandzhieva, S. Recent Developments in enzymatic antioxidant defence mechanism in plants with special reference to abiotic stress. *Biology* **2021**, *10*, 267. [CrossRef]
38. Ottow, E.A.; Brinker, M.; Teichmann, T.; Fritz, E.; Kaiser, W.; Brosché, M.; Kangasjärvi, J.; Jiang, X.; Polle, A. *Populus* euphratica displays apoplastic sodium accumulation, osmotic adjustment by decreases in calcium and soluble carbohydrates, and develops leaf succulence under salt stress. *Plant Physiol.* **2005**, *139*, 1762–1772. [CrossRef]
39. Zhang, S.; Shao, L.; Sun, Z.; Huang, Y.; Liu, N. An atmospheric pollutant (inorganic nitrogen) alters the response of evergreen broad-leaved tree species to extreme drought. *Ecotoxicol. Environ. Saf.* **2020**, *187*, 109750. [CrossRef]
40. Li, S.; Zhou, L.; Addo-Danso, S.D.; Ding, G.; Sun, M.; Wu, S.; Lin, S. Nitrogen supply enhances the physiological resistance of Chinese fir plantlets under polyethylene glycol (PEG)-induced drought stress. *Sci. Rep.* **2020**, *10*, 7509. [CrossRef]

41. Saneoka, H.; Moghaieb RE, A.; Premachandra, G.S.; Fujita, K. Nitrogen nutrition and water stress effects on cell membrane stability and leaf water relations in *Agrostis palustris* Huds. *Environ. Exp. Bot.* **2004**, *52*, 131–138. [CrossRef]
42. Gou, W.; Zheng, P.; Tian, L.; Gao, M.; Zhang, L.; Akram, N.A.; Ashraf, M. Exogenous application of urea and a urease inhibitor improves drought stress tolerance in maize (*Zea mays* L.). *J. Plant Res.* **2017**, *130*, 599–609. [CrossRef] [PubMed]
43. Nabi RB, S.; Tayade, R.; Hussain, A.; Kulkarni, K.P.; Imran, Q.M.; Mun, B.G.; Yun, B.W. Nitric oxide regulates plant responses to drought, salinity, and heavy metal stress. *Environ. Exp. Bot.* **2019**, *161*, 120–133. [CrossRef]
44. Bowler, C.; Montagu, M.V.; Inze, D. Superoxide dismutase and stress tolerance. *Annu. Rev. Plant Biol.* **1992**, *43*, 83–116. [CrossRef]
45. Møller, I.M.; Jensen, P.E.; Hansson, A. Oxidative modifications to cellular components in plants. *Annu. Rev. Plant Biol.* **2007**, *58*, 459–481. [CrossRef]
46. Singh, P.K.; Indoliya, Y.; Agrawal, L.; Awasthi, S.; Deeba, F.; Dwivedi, S.; Chakrabarty DShirke, P.; Pandey, V.; Singh, N.; Dhankher Om, P.; et al. Genomic and proteomic responses to drought stress and biotechnological interventions for enhanced drought tolerance in plants. *Curr. Plant Biol.* **2022**, *29*, 100239. [CrossRef]
47. Warren, C.R.; Aranda, I.; Cano, F.J. Responses to water stress of gas exchange and metabolites in *Eucalyptus* and *Acacia* spp. *Plant Cell Environ.* **2011**, *34*, 1609–1629. [CrossRef]
48. Guimarães, G.F.; Gorni, P.H.; Vitolo, H.F.; Carvalho ME, A.; Pacheco, A.C. Sweetpotato tolerance to drought is associated to leaf concentration of total chlorophylls and polyphenols. *Theor. Exp. Plant Physiol.* **2021**, *33*, 385–396. [CrossRef]
49. Sala, A.; Piper, F.; Hoch, G. Physiological mechanisms of drought-induced tree mortality are far from being resolved. *New Phytol.* **2010**, *186*, 274–281. [CrossRef]
50. Amitrano, C.; Arena, C.; Rouphael, Y.; De Pascale, S.; De Micco, V. Vapour pressure deficit: The hidden driver behind plant morphofunctional traits in controlled environments. *Ann. Appl. Biol.* **2019**, *175*, 313–325. [CrossRef]
51. Fichot, R.; Laurans, F.; Monclus, R.; Moreau, A.; Pilate, G.; Brignolas, F. Xylem anatomy correlates with gas exchange, water-use efficiency and growth performance under contrasting water regimes: Evidence from *Populus deltoides* × *Populus nigra* hybrids. *Tree Physiol.* **2009**, *29*, 1537–1549. [CrossRef]
52. Meng, S.; Wang, S.; Quan, J.; Su, W.; Lian, C.; Wang, D.; Xia, X.; Yin, W. Distinct carbon and nitrogen metabolism of two contrasting poplar species in response to different N supply levels. *Int. J. Mol. Sci.* **2018**, *19*, 2302. [CrossRef] [PubMed]
53. Cohen, D.; Bogeat-Triboulot, M.B.; Vialet-Chabrand, S.; Merret, R.; Courty, P.E.; Moretti, S.; Bizet, F.; Guilliot, A.; Hummel, I. Developmental and environmental regulation of *Aquaporin* gene expression across *Populus* species: Divergence or redundancy? *PLoS ONE* **2013**, *8*, e55506. [CrossRef] [PubMed]
54. Muries, B.; Mom, R.; Benoit, P.; Brunel-Michac, N.; Cochard, H.; Drevet, P.; Petel, G.; Badel, E.; Fumanal, B.; Gousset-dupont, A.; et al. Aquaporins and water control in drought-stressed poplar leaves: A glimpse into the extraxylem vascular territories. *Environ. Exp. Bot.* **2019**, *162*, 25–37. [CrossRef]
55. Chen, J.; Li, J.; Huang, Y.; Li, Y.; Su, C.; Zeng, X. *EuPIP1;2*, a plasma membrane aquaporin gene from *Eucommia ulmoides*, enhances drought and salt tolerance in transgenic tobacco. *Agronomy* **2022**, *12*, 615. [CrossRef]
56. Bae, E.K.; Lee, H.; Lee, J.S.; Noh, E.W. Drought, salt and wounding stress induce the expression of the plasma membrane intrinsic protein 1 gene in poplar (*Populus alba* × *P. tremula var. glandulosa*). *Gene* **2011**, *483*, 43–48. [CrossRef]
57. Grabov, A.; Blatt, M.R. Co-ordination of signalling elements in guard cell ion channel control. *J. Exp. Bot.* **1998**, *49*, 351–360. [CrossRef]
58. Lohse, G.; Hedrich, R. Characterization of the plasma-membrane H^+-ATPase from *Vicia faba* guard cells. *Planta* **1992**, *188*, 206–214. [CrossRef]
59. Song, J.; Wang, Y.; Pan, Y.; Pang, J.; Zhang, X.; Fan, J.; Zhang, Y. The influence of nitrogen availability on anatomical and physiological responses of *Populus alba*× *P. glandulosa* to drought stress. *BMC Plant Biol.* **2019**, *19*, 63. [CrossRef]
60. Mubarik, M.S.; Khan, S.H.; Sajjad, M.; Raza, A.; Hafeez, M.B.; Yasmeen, T.; Rizwan, M.; Ali, S.; Arif, M.S. A manipulative interplay between positive and negative regulators of phytohormones: A way forward for improving drought tolerance in plants. *Physiol. Plant.* **2021**, *172*, 1269–1290. [CrossRef]
61. Uc-Chuc, M.A.; Pérez-Hernández, C.; Galaz-Ávalos, R.M.; Brito-Argaez, L.; Aguilar-Hernández, V.; Loyola-Vargas, V.M. YUCCA-mediated biosynthesis of the auxin IAA is required during the somatic embryogenic induction process in *Coffea canephora*. *Int. J. Mol. Sci.* **2020**, *21*, 4751. [CrossRef]
62. Brunetti, C.; Savi, T.; Nardini, A.; Loreto, F.; Gori, A.; Centritto, M. Changes in abscisic acid content during and after drought are related to carbohydrate mobilization and hydraulic recovery in poplar stems. *Tree Physiol.* **2020**, *40*, 1043–1057. [CrossRef] [PubMed]
63. Li, C.; Wang, Z.; Nong, Q.; Lin, L.; Xie, J.; Mo, Z.; Huang, X.; Song, X.; Malviya, M.K.; Solanki, M.K.; et al. Physiological changes and transcriptome profiling in *Saccharum spontaneum* L. leaf under water stress and re-watering conditions. *Sci. Rep.* **2021**, *11*, 5525. [CrossRef] [PubMed]
64. Duan, H.; Resco de Dios, V.; Wang, D.; Zhao, N.; Huang, G.; Liu, W.; Wu, J.; Zhou, S.; Choat, B.; Tissue, D.T. Testing the limits of plant drought stress and subsequent recovery in four provenances of a widely distributed subtropical tree species. *Plant Cell Environ.* **2022**, *45*, 1187–1203. [CrossRef] [PubMed]
65. Mukarram, M.; Choudhary, S.; Kurjak, D.; Petek, A.; Khan, M.M.A. Drought: Sensing, signalling, effects and tolerance in higher plants. *Physiol. Plant.* **2021**, *172*, 1291–1300. [CrossRef]

66. Becana, M.; Aparicio-Tejo, P.; Irigoyen, J.J.; Sanchez-Diaz, M. Some enzymes of hydrogen peroxide metabolism in leaves and root nodules of Medicago sativa. *Plant Physiol.* **1986**, *82*, 1169–1171. [CrossRef] [PubMed]
67. Ekmekci, Y.; Terzioglu, S. Effects of oxidative stress induced by paraquat on wild and cultivated wheats. *Pestic. Biochem. Physiol.* **2005**, *83*, 69–81. [CrossRef]
68. He, J.L.; Qin, J.J.; Long, L.Y.; Ma, Y.L.; Li, H.; Li, K.; Jiang, X.N.; Liu, T.X.; Polle, A.; Liang, Z.S.; et al. Net cadmium flux and accumulation reveal tissue-specific oxidative stress and detoxification in Populus x canescens. *Physiol. Plant.* **2011**, *143*, 50–63. [CrossRef]
69. Kato, M.; Shimizu, S. Chlorophyll metabolism in higher plants. VII. Chlorophyll degradation in senescing tobacco leaves; phenolic-dependent peroxidative degradation. *Can. J. Bot.* **1987**, *65*, 729–735. [CrossRef]

forests

MDPI

Article

Screening of Key Indices and the Gene Transcriptional Regulation Analysis Related to Salt Tolerance in *Salix matsudana* Seedlings

Yuanxiang Pang [1], Longmei Guo [1], Tiantian Wang [1,2], Wei Liu [1], Peili Mao [1], Xiaonan Cao [1], Ying Geng [1] and Banghua Cao [1,*]

1 Taishan Mountain Forest Ecosystem Research Station, Key Laboratory of State Forestry Administration for Silviculture of the Lower Yellow River, Shandong Agricultural University, Tai'an 271018, China; pangyuanxiang2020@126.com (Y.P.); guolm3180@126.com (L.G.); wtt840602@163.com (T.W.); liucw2022@126.com (W.L.); maopl1979@163.com (P.M.); caoxiaonan_05@163.com (X.C.); aimee_gy@163.com (Y.G.)

2 Dongying Landscaping Center (Shandong), Dongying 257091, China

* Correspondence: caobanghua@126.com

Abstract: Pot experiments were performed to comparatively study the differences in 16 salt tolerance indices between the seedlings of six *Salix matsudana* clones under the stress of various concentrations of NaCl (0, 0.1%, 0.3%, 0.5%, and 0.7%), including the salt injury index, shoot fresh weight, root fresh weight, leaf water content, relative conductivity, malondialdehyde content, and antioxidant enzyme activity. The salt-tolerant clones and key indices of salt tolerance were selected. Transcriptome sequencing analysis was performed on the selected salt-tolerant and salt-sensitive clones under salt stress, and the links between the physiological indices of salt tolerance and gene expression were analyzed. Results: (1) Superoxide dismutase (SOD), peroxidase (POD), chlorophyll, and net photosynthetic rate were closely related to the salt tolerance of *Salix matsudana* at the seedling stage. The regression equation was constructed as follows: salt tolerance index (y) = $0.224x_{10} + 0.216x_{11} + 0.127x_{12} + 0.191x_7 - 0.187$ (x_{10} = chlorophyll, x_{11} = SOD, x_{12} = POD, x_7 = net photosynthetic rate). (2) The number of differentially expressed genes between the seedlings of salt-tolerant and salt-sensitive clones varied with the time of exposure (0 h, 4 h, 12 h, and 24 h) to 200 mmol·L^{-1} NaCl stress. The most differentially expressed genes in Sm172 were detected upon 24 h vs. 4 h of salt treatment, while the most in Sm6 were in the 24 h vs. 0 h comparison. Gene Ontology analysis and Kyoto Encyclopedia of Genes and Genomes analysis showed that several differentially expressed genes were involved in carotenoid biosynthesis and plant mitogen-activated protein kinase signaling pathways. The nine highly expressed transcription factor genes (*Sm172-f2p30-2392*, *Sm172-f2p28-2386*, *Sm6-f8p60-2372*, *Sm6-f2p39-2263*, *Sm6-f16p60-2374*, *Sm6-f3p60-931*, *Sm6-f2p60-1067*, *Sm172-f3p54-1980*, and *Sm172-f3p54-1980*) were closely correlated with the four key indices of salt tolerance. These genes could become genetic resources for salt tolerance breeding of *Salix matsudana*.

Keywords: *Salix matsudana*; NaCl stress; salt tolerance index; salt tolerance gene; transcriptome sequencing

Citation: Pang, Y.; Guo, L.; Wang, T.; Liu, W.; Mao, P.; Cao, X.; Geng, Y.; Cao, B. Screening of Key Indices and the Gene Transcriptional Regulation Analysis Related to Salt Tolerance in *Salix matsudana* Seedlings. *Forests* **2022**, *13*, 754. https://doi.org/10.3390/f13050754

Academic Editor: Claudia Cocozza

Received: 2 April 2022
Accepted: 11 May 2022
Published: 13 May 2022

1. Introduction

The coastal saline-alkali land of the Yellow River Delta suffers frequent natural disasters, soil salinization, and great difficulty in afforestation, which restricts the growth of many tree species [1–3]. The selection of salt-tolerant afforestation tree species is a key measure to improving the outcome of afforestation. Willow (*Salix matsudana*) is a native arbor species naturally distributed in this area that has a certain tolerance to salt. Selection and breeding of improved willow varieties could greatly help with the construction of shelter forests in the Yellow River Delta.

Early studies on salt resistance in willows have focused on their salt resistance ability and physiological characteristics. *Salix matsudana* can grow in a culture medium with a NaCl concentration of 0.1–0.2 g L^{-1}, whereas the salt stress caused by 0.4 g L^{-1} NaCl significantly inhibits its growth [4]. Salt stress damages the stems and leaves of willow [5]; reduces its growth increment, biomass, and leaf water content [6]; and causes ion toxicity, osmotic stress, and secondary oxidative stress to the plant, thereby hindering its photosynthesis, growth, and metabolism [7]. When osmotic stress occurs, osmotic adjustment substances (soluble sugars and soluble proteins) accumulate in plant cells to maintain a higher water potential [8,9]. Osmotic stress and the toxicity of excessive Na^+ will cause the accumulation of reactive oxygen species in plants [10]. The antioxidant enzyme system (including superoxide dismutase (SOD) and peroxidase (POD)) in plants will be activated [11], which scavenges peroxide ions and reduces the oxidative damage to the cells [12,13].

As molecular biology techniques have advanced, research on the salt tolerance of willows has advanced to the molecular level. Qiao et al. analyzed the salt-stress-responsive proteome of *Salix matsudana* [14]. Zhou et al. discovered significantly different miRNA expressions between salt-sensitive and salt-tolerant *Salix matsudana* under various salt stress conditions [15]. Chen et al. identified candidate genes for salt stress response in *Salix matsudana* [16]. Shan et al. found that the physiological and metabolic processes of plants could be regulated by controlling salt-stress-related genes [17]. Yang et al. [18] found that in a high-salt environment, the upregulation of H^+-ATPase gene expression in vacuoles led to cytoplasmic sequestration of Na^+ [19]. Plants also respond to salt stress by regulating the expression of salt-tolerance-related transcription factors such as WRKY, MYB, and basic helix-loop-helix (BHLH) [20–22].

With the gradual maturation of sequencing technology, research based on transcriptome sequencing has expanded from animals to plants. Transcriptomic research in plants has mainly focused on stress resistance, regulation of physiological mechanisms, and nutrient utilization. Studies on stress resistance mainly focus on salt and alkali tolerance, disease resistance, and drought resistance [23–26]. Transcriptome sequencing plays an increasingly important role in mining the salt tolerance genes of plant species, such as *Jatropha curcas*, citrus, cotton, wheat, and rose [27–32]. The application of third-generation single-molecule sequencing technology reduces the difficulty of analyzing the nonparametric transcriptome and allows for the easy sequencing of the complete genome and full-length transcripts, making the in-depth study of transcription mechanisms more convenient. However, there have been few studies on the relationship between the physiological salt-stress-responsive indices and the molecular mechanism of salt tolerance that have used third-generation sequencing technology and have applied the findings to the breeding of salt-tolerant varieties.

The present study examined 16 salt-tolerance-related indices in seedlings of *Salix matsudana*. On this basis, correlation analysis, grey cluster analysis, principal component analysis, and stepwise regression analysis were performed to screen out the key indices, pick the salt-tolerant and salt-sensitive *Salix matsudana* clones, and derive the equation that accurately reflected the salinity resistance of *Salix matsudana*. Transcriptome analysis was carried out on the selected salt-tolerant and salt-sensitive clones. The pathways involved in the salt tolerance of *Salix matsudana* and their relevant genes were identified on the basis of the differentially expressed genes and the key indices of salt tolerance. Eventually, an identification method of salt tolerance that was stable and efficient in both the physiological and molecular aspects was established for *Salix matsudana* at the seedling stage. This study might provide a theoretical and technical basis for the selection and breeding of improved varieties that can tolerate the salinity of coastal saline-alkali land.

2. Materials and Methods

2.1. Experimental Materials

The research objects included a total of five willow clones: the improved salt-tolerant varieties Luliu 2 and Luliu 6, which were registered by the Shandong Academy of Forestry

Sciences, as well as three unapproved clones with good traits (preliminarily named Jinan 1, Jinan 2, and Binzhou 1) in the experimental forest of Jinan, Shandong Province. The widely popular Willow No. 172 was used as the control. We numbered Luliu 2, Luliu 6, Jinan 1, Jinan 2, Binzhou 1, and Willow No. 172 as Sm2, Sm6, SmA, SmB, SmC, and Sm172 for this paper, respectively. The branches used in this study were collected from the experimental forest of Xicang Village, Changqing District, Jinan City, Shandong Province.

2.2. Experimental Methods

(1) Pot experiment: In April 2017, potted seedlings were cultivated with sieved soil. The soil was taken from the Forestry Experimental Station of Shandong Agricultural University. Before loading, it was screened to remove impurities and homogenized. The soil is sandy loam with a pH of 7.05 $\pm$ 0.03, a total nitrogen content of 1.27 $\pm$ 0.14 $g{\cdot}kg^{-1}$, the hydrolyzable nitrogen content of 94.36 $\pm$ 13.89 $mg{\cdot}kg^{-1}$, the available phosphorus content of 32.01 $\pm$ 4.10 $mg{\cdot}kg^{-1}$, the available potassium content of 56.72 $\pm$ 7.83 $mg{\cdot}kg^{-1}$, and the organic matter content of 16.27 $\pm$ 2.39 $g{\cdot}kg^{-1}$. The cuttings of *Salix matsudana* clones Sm6 and Sm172 were selected and planted into greenhouse flowerpots at the Forestry Experimental Station of Shandong Agricultural University. The upper diameter of the flowerpots was 30 cm, the lower diameter was 20 cm, and the height was 25 cm. The weight of the soil in each pot was 10 kg. After planting, each cutting protruded approximately 2 cm above the soil surface. Regular watering management was conducted during the study. After the cuttage seedlings survived, the weak seedlings were eliminated in July 2017, while the seedlings that grew well and had no obvious diseases and pests were selected for experiments. The greenhouse was equipped with a water curtain, fan, and other temperature control equipment. During the experiment, the day and night temperatures in the greenhouse were controlled at (25 $\pm$ 2) °C and (20 $\pm$ 2) °C respectively, and the relative humidity was 65–70%.

Soil salinity was determined by the gravimetric method. A soil salinity gradient (0.1%, 0.3%, 0.5%, and 0.7%) was prepared, and deionized water was used as the blank control. First, the mass of NaCl needed for each stress level was calculated. NaCl solutions with the corresponding concentrations were prepared in deionized water and added three times in equal amounts (irrigation once every 7 days). A tray was placed below each pot to prevent the loss of salt. The physiological and biochemical indices were measured 15 days after the last addition of the salt.

(2) Transcriptome analysis: Transcriptome sequencing was performed using a combination of second-generation and third-generation sequencing technologies. Normally growing Sm6 and Sm172 adult plants (one each) were selected. Several branches with similar growth were cut at a length of approximately 15 cm. The cuttings were placed in beakers filled with deionized water and cultivated in a constant-temperature light incubator. After the branches took root and grew new leaves, various groups of branches were soaked in 200 $mmol{\cdot}L^{-1}$ NaCl solution for 0, 4, 12, and 24 h. The leaves were collected from all groups immediately after the soaking treatment. During sampling, leaves were collected from 3 plants in each group and mixed. Immediately after collection, the samples were placed in liquid nitrogen and then stored in a −80 °C freezer. The samples were saved for procedures such as RNA extraction.

2.3. Determination of Salt Tolerance Indices and Data Analysis

We grouped the indices related to the salt tolerance of *Salix matsudana* into four groups, salt injury indices, physiological and biochemical indices, biomass indices, and photosynthetic indices (Table 1), and used them to establish the standard equation of salt tolerance index. The root fresh weight (RFW), shoot fresh weight (SFW), relative leaf water content (RWC), relative seedling height (RH), and relative diameter (RD) were determined according to the method developed by Chen et al. [33]. The chlorophyll content was measured following Wang et al. [34]. Using a portable photosynthesis system (CIRAS-2), the net photosynthetic rate (Pn), stomatal conductance (GS), and intercellular CO_2

concentration (Ci) were measured between 9:00 a.m. and 11:00 a.m. at a saturated light intensity of 1200 μmol·m^{-2}s^{-1} and a leaf chamber temperature of 25 °C. The relative conductivity (RC) and malondialdehyde (MDA) content were determined as described by Qin et al. [35]. The activities of peroxidase (POD) and superoxide dismutase (SOD) were determined following Li et al., and the contents of soluble sugar (SS) and starch (SST) were determined according to Wang et al. [36,37].

Table 1. Indexes for standard equation construction.

Category	Indexes	Abbreviations	Relative Value Codes
Phenotype	Salt injury index	SII	x1
Biomass	Shoot fresh weight	SFW	x2
	Root fresh weight	RFW	x3
	Water content	RWC	x4
	Seedling height growth	RH	x5
	Ground diameter growth	RD	x6
Photosynthetic index	Net photosynthetic rate	Pn	x7
	Intercellular CO_2 concentration	Ci	x8
	Stomatal conductance	GS	x9
	Chlorophyll	Chl	x10
Physiological and biochemical index	Superoxide dismutase	SOD	x11
	Peroxidase	POD	x12
	Malondialdehyde	MDA	x13
	Conductivity	EC	x14
	Soluble sugar	SS	x15
	Soluble starch	SST	x16

All statistical analyses were conducted using Statistical Product Service Solutions for Windows 22.0 (SPSS, Chicago, IL, USA).

2.4. Sequencing Method and Analysis

Leaf RNA was extracted, and a cDNA library was constructed. Second-generation sequencing was performed on the Illumina HiSeq high-throughput sequencing platform. The third-generation full-length transcriptome sequencing was performed by Anoroad Gene Technology Co., Ltd. (Beijing, China) with the PacBio Sequel sequencer. Sequencing quality was assessed using FastQC. Functional annotation was performed with Trinotate. GO classification was performed with BLAST2GO. COG classification was performed using eggNOG. Metabolic pathway analysis was performed using the KEGG. Differential expression analysis was conducted with DESeq. The GO classification of the differentially expressed Unigene and Pathway enrichment analysis were performed using GOseq.

2.5. Verification of the Differentially Expressed Genes

RNAs were extracted from Sm6 and Sm172 that had received salt treatment for 0, 4, 12, or 24 h. The RNAs were reverse-transcribed, and 12 randomly selected salt-tolerance-related genes of *Salix matsudana* were subjected to real-time fluorescence quantitative verification. The polymerase chain reaction (PCR) system (20 μL) was set up as follows: SYBR Green 10 μL, forward primer (10 mM) 1 μL, reverse primer (10 mM) 1 μL, cDNA 2 μL, and ddH_2O to a total volume of 20 μL. The PCR conditions were as follows: predenaturation at 95 °C for 30 s and 40 cycles of denaturation at 95 °C for 5 s and optimal annealing temperature 60 °C for 30 s.

3. Results

3.1. Comprehensive Evaluation of the Salt Tolerance of *Salix matsudana* Clones at the Seedling Stage

3.1.1. Analysis of Variance of Salt Tolerance Index between NaCl Treatment Concentration and Tree Species

Analysis of variance (Table 2) showed that all indexes showed significant or extremely significant differences under different NaCl concentrations and among different clones. Analysis of variance showed that the data of the indicators selected in this test were reliable.

Table 2. ANOVA analyses of effects of salt and clone treatments and their interaction on the growth, biomass, photosynthetic, physiological, and biochemical of *Salix matsudana*.

Variables	Salt Treatment		Clone Treatment		Salt × Clones	
	F-Value	*p*-Value	F-Value	*p*-Value	F-Value	*p*-Value
SFW	120.91	<0.001	8.793	<0.001	1.935	<0.05
RFW	47.081	<0.001	4.34	<0.01	2.902	<0.001
RWC	3.497	<0.05	3.898	<0.01	2.759	<0.001
RH	555.974	<0.001	16.797	<0.001	9.68	<0.001
RD	558.212	<0.001	4.719	<0.001	6.314	<0.001
Chl	249.154	<0.001	181.623	<0.001	31.436	<0.001
GS	44.091	<0.001	3.91	<0.01	1.944	<0.05
Pn	1293.128	<0.001	91.144	<0.001	11.486	<0.001
Ci	3.931	<0.01	1.539	0.191	1.141	0.336
MDA	532.732	<0.001	811.72	<0.001	40.707	<0.001
POD	97.785	<0.001	100.371	<0.001	16.692	<0.001
SOD	720.173	<0.001	818.479	<0.001	85.472	<0.001
EC	7.072	<0.001	3.505	<0.01	2.766	<0.001
SS	128.457	0 < 0.001	226.815	<0.001	66.527	<0.001
SST	0.518	0.723	4.314	<0.01	1.445	0.138

3.1.2. Analysis of the Correlation between Salt Resistance Capability and Salt Tolerance Index

Correlation analysis of the 16 indices (Table 3) showed that there were various degrees of correlation between the indices in *Salix matsudana* exposed to salt stress. The salt injury index was positively correlated with the intercellular concentrations of CO_2 and malondialdehyde and negatively correlated with stomatal conductance, net photosynthetic rate, chlorophyll, biomass, and growth weight. The range of change varied between the indices, indicating that the results might be one-sided if single indices were used to evaluate the salt tolerance of *Salix matsudana*. The salt tolerance of *Salix matsudana* is a complex comprehensive trait.

3.1.3. Grey Relational Cluster Analysis of the Indices of Various Clones

The grey correlation degree was calculated from the membership function values of the relative values of the 16 indices of the six varieties (Table 4), and the indices were subjected to cluster analysis. At a fixed critical value r ∈ (0,1), the index xi and the index x_j were considered to be the same type of index when $\xi_{ij} \geq r$ ($i \neq j$). In this experiment, the critical value r = 0.57 was used to screen out two types of index clusters: $C_1 = \{x_4, x_8, x_{11}, x_{12}, x_{14}, x_{15}, x_{16}\}$ and $C_2 = \{x_1, x_2, x_3, x_5, x_6, x_7, x_9, x_{10}, x_{13}\}$. In the above classification, leaf water content, intercellular CO_2 concentration, SOD, POD, electrical conductivity, soluble sugar, and soluble starch were classified into one category, while the salt injury index, chlorophyll content, shoot fresh weight, root fresh weight, seedling height growth, ground diameter growth, stomatal conductance, net photosynthetic rate, and malondialdehyde were classified into another category. Table 5 shows that these two types of indices contributed different amounts to the comprehensive evaluation of salt tolerance.

Table 3. Correlation coefficient of every single index.

	SII	GS	PN	CI	RWC	SFW	RFW	SOD	MDA	EC	SS	SST	Chl	POD	RH
GS	−0.760 ***														
PN	−0.911 ***	0.882 ***													
CI	0.477 **	−0.222	−0.467 **												
RWC	0.161	−0.185	−0.268	0.003											
SFW	−0.875 ***	0.744 ***	0.827 ***	−0.378 *	−0.146										
RFW	−0.776 ***	0.458 *	0.714 ***	−0.617 **	−0.139	0.666 ***									
SOD	−0.042	−0.268	−0.133	−0.038	−0.127	−0.097	0.061								
MDA	0.581 **	−0.503 **	−0.606 **	0.537 **	−0.193	−0.561 **	−0.375 *	0.019							
EC	−0.324	0.541 **	0.394 *	−0.07	−0.074	0.246	0.234	−0.505 **	−0.013						
SS	0.32	−0.445 *	−0.421 *	0.14	0.141	−0.103	−0.07	−0.02	0.229	−0.276					
SST	0.045	−0.166	−0.142	−0.071	0.035	0.092	−0.055	0.26	−0.093	−0.165	0.515 **				
Chl	−0.631 **	0.610 **	0.665 ***	−0.235	−0.108	0.658 ***	0.422 *	0.082	−0.519 **	−0.057	−0.374 *	−0.208			
POD	−0.032	−0.136	−0.095	−0.024	0.086	−0.17	0.183	0.363 *	0.303	0.012	−0.25	−0.454 *	−0.019		
RH	−0.926 ***	0.829 ***	0.914 ***	−0.338	−0.172	0.899 ***	0.668 ***	−0.023	−0.591 **	0.361	−0.364 *	−0.072	0.665 ***	−0.109	
RD	−0.948 ***	0.753 ***	0.889 ***	−0.496 **	−0.280	0.887 ***	0.712 ***	−0.018	−0.565 **	0.384 *	−0.36	0.025	0.582 **	−0.071	0.908 ***

Notes: All original data are the average value of three repetitions. Except for the salt damage index, we calculated the relative value of each index (relative value = value of salt treated material/value of control material). Salt damage level: Grade 1: the plant is strong, the leaves are flat, green, and shiny; Grade 2: the plant wilts slightly, the leaf edge turns yellow or loses water slightly; Grade 3: plants wilt, and leaves droop, shrink, and turn yellow; Grade 4: the plant is seriously wilted, and the leaves fall off or wither; Grade 5: all leaves dry or fall off. Salt damage index = Σ (number of damaged plants at all levels) × Corresponding grade value)/(total number of plants investigated × Maximum salt damage level value). Levels of significance are indicated by asterisks: * $p < 0.05$; ** $p < 0.01$; *** $p < 0.001$.

Table 4. Grey correlation degree of every single index.

Evaluation Item	Correlation Degree	Ranking
SOD	0.759	1
POD	0.728	2
EC	0.656	3
RWC	0.627	4
SST	0.623	5
SS	0.613	6
CI	0.574	7
RFW	0.562	8
MDA	0.554	9
RH	0.546	10
RD	0.54	11
Chl	0.537	12
SFW	0.532	13
GS	0.524	14
PN	0.523	15
SII	0.497	16

Table 5. Principal component eigenvalues, contribution rate, and cumulative contribution rate of C_1 and C_2 indicators.

Category	Indexes	Characteristic Value	Contribution Rate (%)	Cumulative Contribution Rate (%)
C_1	Pn	7.734	85.94%	85.94%
	RH	0.431		
	SFW	0.344		
	SII	0.219		
	RD	0.106		
	GS	0.08		
	Chl	0.051		
	RFW	0.027		
	MDA	0.007		
C_2	SOD	5.49	87.15%	87.15%
	POD	1.513		
	SST	1.245		
	SS	1.044		
	RWC	0.545		
	CI	0.354		
	EC	0.189		

3.1.4. Principal Component Analysis of the Two Types of Indices

Since the cumulative variance contribution rates of the two types of indices were both greater than 85% (85.94% and 87.15%, respectively), these indices could be subjected to principal component analysis (Table 6). To evaluate the salt tolerance of the examined varieties, the comprehensive salt tolerance value (S) of each variety was calculated from the scores of the two types of indices obtained by principal component analysis ($S = F_1 \times W_1 + F_2 \times W_2$). The scoring formula for the C_1 comprehensive indices was $F_1 = hc_1 \times yc_1$, while the scoring formula for C_2 comprehensive indices was $F_2 = hc_2 \times yc_2$. The weight coefficient of principal components (h) was generally expressed as the variance contribution rate (namely, $hc_1 = 0.8594$, $hc_2 = 0.8715$). The linear combination of principal components of the C_1 and C_2 indices was obtained through principal component analysis: $yc_1 = 0.113x_3 + 0.116x_{10} + 0.124x_5 + 0.123x_1 + 0.118x_9 + 0.121x_6 + 0.123x_2 + 0.128x_7 + 0.112x_{13}$; $yc_2 = -0.15x_8 - 0.21x_4 + 0.391x_{11} + 0.013x_{14} - 0.262x_{16} - 0.26x_{15} + 0.344x_{12}$. The weight coefficients of the index clusters ($w_1 = 0.561$ and $w_2 = 0.439$) were calculated through the analytic hierarchy process.

Table 6. Comparison of score calculated by principal components in six *Salix matsudana*.

Type	Sm6	Sm2	SmA	Sm172	SmC	SmB
F_1	1.593892412	1.739830574	1.61428782	1.63688199	1.506493542	1.407123631
F_2	−0.005852612	−0.393786624	−0.444860471	−0.619278358	−0.474227168	−0.518359797
S	0.891604346	0.803172624	0.71032172	0.646427597	0.636957151	0.561836407
Sort by salt S values	1	2	3	4	5	6

Notes: F_1: Score of class C_1 index; F_2: Score of class C_2 index; S: Consolidated values.

The S values are given in Table 6. On the basis of a cut-off value of 0.7, Sm6, Sm2, and SmA of the six *Salix matsudana* varieties were salt-tolerant, whereas Sm172, SmC, and SmB were salt-sensitive.

3.1.5. Screening of the Identification Indices of Salt Tolerance of *Salix matsudana* at the Seedling Stage

Using the comprehensive salt tolerance value of the six varieties and the salt-stress-related indices, a regression equation was established to identify the salt tolerance of *Salix matsudana* at the seedling stage. The equation was used to screen out the identification indices of salt tolerance at the seedling stage. An optimal regression equation was established through stepwise regression analysis using the comprehensive S value as the dependent variable and the relative values of the indices as the independent variables: $y = 0.224x_{10} + 0.216x_{11} + 0.127x_{12} + 0.191x_7 - 0.187$ ($r = 0.949$, $R^2 = 0.90$), where y is the S index, x_{10} is chlorophyll, x_{11} is SOD, x_{12} is POD, and x_7 is Pn. The regressors x_{10}, x_{11}, and x_{12} were extremely significantly correlated with y ($p = 0.009$, 0.000, and 0.002, respectively), while x_7 was significantly correlated with y ($p = 0.034$). According to the equation, chlorophyll, SOD, POD, and Pn could be used as key indices to identify the salt tolerance of *Salix matsudana* at the seedling stage.

3.2. *Transcriptome Analysis of Salix matsudana under NaCl Stress*

3.2.1. Classification of the *Salix matsudana* Genes That Were Differentially Expressed in Response to Salt Stress

The second-generation sequencing data were compared and quantified using the full-length transcripts as a reference. The obtained read count was compared between groups, yielding intergroup ratios. The genes with $|\log_2\text{Ratio}| \geq 1$ and $q < 0.05$ were considered significantly differentially expressed genes between two groups. The number of differential genes varied with the duration of salt treatment (Table 7). The differential genes were roughly divided into three major categories on the basis of the clustering results: 12 h salt treatment, 24 h salt treatment, and 0–4 h salt treatment (Figure 1). The results of NT, NR, and BLASTX annotation showed that a greater number of differential genes were annotated at 12 h vs. 0 h, 12 h vs. 4 h, 24 h vs. 0 h, and 24 h vs. 4 h (4362 genes on average). According to the annotation results of the major databases, more than 98.33% of genes in the NT and NR databases were annotated (Table 8).

3.2.2. Pathway Analysis of the Differentially Expressed Genes

All the metabolic pathways in the KEGG were subjected to enrichment analysis using the hypergeometric test. The 20 KEGG metabolic pathways with the most significant enrichment of the differentially expressed genes were identified (Table 9), which gave us an intuitive understanding of the metabolic processes and signal transduction pathways altered in *Salix matsudana* under salt stress. More differentially expressed genes were detected after 12 h and 24 h of stress treatment, and more metabolic pathways were involved at these times, such as amino acid biosynthesis, plant hormone signal transduction, amino acid and nucleotide sugar metabolism, cysteine and methionine metabolism, mitogen-activated protein kinase (MAPK) signaling pathway, and carbon metabolism. These findings indicate that these metabolic pathways might play an important role in the salt stress response of *Salix matsudana*.

Table 7. Summary of differentially expressed genes.

Name	Up	Down	Total	Proportion
Sm172t24h-Sm172t12h	1554	674	2228	7.10%
Sm172t24h-Sm172t4h	2486	1519	4005	12.72%
Sm172t24h-Sm172t0h	1710	1118	2828	9%
Sm172t12h-Sm172t4h	1914	2515	4429	14.10%
Sm172t12h-Sm172t0h	1450	2584	4034	12.80%
Sm172t4h-Sm172t0h	223	195	418	1.30%
Sm6t24h-Sm6t12h	1795	1335	3130	9.90%
Sm6t24h-Sm6t4h	2284	2503	4787	15.20%
Sm6t24h-Sm6t0h	2494	2671	5165	16.40%
Sm6t12h-Sm6t4h	1743	3157	4900	15.60%
Sm6t12h-Sm6t0h	1977	2774	4751	15.10%
Sm6t4h-Sm6t0h	52	10	62	0.10%
Sm6t24h-Sm172t24h	48	66	114	0.30%
Sm6t12h-Sm172t12h	33	1	31	0.10%
Sm6t4h-Sm172t4h	497	253	750	2.30%
Sm6t0h-Sm172t0h	24	24	48	0.10%

Name: variance comparison combination; Up: upregulated gene; Down: downregulated gene; Total: sum of differential genes; Proportion: the proportion of significantly different genes in this group in the total different genes in this group.

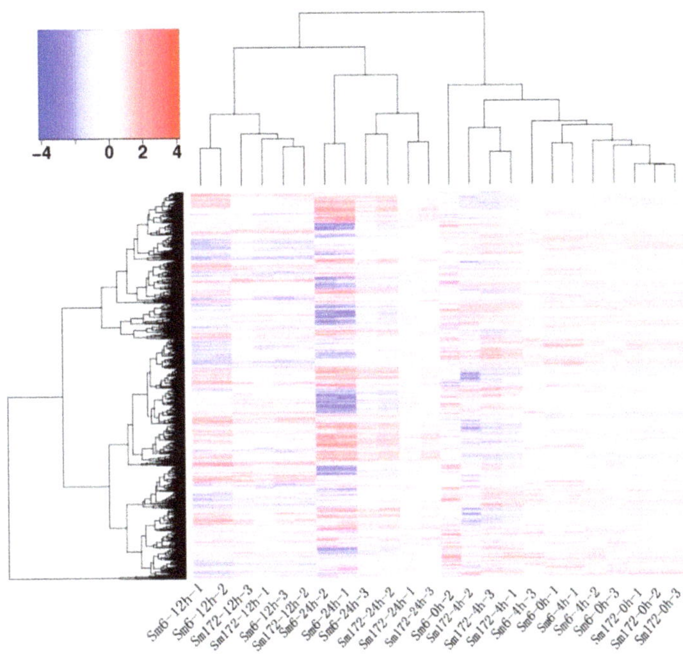

Figure 1. Cluster map of DEGs.

3.2.3. *Salix matsudana* Genes Differentially Expressed in Response to Salt Stress and Expression Verification

Analysis of the transcriptome data revealed that *Salix matsudana* adapted to and resisted salt stress and conducted self-regulation mainly through participating in carotenoid biosynthesis, MAPK signaling, phytohormone signal transduction, flavonoid biosynthesis, starch metabolism, and sucrose metabolism. The differential genes were screened out using a multiple of difference >2 and a corrected $p < 0.01$. Among the nine differential genes with the most significant gene expression difference, *Sm172-f2p30-2392*, *Sm172-f2p28-2386*, *Sm6-*

f8p60-2372, and *Sm172-f2p39-2386* were involved in the synthesis of 9-cis-epoxycarotenoid dioxygenase (NCED) and the regulation of NCED and abscisic acid (ABA) contents. *Sm6-f2p39-2263* participated in the synthesis of asparagine synthase and positively regulated the response of *Salix matsudana* to salt stress. *Sm6-f16p60-2374* was involved in circadian rhythm and photosensitive processes. *Sm172-f3p54-1980* was involved in ABA-activated signaling pathways and was closely related to leaf senescence. *Sm6-f3p60-931*, *Sm6-f2p60-1067*, and *Sm172-f2p30-1863* were involved in MAPK metabolic pathway. *Sm6-f2p60-1067* was related to the PYL gene. It promoted the ABA signal response by inducing PYL overexpression. *Sm172-f2p30-1863* was related to the synthesis of protein phosphatase 2C (PP2C). Silencing of *Sm172-f2p30-1863* promoted the signal response of ABA. As a result of these changes, *Salix matsudana* became more sensitive to ABA and more promptly regulated its own response mechanism in the face of stress.

The expression of the nine selected genes was verified by fluorescence-based quantitative PCR. The genes numbered *Sm172-f2p30-2392*, *Sm172-f2p28-2386*, *Sm6-f8p60-2372*, *Sm6-f2p39-2263*, *Sm6-f16p60-2374*, *Sm6-f3p60-931*, *Sm6-f2p60-1067*, *Sm172-f3p54-1980*, and *Sm172-f3p54-1980* were upregulated. The change trend of the expression of the nine genes was consistent with the results of transcriptome sequencing (Figure 2).

Table 8. Gene annotations in databases.

Comparison Combinations	Annotated Number	The Proportion of Annotated Genes in Databases						
		NT	NR	BLASTX	BLASTP	PFAM	eggNOG	KEGG
Sm6t4h-Sm6t0h	62	98.39%	98.39%	88.71%	83.87%	74.19%	69.35%	41.94%
Sm6t12h-Sm6t0h	4751	99.64%	99.41%	87.77%	79.44%	73.08%	66.93%	34.50%
Sm6t12h-Sm6t4h	4900	99.61%	99.27%	87.98%	79.33%	73.08%	65.88%	33.16%
Sm6t24h-Sm6t0h	5165	99.59%	99.17%	88.02%	79.09%	73.20%	67.76%	35.28%
Sm6t24h-Sm6t4h	4787	99.56%	99.08%	87.61%	78.96%	73.05%	66.79%	33.61%
Sm6t24h-Sm6t12h	3130	99.55%	99.20%	87.73%	79.20%	70.93%	66.90%	30.83%
Sm172t4h-Sm172t0h	418	98.56%	98.33%	87.80%	78.47%	74.64%	59.33%	23.68%
Sm172t12h-Sm172t0h	4034	99.43%	99.23%	88.23%	79.33%	72.43%	66.04%	32.15%
Sm172t12h-Sm172t4h	4429	99.53%	99.32%	87.56%	79.25%	73.40%	65.39%	32.38%
Sm172t24h-Sm172t0h	2828	99.26%	98.76%	86.32%	76.59%	70.47%	65.56%	31.26%
Sm172t24h-Sm172t4h	4005	99.30%	98.90%	86.44%	77.60%	71.16%	65.99%	30.91%
Sm172t24h-Sm172t12h	2228	99.69%	99.15%	88.51%	80.30%	71.99%	66.47%	32.36%

Table 9. The top 20 KEGG pathways with the highest concentration of DEGs.

Metabolic Pathway ID	Number of DEGs in this Pathway with Annotations						Pathway Annotation
Sm6	4 h–0 h	12 h–0 h	12 h–4 h	24 h–0 h	24 h–4 h	24 h–12 h	
ko01230	0	79	67	77	60	0	Biosynthesis of amino acids
ko04075	0	0	61	79	86	55	Plant hormone signal transduction
ko00520	0	39	39	37	40	0	Amino sugar and nucleotide sugar metabolism
ko00270	0	36	32	45	36	0	Cysteine and methionine metabolism
ko04016	0	0	0	55	55	32	MAPK signaling pathway-plant
ko01200	0	0	0	73	64	0	Carbon metabolism
ko00500	0	29	33	0	32	25	Starch and sucrose metabolism
ko04712	0	36	32	0	0	32	Circadian rhythm-plant
ko00941	0	29	31	16	17	0	Flavonoid biosynthesis
ko00280	0	18	19	23	26	0	Valine, leucine, and isoleucine degradation
ko00071	0	17	21	22	24	0	Fatty acid degradation
ko00940	0	27	29	0	26	0	Phenylpropanoid biosynthesis
ko00260	0	26	0	28	26	0	Glycine, serine, and threonine metabolism
ko00561	0	20	21	0	20	16	Glycerolipid metabolism
ko00052	0	15	16	16	14	12	Galactose metabolism
ko01212	0	0	24	25	22	0	Fatty acid metabolism
ko00620	0	0	0	34	31	0	Pyruvate metabolism
ko00053	0	19	22	0	18	0	Ascorbate and aldarate metabolism
ko00592	0	20	20	0	16	0	alpha-Linolenic acid metabolism
ko00906	4	13	14	0	13	0	Carotenoid biosynthesis

Table 9. *Cont.*

Metabolic Pathway ID	Number of DEGs in this Pathway with Annotations						Pathway Annotation
Sm172	4 h–0 h	12 h–0 h	12 h–4 h	24 h–0 h	24 h–4 h	24 h–12 h	
ko04075	0	56	0	80	83	53	Plant hormone signal transduction
ko04712	0	35	32	17	27	29	Circadian rhythm—plant
ko04016	0	0	0	51	54	33	MAPK signaling pathway—plant
ko00500	0	34	27	20	0	22	Starch and sucrose metabolism
ko00940	0	29	24	18	24	0	Phenylpropanoid biosynthesis
ko00941	8	25	27	0	24	9	Flavonoid biosynthesis
ko00270	0	0	37	19	31	0	Cysteine and methionine metabolism
ko01230	0	0	69	0	0	0	Biosynthesis of amino acids
ko00561	0	18	17	15	0	13	Glycerolipid metabolism
ko00073	0	17	19	9	0	9	Cutin, suberine, and wax biosynthesis
ko00906	0	17	14	10	0	10	Carotenoid biosynthesis
ko00592	0	0	14	13	13	9	alpha-Linolenic acid metabolism
ko00280	0	0	16	13	17	0	Valine, leucine, and isoleucine degradation
ko00450	0	11	13	10	11	0	Selenocompound metabolism
ko00196	0	20	15	9	0	0	Photosynthesis—antennal proteins
ko00944	3	10	13	0	11	6	Flavone and flavonol biosynthesis
ko00945	0	11	9	6	8	0	Stilbenoid, diarylheptanoid, and gingerol biosynthesis
ko00330	0	0	0	16	18	0	Arginine and proline metabolism
ko00030	0	0	20	13	0	0	Pentose phosphate pathway
ko04626	0	0	0	0	32	0	Plant–pathogen interaction

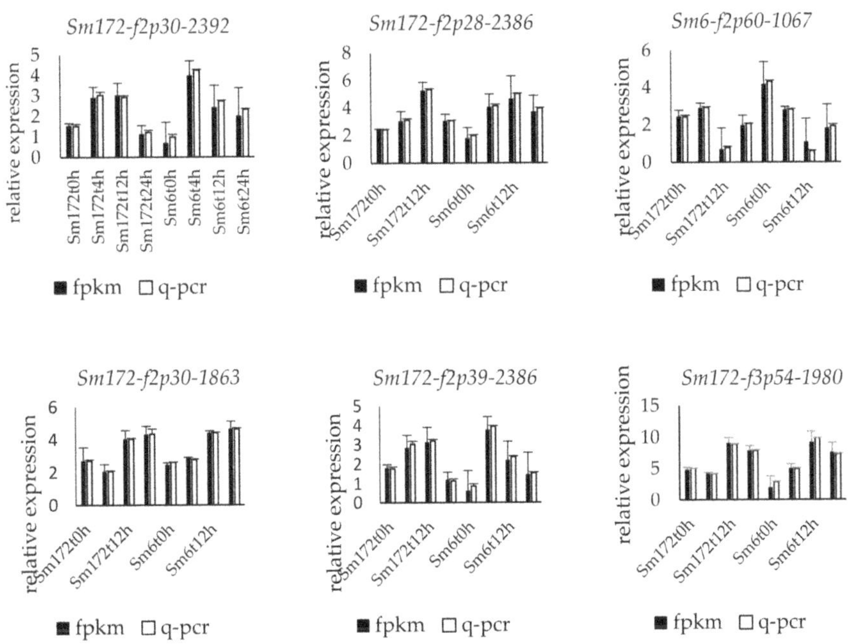

Figure 2. *Cont.*

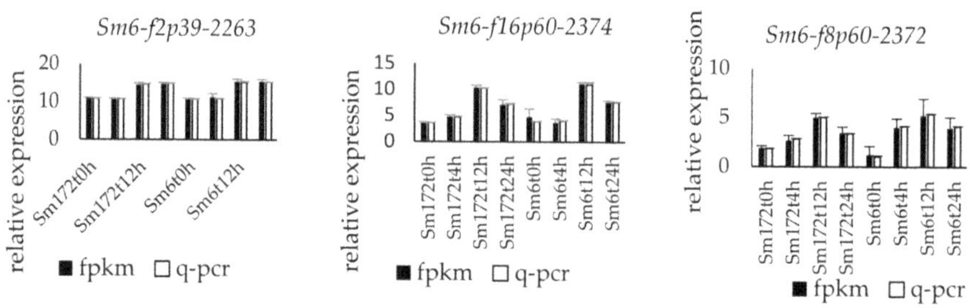

Figure 2. Expression profiles of 9 key DEGs in two poplar cultivars. Note: Sm172 stands for willow 172; Sm6 stands for Luliu 6.

4. Discussion

This study analyzed the chosen indices using a combination of the analytic hierarchy process and principal component analysis, which was more reasonable than directly doing principal component analysis on all indices. SOD, POD, chlorophyll, and net photosynthetic rate may be used as key indices to identify the salt tolerance of *Salix matsudana* at the seedling stage [38–40]. Upon salt stress, the antioxidant enzymes such as SOD and POD were activated in all clones of *Salix matsudana*, which led to the elimination of the toxicity of free radicals. SOD prevents the formation of free radicals and the damage to cells mainly by regulating the content of O^{2-}. It is the first line of defense against cell damage [39,41]. POD plays an important role in protecting cells from H_2O_2 stress [42]. In response to salt stress, photosynthesis is reduced in *Salix matsudana*. The accumulation of photosynthetic products is inhibited, which hinders the growth and development of plants and even causes plant death [43]. As an important photosynthetic pigment, chlorophyll plays an important role in absorption, transmission, and transformation of light energy [3,44,45]. Therefore, the change in chlorophyll content is an important index of plant growth [46]. Salt stress leads to ion toxicity, hinders the absorption of Mg^{2+} by plants, blocks the synthesis of chlorophyll, reduces chlorophyll content, and thus seriously hinders photosynthesis [47–51]. Such findings are consistent with the research results of Gong et al. [52,53].

Under salt stress, reactive oxygen species activate a variety of *MAPK* signaling cascades, inducing specific cellular responses. This study found that in *Salix matsudana* exposed to salt stress, the most strongly expressed *MAPK* signaling pathway was the mitogen-activated protein kinase (*MPK*). This study showed that H_2O_2 phosphorylated mitogen-activated protein kinase kinase 4/5 (*MKK4/5*) by activating *ANP1* upstream of the *MPK* cascade, thereby phosphorylating *MPK3*. This result is consistent with the findings of Kim et al. [54] and Kovtun et al. [55]. H_2O_2 also directly activated *MPK3* by inducing the formation of nucleoside diphosphate kinase 2 (*NDPK2*), in line with the findings of Moon et al. [56]. The serine-threonine protein kinase *OXI1* (oxidative-signal-inducible 1) was also activated by peroxide and responded through the downstream *MPK3*. In the process of cell death caused by reactive oxygen species through the above three pathways, *MPK3* is significantly upregulated, and the level of reactive oxygen species (such as H_2O_2) is increased [57], which eventually leads to cell death. When *Salix matsudana* is exposed to low salt stress or short-term salt stress, the upregulation of *MPK3* and the increase in H_2O_2 lead to elevated activities of SOD and POD in the antioxidant system, which allows the resistance to the damage of reactive oxygen species and maintenance of normal physiological function of cells. Therefore, improving the activity of antioxidant enzymes in plants and increasing the level of plant antioxidant metabolism are important ways to enhance the salt tolerance of plants. Alleviation of the harm of salt stress to *Salix matsudana* could be achieved by inhibiting the expression of the MPK3-related gene *Sm6-f3p60-931*, which suppressed the activity of *MPK3* and inhibited the production of reactive oxygen species. This biological process played a certain positive role in the salt stress resistance of

Salix matsudana and provided evidence that SOD and POD are key indices of salt tolerance in *Salix matsudana*.

In addition, this study showed that *MPK3* of the *MAPK* pathway cascade was significantly upregulated under salt stress, which reduced the function of *SPEECHLESS* (*SPCH*) and inhibited stomatal development. The basic helix-loop-helix (*bHLH*)-type transcription factors that regulate stomatal development include three members: *SPCH*, *MUTE*, and *FAMA*. These proteins are involved in the regulation of the initiation of the M cell lineage, the formation of guard mother cells, and the symmetrical division of guard mother cells [58,59]. *SPCH* has a region between the *bHLH* domain and the C-terminal region that is rich in the acidic amino acids serine and threonine. It mediates the direct phosphorylation of *SPCH* by *MPK3* [60–62], which hinders stomatal development and directly hinders photosynthesis. Therefore, physiological study of *Salix matsudana* showed that, with more concentrated or longer salt treatment, the net photosynthetic rate was reduced in each *Salix matsudana* clone. This finding provides evidence that the net photosynthetic rate is a key index of salt tolerance in *Salix matsudana*.

Moreover, genes with large differences in expression levels are involved in the synthesis of *NCED*. According to the metabolic pathway map of carotenoid biosynthesis, carotenoid biosynthesis is related to ABA synthesis [54,63]. ABA plays a central role in plant responses to stressful environments [64,65]. In the process of ABA synthesis, zeaxanthin epoxidase catalyzes the conversion of carotene to violaxanthin. *NCED* further catalyzes the conversion of 9-cis-flavin to xanthotoxin, which further leads to the production of abscisic aldehyde. Abscisic aldehyde is eventually converted into ABA under the action of an amine oxidase. These findings are consistent with the results of Li et al. [23]. *NCED* is a member of the carotenoid cleavage dioxygenase family. It is the most important rate-limiting enzyme in the process of ABA biosynthesis in higher plants and belongs to a class of key genes involved in the regulation of ABA biosynthesis [66,67]. *NCED* expression level is positively correlated with ABA content in plants [68–70]. Therefore, the goal of plant physiological self-regulation might be achieved through the control of the expression of *NCED* synthesis-related genes such as *Sm172-f2p28-2386*, *Sm172-f2p30-2392*, *Sm172-f2p39-2386*, and *Sm6-f8p60-2372*, which further control the expression of *NCED* and regulate the content of ABA in plants. In future research on the salt tolerance of *Salix matsudana*, ABA might become a key index.

These data provide a scientific basis for future research on the mechanism of *Salix matsudana* salt tolerance, deliver a simpler method for the identification and breeding of salt-tolerant *Salix matsudana* varieties, and can guide the selection of salt-tolerant tree species in the coastal saline-alkali land of the Yellow River Delta.

Author Contributions: Conceptualization, Y.P. and B.C.; formal analysis, P.M.; investigation, Y.G., W.L. and X.C.; data curation, T.W. and L.G.; writing—original draft preparation, Y.P.; writing—review and editing, P.M. and B.C.; visualization, L.G.; supervision, B.C.; funding acquisition, B.C. All authors have read and agreed to the published version of the manuscript.

Funding: This research was funded by the Central Finance Forestry Reform and Development Fund (Forestry Science and Technology Promotion Demonstration Subsidy) ([2020]TG08), the Major Scientific and Technological Innovation in Shandong (2017CXGC0316), and the National Forestry Public Welfare Industry Research Project (201404107).

Institutional Review Board Statement: Not applicable.

Informed Consent Statement: Not applicable.

Data Availability Statement: Not applicable.

Conflicts of Interest: The authors declare no conflict of interest.

References

1. Liu, P.; Bai, J.; Ding, Q.; Shao, H.; Gao, H.; Xiao, R. Effects of Water Level and Salinity on TN and TP Contents in Marsh Soils of the Yellow River Delta, China. *CLEAN–Soil Air Water* **2012**, *40*, 1118–1124. [CrossRef]
2. Liu, S.; Hou, X.; Yang, M.; Cheng, F.; Coxixo, A.; Wu, X.; Zhang, Y. Factors driving the relationships between vegetation and soil properties in the yellow river delta, china. *Catena* **2018**, *165*, 279–285. [CrossRef]
3. Xia, J.; Ren, J.; Zhang, S.; Wang, Y.; Fang, Y. Forest and grass composite patterns improve the soil quality in the coastal saline-alkali land of the yellow river delta, china. *Geoderma* **2019**, *349*, 25–35. [CrossRef]
4. Li, B.; Ouyang, J.; Wang, J.; Wu, H.; Liu, X.; Zou, J. Effects of NaCl on seedling growth and some physiological characteristics of *Salix matsudana Koidz. J. Tianjin Norm. Univ.* **2017**, *37*, 37–42.
5. Dimitriou, I.; Aronsson, P.; Weih, M. Stress tolerance of five willow clones after irrigation with different amounts of landfill leachate. *Bioresour. Technol.* **2006**, *97*, 150–157. [CrossRef]
6. Wang, Y.; Yuan, H.W.; Li, M.; Li, Y.J.; Ma, X.J.; Tan, F.; Zhang, J. Phenotypic and physiological responses of two willow varieties to salt stress. *Isr. J. Plant Sci.* **2013**, *61*, 73–82. [CrossRef]
7. Yang, Y.; Yan, G. Elucidating the molecular mechanisms mediating plant salt-stress responses. *New Phytol.* **2018**, *217*, 523–539. [CrossRef]
8. Hasegawa, P.M. Sodium (Na^+) homeostasis and salt tolerance of plants. *Environ. Exp. Bot.* **2013**, *92*, 19–31. [CrossRef]
9. Munns, R.; Tester, M. Mechanisms of salinity tolerance. *Annu. Rev. Plant Biol.* **2008**, *59*, 651–681. [CrossRef]
10. Zhang, M.; Smith, J.A.C.; Harberd, N.P.; Jiang, C. The regulatory roles of ethylene and reactive oxygen species (ROS) in plant salt stress responses. *Plant Mol. Biol.* **2016**, *91*, 651–659. [CrossRef]
11. Foyer, C.H.; Shigeoka, S. Understanding oxidative stress and antioxidant functions to enhance photosynthesis. *Plant Physiol.* **2011**, *155*, 93–100. [CrossRef] [PubMed]
12. Meloni, D.A.; Oliva, M.A.; Martinez, C.A.; Cambraia, J. Photosynthesis and activity of superoxide dismutase, peroxidase and glutathione reductase in cotton under salt stress. *Environ. Exp. Bot.* **2003**, *49*, 69–76. [CrossRef]
13. Barba-Espín, G.; Clemente-Moreno, M.J.; Álvarez, S.A.; García-Legaz, M.F.; Hernández, J.A.; Díaz-Vivancos, P. Salicylic acid negatively affects the response to salt stress in pea plants. *Plant Biol.* **2011**, *13*, 909–917. [CrossRef] [PubMed]
14. Qiao, G.; Zhang, X.; Jiang, J.; Liu, M.; Han, X.; Yang, H.; Zhuo, R. Comparative proteomic analysis of responses to salt stress in chinese willow (*Salix matsudana koidz*). *Plant Mol. Biol. Rep.* **2014**, *32*, 814–827. [CrossRef]
15. Zhou, J. *Changes of miRNA Expression in Populus Euphratica and Salix Mandshurica under Salt Stress*; Chinese Academy of Forestry: Beijing, China, 2010.
16. Chen, Y.; Jiang, Y.; Chen, Y.; Feng, W.; Liu, G.; Yu, C.; Lian, B.; Zhong, F.; Zhang, J. Uncovering candidate genes responsive to salt stress in *Salix matsudana (koidz)* by transcriptomic analysis. *PLoS ONE* **2020**, *15*, e0236129. [CrossRef]
17. Shan, L.; Zhao, S.; Xia, G. Research Progress on the Identification of Salt-tolerance Related Genes and Molecular Mechanism on Salt Tolerance in Higher Plants. *Mol. Plant Breed.* **2006**, *4*, 15.
18. Yang, H.B.; Dung, C.H.; Xu, X.F.; Wang, Y.; Han, Z.H. Effects of NaCl and Iso-osmotic polyethylene glycol stress on Na^+/H^+ antiport activity of three malus species with different salt tolerance. *J. Integr. Agric.* **2014**, *13*, 1276–1283. [CrossRef]
19. Yang, Y.Q.; Han, X.L.; Ma, L.; Wu, Y.J.; Liu, X.; Fu, H.Q.; Liu, G.Y.; Lei, X.G.; Guo, Y. Dynamic changes of phosphatidylinositol and phosphatidylinositol 4-phosphate levels modulate H^+-atpase and Na^+/H^+ antiporter activities to maintain ion homeostasis in arabidopsis under salt stress. *Mol. Plant* **2021**, *14*, 2000–2014. [CrossRef]
20. Filiz, E.; Kurt, F. Expression and co-expression analyses of WRKY, MYB, bHLH and bZIP transcription factor genes in potato (*Solanum tuberosum*) under abiotic stress conditions: RNA-seq data analysis. *Potato Res.* **2021**, *64*, 721–741. [CrossRef]
21. Das, A.; Pramanik, K.; Sharma, R.; Gantait, S.; Banerjee, J. In-silico study of biotic and abiotic stress-related transcription factor binding sites in the promoter regions of rice germin-like protein genes. *PLoS ONE* **2019**, *14*, e0211887. [CrossRef]
22. Mokhtari, F.; Rafiei, F.; Shabani, L.; Shiran, B. Differential expression pattern of transcription factors across annual medicago genotypes in response to salinity stress. *Biol. Plant.* **2017**, *61*, 227–234. [CrossRef]
23. Li, H.; Li, D.F.; Chen, A.G.; Tang, H.J.; Li, J.J.; Huang, S.Q. RNA-seq for comparative transcript profiling of kenaf under salinity stress. *J. Plant Res.* **2017**, *130*, 365–372. [CrossRef] [PubMed]
24. Okay, S.; Derelli, E.; Unver, T. Transcriptome-wide identification of bread wheat wrky transcription factors in response to drought stress. *Mol. Genet. Genom.* **2014**, *289*, 765–781. [CrossRef] [PubMed]
25. Meng, C.; Quan, T.Y.; Li, Z.Y.; Cui, K.L.; Yan, L.; Liang, Y.; Dai, J.L.; Xia, G.M.; Liu, S.W. Transcriptome profiling reveals the genetic basis of alkalinity tolerance in wheat. *BMC Genom.* **2017**, *18*, 24. [CrossRef] [PubMed]
26. Kumar, S.; Kanakachari, M.; Gurusamy, D.; Kumar, K.; Narayanasamy, P.; Venkata, P.K.; Solanke, A.; Gamanagatti, S.; Hiremath, V.; Katageri, I.S.; et al. Genome-wide transcriptomic and proteomic analyses of bollworm-infested developing cotton bolls revealed the genes and pathways involved in the insect pest defence mechanism. *Plant Biotechnol. J.* **2016**, *14*, 1438–1455. [CrossRef]
27. Zhang, F.; Zhu, G.; Du, L.; Shang, X.; Cheng, C.; Yang, B.; Hu, Y.; Cai, C.; Guo, W. Genetic regulation of salt stress tolerance revealed by RNA-seq in cotton diploid wild species, gossypium davidsonii. *Sci. Rep.* **2016**, *6*, 20582. [CrossRef]
28. Chen, M.-S.; Zhao, M.-L.; Wang, G.-J.; He, H.-Y.; Bai, X.; Pan, B.-Z.; Fu, Q.-T.; Tao, Y.-B.; Tang, M.-Y.; Martínez-Herrera, J.; et al. Transcriptome analysis of two inflorescence branching mutants reveals cytokinin is an important regulator in controlling inflorescence architecture in the woody plant jatropha curcas. *BMC Plant Biol.* **2019**, *19*, 468. [CrossRef]

29. Goyal, E.; Amit, S.K.; Singh, R.S.; Mahato, A.K.; Chand, S.; Kanika, K. Transcriptome profiling of the salt-stress response in *Triticum aestivum* cv. Kharchia local. *Sci. Rep.* **2016**, *6*, 27752. [CrossRef]
30. Amirbakhtiar, N.; Ismaili, A.; Ghaffari, M.R.; Firouzabadi, F.N.; Shobbar, Z.S.; Jain, M. Transcriptome response of roots to salt stress in a salinity-tolerant bread wheat cultivar. *PLoS ONE* **2019**, *14*, e0213305. [CrossRef]
31. Tian, X.; Wang, Z.; Zhang, Q.; Huacong, C.; Wang, P.; Lu, Y.; Jia, G.; Sara, A. Genome-wide transcriptome analysis of the salt stress tolerance mechanism in *Rosa chinensis*. *PLoS ONE* **2018**, *13*, e0200938. [CrossRef]
32. Xie, R.; Pan, X.; Zhang, J.; Ma, Y.; He, S.; Zheng, Y.; Ma, Y. Effect of salt-stress on gene expression in citrus roots revealed by RNA-seq. *Funct. Integr. Genom.* **2017**, *18*, 155–173. [CrossRef] [PubMed]
33. Chen, Y.; Li, C.; Lei, C.; Yi, J.; Gong, M. Identification and Comprehensive Evaluation of Drought Tolerance in Diploid Potato, *S. phureja* Germplasm Resources. *Mol. Plant Breed.* **2019**, *17*, 3416. [CrossRef]
34. Wang, Z.; Li, M.; Hao, R.L.; Chen, L.Q.; Han, Y.Q.; Zhang, B. Comparison of photosynthetic characteristics of broomcorn millet spike mutants. *J. Shanxi Agric. Sci.* **2020**, *48*, 879–883. [CrossRef]
35. Qin, Y. Study on Cold Resistance of Introduced Ornamental Bamboo Species in Shandong Province. Master's Thesis, Shandong Agricultural University, Tai'an, China, 2014.
36. Li, H.S. *Principles and Techniques of Plant Physiological and Biochemical Experiments*; Higher Education Press: Beijing, China, 2000; pp. 164–165.
37. Wang, X.Y.; Wang, S.L.; Tang, Y.; Zhou, W.M.; Zhou, L.; Zhong, Q.L.; Dai, L.M.; Yu, D.P. Characteristics of non-structural carbohydrate reserves of three dominant tree species in broadleaved korean pine forest in Changbai mountain, China. *Chin. J. Appl. Ecol.* **2019**, *30*, 1608–1614. [CrossRef]
38. Yu, C.W.; Guan, Z.Q.; Hong, Y.L.; Wu, Y.J.; Wang, C.; Liu, G.F.; Yang, C.P. Enhanced salt tolerance of transgenic poplar plants expressing a manganese superoxide dismutase from tamarix androssowii. *Mol. Biol. Rep.* **2010**, *37*, 1119–1124. [CrossRef]
39. Foyer, C.H.; Noctor, G. Redox signaling in plants. *Antioxid. Redox Signal.* **2013**, *18*, 2087–2090. [CrossRef]
40. Kravchik, M.; Bernstein, N. Effects of salinity on the transcriptome of growing maize leaf cells point at cell-age specificity in the involvement of the antioxidative response in cell growth restriction. *BMC Genom.* **2013**, *14*, 24. [CrossRef]
41. Amor, N.B.; Jiménez, A.; Megdiche, W.; Lundqvist, M.; Sevilla, F.; Abdelly, C. Response of antioxidant systems to nacl stress in the halophyte cakile maritima. *Physiol. Plant.* **2010**, *126*, 446–457. [CrossRef]
42. Chaparzadeh, N.; D'Amico, M.L.; Khavari-Nejad, R.A.; Izzo, R.; Navari-Izzo, F. Antioxidative responses of *Calendula officinalis* under salinity conditions. *Plant Physiol. Biochem.* **2004**, *42*, 695–701. [CrossRef]
43. Yang, J.Y.; Zheng, W.; Tian, Y.; Wu, Y.; Zhou, D.W. Effects of various mixed salt-alkaline stresses on growth, photosynthesis, and photosynthetic pigment concentrations of medicago ruthenica seedlings. *Photosynthetica* **2011**, *49*, 275–284. [CrossRef]
44. Wu, Y.; Jin, X.; Liao, W.; Hu, L.; Dawuda, M.M.; Zhao, X.; Tang, Z.; Gong, T.; Yu, J. 5-aminolevulinic acid (ALA) alleviated salinity stress in cucumber seedlings by enhancing chlorophyll synthesis pathway. *Front. Plant Sci.* **2018**, *9*, 635. [CrossRef] [PubMed]
45. Huihui, Z.; Yue, W.; Xin, L.; Guoqiang, H.; Yanhui, C.; Zhiyuan, T.; Jieyu, S.; Nan, X.; Guangyu, S. Chlorophyll synthesis and the photoprotective mechanism in leaves of mulberry (*Morus alba* L.) seedlings under NaCl and $NaHCO_3$ stress revealed by tmt-based proteomics analyses. *Ecotoxicol. Environ. Saf.* **2020**, *190*, 110164. [CrossRef] [PubMed]
46. Koldobika, H.; María, B.J.; Isabel, F.; Marta, P.; Ignacio, G.-P.J. Functional role of red (*retro*)-carotenoids as passive light filters in the leaves of *Buxus sempervirens* L.: Increased protection of photosynthetic tissues? *J. Exp. Bot.* **2005**, *56*, 2629–2636. [CrossRef]
47. Mukhtar, E.; Siddiqi, E.H.; Bhatti, K.H.; Nawaz, K.; Hussain, K. Gas exchange attributes can be valuable selection criteria for salinity tolerance in canola cultivars (*Brassica napus* L.). *Pak. J. Bot.* **2013**, *45*, 35–40.
48. Ye, L.; Zhao, X.; Bao, E.C.; Cao, K.; Zou, Z.R. Effects of arbuscular mycorrhizal fungi on watermelon growth, elemental uptake, antioxidant, and photosystem II activities and stress-response gene expressions under salinity-alkalinity stresses. *Front. Plant Sci.* **2019**, *10*, 863. [CrossRef]
49. Guo, R.; Zhou, J.; Ren, G.X.; Hao, W. Physiological responses of linseed seedlings to iso osmotic polyethylene glycol, salt, and alkali stresses. *Agron. J.* **2013**, *105*, 764–772. [CrossRef]
50. Guo, R.; Zhou, J.; Hao, W.P.; Gu, F.X.; Liu, Q.; Li, H.R.; Xia, X.; Mao, L.L. Germination, growth, chlorophyll fluorescence and ionic balance in linseed seedlings subjected to saline and alkaline stresses. *Plant Prod. Sci.* **2014**, *17*, 20–31. [CrossRef]
51. He, Q. The Ionic Response and Physiological Ecological Changes of *Ph.Praecox* under Salt Stress. Ph.D. Thesis, Chinese Academy of Forestry, Beijing, China, 2011.
52. Gong, B.; Wen, D.; Vandenlangenberg, K.; Wei, M.; Yang, F.; Shi, Q.; Wang, X. Comparative effects of NaCl and $NaHCO_3$ stress on photosynthetic parameters, nutrient metabolism, and the antioxidant system in tomato leaves. *Entia Hortic.* **2013**, *157*, 1–12. [CrossRef]
53. Hui-Hui, Z.; Guang-Liang, S.; Jie-Yu, S.; Xin, L.; Ma-Bo, L.; Liang, M.; Nan, X.; Guang-Yu, S. Photochemistry and proteomics of mulberry (*Morus alba* L.) seedlings under NaCl and $NaHCO_3$ stress. *Ecotoxicol. Environ. Saf.* **2019**, *184*, 109624. [CrossRef]
54. Kim, S.H.; Woo, D.H.; Kim, J.M.; Lee, S.Y.; Chung, W.S.; Moon, Y.H. Arabidopsis MKK4 mediates osmotic-stress response via its regulation of MPK3 activity. *Biochem. Biophys. Res. Commun.* **2011**, *412*, 150–154. [CrossRef]
55. Kovtun, Y.I.; Chiu, W.L.; Tena, G.; Sheen, J. Functional analysis of oxidative stress-activated mitogen-activated protein kinase cascade in plants. *Proc. Natl. Acad. Sci. USA* **2000**, *97*, 2940–2945. [CrossRef]

56. Moon, H.; Lee, B.; Choi, G.; Shin, S.; Prasad, D.T.; Lee, O.; Kwak, S.S.; Kim, D.N.J.; Bahk, J.; Hong, J.C. NDP kinase 2 interacts with two oxidative stress-activated MAPKs to regulate cellular redox state and enhances multiple stress tolerance in transgenic plants. *Proc. Natl. Acad. Sci. USA* **2003**, *100*, 358–363. [CrossRef] [PubMed]
57. Su, J.; Yang, L.; Zhu, Q.; Wu, H.; He, Y.; Liu, Y.; Xu, J.; Jiang, D.; Zhang, S. Active photosynthetic inhibition mediated by MPK3/MPK6 is critical to effector-triggered immunity. *PLoS Biol.* **2018**, *16*, e2004122. [CrossRef] [PubMed]
58. Macalister, C.A.; Ohashi-Ito, K.; Bergmann, D. Transcription factor control of asymmetric cell divisions that establish the stomatal lineage. *Nature* **2007**, *445*, 537–540. [CrossRef] [PubMed]
59. Morales-Navarro, S.; Pérez-Díaz, R.; Ortega, A.; Marcos, A.D.; Mena, M.; Fenoll, C.; González-Villanueva, E.; Ruiz-Lara, S. Overexpression of a sdd1-like gene from wild tomato decreases stomatal density and enhances dehydration avoidance in arabidopsis and cultivated tomato. *Front. Plant Sci.* **2018**, *9*, 940. [CrossRef] [PubMed]
60. Lampard, G.R.; Lukowitz, W.; Ellis, B.E.; Bergmann, D.C. Novel and expanded roles for mapk signaling in arabidopsis stomatal cell fate revealed by cell type-specific manipulations. *Plant Cell* **2009**, *21*, 3506–3517. [CrossRef]
61. Zhang, Y.; Guo, X.; Dong, J. Phosphorylation of the polarity protein BASL differentiates asymmetric cell fate through MAPKS and SPCH. *Curr. Biol.* **2016**, *26*, 2957–2965. [CrossRef]
62. Samakovli, D.; Ticha, T.; Vavrdova, T.; Ovecka, M.; Luptovciak, I.; Zapletalova, V.; Kucharova, A.; Krenek, P.; Krasylenko, Y.; Margaritopoulou, T. Yoda-hsp90 module regulates phosphorylation-dependent inactivation of speechless to control stomatal development under acute heat stress in arabidopsis. *Mol. Plant Engl. Version* **2020**, *13*, 612–633. [CrossRef]
63. Wang, Y.H.; Que, F.; Li, T.; Zhang, R.R.; Khadr, A.; Xu, Z.S.; Tian, Y.S.; Xiong, A.S. DcABF3, an ABF transcription factor from carrot, alters stomatal density and reduces ABA sensitivity in transgenic *arabidopsis*. *Plant Sci.* **2021**, *302*, 110699. [CrossRef]
64. Hasegawa, P.M.; Bressan, R.A.; Zhu, J.K.; Bohnert, H.J. Plant cellular and molecular responses to high salinity. *Annu. Rev. Plant Physiol. Plant Mol. Biol.* **2000**, *51*, 463–499. [CrossRef]
65. Finkelstein, R.R.; Gampala, S.; Rock, C.D. Abscisic acid signaling in seeds and seedlings. *Plant Cell* **2002**, *14* (Suppl. S1), S15–S45. [CrossRef] [PubMed]
66. Kitahata, N.; Ito, S.; Kato, A.; Ueno, K.; Nakano, T.; Yoneyama, K.; Yoneyama, K.; Asami, T. Abamine as a basis for new designs of regulators of strigolactone production. *J. Pestic. Sci.* **2011**, *36*, 53–57. [CrossRef]
67. Zhao, J.; Li, J.; Zhang, J.; Chen, D.; Qin, G. Genome-wide identification and expression analysis of the carotenoid cleavage oxygenase gene family in five rosaceae species. *Plant Mol. Biol. Rep.* **2021**, *39*, 739–751. [CrossRef]
68. Li, Q.H.; Yu, X.T.; Chen, L.; Zhao, G.; Li, S.Z.; Zhou, H.; Dai, Y.; Sun, N.; Xie, Y.F.; Gao, J.S.; et al. Genome-wide identification and expression analysis of the NCED family in cotton (*Gossypium hirsutum* L.). *PLoS ONE* **2021**, *16*, e0246021. [CrossRef]
69. Milosavljevic, A.; Prokic, L.; Marjanovic, M.; Stikic, R.; Sabovljevic, A. The effects of drought on the expression of TAO1, NCED and EIL1 genes and ABA content in tomato wild-type and flacca mutant. *Arch. Biol. Sci.* **2012**, *64*, 297–306. [CrossRef]
70. Awan, S.Z.; Chandler, J.O.; Harrison, P.J.; Sergeant, M.J.; Bugg, T.D.H.; Thompson, A.J. Promotion of germination using hydroxamic acid inhibitors of 9-cis-epoxycarotenoid dioxygenase. *Front. Plant Sci.* **2017**, *8*, 357. [CrossRef] [PubMed]

MDPI
St. Alban-Anlage 66
4052 Basel
Switzerland
Tel. +41 61 683 77 34
Fax +41 61 302 89 18
www.mdpi.com

Forests Editorial Office
E-mail: forests@mdpi.com
www.mdpi.com/journal/forests

www.ingramcontent.com/pod-product-compliance
Lightning Source LLC
Chambersburg PA
CBHW080937240926
51849CB00004BA/477
* 9 7 8 3 0 3 6 5 8 4 9 5 9 *